AF681889

CLASSICAL AND MODERN MECHANISMS FOR ENGINEERS AND INVENTORS

MECHANICAL ENGINEERING

A Series of Textbooks and Reference Books

Editor: L.L. FAULKNER Columbus Division, Battelle Memorial Institute, and Department of Mechanical Engineering, The Ohio State University, Columbus, Ohio

Associate Editor: S.B. MENKES Department of Mechanical Engineering, The City College of the City University of New York, New York

1. Spring Designer's Handbook, *by Harold Carlson*
2. Computer-Aided Graphics and Design, *by Daniel L. Ryan*
3. Lubrication Fundamentals, *by J. George Wills*
4. Solar Engineering for Domestic Buildings, *by William A. Himmelman*
5. Applied Engineering Mechanics: Statics and Dynamics, *by G. Boothroyd and C. Poli*
6. Centrifugal Pump Clinic, *by Igor J. Karassik*
7. Computer-Aided Kinetics for Machine Design, *by Daniel L. Ryan*
8. Plastics Products Design Handbook, Part A: Materials and Components; Part B: Processes and Design for Processes, *edited by Edward Miller*
9. Turbomachinery: Basic Theory and Applications, *by Earl Logan, Jr.*
10. Vibrations of Shells and Plates, *by Werner Soedel*
11. Flat and Corrugated Diaphragm Design Handbook, *by Mario Di Giovanni*
12. Practical Stress Analysis in Engineering Design, *by Alexander Blake*
13. An Introduction to the Design and Behavior of Bolted Joints, *by John H. Bickford*
14. Optimal Engineering Design: Principles and Applications, *by James N. Siddall*
15. Spring Manufacturing Handbook, *by Harold Carlson*
16. Industrial Noise Control: Fundamentals and Applications, *edited by Lewis H. Bell*
17. Gears and Their Vibration: A Basic Approach to Understanding Gear Noise, *by J. Derek Smith*

18. Chains for Power Transmission and Material Handling: Design and Applications Handbook, *by the American Chain Association*
19. Corrosion and Corrosion Protection Handbook, *edited by Philip A. Schweitzer*
20. Gear Drive Systems: Design and Application, *by Peter Lynwander*
21. Controlling In-Plant Airborne Contaminants: Systems Design and Calculations, *by John D. Constance*
22. CAD/CAM Systems Planning and Implementation, *by Charles S. Knox*
23. Probabilistic Engineering Design: Principles and Applications, *by James N. Siddall*
24. Traction Drives: Selection and Application, *by Frederick W. Heilich III and Eugene E. Shube*
25. Finite Element Methods: An Introduction, *by Ronald L. Huston and Chris E. Passerello*
26. Mechanical Fastening of Plastics: An Engineering Handbook, *by Brayton Lincoln, Kenneth J. Gomes, and James F. Braden*
27. Lubrication in Practice, Second Edition, *edited by W. S. Robertson*
28. Principles of Automated Drafting, *by Daniel L. Ryan*
29. Practical Seal Design, *edited by Leonard J. Martini*
30. Engineering Documentation for CAD/CAM Applications, *by Charles S. Knox*
31. Design Dimensioning with Computer Graphics Applications, *by Jerome C. Lange*
32. Mechanism Analysis: Simplified Graphical and Analytical Techniques, *by Lyndon O. Barton*
33. CAD/CAM Systems: Justification, Implementation, Productivity Measurement, *by Edward J. Preston, George W. Crawford, and Mark E. Coticchia*
34. Steam Plant Calculations Manual, *by V. Ganapathy*
35. Design Assurance for Engineers and Managers, *by John A. Burgess*
36. Heat Transfer Fluids and Systems for Process and Energy Applications, *by Jasbir Singh*
37. Potential Flows: Computer Graphic Solutions, *by Robert H. Kirchhoff*
38. Computer-Aided Graphics and Design, Second Edition, *by Daniel L. Ryan*
39. Electronically Controlled Proportional Valves: Selection and Application, *by Michael J. Tonyan, edited by Tobi Goldoftas*
40. Pressure Gauge Handbook, *by AMETEK, U.S. Gauge Division, edited by Philip W. Harland*
41. Fabric Filtration for Combustion Sources: Fundamentals and Basic Technology, *by R. P. Donovan*
42. Design of Mechanical Joints, *by Alexander Blake*
43. CAD/CAM Dictionary, *by Edward J. Preston, George W. Crawford, and Mark E. Coticchia*

44. Machinery Adhesives for Locking, Retaining, and Sealing, *by Girard S. Haviland*
45. Couplings and Joints: Design, Selection, and Application, *by Jon R. Mancuso*
46. Shaft Alignment Handbook, *by John Piotrowski*
47. BASIC Programs for Steam Plant Engineers: Boilers, Combustion, Fluid Flow, and Heat Transfer, *by V. Ganapathy*
48. Solving Mechanical Design Problems with Computer Graphics, *by Jerome C. Lange*
49. Plastics Gearing: Selection and Application, *by Clifford E. Adams*
50. Clutches and Brakes: Design and Selection, *by William C. Orthwein*
51. Transducers in Mechanical and Electronic Design, *by Harry L. Trietley*
52. Metallurgical Applications of Shock-Wave and High-Strain-Rate Phenomena, *edited by Lawrence E. Murr, Karl P. Staudhammer, and Marc A. Meyers*
53. Magnesium Products Design, *by Robert S. Busk*
54. How To Integrate CAD/CAM Systems: Management and Technology, *by William D. Engelke*
55. Cam Design and Manufacture, Second Edition; with cam design software for the IBM PC and compatibles, disk included, *by Preben W. Jensen*
56. Solid-State AC Motor Controls: Selection and Application, *by Sylvester Campbell*
57. Fundamentals of Robotics, *by David D. Ardayfio*
58. Belt Selection and Application for Engineers, *edited by Wallace D. Erickson*
59. Developing Three-Dimensional CAD Software with the IBM PC, *by C. Stan Wei*
60. Organizing Data for CIM Applications, *by Charles S. Knox, with contributions by Thomas C. Boos, Ross S. Culverhouse, and Paul F. Muchnicki*
61. Computer-Aided Simulation in Railway Dynamics, *by Rao V. Dukkipati and Joseph R. Amyot*
62. Fiber-Reinforced Composites: Materials, Manufacturing, and Design, *by P. K. Mallick*
63. Photoelectric Sensors and Controls: Selection and Application, *by Scott M. Juds*
64. Finite Element Analysis with Personal Computers, *by Edward R. Champion, Jr. and J. Michael Ensminger*
65. Ultrasonics: Fundamentals, Technology, Applications, Second Edition, Revised and Expanded, *by Dale Ensminger*
66. Applied Finite Element Modeling: Practical Problem Solving for Engineers, *by Jeffrey M. Steele*
67. Measurement and Instrumentation in Engineering: Principles and Basic Laboratory Experiments, *by Francis S. Tse and Ivan E. Morse*

68. Centrifugal Pump Clinic, Second Edition, Revised and Expanded, *by Igor J. Karassik*

69. Practical Stress Analysis in Engineering Design, Second Edition, Revised and Expanded, *by Alexander Blake*

70. An Introduction to the Design and Behavior of Bolted Joints, Second Edition, Revised and Expanded, *by John H. Bickford*

71. High Vacuum Technology: A Practical Guide, *by Marsbed H. Hablanian*

72. Pressure Sensors: Selection and Application, *by Duane Tandeske*

73. Zinc Handbook: Properties, Processing, and Use in Design, *Frank Porter*

74. Thermal Fatigue of Metals, *by Andrzej Weronski and Tadeusz Hejwowski*

75. Classical and Modern Mechanisms for Engineers and Inventors, *by Preben W. Jensen*

Additional Volumes in Preparation

Mechanical Engineering Software

Spring Design with an IBM PC, *by Al Dietrich*

Mechanical Design Failure Analysis: With Failure Analysis System Software for the IBM PC, *by David G. Ullman*

CLASSICAL AND MODERN MECHANISMS FOR ENGINEERS AND INVENTORS

PREBEN W. JENSEN

Mankato State University
Mankato, Minnesota

Marcel Dekker, Inc. New York • Basel • Hong Kong

Library of Congress Cataloging-in-Publication Data

Jensen, Preben W.
Classical and modern mechanisms for engineers and inventors /
Preben W. Jensen.
p. cm. -- (Mechanical Engineering)
Includes bibliographical references and index.
ISBN 0-8247-8547-4 (acid-free paper)
1. Mechanical movements. 2. Machinery, Kinematics of. I. Title.
II. Series: Mechanical engineering (Marcel Dekker, Inc.)
TJ181.J44 1991
621.3--dc20 91-19847
CIP

This book is printed on acid-free paper.

Copyright © 1991 by MARCEL DEKKER, INC. All Rights Reserved

Neither this book nor any part may be reproduced or transmitted in any form or by any means, electronic or mechanical, including photocopying, microfilming, and recording, or by any information storage and retrieval system, without permission in writing from the publisher.

MARCEL DEKKER, INC.
270 Madison Avenue, New York, New York 10016

Current printing (last digit):
10 9 8 7 6 5 4 3 2 1

PRINTED IN THE UNITED STATES OF AMERICA

Preface

Over a long period, I have written a number of articles on application and theory of mechanisms in various periodicals in the United States, England, Germany, France, Switzerland, and Denmark. It was the purpose of these articles to present the design engineer with a variety of solutions to mechanical motion problems. The source of the material for those articles as well as for this book consists primarily of publications from the United States and Germany. It has often been a delight of mine to look through old and newer magazines to find inspiration when designing mechanisms.

This book is oriented toward the mechanical engineer working in industry, often under time pressure, who must come up with solutions to new mechanical motion problems as well as create new solutions to familiar problems.

The bibliographies in the back of the book should prove helpful for those who want to study further the world of mechanisms. Only a few cam mechanisms are included in the present book. The reader who wants more information on this subject is referred to my book *Cam Design and Manufacture,* Second Edition (Marcel Dekker, Inc.).

I dedicate this book to the memory of the untold thousands of inventors who have contributed to the foundation of kinematics but often have not received the credit they deserve.

Preben W. Jensen

Contents

1

Five Basic Mechanisms

The design engineer is often faced with the problem of finding a mechanism for a given purpose or of changing a mechanism to make it more suitable for the purpose. It is his primary objective to find the simplest possible solutions based on the three criteria: space, speed, and savings.

An introduction to mechanisms begins with a review of the foundation of kinematics, namely, the concept of joints. A joint connects two links. There are five joints with one degree of freedom. One degree of freedom means that one coordinate describes the position of one link relative to the other link. The five basic joints are

Turning joint: The relative motion between the two links is a rotation.

Sliding joint: The relative motion between the two links is a translation. Theoretically the bar and the hole could have constant radius of curvature along the axis, but this kind of sliding joint is rarely used in practice.

Rolling joint: This type of joint permits the two links to roll on each other without sliding. The joint takes on two forms: the gear joint and the band joint, where the connection is maintained by two bands, each band being wrapped around both links. Both gear and band

Table 1.1

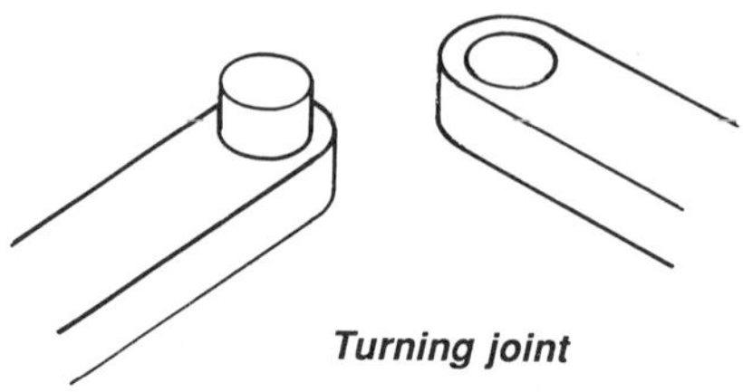

Turning joint

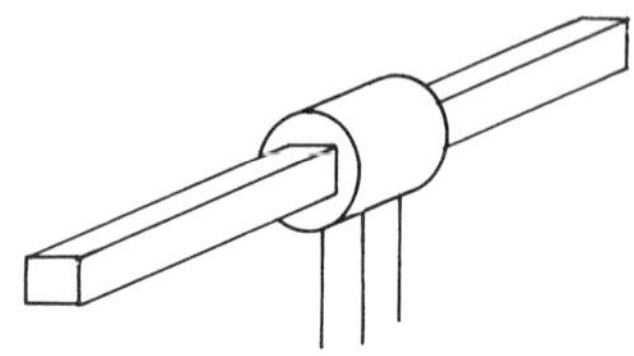

Sliding joint

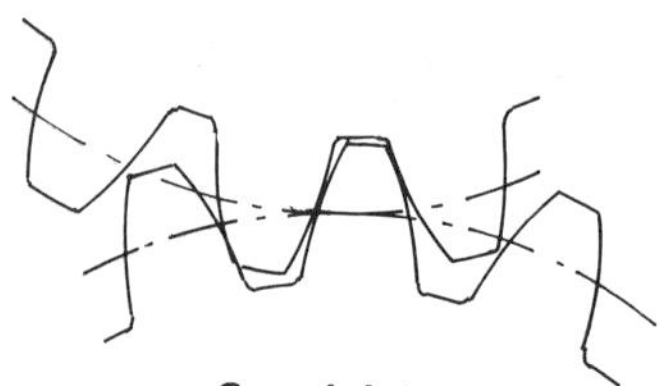

Gear joint

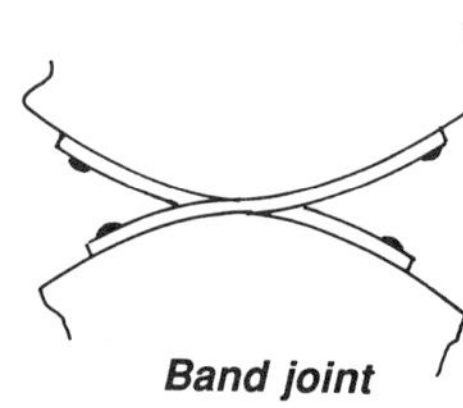

Band joint

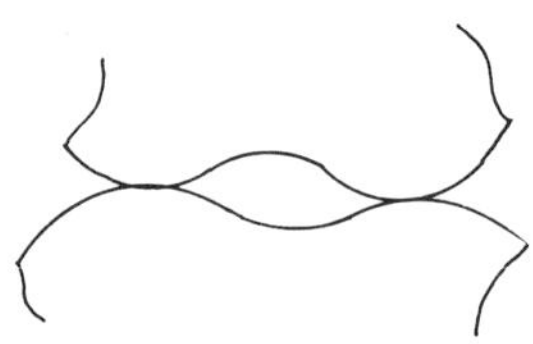

Roll-slide joint

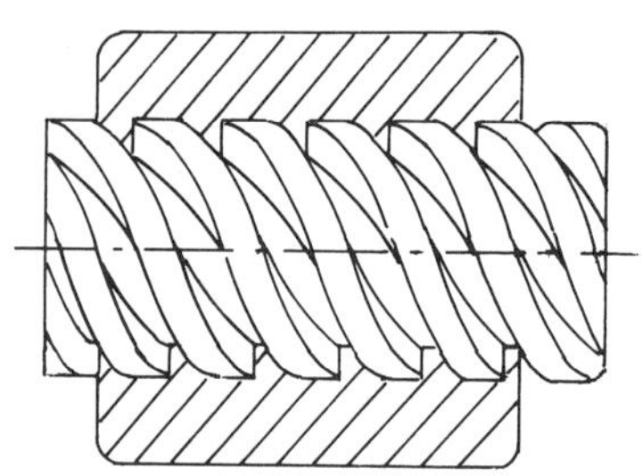

Screw joint

joints include noncircular surfaces. The joints can be interchanged without resulting change in motion.

Roll-slide joint: The two links are kept in contact by force closure. The joint has two points of contact. For special proportions form closure can be achieved (see Fig. 1.6).

Screw joint: This is the well-known screw and nut. The relative motion between the links is a combined turning and translating motion.

In general the four-bar linkage and mechanisms that can be derived from the four-bar linkage are considered the simplest mechanisms. This assumption, however, does not always lead to the best mechanism design solution. In the literature joints are defined as connecting links of a mechanism. What is not recognized is that some of them should be considered mechanisms in their own right. Examples will be shown of how joint mechanisms can replace far more complicated mechanisms and result in patentable devices. The turning-joint mechanism, described later is especially impressive because of its simplicity and speed, which could not be achieved with a more complex mechanism having more links.

THE SCREW MECHANISM

Archimedes (287–212 B.C.) had an inclination for practical applications of mathematical principles.* A short time after his arrival in Egypt, Archimedes invented the screw pump. He had been surprised with the amount of work people and animals performed in order to pump water from the Nile onto the fields. He decided to improve the impractical pumps that had been used for thousands of years, and he designed a wooden screw, as shown in Fig. 1.1. When he dipped this device into the Nile and started rotating it, it immediately started pumping. A paddle wheel was used as the driving means, so that the river provided power for the pumping action itself.

THE SLIDING-JOINT MECHANISM

A sliding joint is an irregular-shaped shaft sliding in a hole with identical cross section. In the literature this joint is designated a sliding joint and is not considered a mechanism. For practical purposes it can be a piston in a cylinder. An application of this joint is described later in the chapter.

*Archimedes intuitively solved one of the greatest problems in mathematics of the day, namely the calculation of the volume of a cone, a cylinder, and a sphere. The Greek mathematicians had not yet developed the tools to calculate the volume of these bodies when their dimensions were known. Archimedes guessed that the ratio of the volumes was as 1:2:3. Because he couldn't find a mathematical proof, he compared the weight of these bodies. They all had the same circular base and the same height. He decided to give a talk to the mathematical society in Egypt about his discovery. His lecture, entitled "On the Volumes of Round Bodies," was ill received because it was not pure mathematics. He left Egypt and returned to his homeland where he continued his brilliant career as a mathematician and inventor, discovering the principle of leverage ("Give me a place to stand and I will move the earth") as well as finding an approximation to the number π.

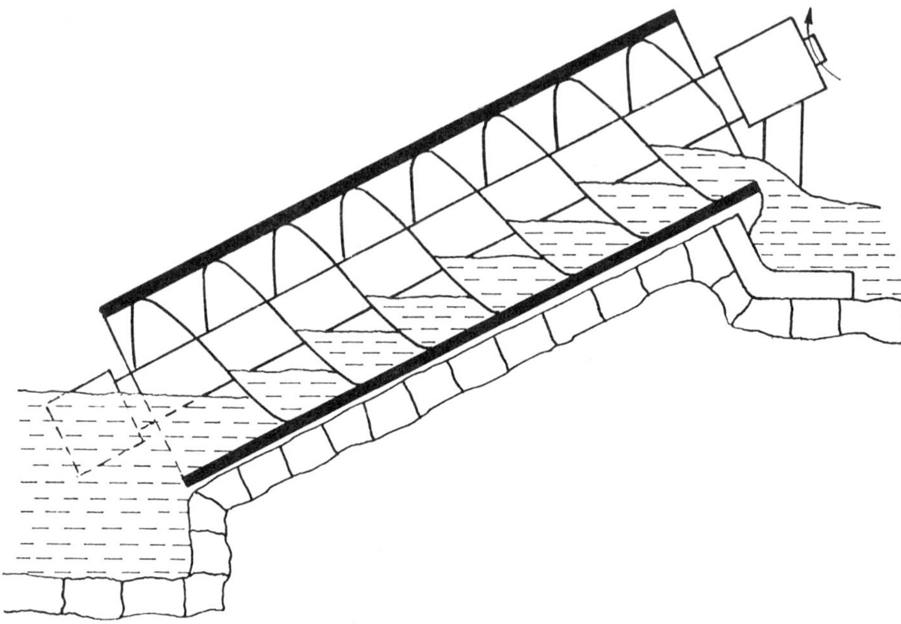

Figure 1.1 Archimedes' water pump.

THE TURNING-JOINT MECHANISM

The turning-joint mechanism is, in principle, just a pin on one link and a corresponding hole in the neighboring link. In the literature it is called a turning joint, or revolute pair, and is not referred to as a mechanism. First, a complex solution to cutting material on-the-fly will be shown. This device was invented many years ago. Then, a device in the form of a turning-joint mechanism will be shown that does the same but is much simpler and much faster.

Cutoff Device for Cigarettes

Continuously moving material can be cut without stopping the material, thus speeding production. Conventionally, this type of cutting is done while moving the tool at the same velocity as the material and then withdrawing the cutter and returning it to the starting position and repeating the operation.

The principle is illustrated by the device in Fig. 1.2. The material to be cut moves at constant velocity. A circular knife on a shaft is supported by a cutter carrier, which swings in a circular path on a pair of equal-length cranks. The cranks are driven at the same angular velocity by one or both of the connected crankshafts. When the right-hand crank is rotated, a bevel

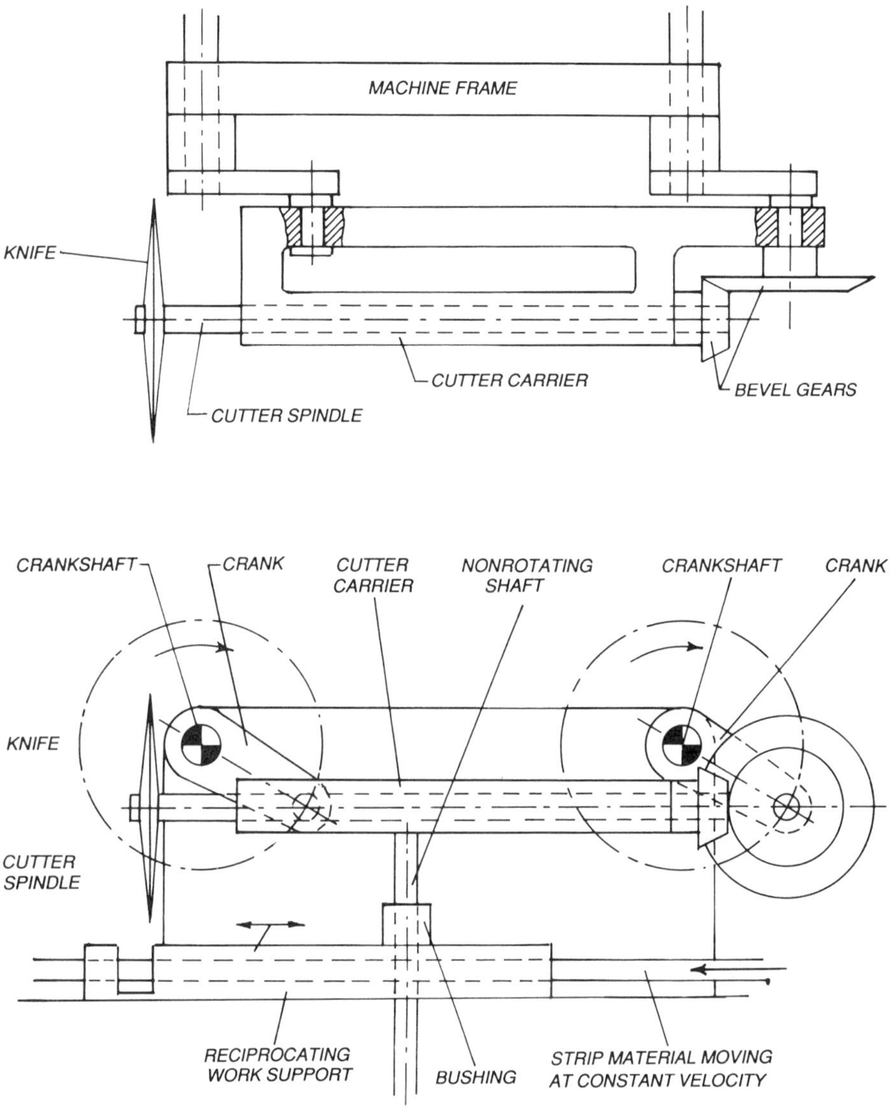

Figure 1.2 Device for cutting material on-the-fly.

gear fixed to the crank drives a second gear fixed to the cutter spindle and revolves the knife. The cutter carrier has a nonrotating shafts, which imparts a reciprocating motion through a sleeve bushing to a slide supporting the material while it is being cut. Since the cutter and work move in the same direction at the same speed, it is not possible to change the length of the cutoff piece without reproportioning of the device. Because of the large reciprocating masses, the device can run at moderate speed only.

Simplified and Improved Cutoff Device for Cans and Cigarettes Using the Turning-Joint Mechanism

An improved version of a cutoff mechanism was invented by the author, U.S. Patent #3,183,754. Length of the cutoff piece can be varied using the turning-joint mechanism illustrated in Fig. 1.3. The device consists of a cutter blade attached to a circular disc on a shaft supported by two bearing

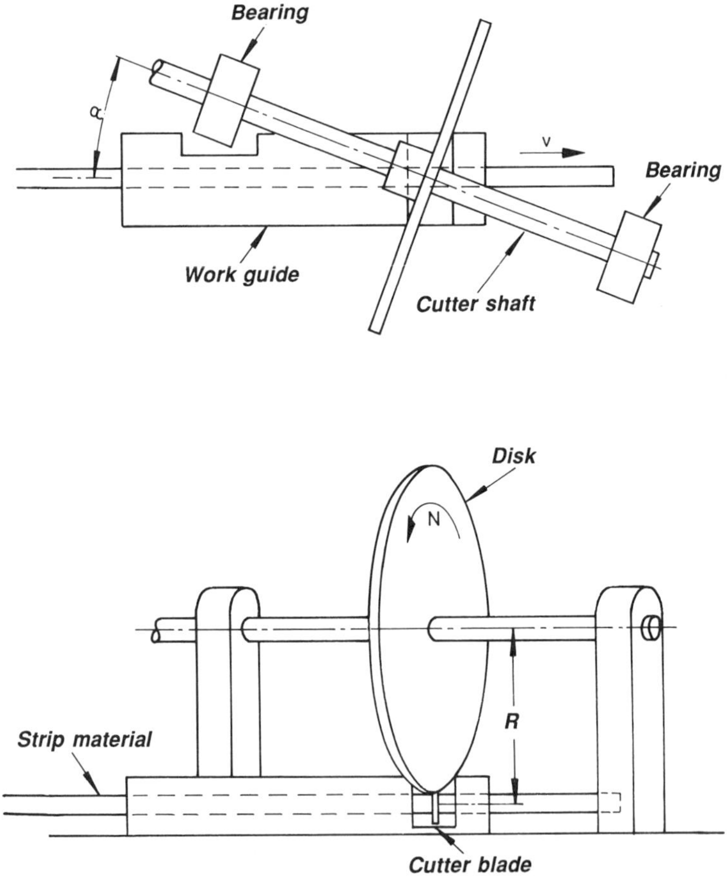

Figure 1.3 Device for cutting material on-the-fly but greatly simplified in comparison with the design in Fig. 1.2.

brackets. The cutter blade is oriented so that when it is in its lowest position, as shown, it is perpendicular to the center line of the material being cut. The strip is supported by a work guide.

Angle α between the center line of the material and that of the cutter shaft can be determined by the equation

$$2RN \sin \alpha/60 = V$$

or

$$\alpha = \sin^{-1} (30V/RN)$$

where R is the active cutting radius of the blade in inches, N is the rotary speed of the disc in rpm, and V is the velocity of the material in inches per second. In order to cut various lengths of strip material at different strip velocities V, the following principle is applied: the knife is rotated with a speed N equal to the number of pieces to be cut per minute. Angle α is then adjusted according to the above equation and the blade angle is reset to cut perpendicular to the moving strip.

The device has been successful in cutting pre-slit tubes for can manufacture at a production rate of 1000–2000 cans per minute. Actual deviation from an exact perpendicular cut is about 0.002 in. for an active cutting radius R of 2.5 in. It is used today in high speed cigarette-making machines, where it cuts cigarettes on-the-fly with a speed of 2000 cigarettes per minute. When two knives on the same disc are employed, the device cuts 4000 cigarettes per minute. The device can be completely dynamically balanced.

Figure 1.4 shows four sliding-joint mechanisms in series, each mechanism in the form of a cylinder and a piston. Due to the series connection of these mechanisms, the stroke of each cylinder is added to the output motion. Cylinder 1 has a stroke of 1 mm, cylinder 2 a stroke of 2 mm, cylinder 3 a stroke of 4 mm, and cylinder 4 a stroke of 8 mm. Because the strokes are all added linearly, the output member can be positioned in any one of 16 position. The mechanism has four degrees of freedom and is an open kinematic chain. If a link mechanism were to be used instead, based on prevailing kinematic literature, the result would be a monster of links and would not even be accurate.

THE ROLLING-JOINT MECHANISM

I have been told that the inventor of the Spirograph drawing toy has earned millions. The spirograph in its simplest form is made of two gears, where one gear is fixed and the other gear is made to roll on the fixed gear by means of a pencil that presses the two gears against each other. The pencil inserted through a hole in the moving gear, at the same time traces a curve.

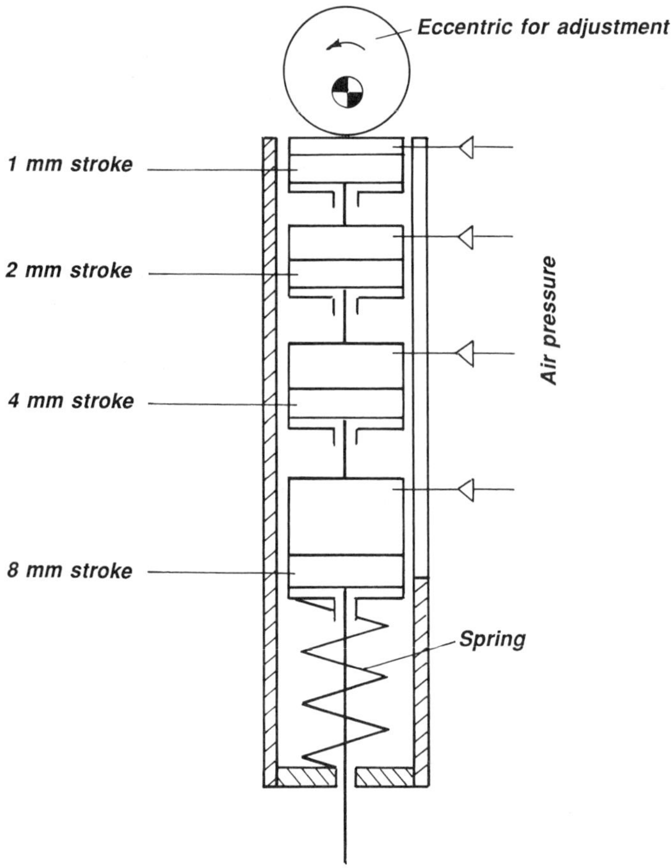

Figure 1.4 Mechanism with four inputs (air pressure). The output member can be positioned in any one of 16 positions.

The position of the pencil on the moving gear can be changed by using another of a number of holes. In this way intricate patterns are generated.

Figure 1.5a shows a double-crank linkage A_0ABB_0, which can be replaced by the band mechanism in Fig. 1.5b, where the moving link AB makes exactly the same motion as the coupler AB in Fig. 1.5a. Therefore, the rolling mechanism is also a mechanism and not only a joint.

THE ROLL-SLIDE-JOINT MECHANISM

Figure 1.6 shows a roll-slide joint mechanism. The triangular-shaped polygon, a constant-diameter cam, rotates inside the rhombus with three points

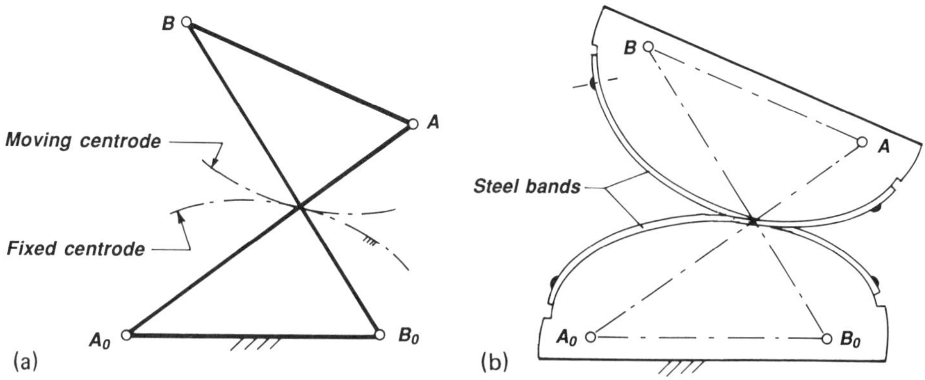

Figure 1.5 (a) A double-crank linkage A_0ABB_0 and its associated moving and fixed centrode. (b) The centrodes have been replaced by steel bands.

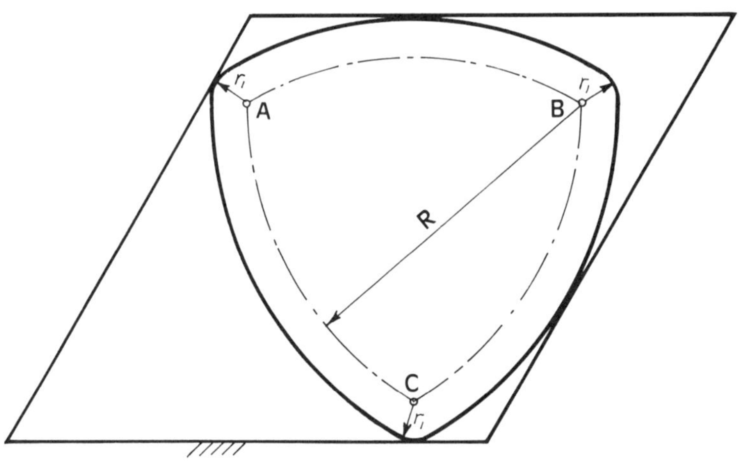

Figure 1.6 A roll-slide mechanism.

of contact. Each point on the polygon traces different curves. To construct the shape of the triangular-shaped member, draw a circle with A and B as centers, where the distance AB = R is arbitrary. The point of intersection of the two circles is C. With A, B, and C as centers, circular arcs are drawn with r_1 as radius. Finally, circles with A, B, and C as centers and $R + r_1$ as radius are drawn and yield the profile shown. The fixed frame has opposite sides parallel (rhomboid). The mechanism can be used for pumps.

Only a few examples of the five basic mechanisms have been shown. It is possible to find delightfully simple solutions to motion problems in everyday life, employing one or more of the five basic mechanisms. Just take, for instance, a trip to your local hardware store and look in the tool section. You will find many turning-joint mechanisms.

2

Motion and Force Transmission in Linkages

When considering a mechanism for practical purposes, it is important to evaluate the mechanism's ability to transmit forces from the input to the output member. In this chapter the main topic is four-bar linkages.

A four-bar linkage is shown in Fig 2.1. It consists of the crank A_0A, which can make a full revolution, the coupler AB which transmits motion from the crank to the rocker B_0B which oscillates back and forth. The crank and the rocker are supported by link A_0B_0 which is designated the frame.

The motion of the four-bar linkage is not influenced by the shape of the links. The motion is determined only by the lengths of the links. It is customary to represent a link with a heavy line as shown in Fig. 2.2. The type of four-bar linkage that is possible is discussed below.

GRASSHOF'S INEQUALITY

Depending on which link is fixed, various types of linkages result: if the sum of the lengths of the shortest and longest link is less than the sum of the two other links, the result is a crank-and-rocker, Fig. 2.1; a double crank or drag-link, Fig. 2.2, where both links b and d can make complete revolutions; or a double rocker, Fig. 2.3, where links b and d can only oscillate.

If the sum of two neighboring links is equal to the sum of the two other

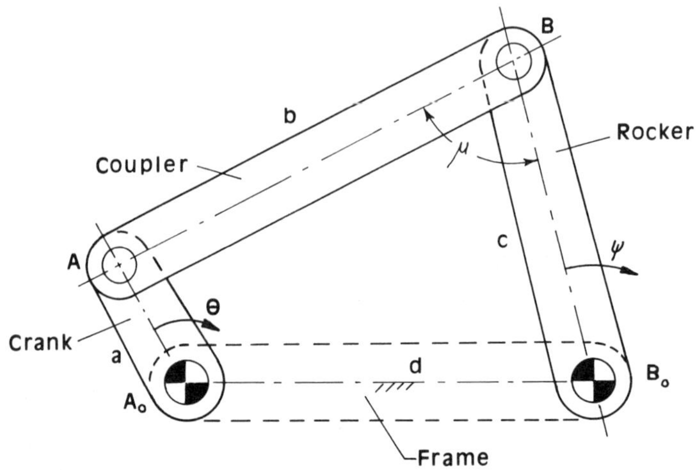

Figure 2.1 Crank-and-rocker linkage (oscillating output).

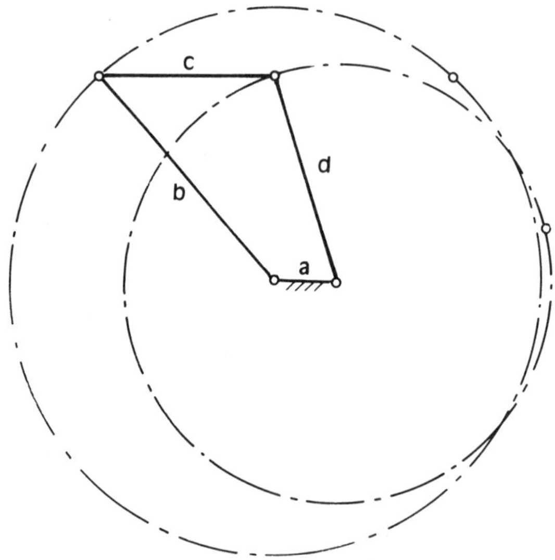

Figure 2.2 Double crank or drag-link linkage. Cranks b and d make full revolutions.

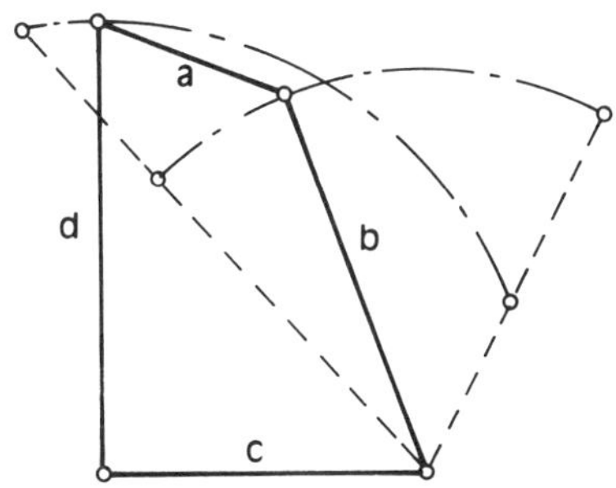

Figure 2.3 Double rocker. Links b and d oscillate, but link can make a full revolution.

links, then all the links in a certain position lie on a straight line. Consider the parallelogram $A_0A_1B_1B_0$, Fig. 2.4, where the above condition is fulfilled. If the crank A_0A moves from position A_0A_1 to position A_0A_2, then crank B_0B moves from position B_0B_1 to B_0B_2, i.e., all links lie on a straight line. If the crank A_0A moves further, crank B_0B can move up or down. If it moves up, the result is a so-called antiparallelogram linkage A_0ABB_0, shown in Fig. 2.5. To overcome this uncertainty position, a double-parallelogram linkage can be used, Fig. 2.6.

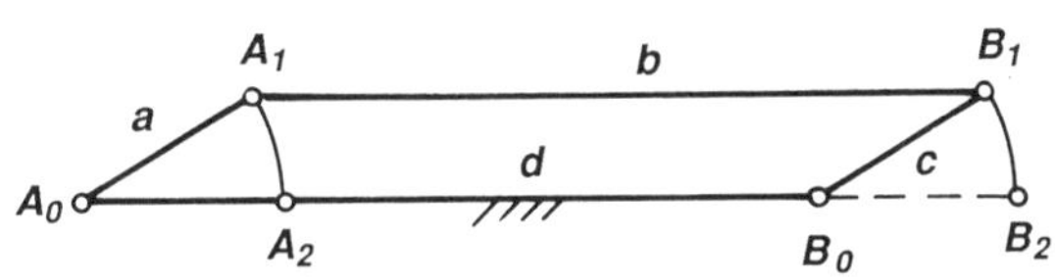

Figure 2.4 Parallelogram linkage.

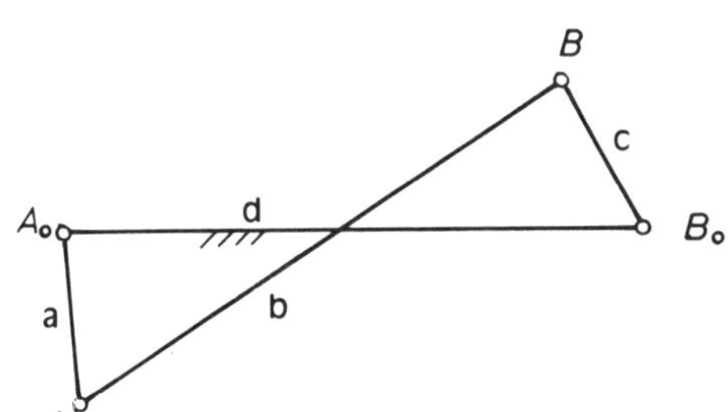

Figure 2.5 Antiparallelogram linkage.

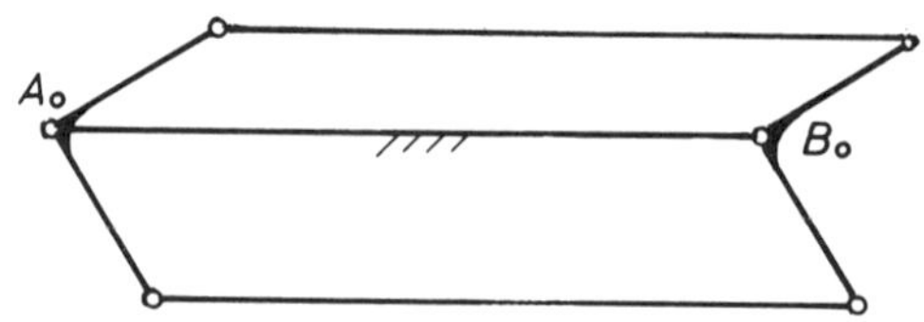

Figure 2.6 Double parallelogram linkage to overcome dead-center positions.

If the sum of the lengths of the shortest and longest links is greater than the sum of the two other links, the result is a double-rocker, Fig. 2.7a and b, where no link can make a full revolution.

THE TRANSMISSION ANGLE

Figure 2.8 shows a four-bar linkage, A_0ABB_0. The transmission angle μ is defined as the smallest angle between the coupler AB and the rocker B_0B,

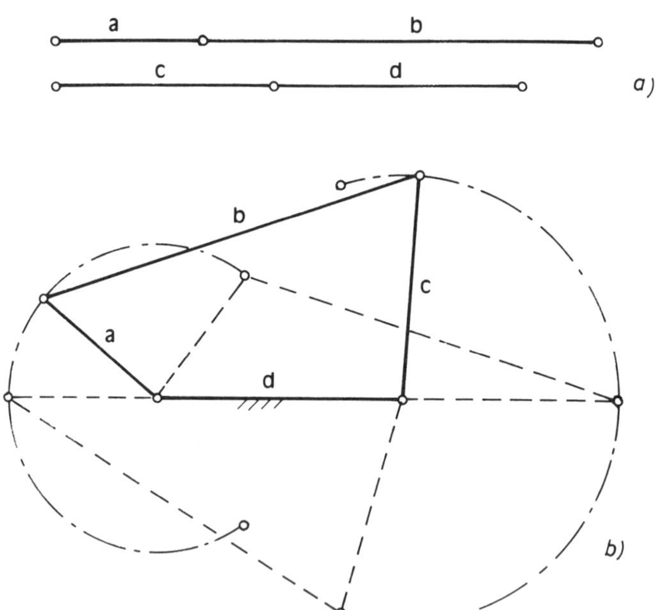

Figure 2.7 The sum of the lengths of the shortest link and the longest link is greater then the sum of the lengths of the two other links. Links a and c can only oscillate.

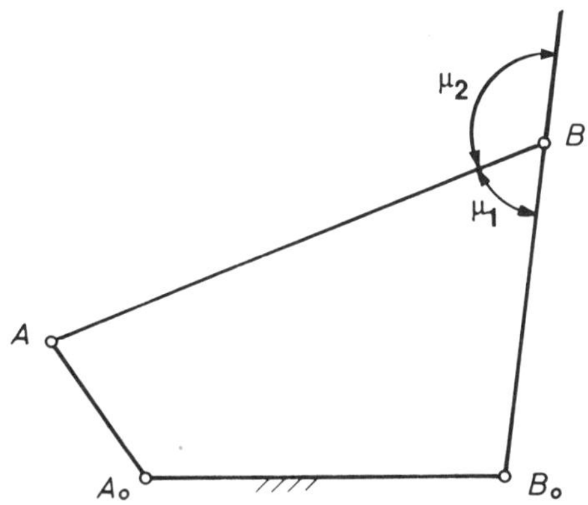

Figure 2.8 Four-bar linkage and associated transmission angle μ.

provided that crank A_0A is the driving member. The smallest angle is μ_1. The same holds for the double crank Fig. 2.9.

Two four-bar linkages in series are shown in Fig. 2.10; here, the transmission angle now occurs at two places, B and D. In Fig. 2.11 D_0D is driven by the coupler of the four-bar linkage A_0ABB_0. There is one transmission angle at B and one at D. Both transmission angles must fulfill certain conditions with respect to minimum value (see later in the chapter).

Figure 2.12, which is actually a reversal of input and output of Fig. 2.11, shows that μ is found by first finding O, the instant center for the motion of link CD. The direction of motion of B is perpendicular to OB. The angle μ is the angle between OB and AB.

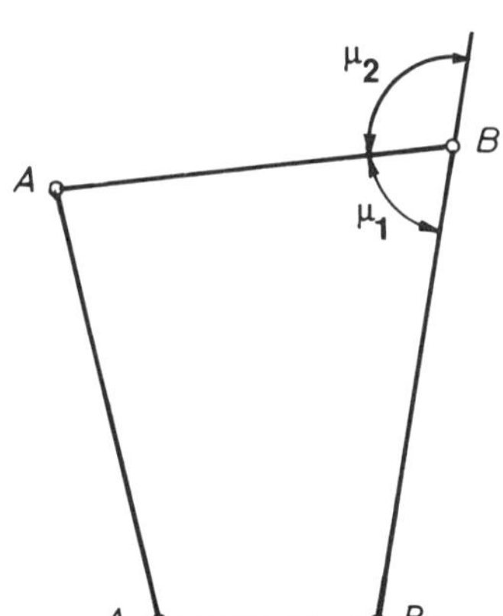

Figure 2.9 Double-crank linkage and associated transmission angle μ.

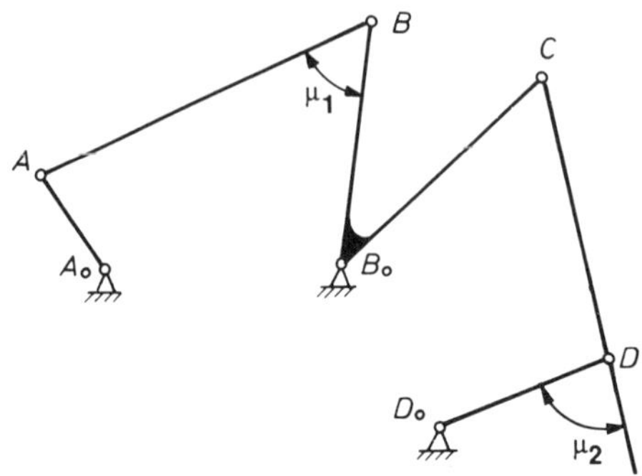

Figure 2.10 Two four-bar linkages in series and their associated transmission angles μ_1 and μ_2.

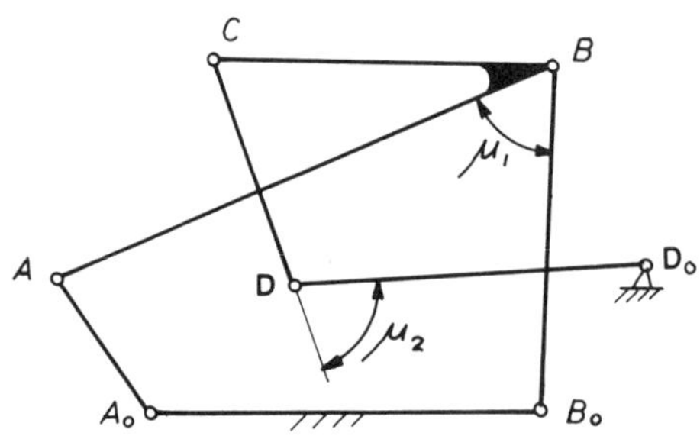

Figure 2.11 Link D_0D driven through link CD where C is a coupler point of the four-bar linkage A_0ABB_0, and associated transmission angles μ_1 and μ_2.

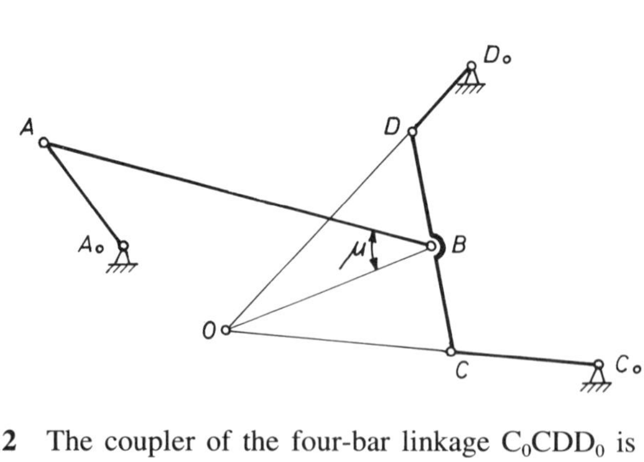

Figure 2.12 The coupler of the four-bar linkage C_0CDD_0 is driven from crank A_0A through link AB.

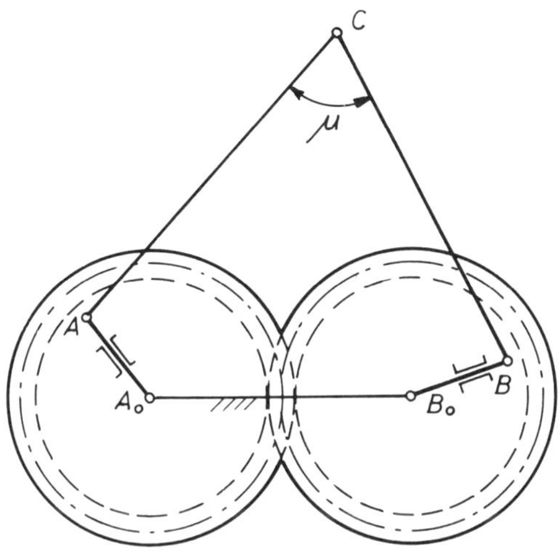

Figure 2.13 Geared double-crank linkage with its associated transmission angle.

Figure 2.13 shows a geared double-crank mechanism, and μ is defined as the smallest angle between links AC and BC. However, this is a purely geometrical definition and often does not correspond to the actual force conditions.

HOW TO DETERMINE OPTIMAL PROPORTIONS FOR A FOUR-BAR LINKAGE*

A common function of a four-bar linkage is to transform rotary into oscillating motion. Frequently in such applications a large force must be transmitted, or force must be converted at high speed. It is then that the transmission angle becomes of paramount importance.

The transmission angle, angle μ in Fig. 2.14, is comparable to the pressure angle in cams. For best results, μ should be as close to 90° as possible during the entire rotation of the crank. This will reduce bending in the links and will produce the most favorable force-transmission conditions. When μ becomes small, a large force is required to drive the rocker arm, and the force fluctuations increase. Charts showing optimal transmission angles make it easier to find the best force-transmission linkage in a wide range of possible selections. Subsequent examples will show how to apply the charts.

*Volmer, J. Four-Bar Power Linkages, *Product Engineering*, Nov. 12, 1962, pp. 71–76.

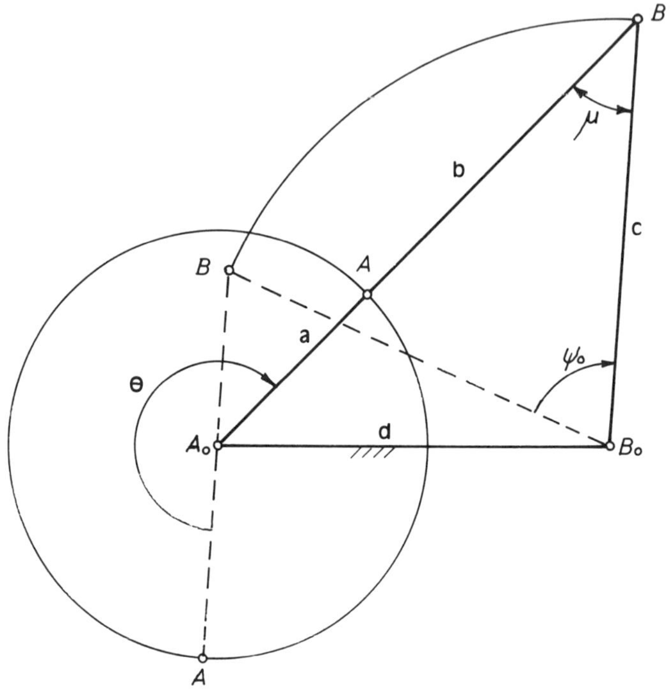

Figure 2.14 A four-bar linkage in its two extreme positions (the heavy- and dotted-line positions).

The Linkage Family

When a four-bar linkage is designed to operate, for example, as a quick-return mechanism, angular displacement of the output link (the rocker) is usually prescribed in relationship to that of the input link (the crank). These angles are measured from the dead-center positions of the rocker, where the rocker reverses its direction of rotation. Thus, in Fig. 2.14, ψ_0 is the angle between the two dead-center positions of the rocker, and θ_0 is the corresponding angle of the crank. Both angles are measured CW from the outside dead-center position, where crank and coupler are superimposed, to the inside dead-center position.

When values are given for ψ_0, θ_0, and for the distance A_0B_0, a family of linkages that meet these conditions can be evolved by a geometrical method. This method, illustrated in Fig. 2.15, is the first phase of the design procedure:

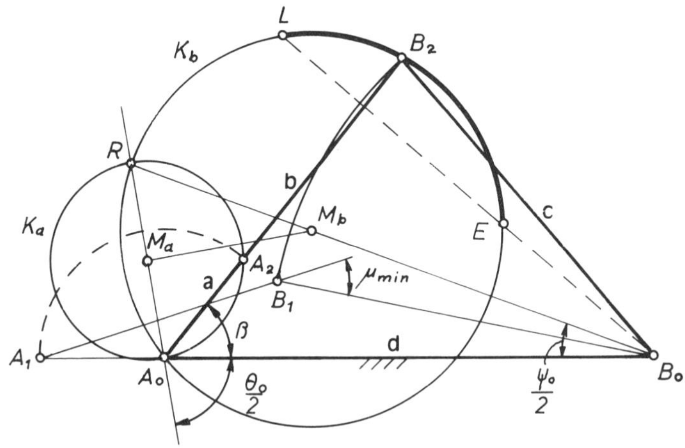

Figure 2.15 Geometrical construction when $\theta_0 < 180°$.

1. Lay out the line of fixed centers, A_0B_0.
2. Construct angles $\theta_0/2$ with center at A_0, and $\psi_0/2$ with center at B_0. Both angles are measured in the same direction. This locates point R.
3. Draw M_aM_b, the midnormal to A_0R. Locate points M_a and M_b.
4. Draw circles K_a and K_b through point R with M_a and M_b as centers, respectively.
5. On circle K_b make $RL = A_0R$ and connect point L with B_0 to locate point E on circle K_b.
6. Choose any point on the circular arc LE as point B, which is the center of the moving joint. This will be one of a family of linkages.
7. Construct line A_0B to intersect circle K_a. This locates crank pivot point A, thus defining all dimensions of the required four-bar linkage.

This method, however, stops short of determining which of all possible mechanisms obtained from the construction is the best power linkage.

Optimal Transmission Angles

The question is, which linkage from the above family has μ_{min} closest to 90°? Since angle μ can be either smaller or larger than 90°, it is useful to define it as the angle between AB or BB_0—or between the extension of lines AB and BB_0. Thus μ is always taken as $\mu < 90°$. For example, if μ

= 120°, it is taken as μ = 60°. A linkage in which μ varies from 75° to 120° (μ_{min} = 60°) is more desirable than a linkage where μ varies from 45° to 90° (μ_{min} = 45°)

For $\theta_0 < 180°$, μ_{min} occurs when crank pin is at A_1 (see Fig. 2.15); for $\theta_0 > 180°$, when at A_2 (see Fig. 2.16); when $\theta_0 = 180°$, the minima of μ at A_1 and A_2 are equal (see Fig. 2.17). The linkage with $\theta_0 = 180°$ is called a centric crank-and-rocker mechanism.

Optimal Transmission Angle Design Charts

Here is how to use the charts to find the linkage with the largest μ_{min} value from the linkage family in Fig. 2.15:

For the given values of θ_0 and ψ_0, find the value for angle β from Chart 2.1 (dashed lines). Angle β = angle AA_0B_0. This locates point B and thus defines the mechanism. Read also the value for max μ_{min} (solid lines). No better transmission angle can be obtained for the given conditions—every other linkage for the same dead-center angles possess a lower value of μ_{min}.

Interpolation can be employed between the curves for β and μ_{min}. How-

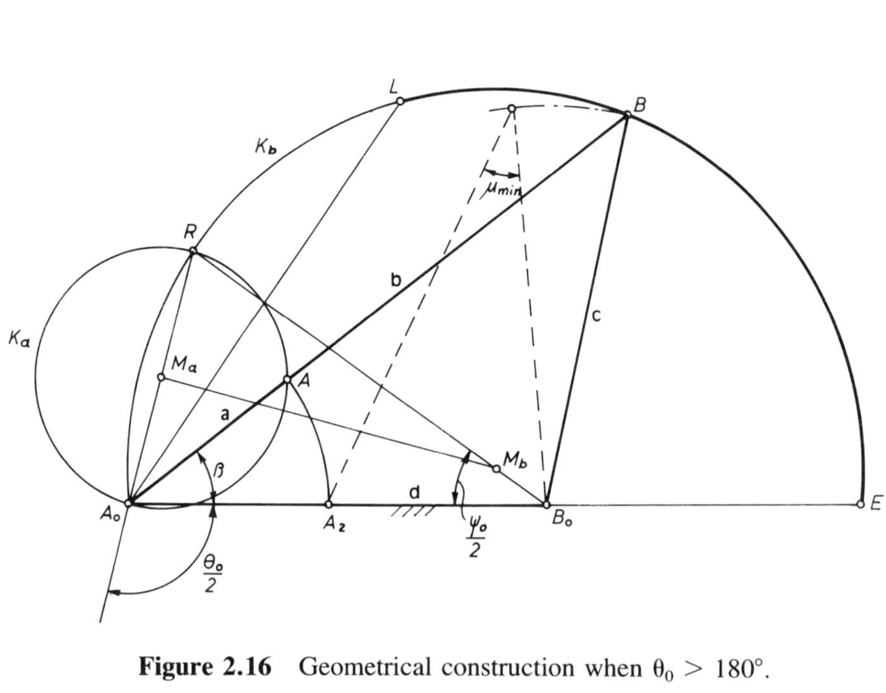

Figure 2.16 Geometrical construction when $\theta_0 > 180°$.

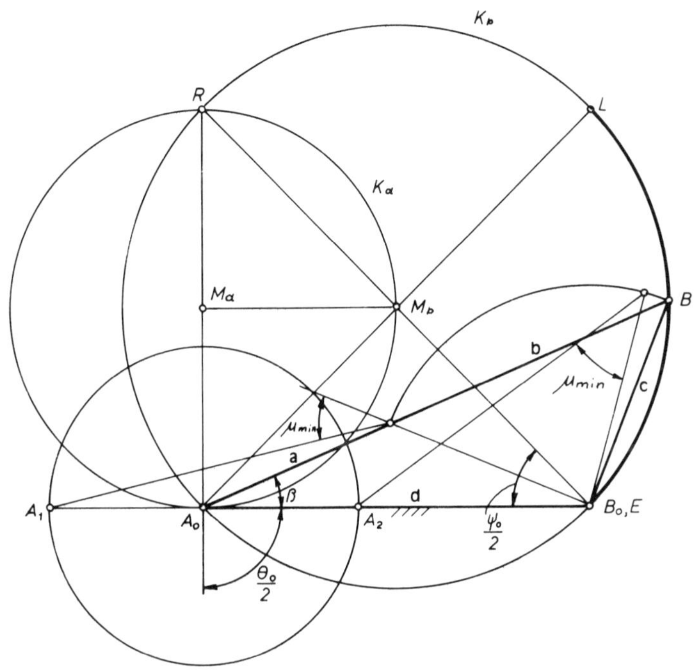

Figure 2.17 Geometrical construction when $\theta_0 = 180°$.

ever, when more accurate values are needed, use the equations (by assuming values for β and d):

$$\frac{a}{d} = -\frac{\sin\frac{\psi_0}{2}\cos\left(\frac{\theta_0}{2}+\beta\right)}{\sin\left(\frac{\theta_0}{2}-\frac{\psi_0}{2}\right)}$$

$$\frac{b}{d} = \frac{\sin\frac{\psi_0}{2}\sin\left(\frac{\theta_0}{2}+\beta\right)}{\cos\left(\frac{\theta_0}{2}-\frac{\psi_0}{2}\right)}$$

$$c_2 = (a+b)^2 + d^2 - 2(a+b)d\cos\beta$$

where $a = AA_0$, $b = AB$, $c = BB_0$, and $d = A_0B_0$.

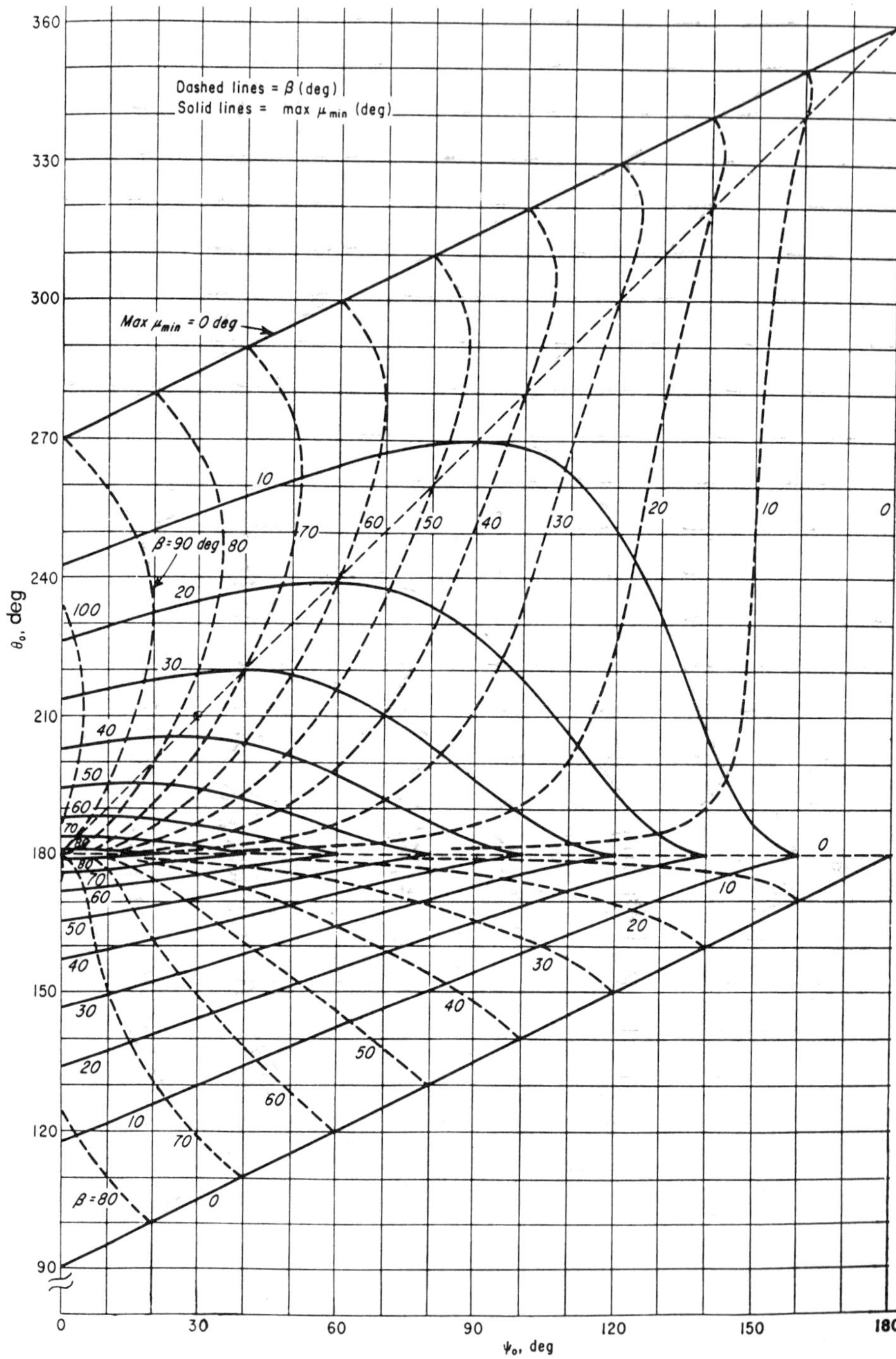

Chart 2.1 Optimal transmission angles: general case.

The design chart in Fig. 2.2 is for the special case where $\theta_0 = 180°$. The following examples illustrate use of the charts.

Example I: Typical Design

The dead-center position construction for $\theta_0 = 160°$ and $\psi_0 = 40°$ is shown in Fig. 2.15. From the Chart 2.1, $\beta = 50.5°$ and max $\mu_{min} = 32°$. Although there is no linkage with more favorable force-transmission characteristics, a linkage with such a low minimum value of transmission angle is not capable of running at high speeds or transmitting great forces.

Example II: When $\theta_0 = \psi_0 + 180°$

When θ_0 and ψ_0 are chosen so that $\theta_0 = \psi_0 + 180°$ at the dead-center position construction, then angle A_0RB_0 becomes a right angle, and circle K_b degenerates into the line A_0R. Figure 2.18 shows the construction for $\theta_0 = 210°$ and $\psi_0 = 30°$. For all linkages within this family, the crank-pin center A coincides with point R, and line A_0R extended beyond point L is the locus of point B, where $A_0R = RL$. Therefore, angle β is useless for finding the optimum linkage.

To obtain this linkage, locate A ($A_1A_0 = A_0A$). Draw circle t, with A_1B_0 as the diameter, to intersect with a line from A_0 perpendicular to A_0B_0. This locates point B_1. The required optimal linkage is $A_0A_1B_1B_0$. Angle $A_2B_1B_0$ = max μ_{min}. All linkages of this family lie on the diagonal dashed line in the Chart 2.1. For this example, the chart shows that max $\mu_{min} = 36°$, and that $\beta = 76°$.

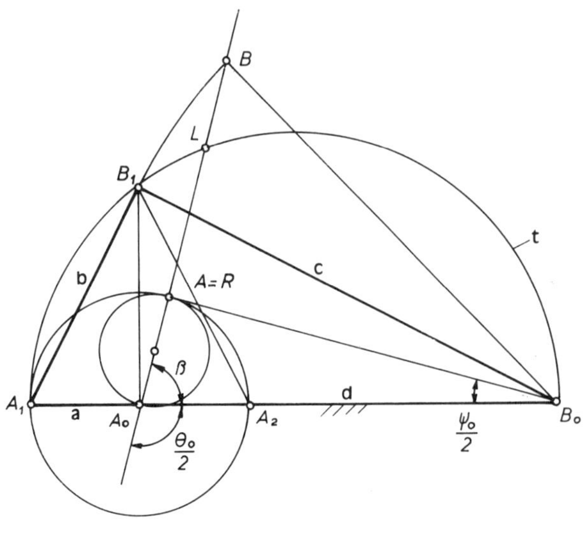

Figure 2.18 Geometrical construction when $\theta_0 = \psi_0 + 180°$.

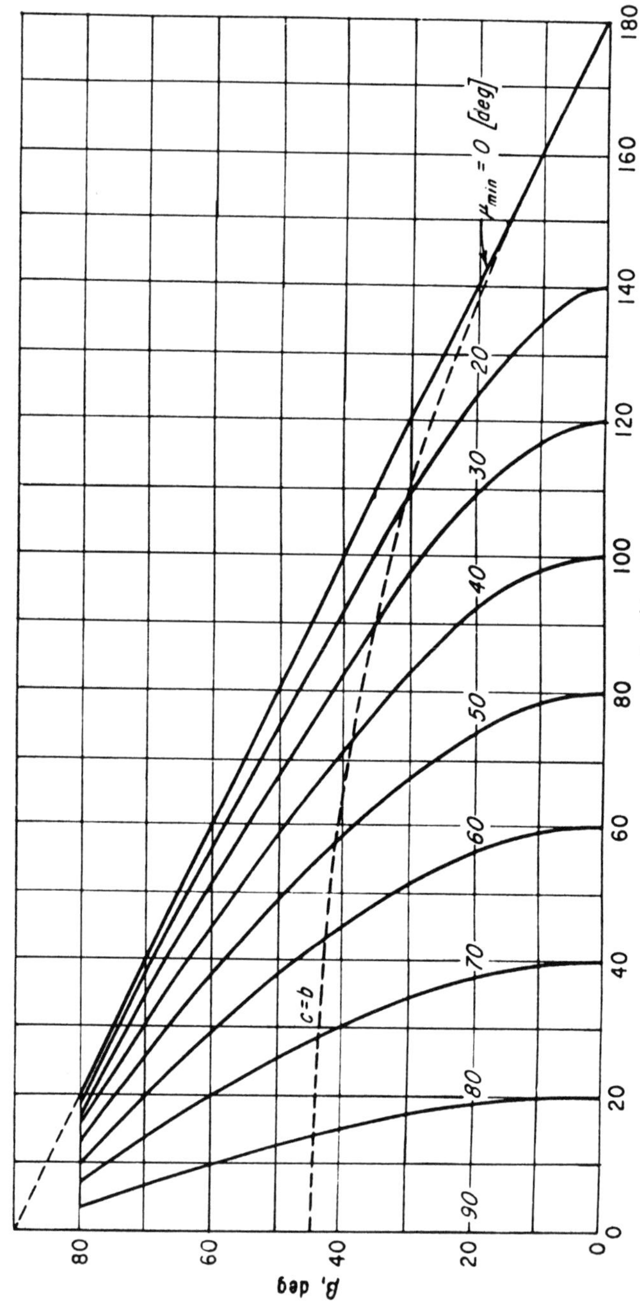

Chart 2.2 Optimal transmission angles when $\theta_0 = 180°$.

Example III: Centric Linkage Design

The centric four-bar linkage where the crank rotates 180° between the two dead-center positions ($\theta_0 = 180°$) provides the greatest amplitudes ψ_0 of the rocker with most favorable transmission angle. However, Chart 2.1 is not applicable for this case because it indicates that $\beta = 0°$ for any desired rocker angle ψ_0. This means that the length of crank and rocker must be zero—a solution without practical significance. Thus, Chart 2.2 is used. For example, if a total rocker displacement of $\psi_0 = 90°$ is required, a linkage with $\mu_{min} = 40°$ will be constructed with angle of $\beta = 23°$, and a linkage with $\mu_{min} = 30°$ with an angle of $\beta = 35°$. The dead-center position construction for $\beta = 23°$ is shown in Fig. 2.17. Here, other conditions can also be prescribed, for instance, a desired length of crank, rocker, or coupler.

ROBERTS' LAW

One of the most remarkable theorems in kinematics was discovered by Roberts,* an English mathematician. Roberts' theorem states that there are three and only three different four-bar linkages that will generate exactly the same coupler curve.

A four-bar linkage A_0ABB_0 with the coupler point C describes the coupler curve shown in Fig. 2.19. As the crank A_0A rotates, the rocker B_0B oscillates back and forth. Through the combined motion of A and B, the coupler point C describes the coupler curve shown. In Fig. 2.19 the two angles by A and B are designated α and β, respectively.

To find the proportions of the two other four-bar linkages that will trace exactly the same coupler curve, a Cayley diagram is used. This diagram is shown in Fig. 2.20 and is constructed by placing the links A_0A, AB, and BB_0 on a straight line and laying out angles α and β as shown by A and B. Draw a parallel through A_0 to AC and through B_0 to BC and lay out the angles α and β at C as shown. This determines points D, E, F, G, and H_0.

To find the actual linkage, fold back the links of Fig. 2.20 to their original position as shown in Fig. 2.21. When doing so, recall that parallel lines in parallelograms remain parallel. Draw parallelogram A_0ACD and B_0BCF, whereby points D and F are found. Make triangle CDE similar to triangle BAC by laying out angle α and β as shown. Do the same for triangles FCG and BAC. Finally the parallelogram $GCEH_0$ determines H_0.

---

*Roberts, S. *Three Bar Motion in Plane Space. Proceedings of the London Mathematical Society*, vol. 7, 1875, p. 14.

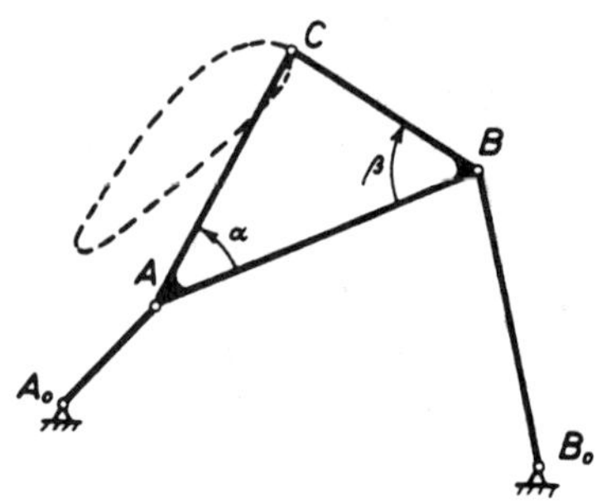

Figure 2.19 Four-bar linkage and coupler curve.

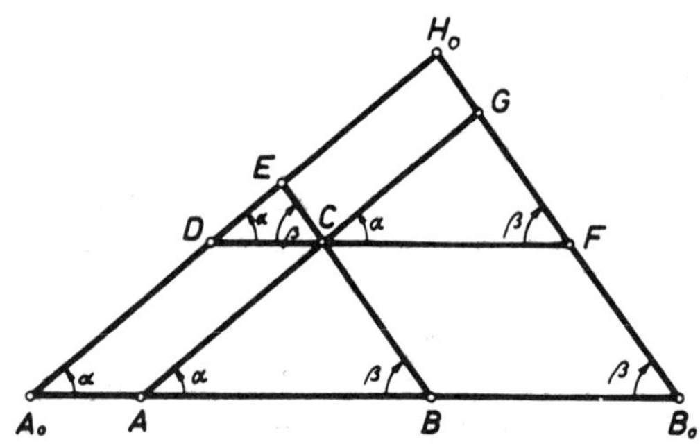

Figure 2.20 Cayley diagram to obtain the two other four-bar linkages related by Roberts' theorem.

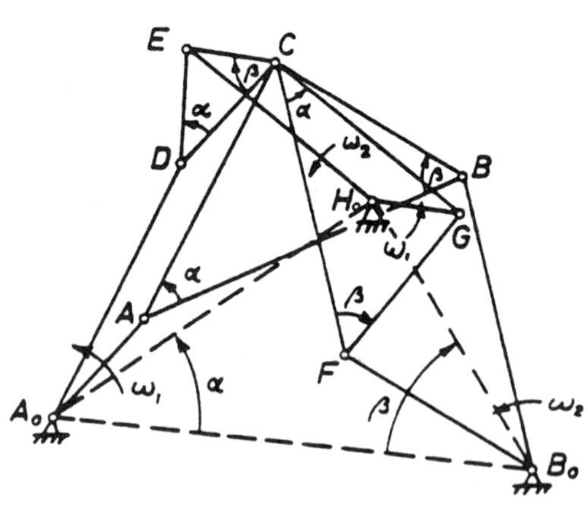

Figure 2.21 Construction relating the three Roberts' linkages.

The three four-bar linkages that traces identical coupler curves are

A_0ABB_0 with the coupler point C (Fig. 2.19)
A_0DEH_0 with the coupler point C (Fig. 2.22)
B_0FGH_0 with the coupler point C (Fig. 2.23)

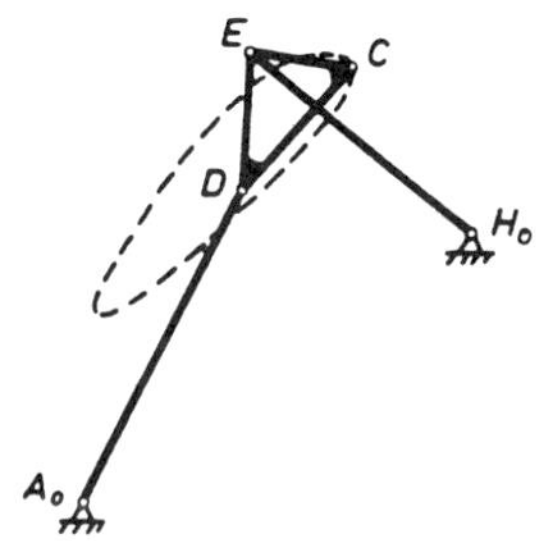

Figure 2.22 The coupler curve shown is identical to the one shown in Fig. 2.19.

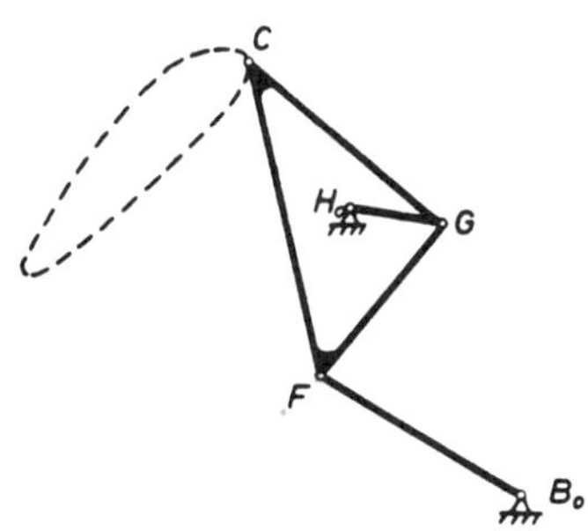

Figure 2.23 The coupler curve shown is identical to the one shown in Fig. 2.19

Roberts' law can be applied, for instance, if the linkage in question is a crank-and-rocker, because then one of Roberts' linkages is another crank-and-rocker where the optimal transmission angle might be closer to 90°.

3
The Slider Crank

INTRODUCTION

The slider crank is used extensively in internal combustion engines, where it converts the gas pressure on a piston to a moment on the crankshaft to power automobiles, lawn mowers, earth-moving equipment, etc., as well as pumps; it is also used widely as a motion converter (or mechanism). Slider cranks can be characterized as centric or offset. An offset slider crank is shown in Fig. 3.1. If $e = 0$, then the slider crank is called a centric slider crank.

The process of converting gas pressure on a piston to a moment on a rotating crankshaft can be inverted; that is a moment exerted on the crankshaft can be used to let the piston compress air (a compressor) or to pump liquid. The use of the slider crank as a motion converter is comprehensive, and quite a variety of motions can be obtained from it. The slider crank can also be used in series with other mechanisms.

A slider crank is made up of four parts, namely, the crank A_0A, the connecting rod AB, the slider at B_0, and the frame (Fig. 3.1). The joints by A_0 and A are turning joints, the joint by B is called a sliding joint. If the motion of B, as it moves back and forth, passes through A_0, then the

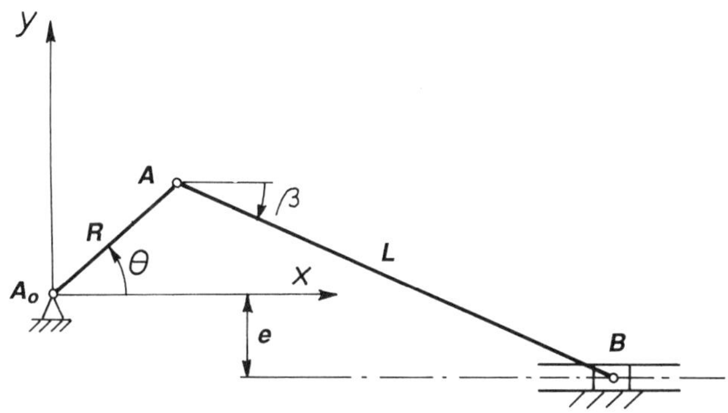

Figure 3.1 Eccentric, or offset, slider crank.

mechanism is called a centric slider crank, otherwise an offset slider crank. A so-called pin enlargement is shown in Fig. 3.2, where a circular disc has been fastened to the crank A_0. A is the center of the disc, and if $A_0A = R$ is the same as in Fig. 3.1, then the two mechanisms are exactly equivalent kinematically because the relative motions of the links are the same.

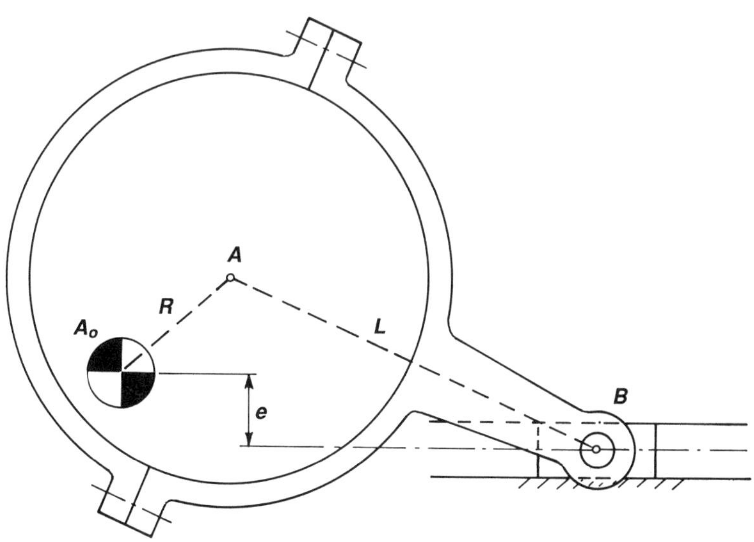

Figure 3.2 Principle of pin enlargement.

DISPLACEMENT, VELOCITY, AND ACCELERATION

The displacement of the slider (S) is calculated from the position where the crank is on the extension of the connecting rod (A_0AB form a straight line). (Symbols are defined at the end of this chapter.)

$$S = \sqrt{(R + L)^2 - e^2} - R \cos \theta - L \cos \beta \tag{3.1}$$

For e = 0

$$S = R + L - R \cos \theta - L \cos \beta \tag{3.1a}$$

The velocity of the slider (V_B) is found using unit vectors:

$$V_B = V_A + V_{B/A}$$

$$V_B e^{io} = iR\omega e^{i\theta} + iL\omega_1 e^{i\beta} \tag{3.2}$$

$$V_B(\cos 0° + i \sin 0°) = iR\omega(\cos \beta + i \sin \beta) + iL\omega_1(\cos \beta_1 + i \sin \beta_1) \tag{3.3}$$

From eq. (3.3) two equations can be written, namely

$$V_B + L\omega_1 \sin \beta = -R\omega \sin \theta \tag{3.4}$$

$$-L\omega_1 \cos \beta = R\omega \cos \theta \tag{3.5}$$

Solving for V_B and ω_1:

$$V_B = \frac{R\omega \sin (\beta - \theta)}{\cos \beta} \tag{3.6}$$

$$\omega_1 = -\frac{R\omega \cos \theta}{L \cos \beta} \tag{3.7}$$

The value ω_1 is positive when rotation is CCW, otherwise it is negative. V_B is positive when directed away from A_0 (to the right in Fig. 3.1).

The acceleration of the slider (A_B) is

$$A_B = A^n_A + A^t_A + A^n_{B/A} + A^t_{B/A}$$

$$A_B e^{io} = -R\omega e^{i\theta} + iR\alpha e^{i\theta} - L\omega_1^2 e^{i\beta_1} + iL\alpha_1 e^{i\beta_1} \tag{3.8}$$

or

$$A_B + L\alpha_1 \sin \beta = -R\omega^2 \cos \theta - R\alpha \sin \theta - L\omega^2{}_1 \cos \beta \tag{3.9}$$

$$-L\alpha_1 \cos \beta = -R\omega^2 \sin \theta + R\alpha \cos \theta - L\omega^2{}_1 \sin \beta \tag{3.10}$$

$$A_B = R\omega^2\left[\frac{\cos(\theta - \beta)}{\cos \beta} - \lambda \frac{\cos^2 \theta}{\cos^3 \beta}\right] + \frac{R\alpha}{\cos \beta} \sin (\theta - \beta) \tag{3.11}$$

$$\text{where } \lambda = \frac{R}{L} \tag{3.12}$$

$$\alpha_1 = R\omega^2\left(\frac{\sin\theta}{L\cos\beta} + \frac{R\cos^2\theta\sin\beta}{L^2\cos^3\beta}\right) - \frac{R\alpha\cos\theta}{L\cos\beta} \tag{3.13}$$

Eqs. (3.6) and (3.11) are valid for both centric and offset slider cranks. The angle between the connecting rod and the motion of the slider (β) can be found from

$$e = R\sin\theta + L\sin\beta$$

or

$$\beta = \sin^{-1}\left(\frac{e - R\sin\theta}{L}\right) \tag{3.14}$$

where e is negative, if B lies below the x axis. For $e = 0$, the slider crank is a centric slider crank:

$$\beta = -(R/L)\sin\theta = -\lambda\sin\theta \qquad (e = 0) \tag{3.14a}$$

Design Charts

For the centric slider crank, the Charts 3.1, 3.2, and 3.3 were calculated using eqs. (3.1), (3.6), (3.11), and 3.14a. It is assumed that $R = 1$ (unit) and ω is constant ($\alpha = 0$).

Example

$R = 100$ mm, $L = 200$ mm, $\theta = 45°$, $N = 3000$ rpm counter-clockwise (CCW). Find the displacement, velocity, and acceleration of the slider.

Solution

From Chart 3.1 $\quad S' = 1 - 0.58 = 0.42$
From Chart 3.2 $\quad V'_B = 0.97$
From Chart 3.3 $\quad A'_B = 0.75$

The value of S', Chart 3.1, was found at the intersection of the $L/R = 2.0$ value and $\theta = 45°$. One must remember that the value 1.0 corresponds to the nearest position. Therefore the values of the displacement of the slider from the farthest away position are found by subtracting the value read from 1.0. The values found must be multiplied by the factors shown in Table 3.1.

For $R = 100$ mm and $\omega = n\,2\pi/60 = 3000\,2\pi/60 = 314/\text{s}$

$$S = S'R = 0.42 \times 100 = 42 \text{ mm}$$

$$V_B = V'_B R\omega = \frac{0.97(100)(314)}{1000} = 30.5 \text{ m/s}$$

$$A_B = A'_B R\omega^2 = \frac{0.75(100)(314)^2}{1000} = 7395 \text{ m/s}^2$$

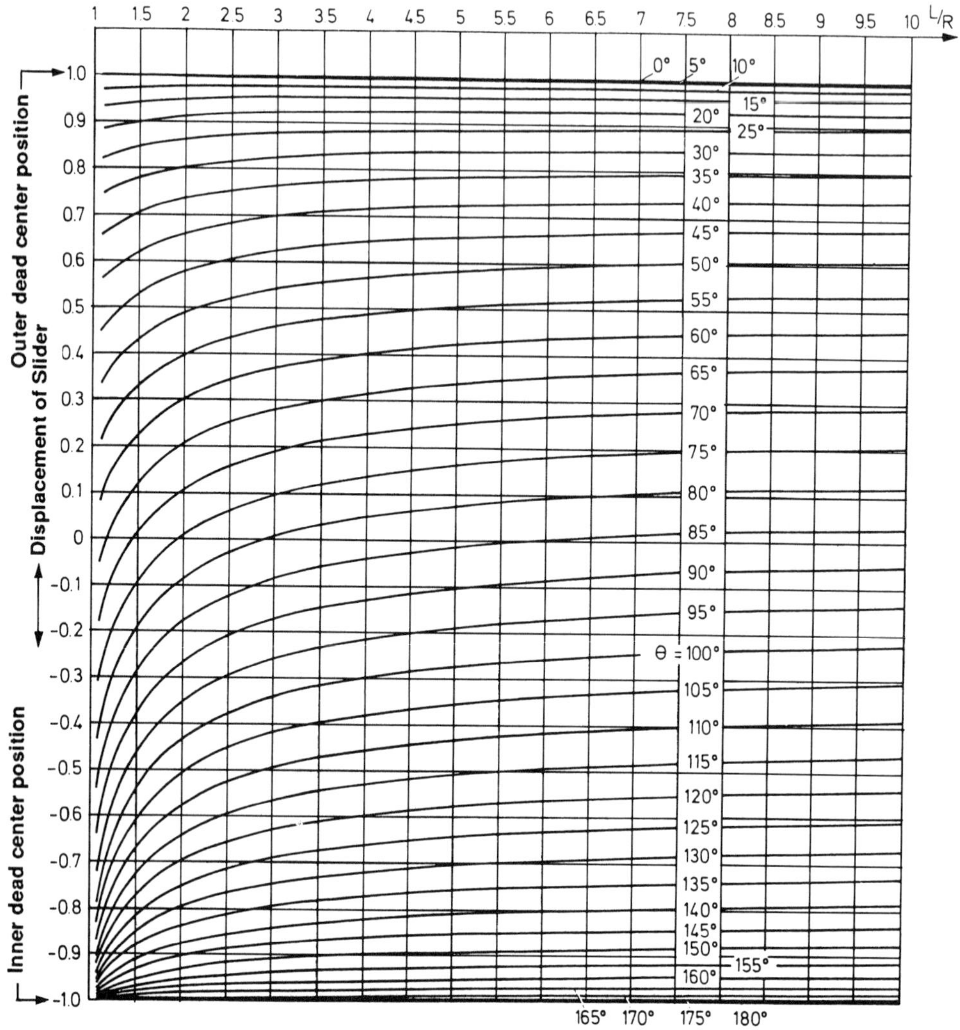

Chart 3.1 Displacement of slider as a function of θ and L/R (centric slider crank).

MOTION CHARACTERISTICS OF THE SLIDER CRANK

In Fig. 3.3 the motion of the slider of the eccentric slider crank shown in Fig. 3.1 is considered in more detail. As the crank A_0A rotates, the slider moves back and forth. In the positions where it reverses its motion the ve-

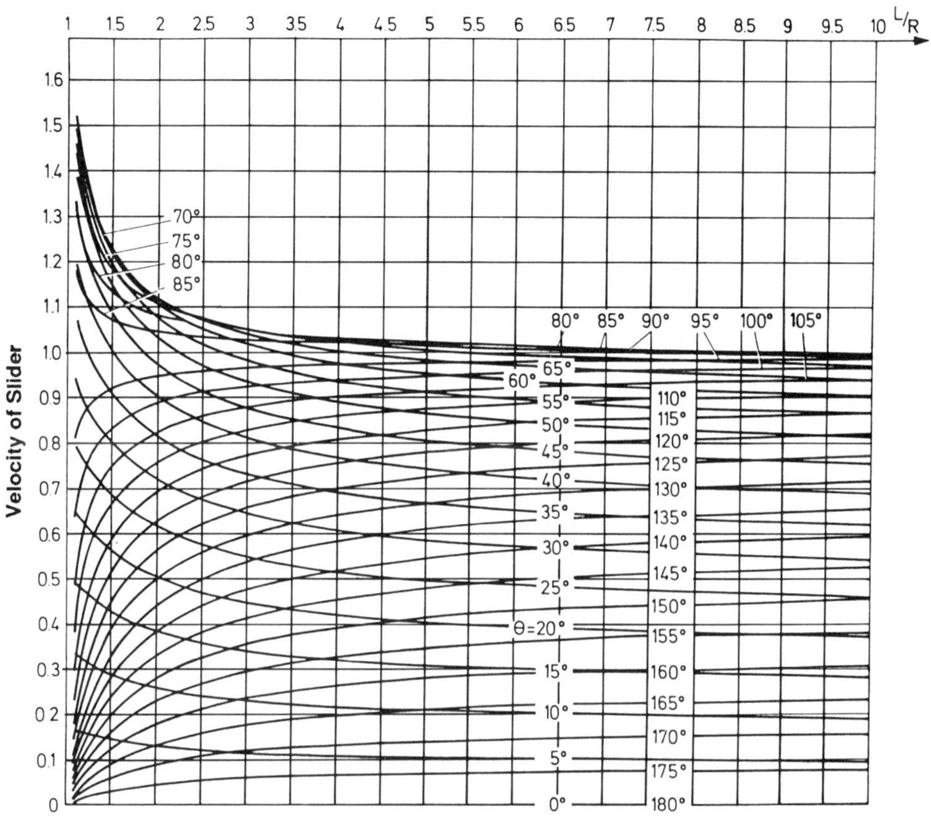

Chart 3.2 Velocity of slider as a function of θ and L/R.

locity of the slider must be zero. These two positions occur when the crank and the connecting rod lie on top of each other (position $A_0A_1C_1$), and when the crank is colinear with the extension of the rod (position $A_0A_2C_2$). The distance between the two points C_1 and C_2 is equal to the stroke S.

For given R, L, and e (Fig. 3.4), the positions C_1 and C_2 are found as follows: a horizontal line is drawn at a distance e from A_0. The value e is considered positive when B is above A_0 and negative when below. A circle with A_0 as center and L + R as radius locates C_2 on the line of eccentricity, and a circle with radius L − R locates C_1.

Often the crank angle, when the crank rotates from position A_1 to A_2 and the slider moves from C_1 to C_2, is given. Suppose the crank angle ϕ between dead-center positions (Fig. 3.3) is 210° and C_1C_2 = S = 60 mm.

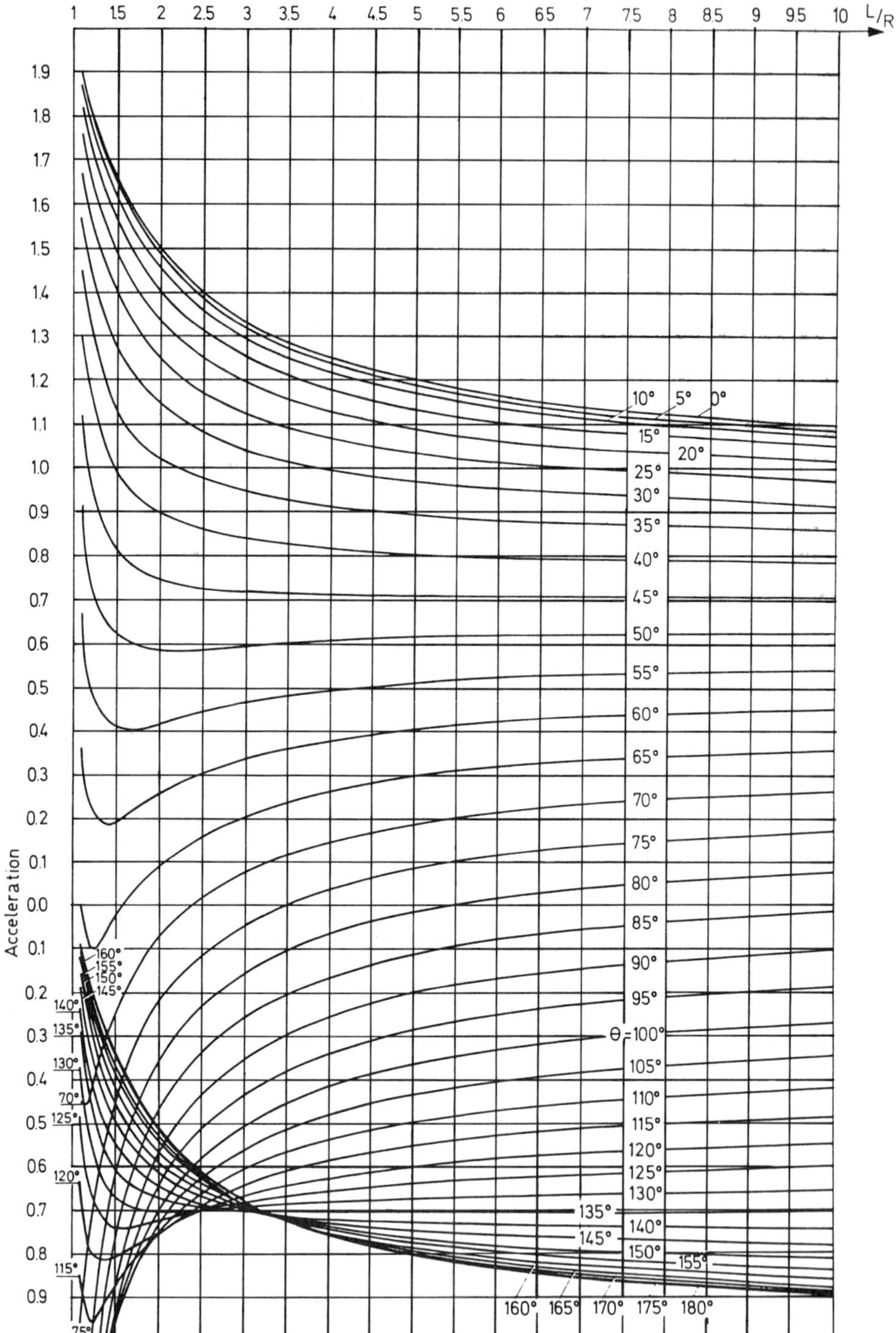

Chart 3.3 Acceleration of slider as a function of θ and L/R.

Table 3.1 Slider Displacement

Value read	Correction factor
S'	R
V'_R	$R\omega$
A'_B	$R\omega^2$

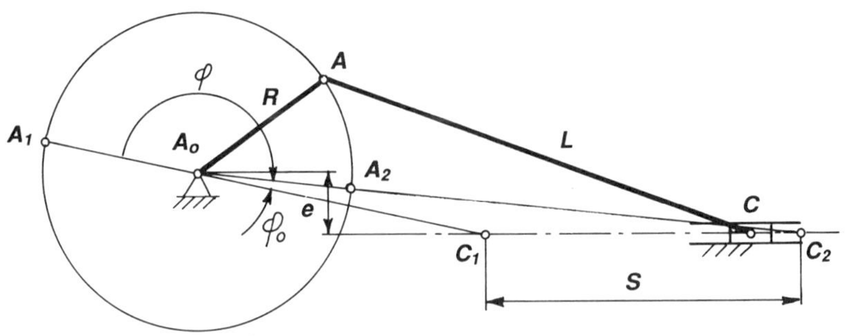

Figure 3.3 The extreme positions of a slider crank.

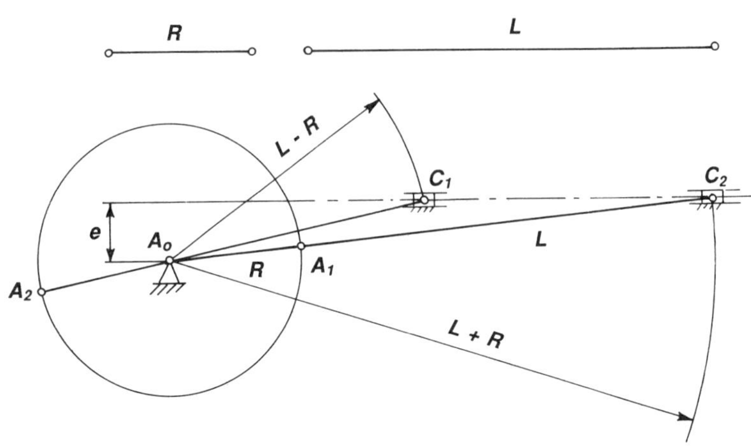

Figure 3.4 Finding A_0, C_1, and C_2 for given e, R, and L.

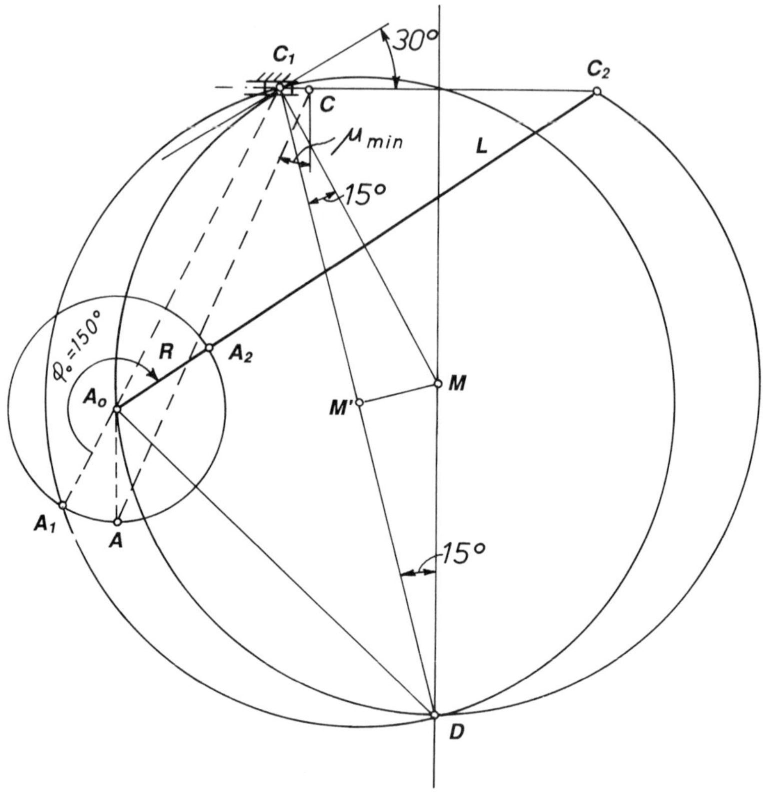

Figure 3.5 Locus for the position of A_0 for given crank angle ϕ_0.

Figure 3.5 shows the construction that locates A_0. First a line subtending an angle of 210° − 180° is 30° with C_1C_1 is drawn. A perpendicular to this line at C_1 locates M on the midnormal to C_1C_2. A circle with M as center and passing through C_1 (and C_2) is the locus for the crankshaft center A_0. Crank radius R and rod length L are calculated from

$$L + R = A_0C_2$$

$$L - R = A_0C_1$$

$$\overline{}$$

$$L = 1/2(A_0C_2 + A_0C_1)$$

$$R = 1/2(A_0C_2 - A_0C_1)$$

A circle with center at the midpoint M_1 of C_1D passing through C_1 is the locus for crank points A_1. However, not all points on the A_0 circle yield useable solutions because the minimum transmission angle μ_{min}, which is shown in Fig. 3.5, may be too small, so that motion cannot be transmitted from the crank to the slider because of the additional friction forces that are created. In general, this angle should be greater than 40° by the slider crank. The minimum transmission angle occurs when point A is farthest away from the line of motion of the slider, which occurs when A_0A is perpendicular to C_1C_2. As the position of A_0 on the locus circle changes, the minimum transmission angle changes, and at some point this angle has its greatest value, which is designated optimum of μ_{min}.

Figure 3.6 shows how the optimum μ_{min} can be achieved. Here, circles indicate the possible location of the crankshaft center, for given crank angles have been drawn. An example will illustrate its use.

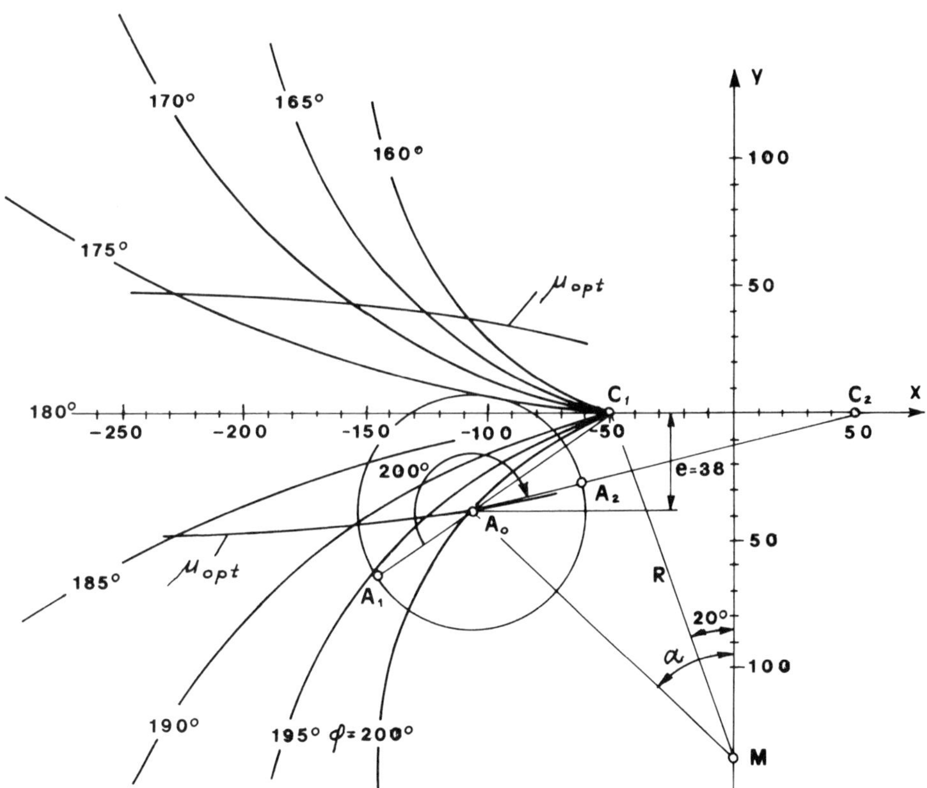

Figure 3.6 Finding optimal proportions of a slider crank with optimal value of μ_{mim}.

Example

Find the proportions of a slider crank for which $C_1C_2 = S = 100$ mm and $\phi = 200°$ (CW rotation).

Solution

A_0 is located at the intersection of the circle $\phi = 200°$ and the μ_{opt} curve. The eccentricity e can be read on the vertical axis; here e = 38 mm. In Fig. 3.7 the resultant mechanism is shown.

The exact values based on an eccentricity of e = 38 mm can be calculated as follows:

$$R = \frac{\frac{1}{2}S}{\sin(\varphi - 180°)} = \frac{50}{\sin 20°} = 146.2 \qquad (3.15)$$

$$R \cos 20° - R \cos \alpha = e$$

$$\alpha = \cos^{-1}\frac{R \cos 20° - e}{R} = \cos^{-1}\frac{146.2 \cos 20° - 38}{146.2} = 42.26° \qquad (3.16)$$

Coordinates of A_0 (x_0, y_0):

$$x_0 = -R \cos \alpha = -146.2 \cos 42.26° = -108.2 \text{ mm} \qquad (3.17)$$

$$y_0 = e = -38 \text{ mm} \qquad (3.18)$$

$$L - R = A_0C_1 = \sqrt{\left[x_0 - \left(-\frac{S}{2}\right)\right]^2 + e^2}$$

$$= \sqrt{[-108.2 - (-50)]^2 + 38^2} = 69.5 \text{ mm} \qquad (3.19)$$

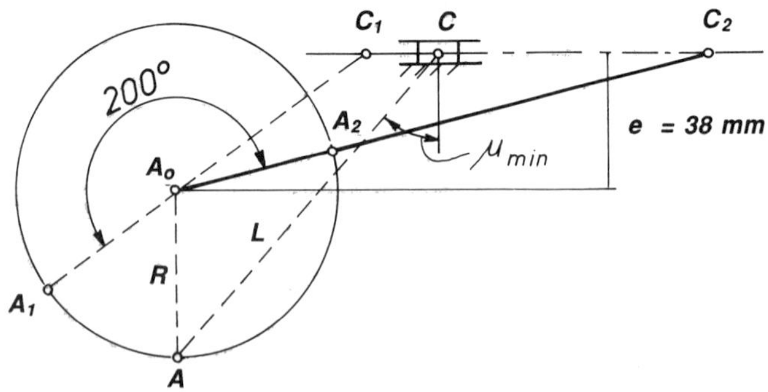

Figure 3.7 Finding the value of μ_{mim}.

$$L + R = A_0C_2 = \sqrt{\left(x_0 - \frac{S}{2}\right)^2 + e^2}$$

$$= \sqrt{(-108.2 + 50)^2 + 38^2} = 162.7 \text{ mm} \quad (3.20)$$

From eqs. (3.19) and (3.20):

$$L = 116.1 \text{ mm}$$
$$R = 46.6 \text{ mm}$$

The value μ_{min} occurs in the position where the crank in A is farthest away from the path of the slider.

$$\mu_{min} = \cos^{-1}\frac{R + e}{L} = \cos^{-1}\frac{46.6 + 38}{116.1} = 43.2° \quad (3.21)$$

The above-calculated values of L and R result in a slider crank where $\phi = 200°$ but where the calculated μ_{min} might be off the theoretical optimal value of μ_{min} because e = 38 mm was read from the diagram; however, repeated calculations of μ_{min} for values of e around the value of 38 mm will yield the exact optimal value.

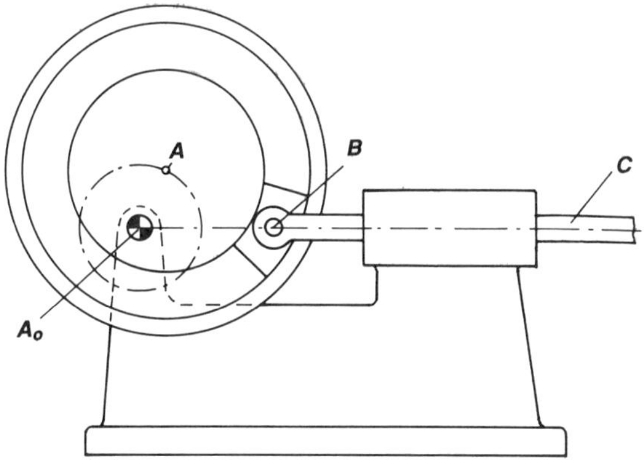

Figure 3.8 Pin-enlarged slider crank. The eccentric with a groove (cam) and center A is rotating around A_0. A slider formed as a circular arc slides in the groove. The slider is attached through a joint B to another slider C. A has the same distance from A_0 at all times. Therefore the mechanism shown is exactly kinematic equivalent to a slider crank A_0AB because the output motion of slider C is the same in both cases. The crank A_0A has been done away with and instead of a crankshaft, a straight shaft can be used. The crank A_0A has actually been enlarged so that the crank is a cam.

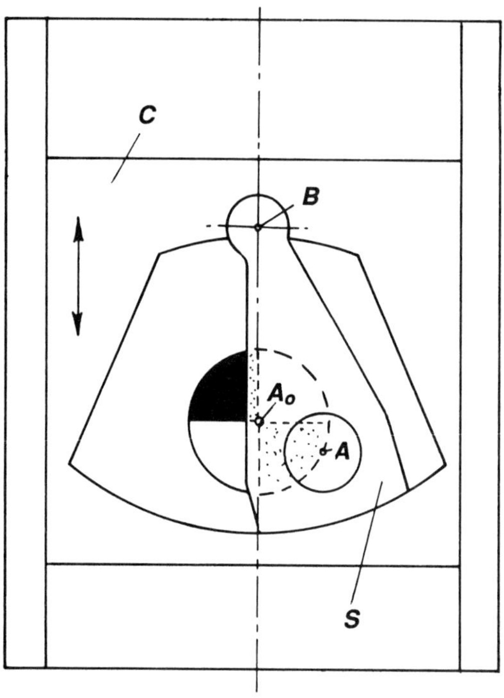

Figure 3.9 Instead of enlarging the crank as in Fig. 3.8, a pin enlargement can be made with respect to the connecting rod AB. AB is supported in the slider C that slides up and down. The arc-shaped surface C of the connecting rod has its center at B. Rod AB can therefore make an oscillating motion relative to the slider C. F. Reuleaux, a German kinematician, is the originator of this compact design.

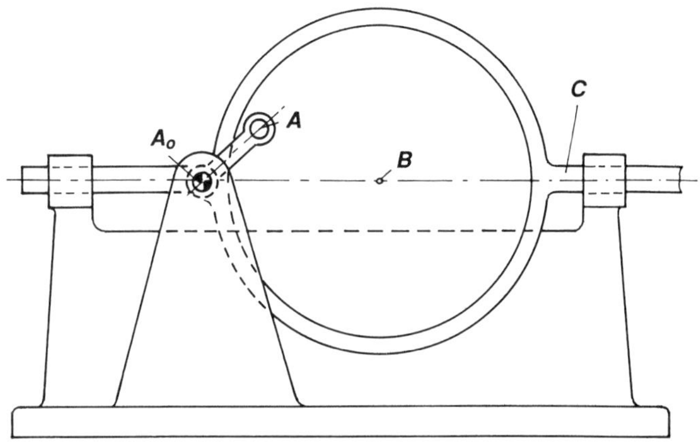

Figure 3.10 A pin enlargement has been made with respect to the connecting rod AB. AB is supported in the slider C that moves back and forth horizontally. The circular-shaped surface C of the connecting rod has its center at B. Rod AB can therefore make an oscillating motion relative to the slider C.

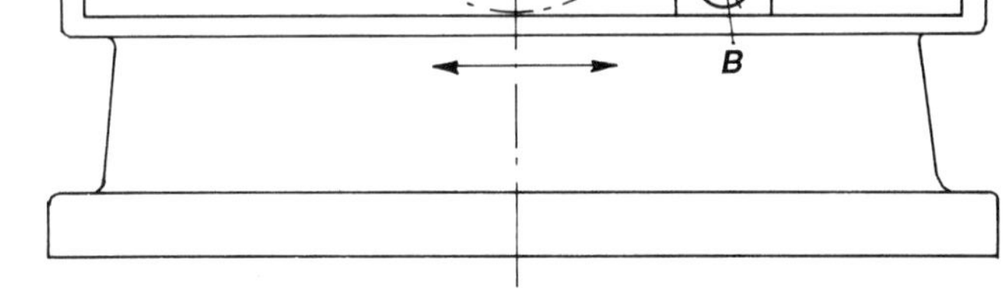

Figure 3.11 The eccentric slider crank is proportioned so that the crank can make a complete revolution. The sum of the length of A_0A and the eccentricity e is equal to the length of the connecting rod AB. Slider B must be pushed near its center position to overcome the dead-center position and move to the other side when B is right below A_0.

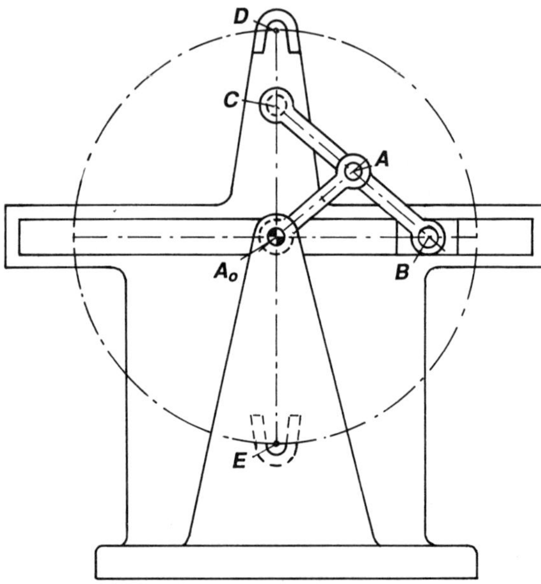

Figure 3.12 The crank A_0A drives link ABC. When A is right above A_0, B is at A_0. The transmission angle in this position is zero, and theoretically no motion can take place. This dead-center position is overcome by pin C that enters guides D and E alternately.

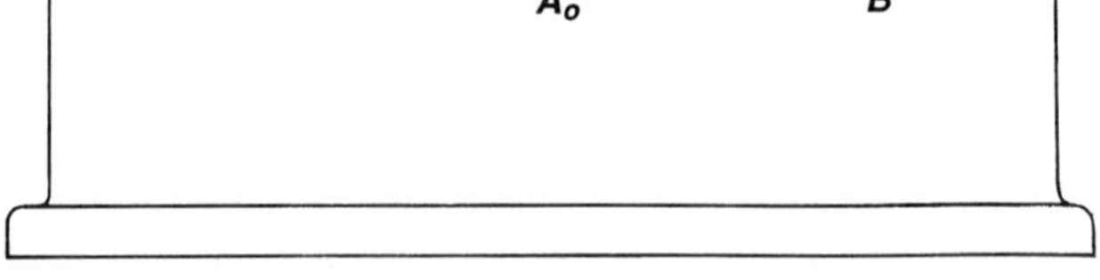

Figure 3.13 The centric slider-crank A_0AB has an extended slider with a slot that is guided by the square-shaped member at A_0, resulting in a very good guidance ratio for the slider and also improved frictional forces (guidance ratio = length of slider/diameter of slider).

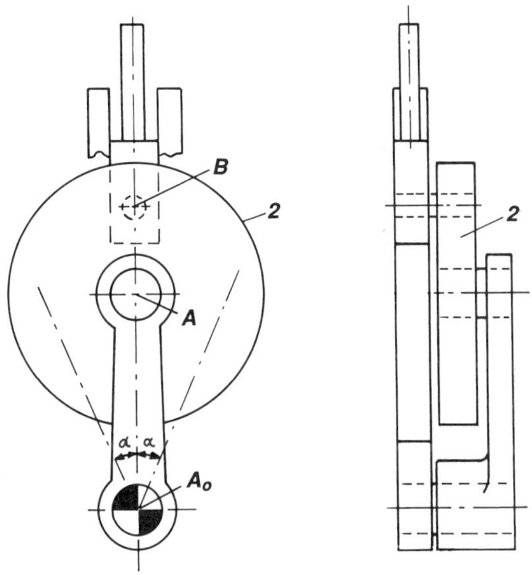

Figure 3.14 The rotary motion of the fly-wheel-like disc 2 is converted to an oscillating motion of the slider B. The basic mechanism is a slider crank A_0AB with slider at B. Crank A_0A is larger than the connecting rod AB, so rotary input cannot be link A_0A. Instead, rotary input is to the fly-wheel 2.

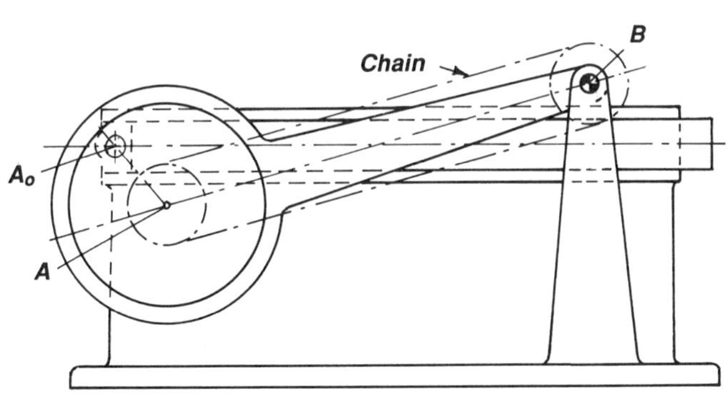

Figure 3.15 Kinematic inversion of a slider crank. The basic mechanism is the slider crank A_0AB, but instead of making B slide back and forth and A_0 a fixed center of rotation, B is made stationary and A_0 slides back and forth. Input is at the fixed center B to the chain. The motion of the chain is transmitted to a sprocket at A. The sprocket is fixed to the crank A_0A, which makes a rotary motion. A_0 slides back and forth.

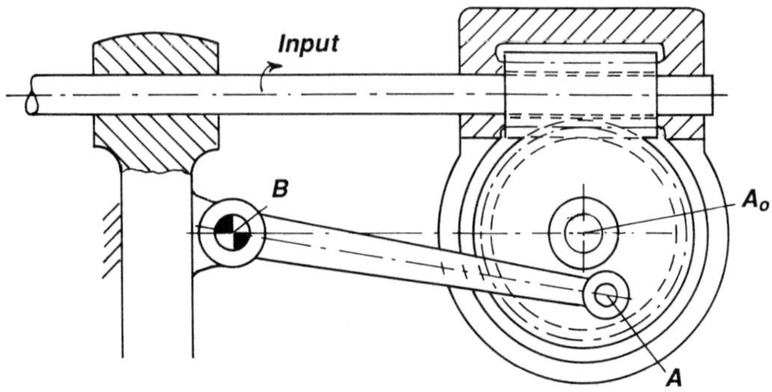

Figure 3.16 The input shaft rotates and is imparted an oscillating motion through the worm and worm gear. A_0A rotates around A_0. This makes A_0 move back and forth along a horizontal line because A is connected to B, which is stationary, causing the input shaft to oscillate back and forth.

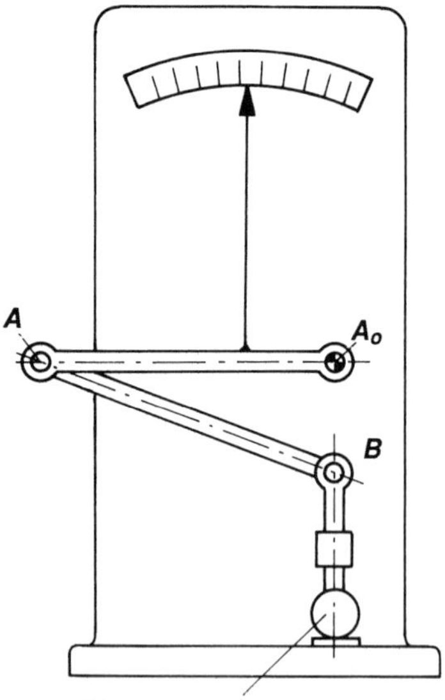

Figure 3.17 Slider crank A_0AB with slider at B is employed to measure thickness of work specimen.

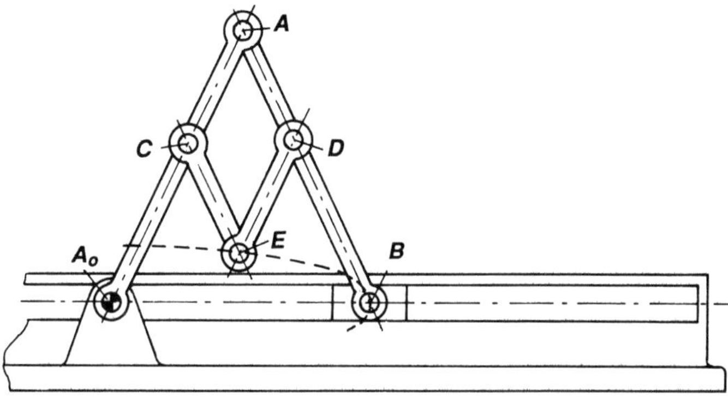

Figure 3.18 The slider crank A_0AB is provided with links at C and D. If A_0A = BDA, and AC = CE = AD = DE, then E traces an ellipse. The equation for the ellipse is $x^2/I^2 + y^2/(I + 2a)^2 = 1$, where A_0A = AB = I and AC = AD = CE = DE = a.

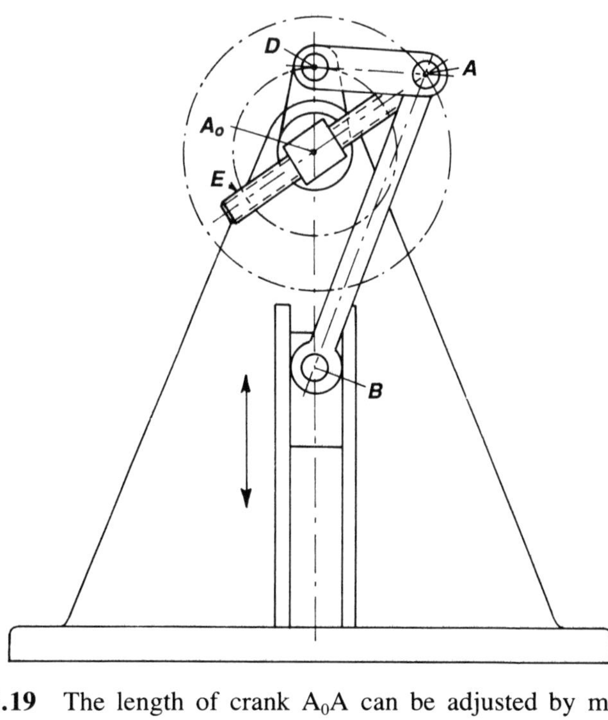

Figure 3.19 The length of crank A_0A can be adjusted by means of the screw E to obtain a variable stroke of slider B.

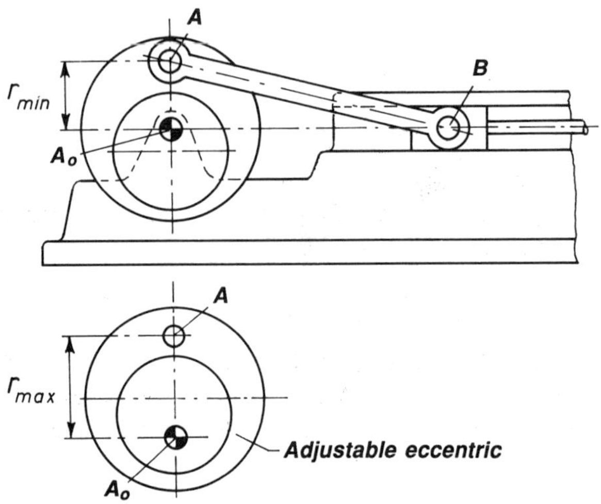

Figure 3.20 The adjustable eccentric A_0A imparts a stroke to the slider B. The stroke is made variable by the double eccentric, the eccentricity of which is changed by changing the position of the outer eccentric relative to the position of the eccentric that is fastened to the cam.

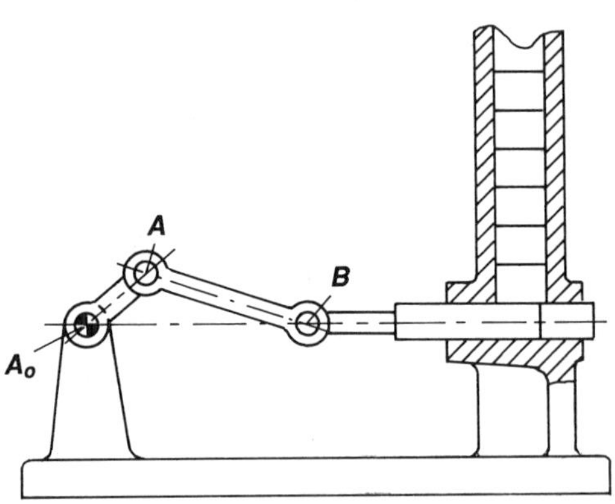

Figure 3.21 The slider crank A_0AB imparts an oscillating motion to the slider. The slider, in turn, feeds a carton and, at the same time, it prevents the stored cartons from moving down.

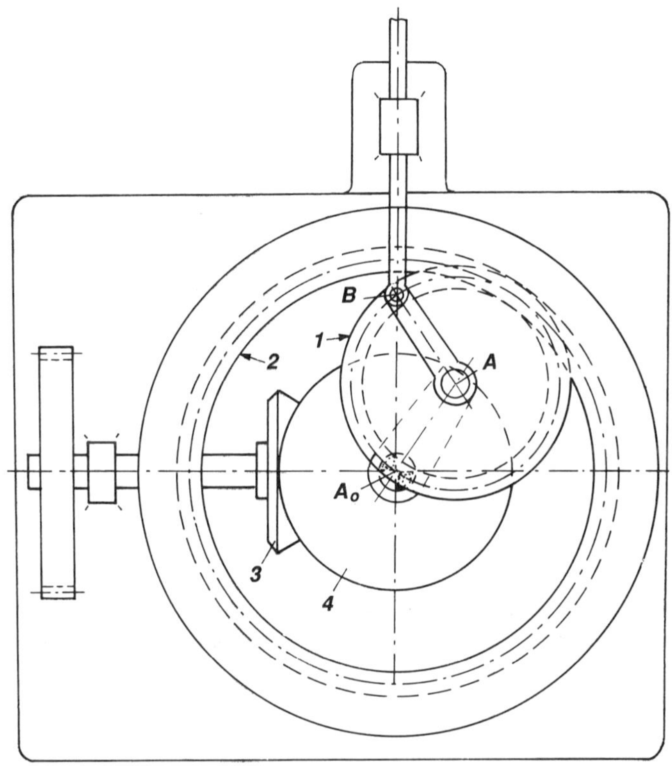

Figure 3.22 Pin B on the pitch circle of the small gear 1 traces an exact straight line when the small gear rolls inside the large gear 2 with twice the diameter. Crank A_0A is fastened to a bevel gear 4, which is driven from the input bevel gear 3. This principle was once used in a high-speed printing press made by the company Koenig & Bauer. Compared with the total stroke of the slider, the mechanism itself builds very small.

SYNTHESIS OF CENTRIC SLIDER CRANKS USED AS AN APPROXIMATE STRAIGHT-LINE LINKAGE

Figures 3.23–3.27 show a centric slider crank A_0AB with the coupler point C. As the crank A_0A rotates through angle ϕ_0, the coupler point traces a coupler curve with a maximum deviation Δx from the ideal straight line (Δx is shown exaggerated). The coupler curve of a slider crank can have a maximum of four points common with a straight line, or the coupler curve can have a straight line as tangent no more than two times. With the symbols

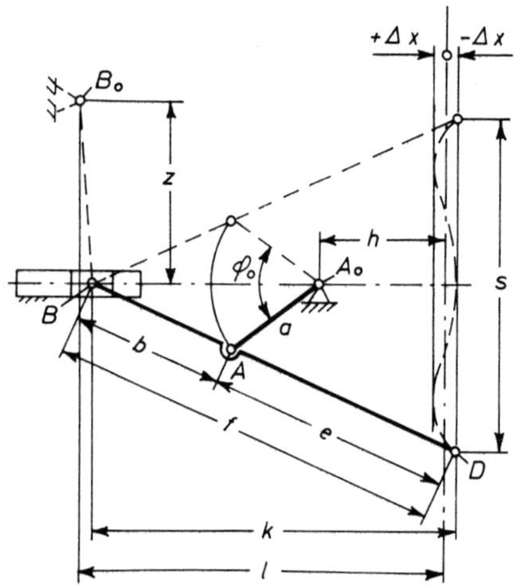

Figure 3.23 The centric slider crank A_0AB with the coupler point C traces an approximate straight line. The deviation Δx from the straight line is shown exaggerated.

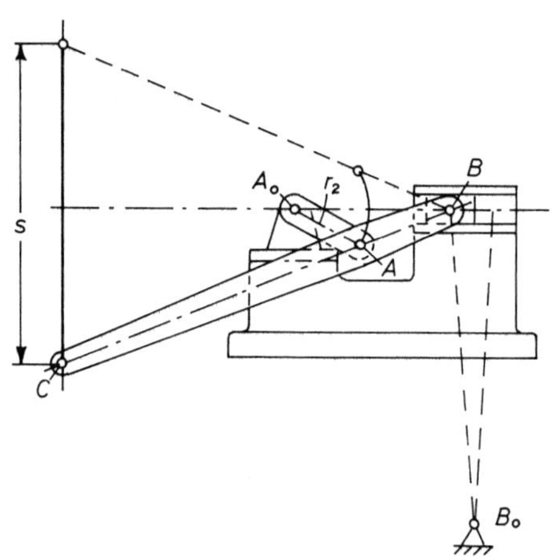

Figure 3.24

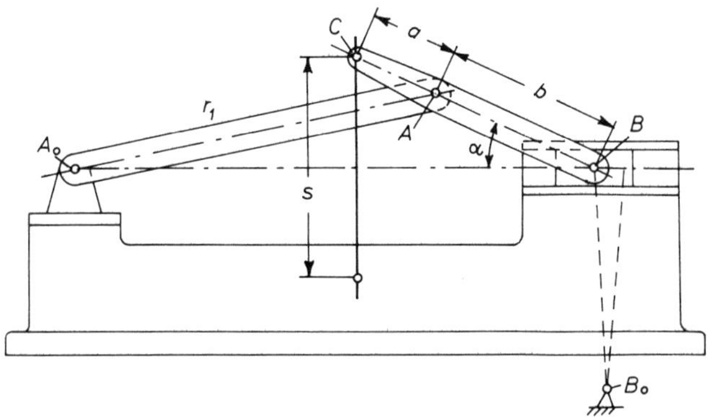

Figure 3.25

of Fig. 3.23, the following equations describing the coupler curve can be written:

$$x = h + a\sqrt{1 - \left(\frac{b}{a}\right)^2\left(\frac{y}{f}\right)^2} - e\sqrt{1 - \left(\frac{y}{f}\right)^2} \tag{3.22}$$

$$m = a + b \tag{3.23}$$

$$k = \sqrt{f^2 - \left(\frac{S}{2}\right)^2} \tag{3.24}$$

$$f = b + e \tag{3.25}$$

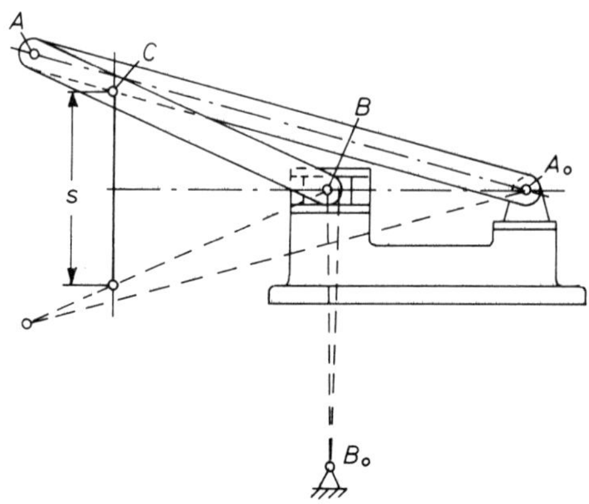

Figure 3.26

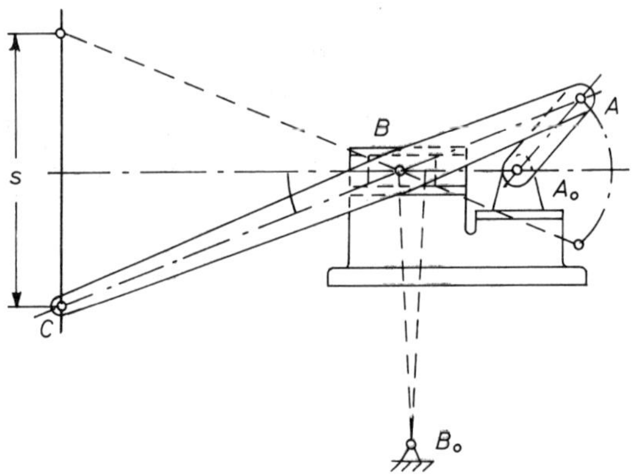

Figure 3.27

$$\Delta x = (f - m)\left[\frac{1}{2} - \frac{\sqrt{m(k - f + m)}}{|2m - f + k|}\right] \tag{3.26}$$

$$h = f - m - \Delta x \tag{3.27}$$

$$b = \frac{f}{2}\left(1 - \frac{f - m}{k + m}\right) \tag{3.28}$$

$$S = \frac{4}{e - b}\sqrt{e(b^2 - ae)(a - e)} \tag{3.29}$$

$$\frac{\varphi_0}{2} = \sin^{-1}\left|\frac{bS}{2af}\right| \tag{3.30}$$

The variables are S, Δx, a, b, f, k, and ϕ_0. Any three determine the final proportions, but it is advantageous to select S, Δx, and f. In the above equations, a, b, and e, are positive or negative according to Table 3.2.

Table 3.2[a]

	Fig. 24	Fig. 25	Fig. 26	Fig. 27
Sign of a	+	+	−	+
Sign of b	+	+	+	−
Sign of e	+	+	−	+

[a]Based on work by J. Volmer, *Getriebetechnik*, Verlag Technik, Berlin, 1987, pp. 498–503.

Example

Let

$S = 275$ mm
$f = 380$ mm
$m = 270$ mm

Solution

$$k = \sqrt{380^2 - \left(\frac{275}{2}\right)^2} = 354.251 \text{ mm} \tag{3.24}$$

$$b = \frac{380}{2}\left(1 - \frac{380 - 270}{354.251 + 270}\right) = 156.520 \text{ mm} \tag{3.28}$$

$$a = 270 - 156.520 = 113.48 \text{ mm} \tag{3.23}$$

$\Delta x =$

$$(380 - 270)\left[\frac{1}{2} - \frac{\sqrt{270(354.251 - 380 + 270)}}{2 \cdot 270 - 380 + 354.251}\right] = 0.069 \text{ mm} \tag{3.26}$$

$$h = 380 - 270 - 0.069 = 109.931 \text{ mm} \tag{3.27}$$

$$\varphi_0 = 2\sin^{-1}\left|\frac{156.52 \cdot 275}{2 \cdot 113.48 \cdot 380}\right| = 59.88^\circ \tag{3.30}$$

If it is desired to replace the slider by B with a link B_0B of finite length to improve friction conditions, then (Fig. 3.23)

$$l = \frac{1}{2}(k + f) = \frac{1}{2}(354.251 + 380) = 367.13 \text{ mm} \tag{3.31}$$

$$B_0B = z + \frac{(f - k)^2}{16z} \tag{3.32}$$

and choosing $z = 250$ mm

$$B_0B = 250 + \frac{(380 - 354.251)^2}{16 \cdot 250} = 250.166 \text{ mm}$$

PUMPS AND ENGINES

Pumps and engines differ from each other in their use of the slider crank: in pumps input power is to the crank(s); in engines input power is to the piston(s). In general, the cylinder is stationary for both pumps and engines.

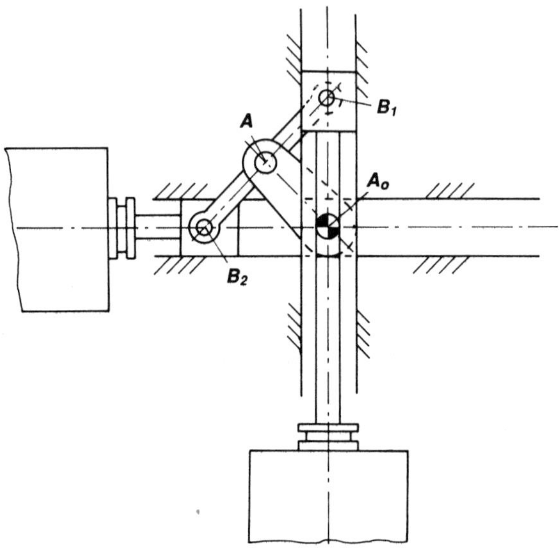

Figure 3.28 The slider cranks A_0AB_1 and A_0AB_2, where crank A_0A is common to both slider cranks, drive the two crossheads B_1 and B_2. The direction of motion of the two crossheads is shown as a 90° angle with respect to each other, but any other value can be used.

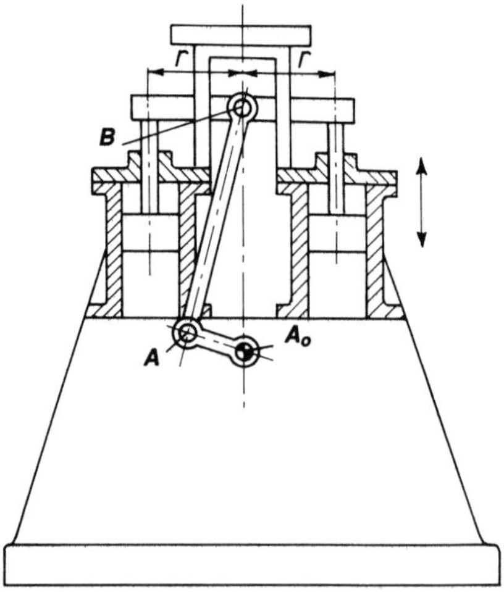

Figure 3.29 A battery of pistons operating with their cylinders arranged in a circle with radius r. The pistons are driven by the slider crank A_0AB.

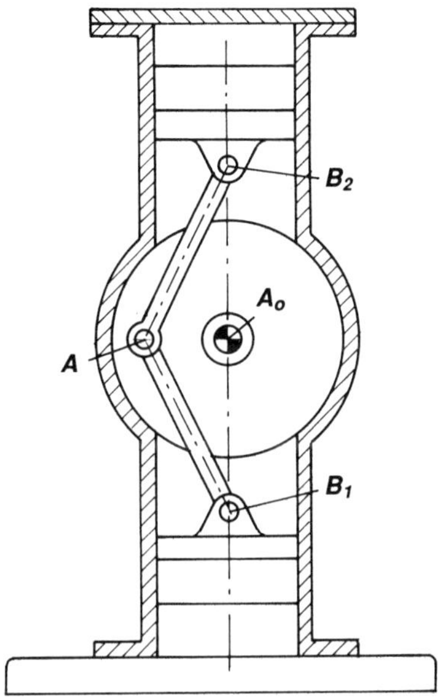

Figure 3.30 A heavy fly-wheel with center A_0 is guided by the cylinder walls. A crank A_0A is attached to the fly-wheel and drives two pistons B_1 and B_2, 180° out of phase with each other.

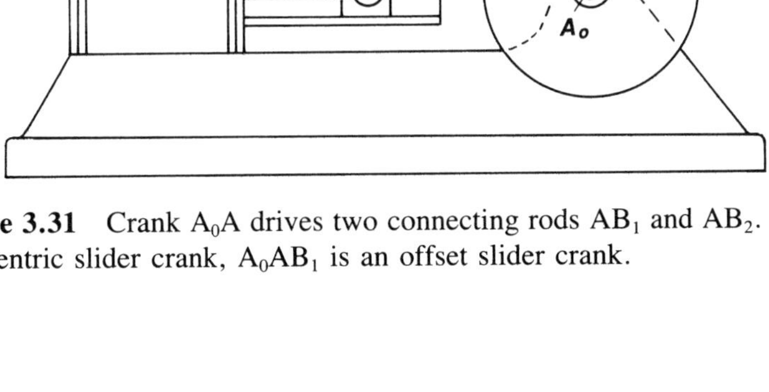

Figure 3.31 Crank A_0A drives two connecting rods AB_1 and AB_2. A_0AB_2 is a centric slider crank, A_0AB_1 is an offset slider crank.

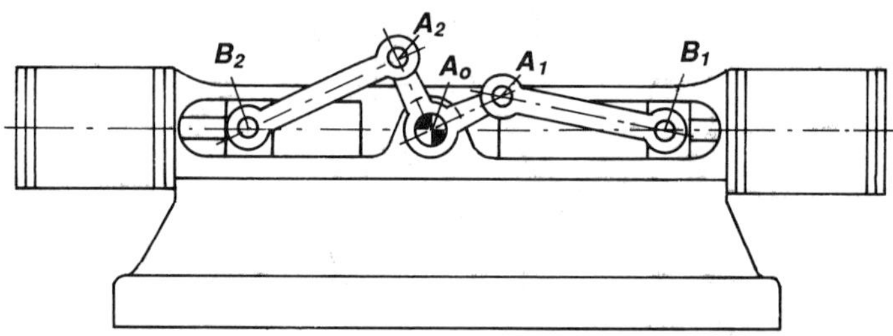

Figure 3.32 The double crank $A_1A_0A_2$ drives two slider cranks $A_0A_1B_1$ and $A_0A_2B_2$. The two slider cranks are in line.

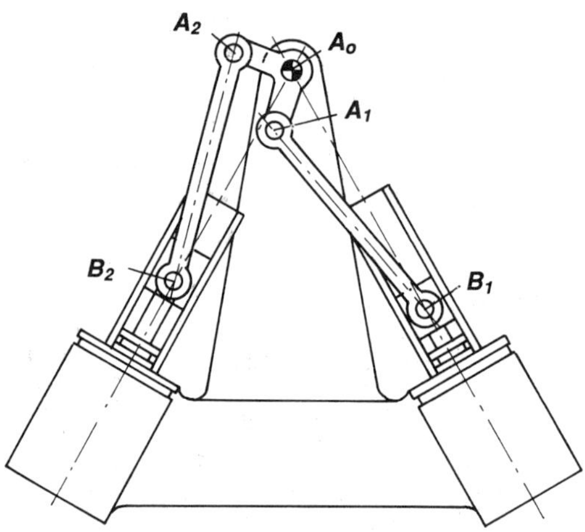

Figure 3.33 The double crank $A_1A_0A_2$ drives two slider cranks $A_0A_1B_1$ and $A_0A_2B_2$. The two centric slider cranks have different lines of motion.

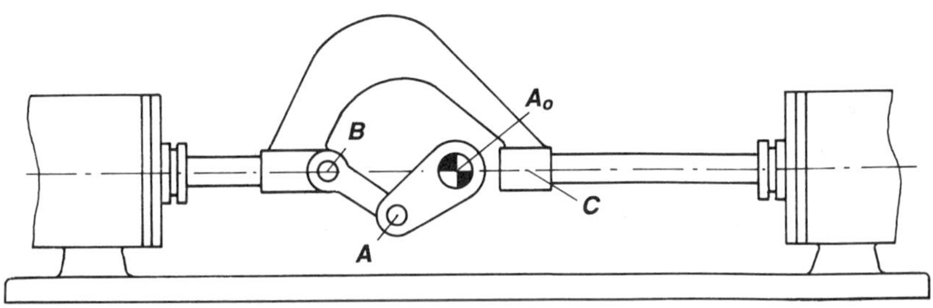

Figure 3.34 The slider crank A_0AB drives connecting rod AB, which drives the two integral piston rods B and C. The pistons move 180° out of phase with each other.

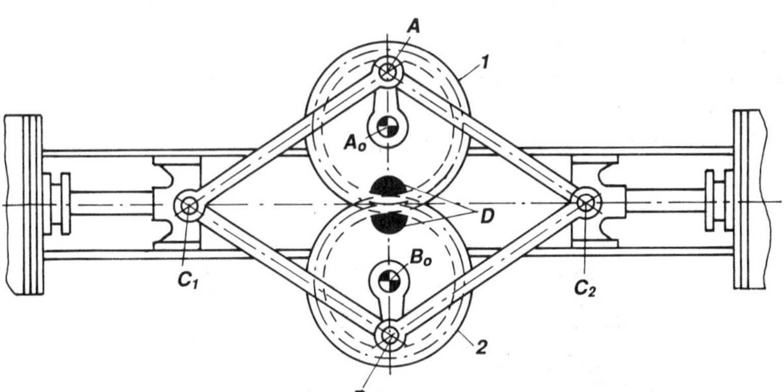

Figure 3.35 The twin-geared double-crank linkage $A_0AC_1BB_0$ and $A_0AC_2BB_0$ drives the two crossheads C_1 and C_2. Counterbalance is provided by the two weights D.

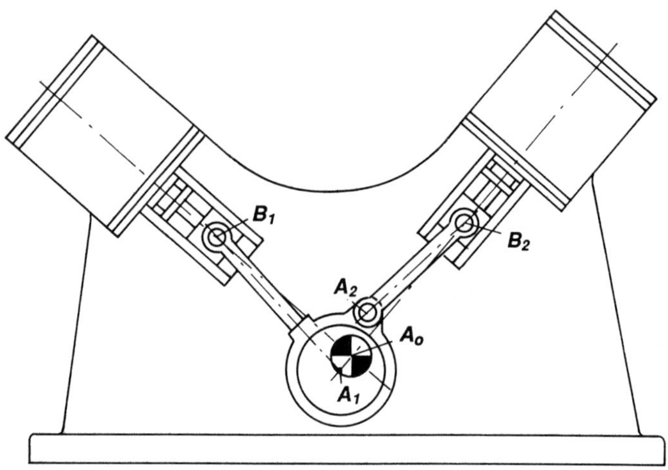

Figure 3.36 The centric slider crank $A_0A_1B_1$ drives the crosshead at B_1. A_0A_1 is an eccentric fixed to the input shaft. The connecting rod A_2B_2 is hinged to connecting rod A_1B_1 and drives crosshead B_2. A double crank or double eccentric is avoided, thus resulting in a less complicated design.

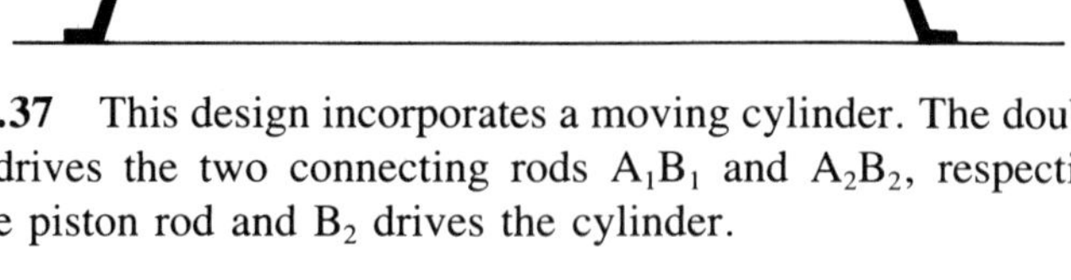

Figure 3.37 This design incorporates a moving cylinder. The double crank $A_1A_0A_2$ drives the two connecting rods A_1B_1 and A_2B_2, respectively. B_1 drives the piston rod and B_2 drives the cylinder.

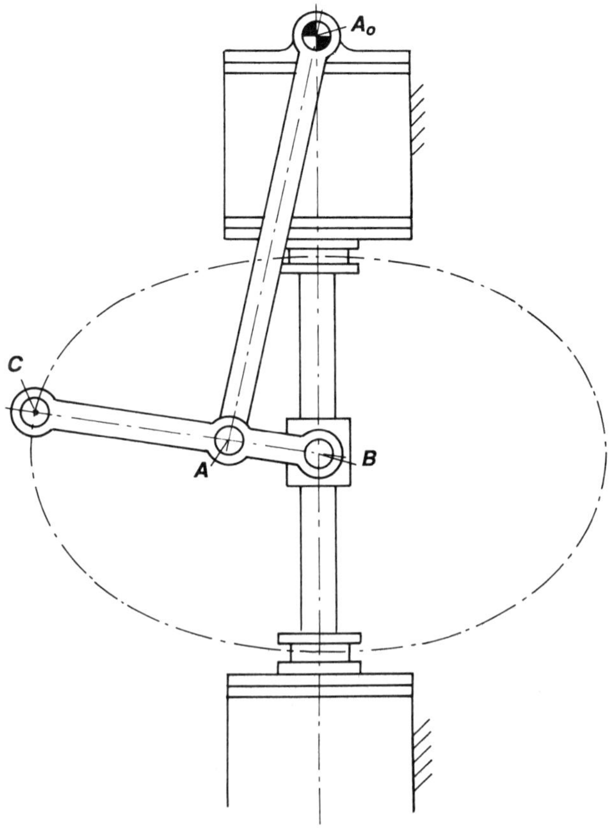

Figure 3.38 Hand-operated double-acting pump. The connecting rod A_0A of the centric slider crank A_0AB can only make an oscillating motion. The crank is being driven from the crank AB with the coupler point C that traces the oval-shaped coupler curve. The handle is located at C, and in the horizontal position of the rod, where the pistons are close to their maximum velocities, the mechanical advantage is greater. The motion of the pistons are 180° out of phase so that only one piston pumps at a time.

A classification is made in Table 3.3.

Table 3.3 Slider Crank Classification

Number of Cranks	Number of Pistons	Number of Connecting Rods	Cylinder Stationary	Fig. No.
1	1	1	yes	3.28
1	Several	1	yes	3.29
1	2	2	yes	3.30
1	2	2	yes	3.31
2	2	2	yes	3.32
2	2	2	yes	3.33
1	2[b]	2	yes	3.34
2[a]	2	4	yes	3.35
1	2	1 + 1	yes	3.36
2	1	1	no[c]	3.37
1	1	1	yes	3.38

[a]Geared together.
[b]Integral.
[c]Cylinder and piston both moving.

ANALYSIS AND SYNTHESIS OF SLIDER CRANKS IN SERIES WITH GEAR MECHANISMS

When a gear is fixed to one or more neighboring members of a mechanism and meshing of these gears with the output gear, some interesting motions can be obtained. The analysis and synthesis of some of these gear drives can be developed from a centric slider crank with help of charts.

Slider Crank in Series with a Gear-Rack Mechanism

Figure 3.39. The slider crank A_0AB drives the rack. The rack is in meshed with a gear of radius r. The oscillating motion of the rack is converted to the oscillating motion of the gear. The angle Ψ through which the gear oscillates, can be calculated from

$$\Psi = S/(2\pi r) \times \frac{180}{\pi} \tag{3.33}$$

where Ψ is in degrees, S is the maximum stroke of the slider, and r is the pitch radius of the gear. With this type of mechanism the angle of oscillation can be greater than 360°. If Ψ, the total angle of oscillation, is given and it is also required that a certain position of the crank correlate with a certain angular position of the gear, then Chart 3.1 can be employed.

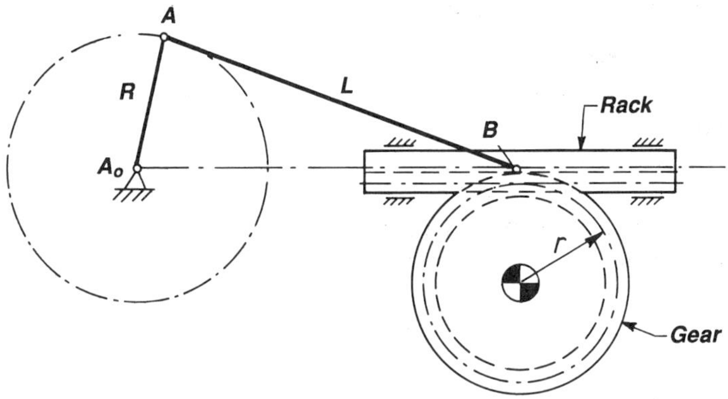

Figure 3.39 Centric slider crank drives rack and imparts an oscillating motion to the gear.

Example

Let r = 25 mm and Ψ = 360°, and for θ = 40° the gear must have rotated 60° from its extreme position. Determine the length of the connecting rod L.

Solution

From equation (3.12)

$$r = \frac{R(360)}{\psi\pi} = \frac{25(360)}{360\pi} = 7.69 \text{ mm}$$

The angle Ψ = 360° corresponds to two limits on the vertical axis of Chart 3.1. Therefore, the rotation of 60° corresponds to

$$\frac{2(60)}{360} = 1/3 \text{ unit}$$

This value corresponds to a value

$$S' = 1 - 1/3 = 2/3$$

on the vertical axis, and the intersection of a horizontal through this value and θ = 40° yields L/E = 2.3. Therefore, the length of the connecting rod is

$$L = R \times 2.3 = 25 \times 2.3 = 57.5 \text{ mm}$$

Slider Crank in Combination with Gears

The mechanism shown in Fig. 3.40 consists of a crank A_0A and a connecting rod AB. Gear 3 is fastened to the connecting rod, and gear 4, which is in mesh with gear 3, is free to rotate about A_0.

When the crank rotates, the output gear 4 will move in a certain manner. To find a relationship between θ, the angular displacement of the crank, and ω, the angular displacement of gear 4, use Swamp's tabular method (Table 3.4), starting out from the position when $\theta = 0°$ (Fig. 3.41).

$$\gamma = \theta + \frac{r_3}{r_4}(\theta + \beta) \tag{3.34}$$

θ and β are related by

$$\sin \beta = \frac{R}{L} \sin \theta \tag{3.35}$$

Differentiating eq. (3.35) with respect to t:

$$\cos \beta \, \dot{\beta} = \frac{R}{L} \cos \theta \, \dot{\theta} = \frac{R}{L} \cos \theta \cdot \omega \tag{3.36}$$

$$\frac{\dot{\beta}}{\omega} = \frac{R \cos \theta}{L \cos \beta}$$

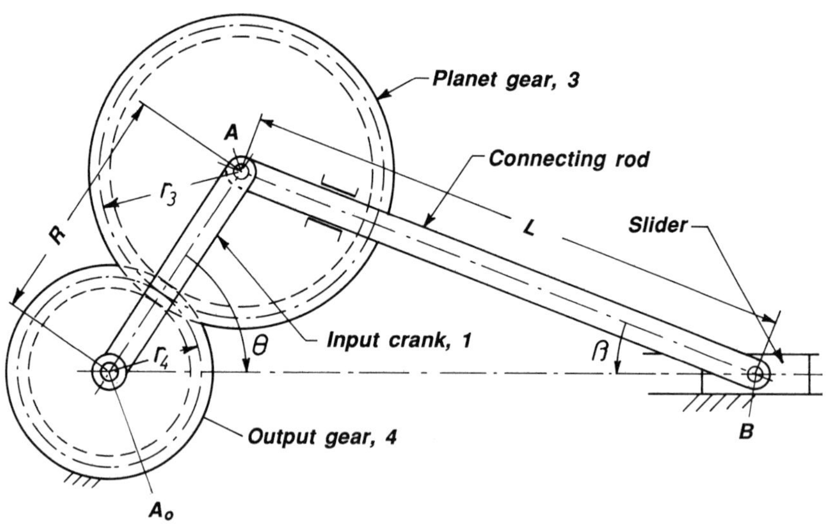

Figure 3.40 The input crank A_0A is part of a centric slider crank A_0AB. Planet gear 3 is fixed to connecting rod. Output is to gear 4.

Table 3.4

	Arm 1	Rod 2 and Gear 3	Gear 4
Gears 3 and 4 and arm 1 interlocked	θ	θ	θ
Arm 1 locked and gear 3 rotated	0	$-(\theta + \beta)$	$\frac{r_3}{r_4}(\theta + \beta)$
Resultant motion	θ	$-\beta$	$\theta + \frac{r_3}{r_4}(\theta + \beta)$

and from (3.35)

$$\cos^2 \beta = 1 - \left(\frac{R}{L}\right)^2 \sin^2 \theta \tag{3.37}$$

substituting eq. (3.37) into eq. (3.36) gives

$$\frac{\dot{\beta}}{\omega} = \frac{R}{L} \frac{\cos \theta}{\left[1 - \left(\frac{R}{L}\right)^2 \sin^2 \theta\right]^{1/2}} \tag{3.38}$$

Differentiating once more and collecting terms:

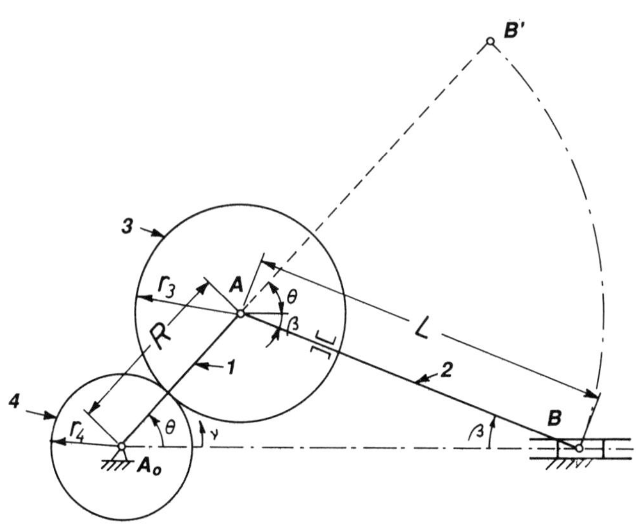

Figure 3.41 Designation for two-gear drive of fig. 3.40.

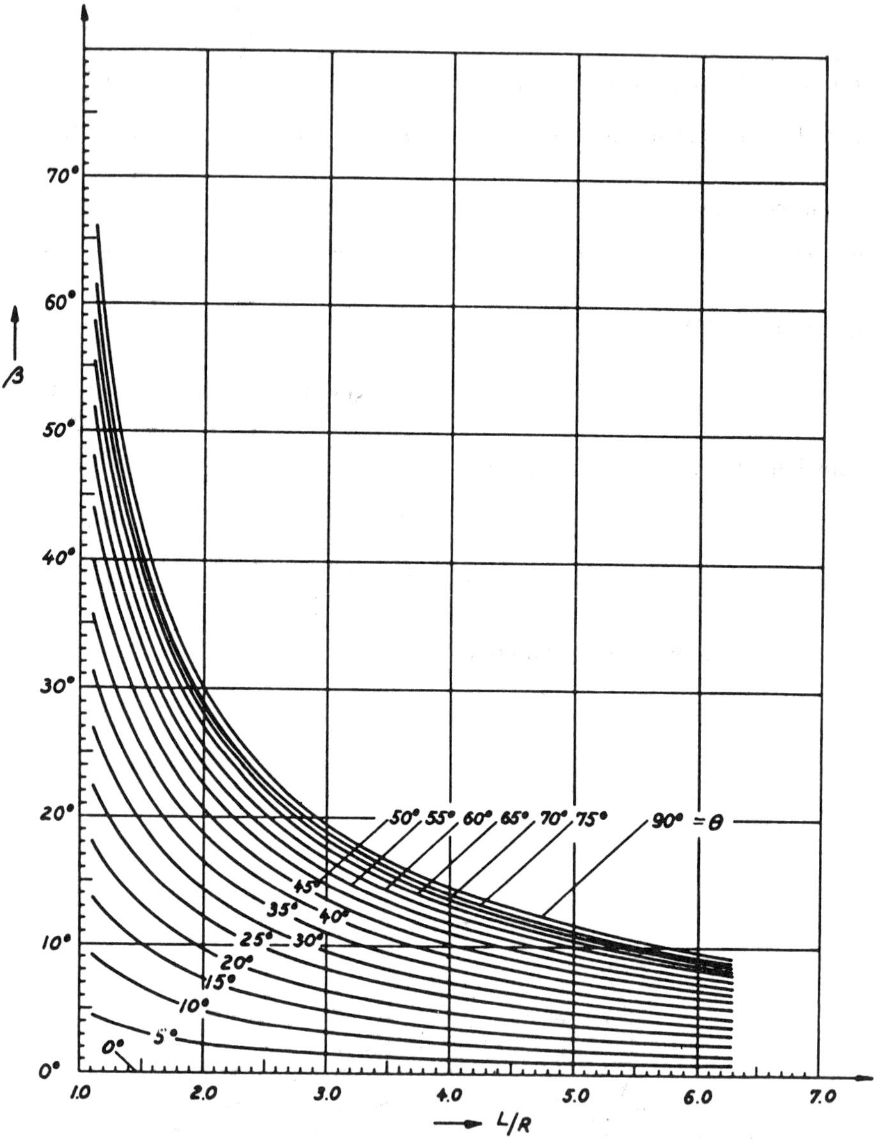

Chart 3.4 Angular displacement diagram of connecting rod for L/R = 1.1 [.1] 6.3.

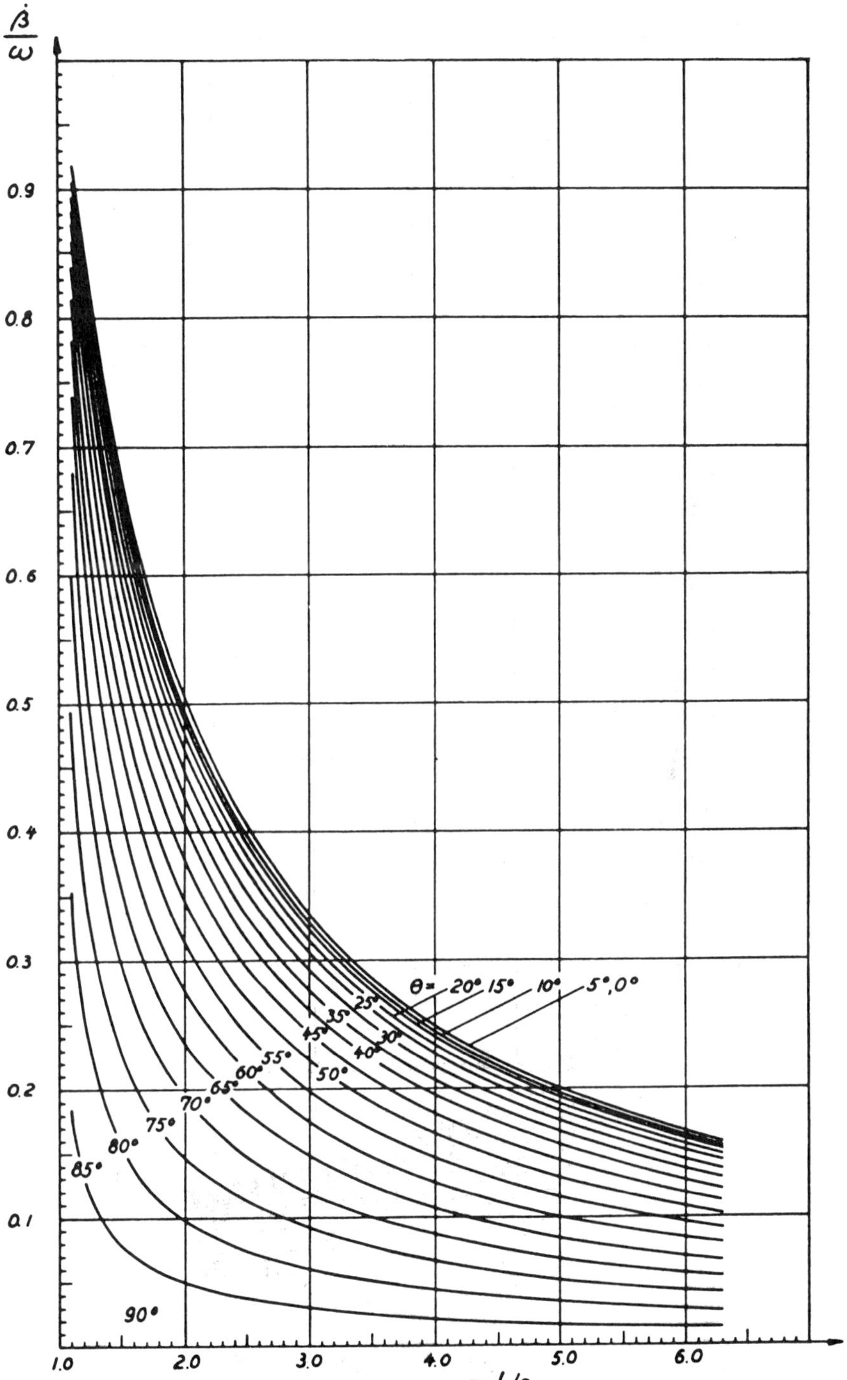

Chart 3.5 Angular velocity of connecting rod for L/R = 1.1 [.1] 6.3.

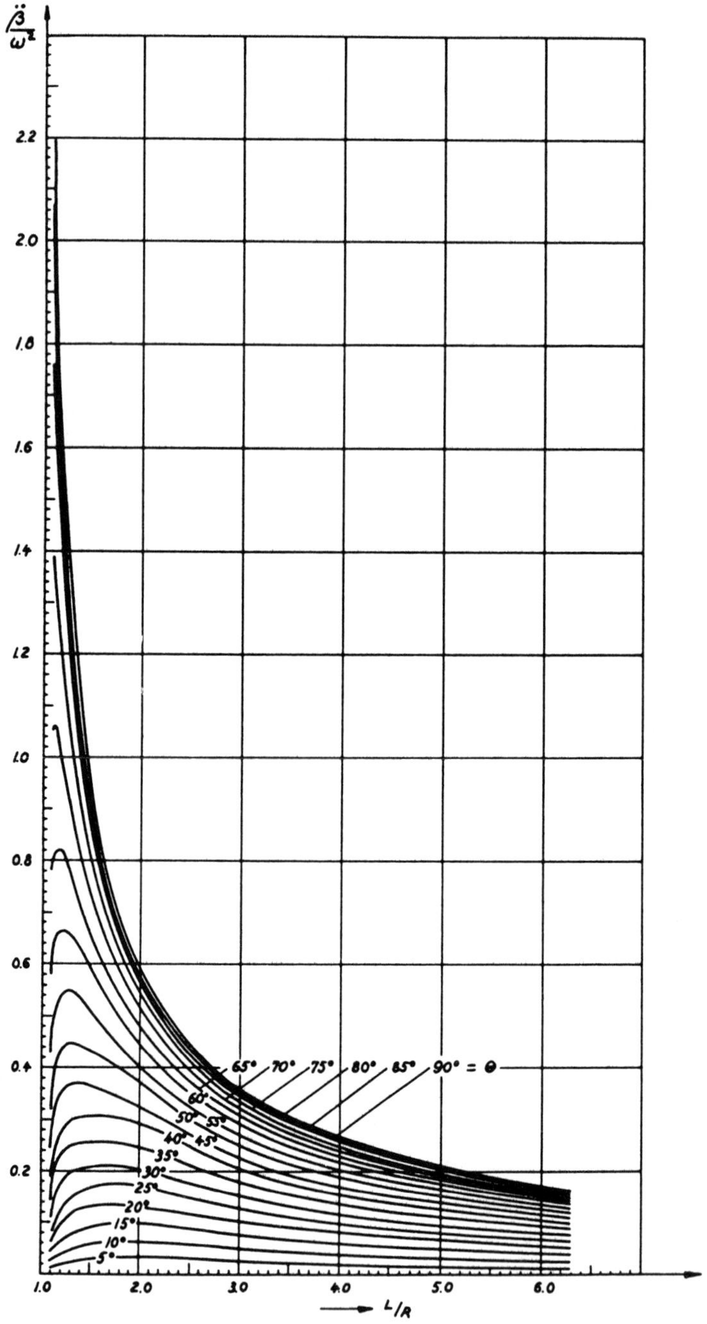

Chart 3.6 Angular acceleration of connecting rod for L/R = 1.1 [.1] 6.3.

$$\frac{\ddot{\beta}}{\omega^2} = \frac{R}{L} \frac{\sin\theta\left[\left(\frac{R}{L}\right)^2 - 1\right]}{\left[1 - \left(\frac{R}{L}\right)^2 \sin^2\theta\right]^{3/2}} \tag{3.39}$$

Equations (3.35), (3.36), and (3.37) were used to make the Charts 3.4, 3.5, and 3.6.

Differentiating eq. (3.13) with respect to time:

$$\dot{\gamma} = \omega + \frac{r_3}{r_4}(\omega + \ddot{\beta}) \tag{3.40}$$

$$\ddot{\gamma} = \alpha + \frac{r_3}{r_4}(\alpha + \ddot{\beta}) \tag{3.41}$$

For constant angular velocity, eq. (3.41) becomes:

$$\ddot{\gamma} = \frac{r_3}{r_4}\ddot{\beta} \tag{3.41a}$$

Equations (3.34), (3.40) and (3.41a) give the angular velocity and acceleration of the output gear in terms of r_3, r_4, $\dot{\beta}$, and $\ddot{\beta}$. These equations together with the Charts 3.4–3.6 will determine the output motion.

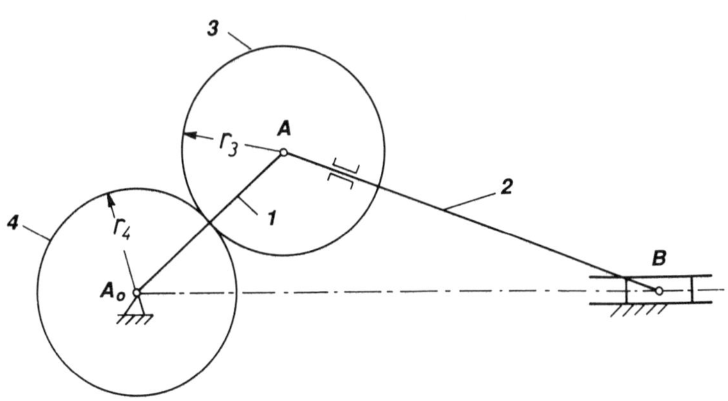

Figure 3.42 The two-gear drive from Fig. 3.40, also called Watt's crank mechanism, where $r_3 = r_4$ and in which the output gear 4 makes two revolutions for each revolution of the crank. Superimposed on the rotation of gear 4 is the oscillating motion of gear 3. Only in the limiting case when L = R, which is of little practical use, does gear 4 make only one revolution. This is all shown in Chart 3.7.

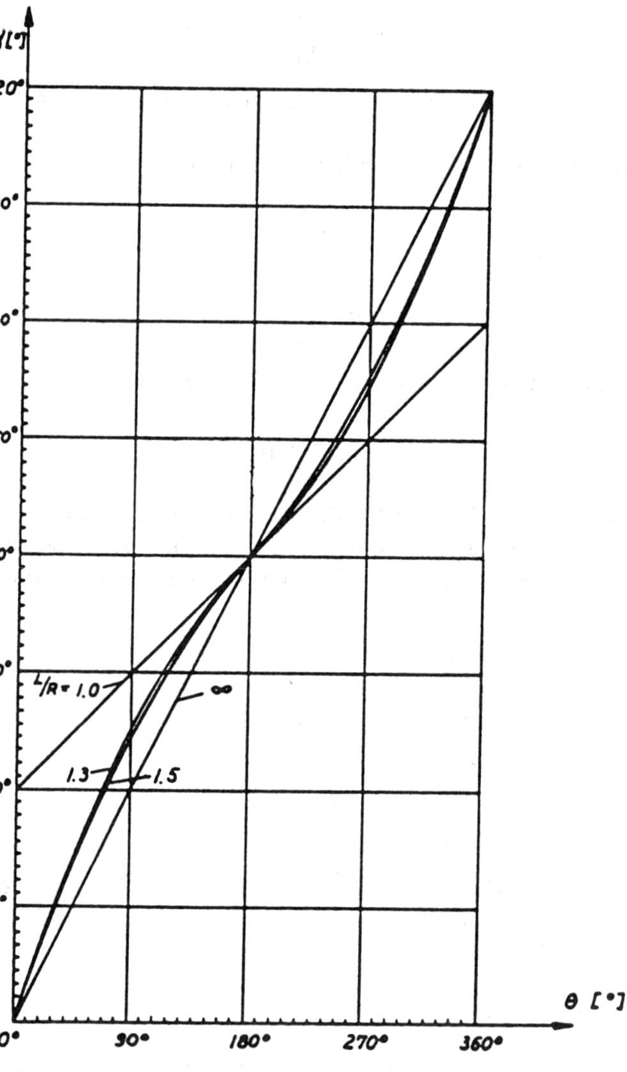

Chart 3.7 Angular displacement diagram for the drive of Fig. 3.41 for different values of L/R ($r_3 = r_4$).

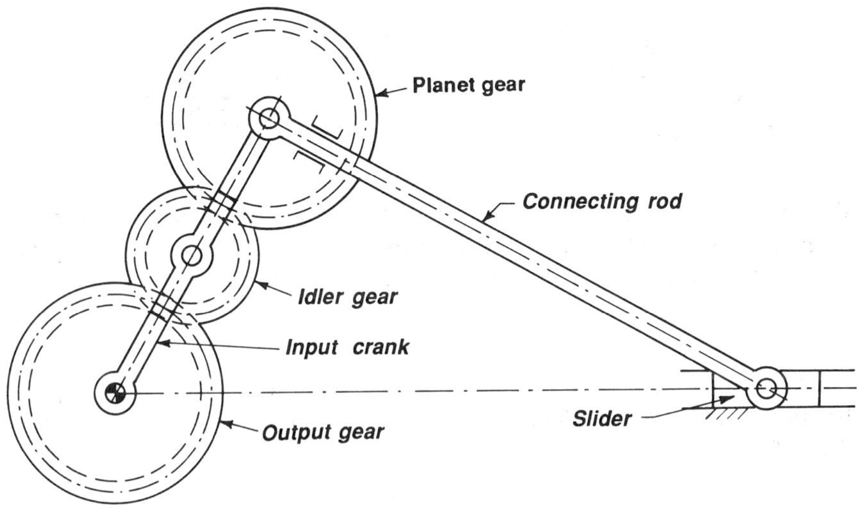

Figure 3.43 Gear drive where an idler gear has been interposed between the planet gear and the output gear.

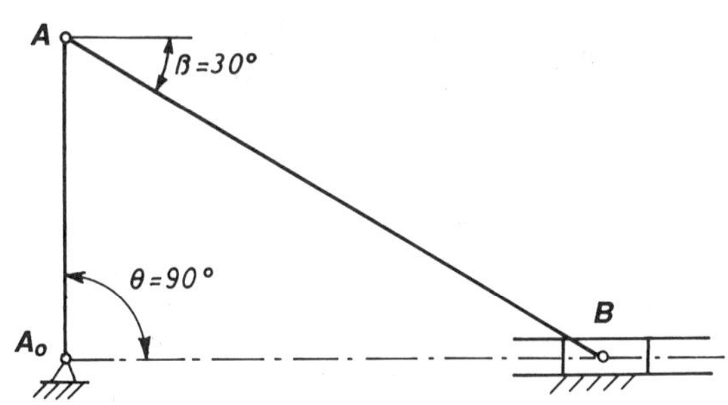

Figure 3.44 Simple geometrical construction used to find L/R when θ and γ are given, is as follows: given $\theta = 90°$, $\gamma = 210°$, and $r_3 = r_4$, the angle β is, from eq. (3.14), $210 - 90 - 90 = 30°$. First $\theta = 90°$ is laid out and an arbitrary distance chosen for $A_0A = R$. Then $\beta = 30°$ is laid out, determining $AB = L$. For the values given, $L/R = 2$.

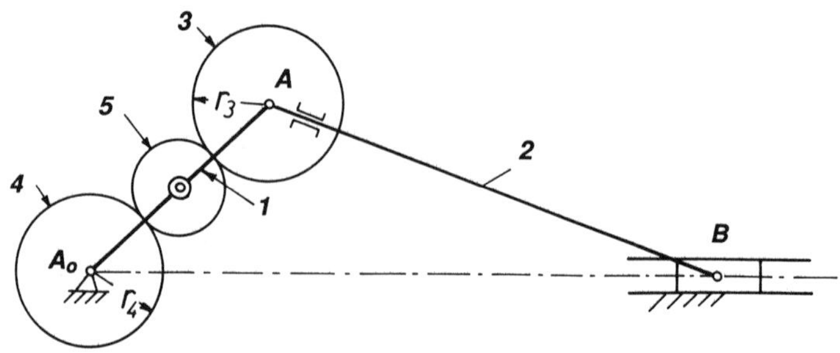

Figure 3.45 Another special case of Watt's crank mechanism, where $r_3 = r_4$ and gear 5 is used as an idler gear. This special arrangement will let gear 4 oscillate exactly as the oscillation of the connecting rod.

Figure 3.46 shows the mechanism of Fig. 3.42 when the output gear has the lowest angular velocity. Let ω_1 be the angular velocity of the crank A_0A. Then

$$\omega_1 = \frac{V_A}{r_3 r_4}; \; \omega_4 = \frac{V_4}{r_4}; \; \frac{V_A}{\omega_4} = \frac{L}{L - r_3}$$

$$\omega_4 = \frac{V_4}{r_4} = \frac{(L - r_3)\, V_A}{r_4 \cdot L} = \frac{(L - r_3)\, \omega_1 (r_3 + r_4)}{r_4 \cdot L}$$

$$R_{min} = \left(\frac{\omega_4}{\omega_1}\right)_{\theta = 180^\circ} = \frac{R}{L}\,\frac{L - r_3}{R - r_3} \tag{3.42}$$

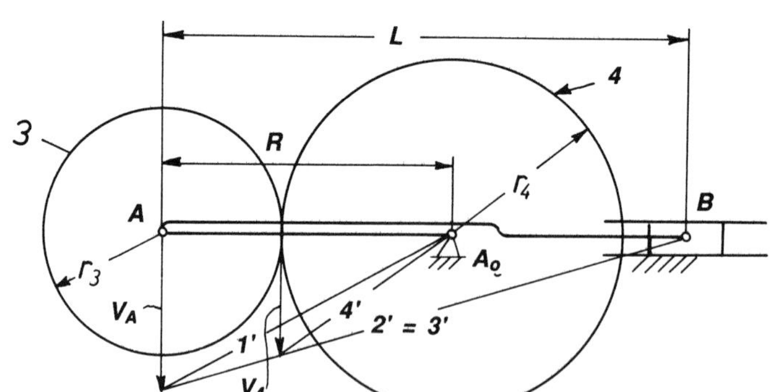

Figure 3.46 Geometrical construction to find angular velocity of output gear, when the slider crank is in its one extreme position.

Drawing the mechanisms in the other dead-center position gives

$$R_{max} = \left(\frac{\omega_4}{\omega_1}\right)_{\theta=0°} = \frac{R(L - R + r_3)}{(R - r_3)(L - R)} \tag{3.43}$$

It is interesting to note that the average gearing ratio is

$$i_{avg} = \left[\frac{\gamma}{\theta}\right]_{\theta=0°\rightarrow 360°} = 1 + \frac{r_3}{r_4} \tag{3.44}$$

It is not possible to obtain an instantaneous dwell of gear 4 of the mechanism of Fig. 3.42.

Figures 3.47 and 3.48 show how by replacing one of the external gears with an internal one, other types of mechanisms are obtained. The mechanism in Fig. 3.48 is interesting because it can be proportioned to give either an instantaneous dwell or a progressive oscillation.

Figure 3.49 is used to develop an expression for the output motion of gear 4 ($\theta = 180°$):

$$V_A = \omega_1 R; \frac{V_A}{V_4} = \frac{L}{L - r_3} \tag{3.45}$$

$$\omega_4 = -\frac{V_4}{r_4} = -\frac{(L - r_3)\omega_1 R}{r_4 \cdot L} = -\frac{(L - R - r_4)R}{r_4 \cdot L}\omega_1$$

Figure 3.50 shows the case where $\omega_4 = 0$, when $r_4 = L - R$.

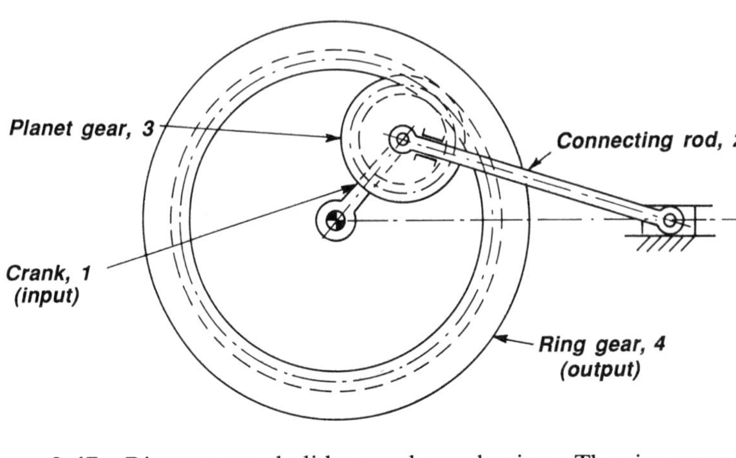

Figure 3.47 Ring gear and slider crank mechanism. The ring gear is the output and replaces the center gear in Fig. 3.40.

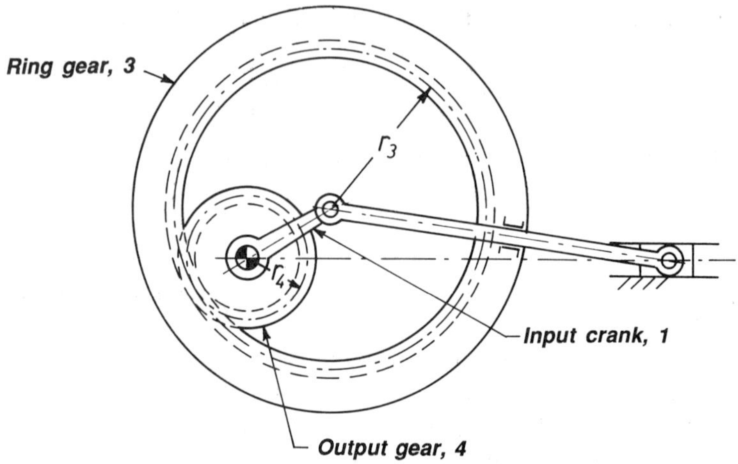

Figure 3.48 Same mechanism as Fig. 3.47 but with output from the smaller gear.

Figure 3.51 shows how to obtain a progressive oscillation, i.e., gear 4 reverses its rotation for a short time interval; r_4 must be larger than L − R.

If gear 4 turns back and then starts moving forward again Fig. 3.52, there must be two positions where the motion of gear 4 is zero. The two positions are symmetrical with respect to A_0B. Gear 4 is in a dwell position

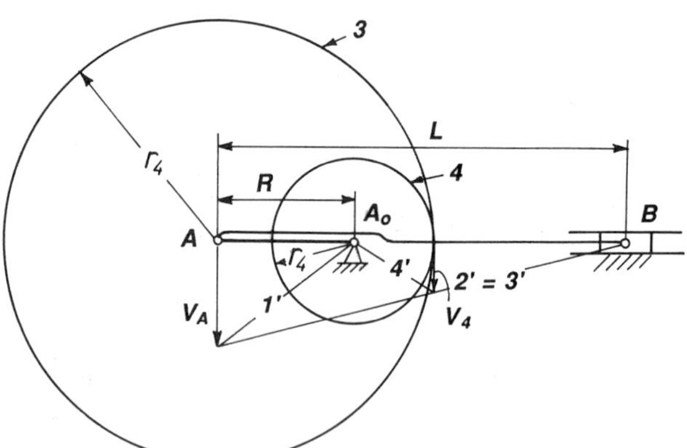

Figure 3.49 Geometrical construction to find angular velocity of output gear in the position where it has its lowest angular velocity.

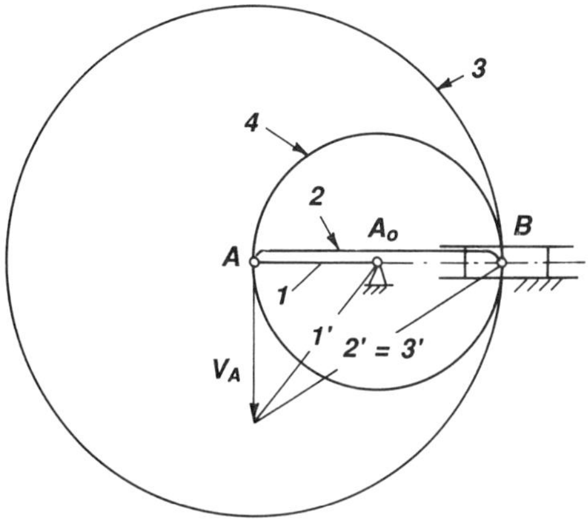

Figure 3.50 A two-gear drive proportioned to give instantaneous dwell.

when the pitch point P of gears 3 and 4 also is the instant center of rotation for connecting rod 2. The instant center of rotation is found at the intersection of A_0A with a perpendicular to A_0B at B. From triangle APB we have

$$(R + r_4)^2 \cos^2 \theta' = L^2 \cos^2 \beta'$$
$$R^2 \sin^2 \theta' = L^2 \sin^2 \beta'$$

and

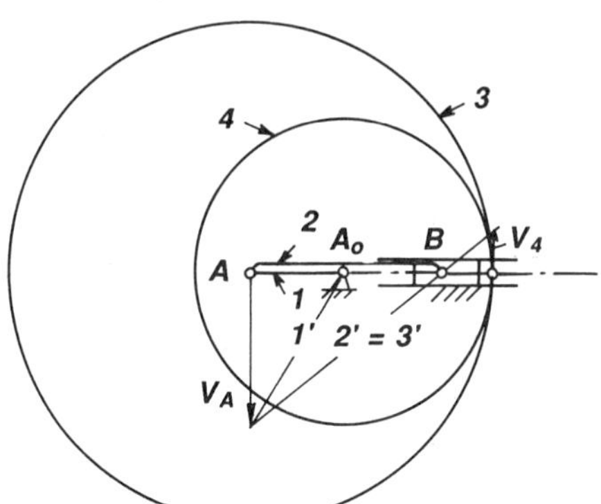

Figure 3.51 A two-gear drive proportioned to give a progressive oscillation.

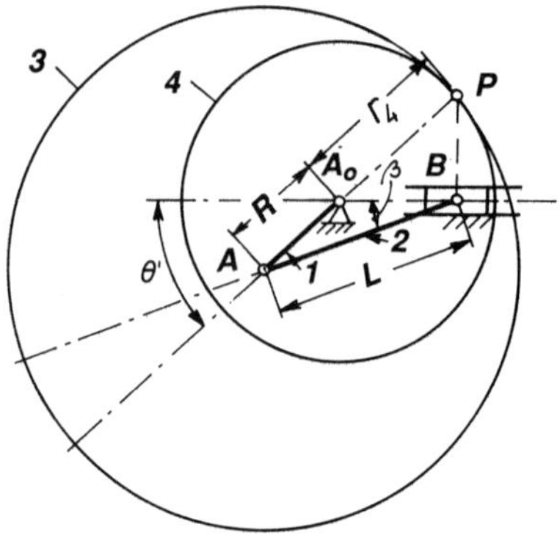

Figure 3.52 A two-gear drive with progressive oscillation shown in the position where the output motion is zero (instantaneous dwell).

$$(R + r_4)^2\cos^2\theta' = L^2\left[1 - \left(\frac{R}{L}\right)^2 \sin^2\theta'\right]$$

$$R^2\cos^2\theta' + 2Rr_4\cos^2\theta' + r^2_4\cos^2\theta' = L^2 - R^2\sin^2\theta'$$

$$\cos\theta' = \left[\frac{L^2 - R^2}{r_4(2R + r_4)}\right]^{1/2} \tag{3.46}$$

Letting θ_0 = crank-angle rotation during which the output gear reverses its motion, and φ = the angle through which gear 4 rotates back, Swamp's tabular method gives values as in Chart 3.8 and Table 3.5 in which

$$\theta_0 = 2\theta' \tag{3.47}$$

Figure 3.53 shows a two-gear drive equivalent to that in Fig. 3.50. It has the advantage that the output motion can be changed from progressive oscillation to instantaneous dwell or nonuniform CW or CCW rotation by changing the position of the pin, which acts as the sliding piece of the centric slider crank. It is, of course, also possible to use an eccentric slider crank, a four-bar linkage, or a sliding-block linkage as the basic mechanism.

Two mechanisms in series will give an output with either a prolonged approximate dwell or two separate dwells. The angle between the separated dwells can be adjusted during the time of operation by interposing a gear differential, so that the position of the output shaft of the first mechanism can be changed relative to the input shaft of the second mechanism.

Table 3.5

	Arm 1	Rod 2 and Gear 3	Gear 4
Gears 3 and 4 and arm 1 interlocked	θ'	θ'	θ'
Arm 1 and gear 3 rotated	0	$-(\theta' - \beta')$	$-\frac{r_3}{r_4}(\theta' - \beta')$
Resultant motion	θ'	β'	$\theta' - \frac{r_3}{r_4}(\theta' - \beta')$

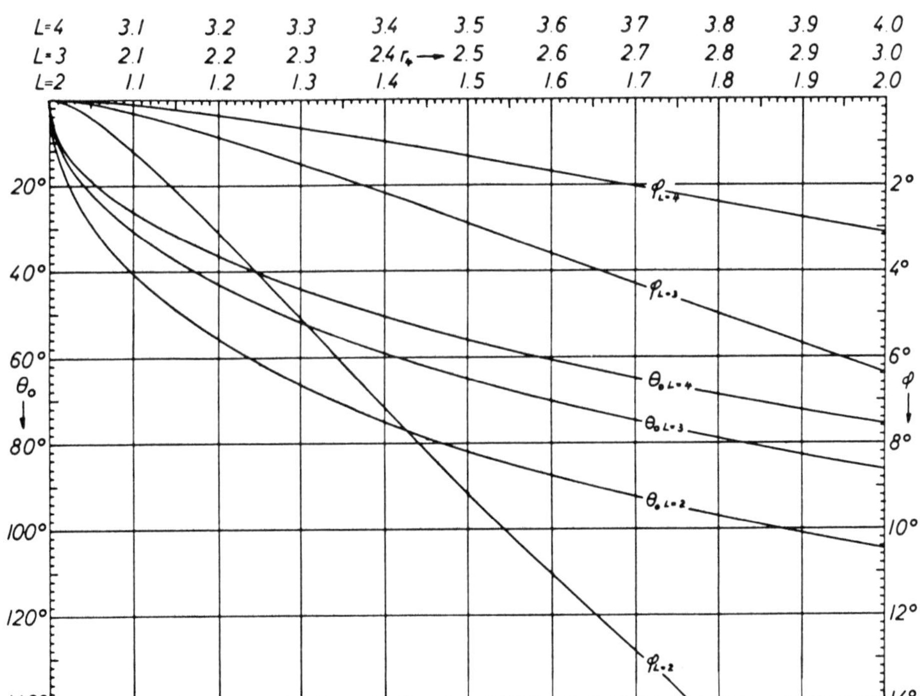

Chart 3.8 Proportioning diagram for two-gear drive.

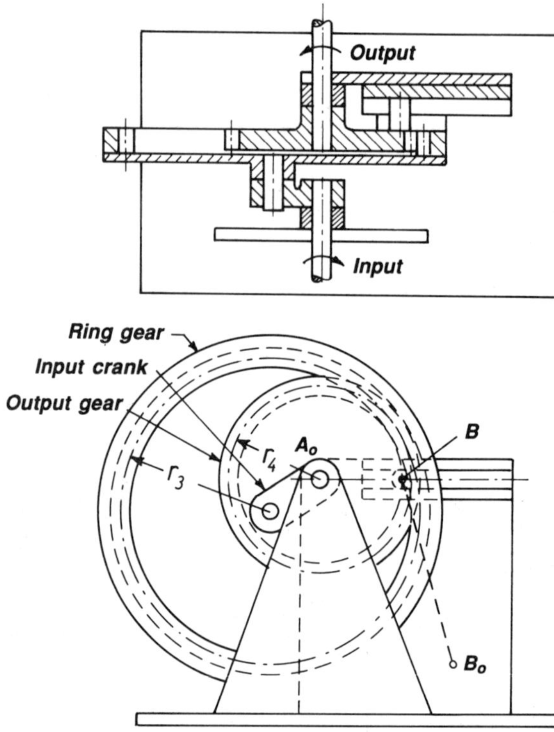

Figure 3.53 Model of a two-gear drive (author's invention).

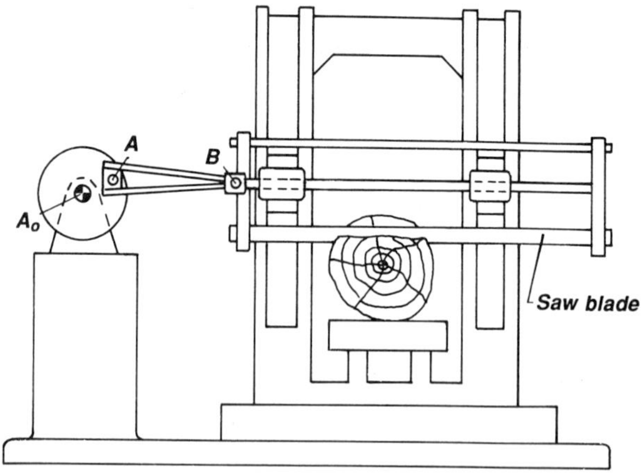

Figure 3.54 Slider crank A_0AB used to drive a saw.

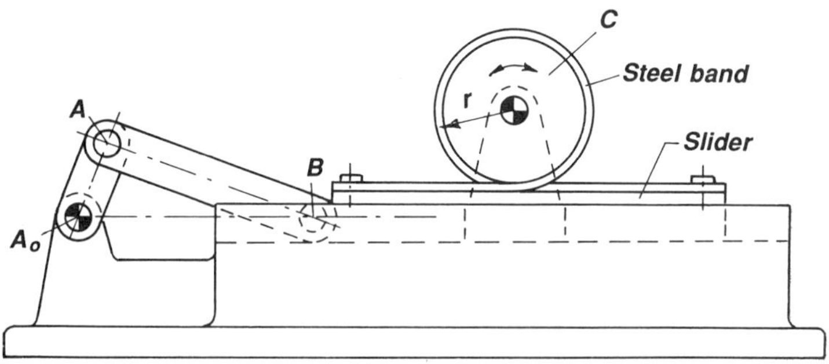

Figure 3.55 Mechanism similar to that in Fig. 3.39, but instead of using a rack and a gear in mesh with each other, a steel band is employed to force a rolling motion between the slider B of the centric slider crank A_0AB and the disc C. The angle Ψ through which the gear oscillates, can be calculated from $\Psi = S/(2\pi r) \times 180/\pi$ where Ψ is in degrees, S is the maximum stroke of the slider, and r is the radius of the disc, calculated to the middle of the steel band. With this type of mechanism the angle of oscillation can be greater than 360°. (See the example given with Fig. 3.39 for more information.)

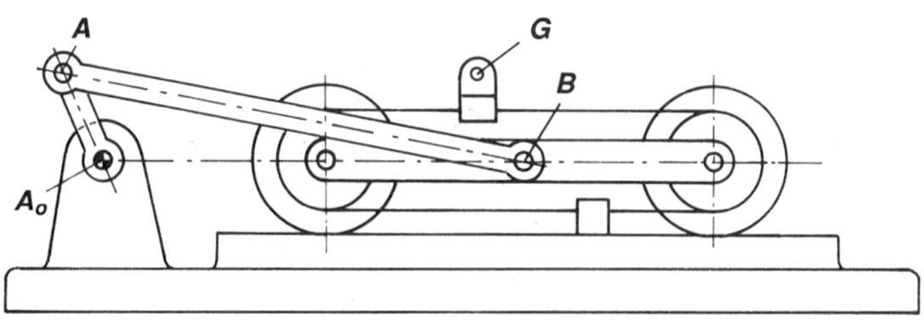

Figure 3.56 Motion amplifier. A band is wrapped around two cylinders. The centers of the two cylinders are on the slider of a slider crank A_0AB. The cylinders roll on a stationary guide, and the band is fixed to the stationary guide. The resultant motion of G is four times the length of the crank A_0A.

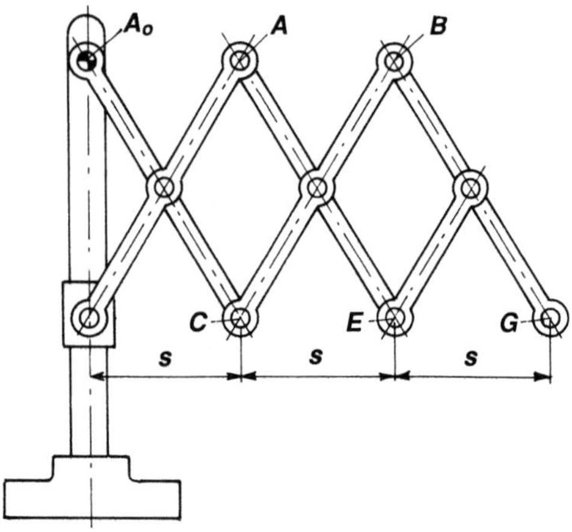

Figure 3.57 Known as *lazy tongs* or *Nuremberg scissors*. Points C, E, and G move on a straight line, but the movement of E is double that of C, and the movement of G is triple that of C.

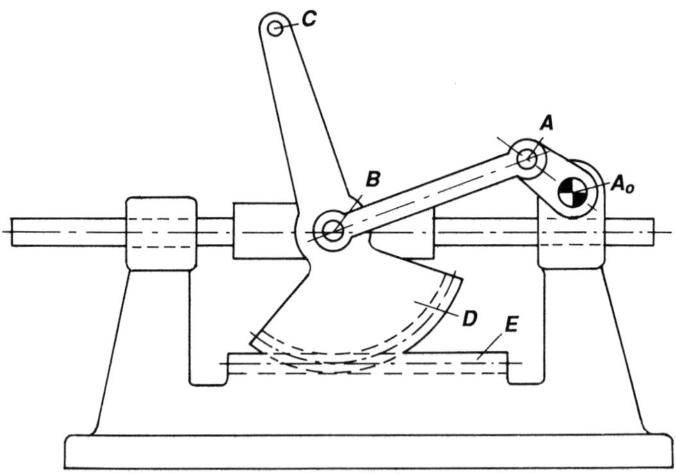

Figure 3.58 The slider crank A_0AB drives the gear segment D, that rolls on a rack E. The resultant motion of point C is an amplified motion because the rolling motion of gear segment D is superimposed on the sliding motion of B.

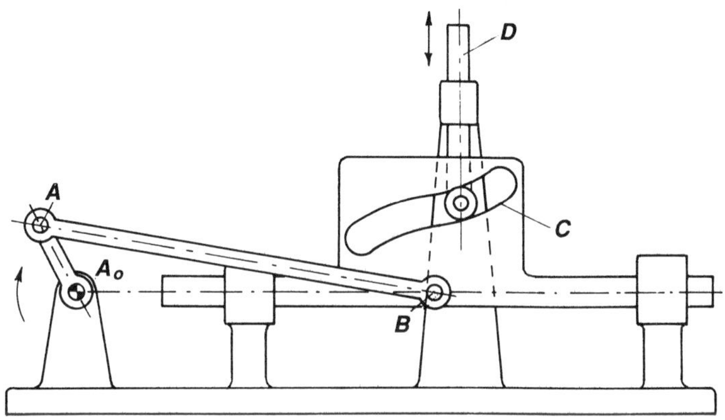

Figure 3.59 A groove C is cut into the slider of a slider crank A_0AB. The groove (or track) moves the roller of slider D up and down.

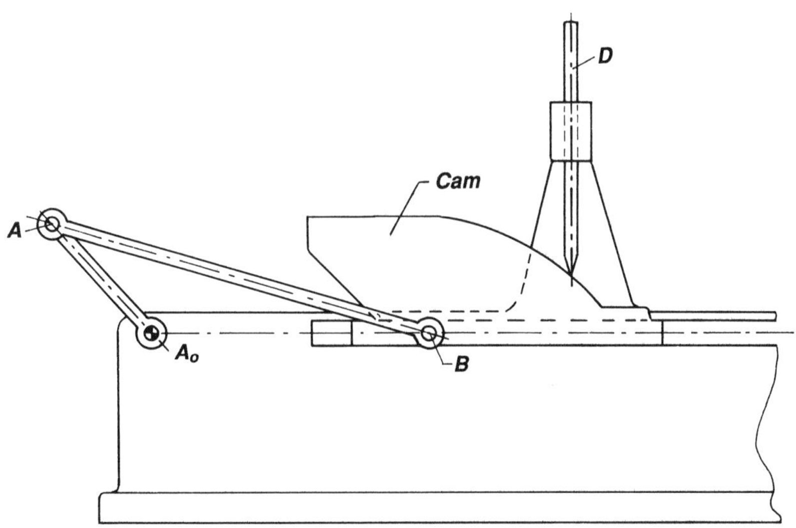

Figure 3.60 A knife-edged translating follower member is imparted an oscillating motion by the translating slider of a centric slider crank A_0AB.

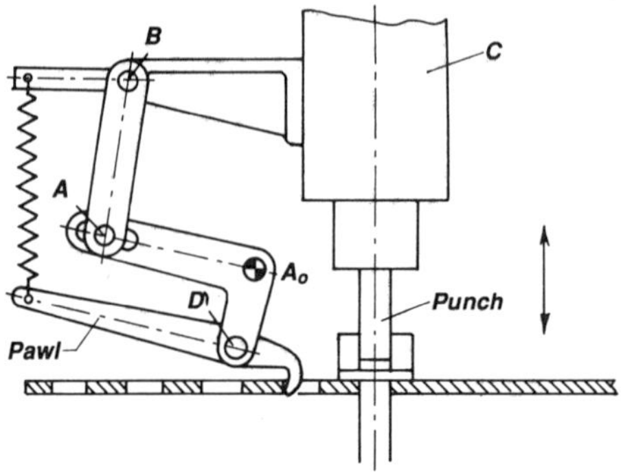

Figure 3.61 The reciprocating motion of the slider C of a slider crank A_0AB oscillates the bell crank AA_0D. At D is mounted a spring-loaded hook. When the slider punches (or pierces) the strip material, the hook moves from left to right without engaging the strip. When the punch (the slider) moves upward, the hook transports the material to the left.

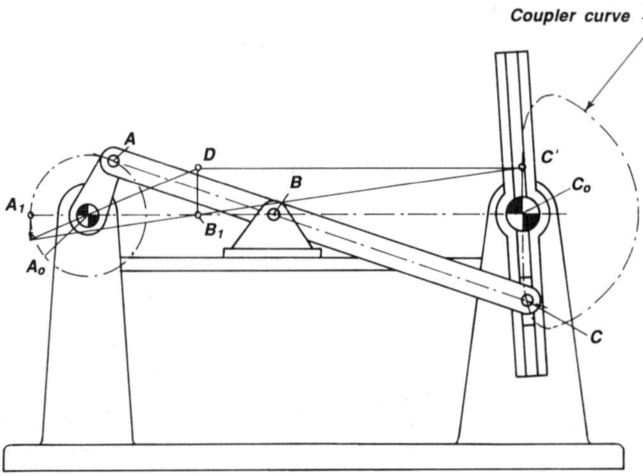

Figure 3.62 The coupler curve of a centric slider crank A_0AB with the coupler point C traces an approximate straight line. For each revolution of crank A_0A, the Geneva wheel is rotated 180°. To find the *proportions* of such a slider crank, employ a geometrical method: lay out an arbitrary vector from A_1 perpendicular to A_0A_1. Connect the head of the vector with B_1, where B_1 is the position of the slider B when the crank is at A_1. Draw another line from the head of the vector and through A_0 to intersect a perpendicular at B_1 to A_0A_1. Point of intersection is D. Draw a parallel to A_0A through D to intersect the line though B_1 at C′. Project C′ onto A_0A_1 to locate C_1. The slider crank $A_0A_1B_1C_1$ is proportioned so that C_1 traces an approximate straight line around the center position.

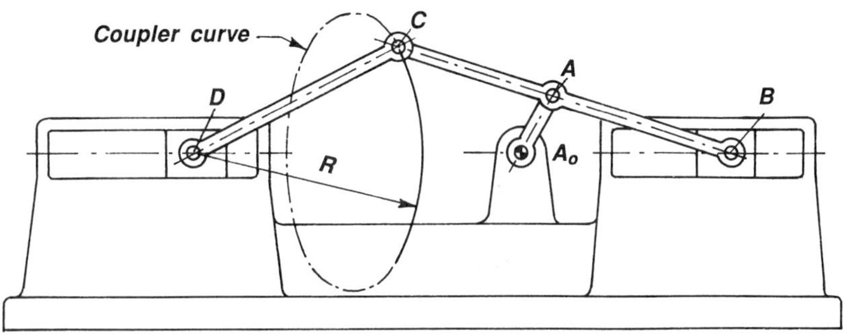

Figure 3.63 The coupler point C of a centric slider crank A_0AB traces an approximate circular arc with radius R. As long as C traces the approximate circular arc with center at D, the output member D will remain in an approximate dwell position.

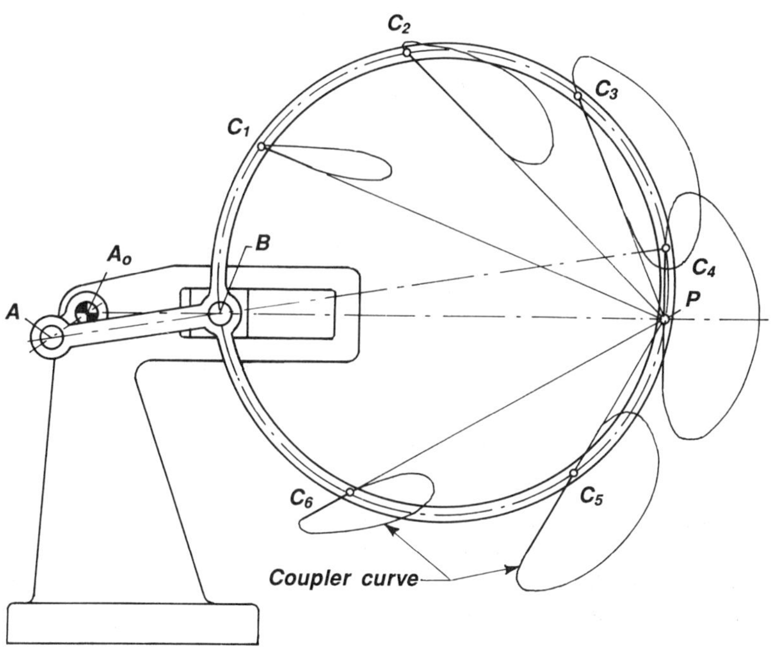

Figure 3.64 The coupler points C_1, C_2, C_3, etc., of a centric slider crank. A_0AB traces approximate straight lines over a limited range. Find the circle by using the geometrical method of Fig. 3.62, and then draw a circle with BC as diameter.

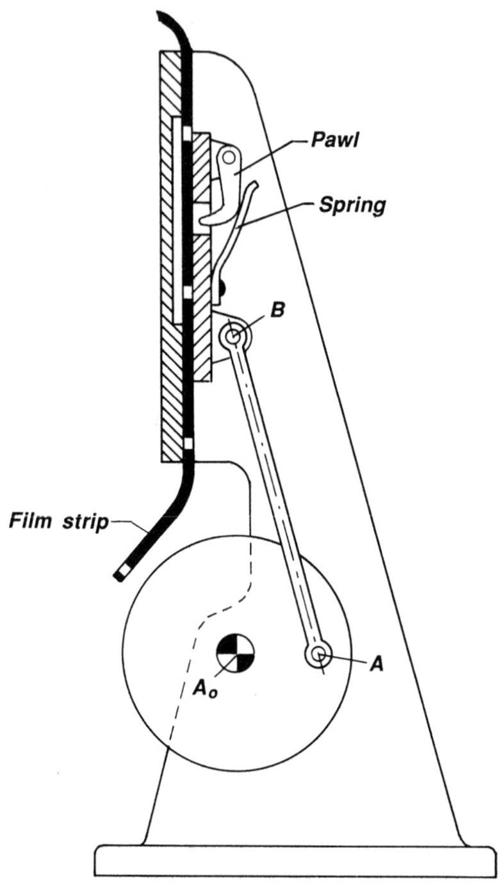

Figure 3.65 A slider crank in series with a ratchet-type mechanism. The slider B of the slider crank A_0AB carries a spring-loaded pawl. When B moves upwards, the pawl remains out of engagement with the film strip. On the downwards stroke, the pawl engages the film strip and moves it downwards.

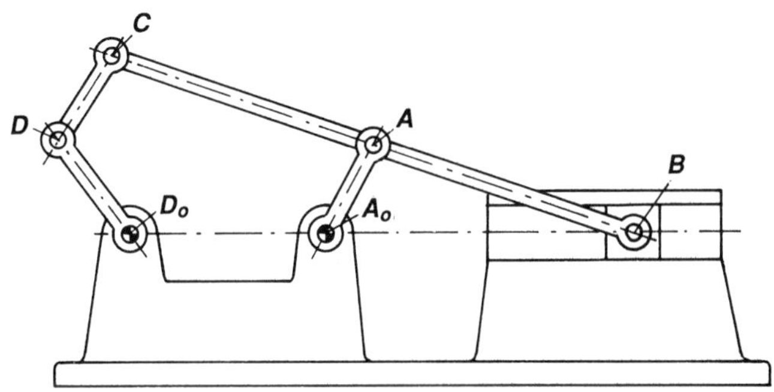

Figure 3.66 The slider crank A_0AB with the coupler point C traces a coupler curve around D_0. Links CD and DD_0 are connected to each other and to C and the frame. When the crank A_0A rotates, link DD_0 will make a complete revolution with varying angular velocity.

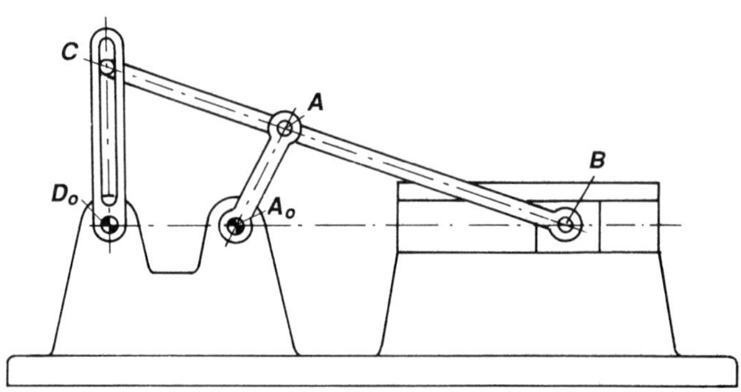

Figure 3.67 The slider crank A_0AB with the coupler point C traces a coupler curve around D_0. The slotted link rotates around D_0. When the crank A_0A rotates, the slotted link will make a complete revolution with varying angular velocity.

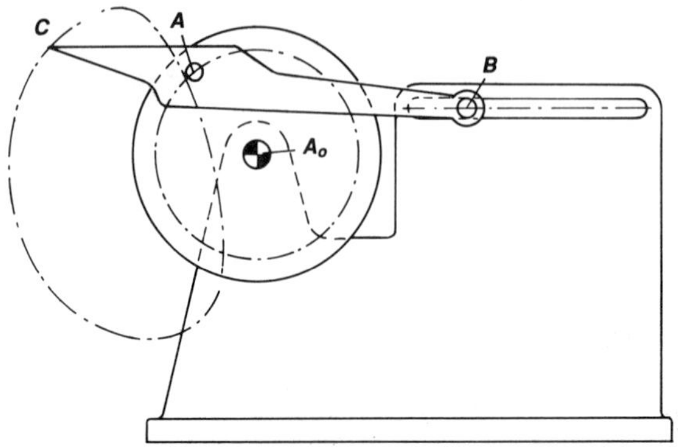

Figure 3.68 The eccentric slider crank A_0AB with the coupler point C traces the coupler curve shown. Point C moves into the perforation of the film strip (not shown), advances the film, and then moves out of engagement.

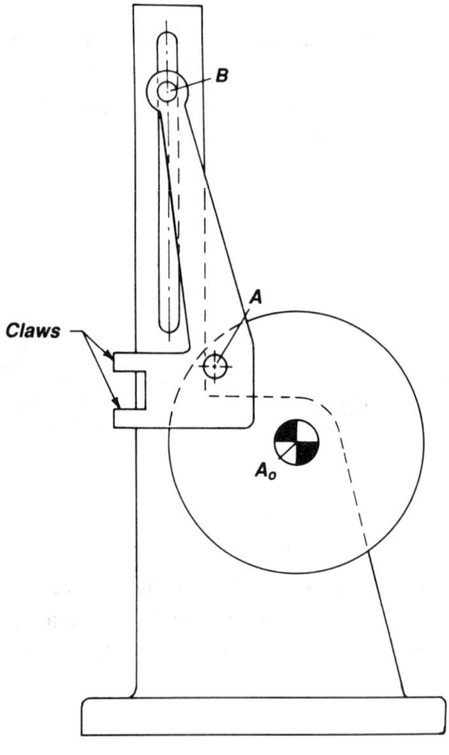

Figure 3.69 The eccentric slider crank A_0AB has a connecting rod with two claws that move in and out of the perforations of a film strip (not shown). When the claws are engaged, they move the film strip. Using two claws lessens the contact pressure between the claws and the film strip.

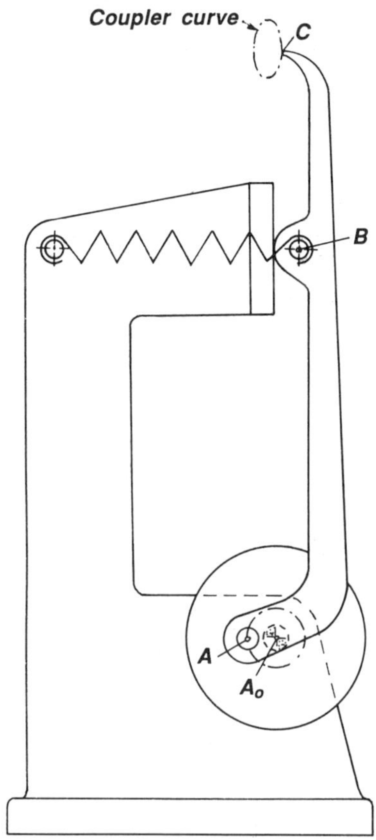

Figure 3.70 The eccentric slider crank A_0AB with the coupler point C traces the coupler curve shown. Point C moves into the perforation of the film strip (not shown), advances the film, and then moves out of engagement. The constrained motion of the rod is obtained here by the use of a spring.

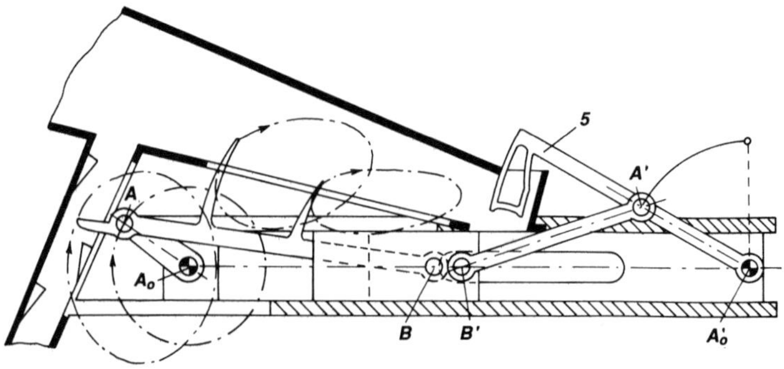

Figure 3.71 One crank drives separate linkages of this hay compressor. The slider crank A_0AB has tines on the connecting rod that serves to transport the hay to the right. A second slider crank $A_0'A'B'$, where the sliders B and B′ are integral, is driven by the motion of slider B. This motion causes link 5 (the crank) to move up and down, thereby pressing the hay into a duct (not shown).

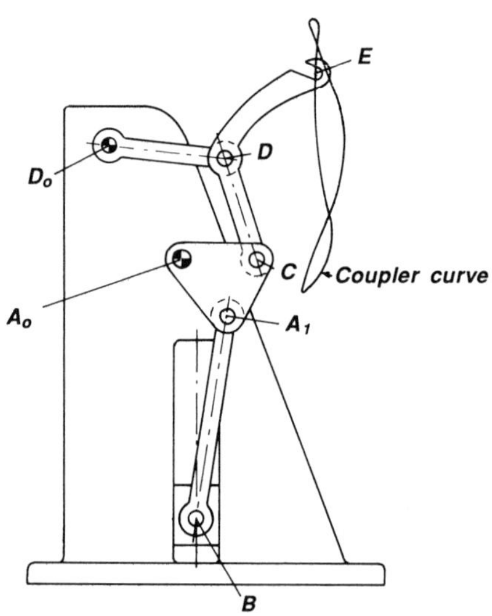

Figure 3.72 The slider crank A_0AB drives the needle of a sewing machine up and down. The needle is fixed to the slider B. The crank A_0A is integral with crank A_0C of the four-bar linkage A_0CDD_0 with a coupler point E, which traces the coupler curve shown. The eye E, through which the thread to the needle passes, makes a motion that is synchronized to the motion of the point of the needle.

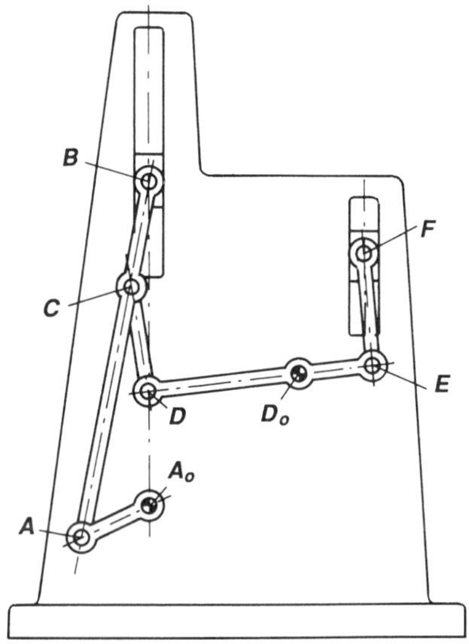

Figure 3.73 The centric slider crank A_0AB drives the slider B. Link CD, attached to the rod AB, oscillates link DD_0E around D_0, causing the slider F to move up and down, approximately 180° out of phase with slider B.

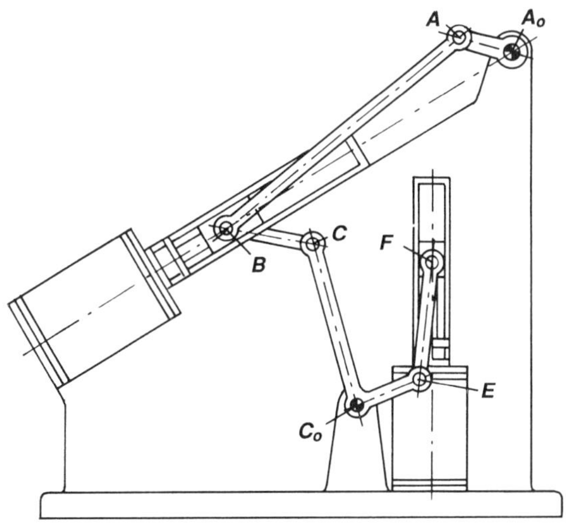

Figure 3.74 The slider crank A_0AB moves the crosshead at B. A link BC is attached at B and oscillates the bell crank BC_0E around C_0, causing slider F to move up and down.

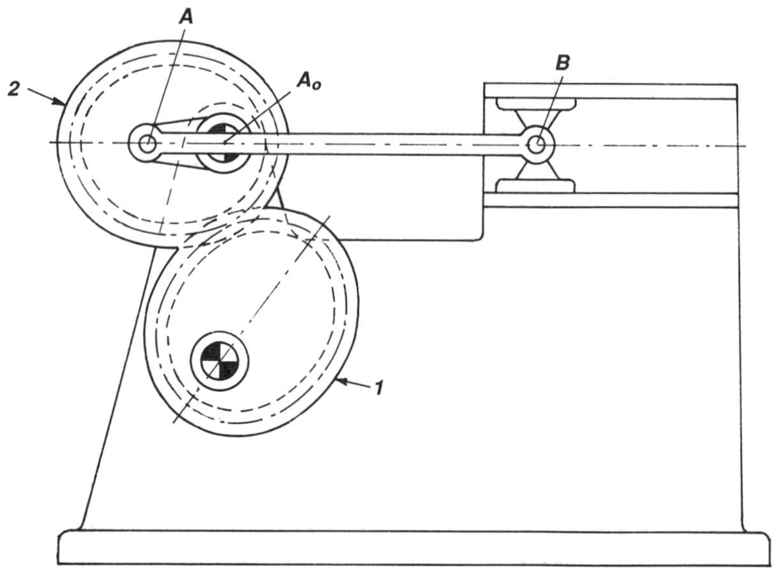

Figure 3.75 The elliptical gear 1 drives the elliptical gear 2, which in turn drives the crank A_0A of the slider crank A_0AB. The motion of the slider is dependent on the proportions of the slider crank and on the motion of the elliptical gears 1 and 2.

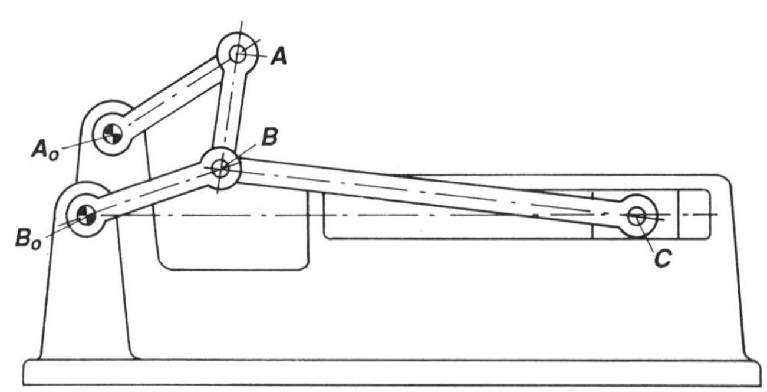

Figure 3.76 The crank A_0A of the double crank A_0ABB_0 imparts a rotary motion to link B_0B, which acts as crank B_0B of the slider crank B_0BC. The motion of the slider is now dependent on the proportions of the double crank and on the proportions of the slider crank.

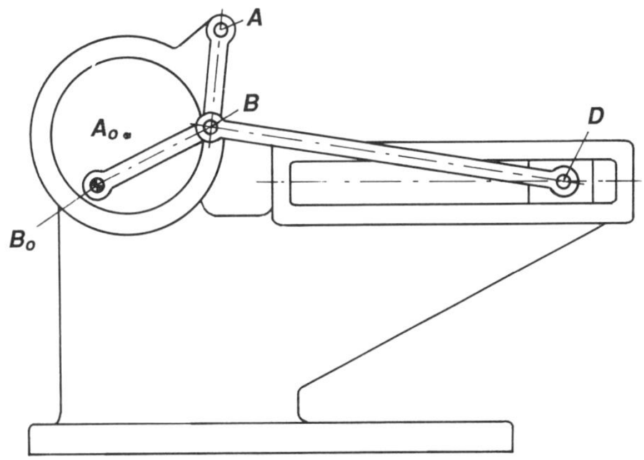

Figure 3.77 This mechanism is kinematically equivalent to the mechanism in Fig. 3.76, and if link lengths are identical, the motion of B will remain the same. Here the principle of pin enlargement is used.

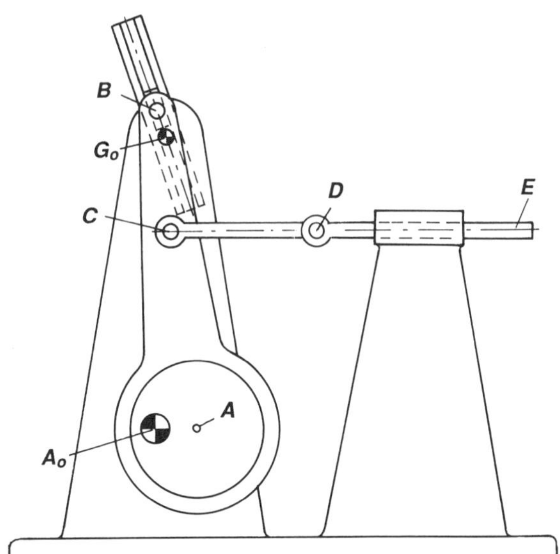

Figure 3.78 The slider crank A_0AB drives the slider DE through link CD. The direction of motion of the slider is made adjustable so that a varying output motion can be obtained.

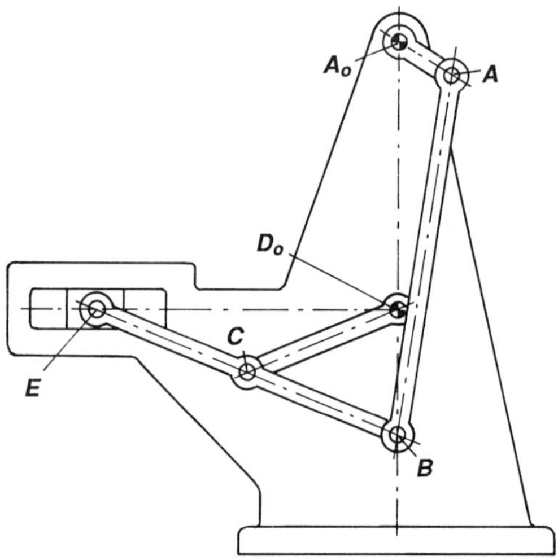

Figure 3.79 Straight-line motion of point B. If the direction of motion of the slider E passes through D_0, and CE = CD = CB, then B moves on a straight line (see Fig. 3.22), as is true here. Principle is useful if, for some reason, a sliding motion is not possible near B.

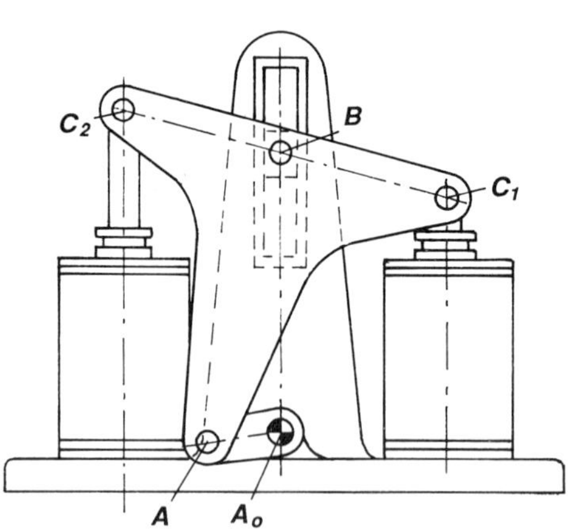

Figure 3.80 The slider crank A_0AB drives the pistons which are connected to joints C_1 and C_2. The mechanism is overconstrained, i.e., clearance must be provided somewhere in order that it can move satisfactorily.

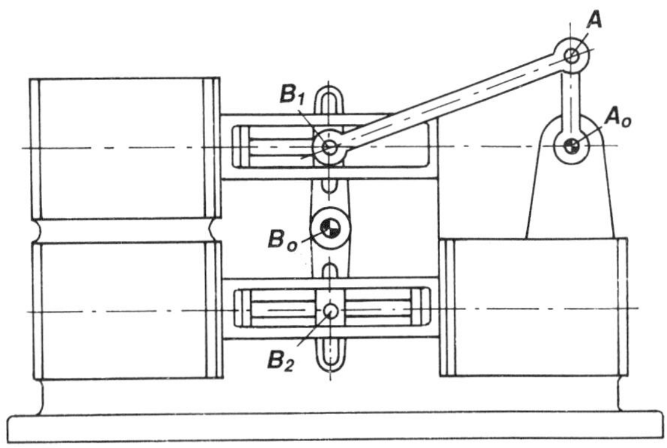

Figure 3.81 The slider crank A_0AB_1 drives the upper piston. Slider B_1 oscillates the slotted member $B_1B_0B_2$, thereby driving the piston rod that is connected to the pistons of the two lower cylinders. One of the lower pistons is out of phase with the upper one.

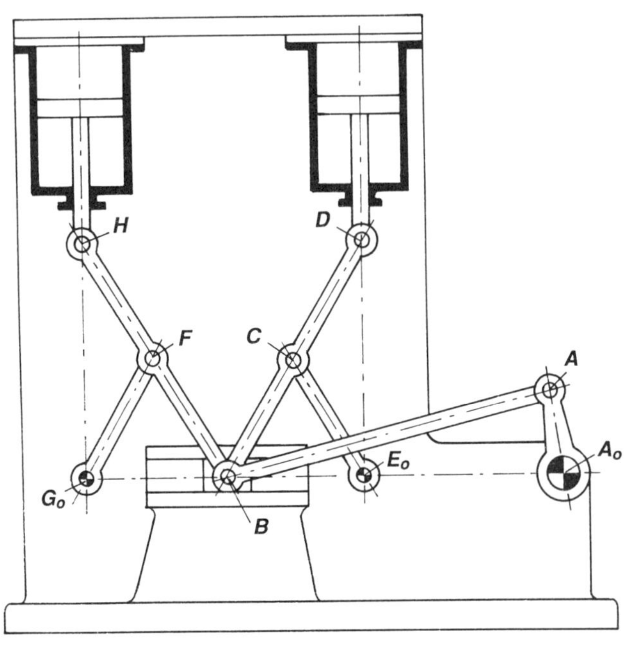

Figure 3.82 The slider crank A_0AB drives the slider B back and forth along a straight line. Because the length of link FB = FG = FH, and CB = CD = CE, points D and H move on straight lines and drive the pistons (see Fig. 3.22).

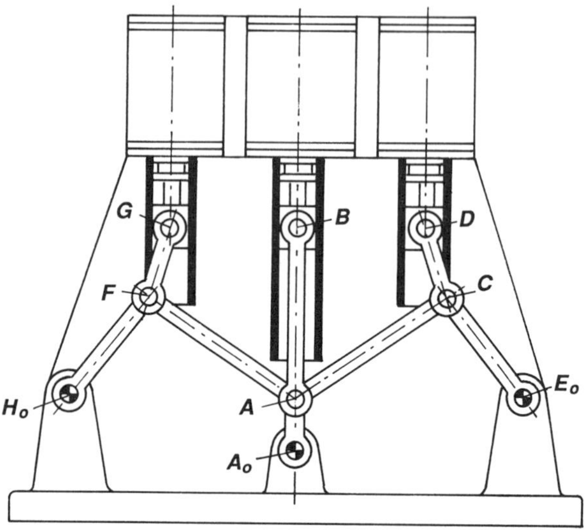

Figure 3.83 The slider B of the slider crank A_0AB moves the center piston. The two four-bar linkages A_0ACE_0 and A_0AFH_0 drive the right and left piston through the rods CD and FG, respectively.

SYMBOLS

α	angular acceleration of crank, s^{-2}
α_1	angular acceleration of connecting rod, s^{-2}
A_B	acceleration of the slider, in./s^2 or m/s^2
β	angle between the connecting rod and the motion of the slider, degrees
ϕ	angle through which the output gear rotates back, degrees
γ	output rotation, degrees
L	length of connecting rod, in. or mm
λ	R/L
ω	angular velocity of crank, s^{-1}
ω_1	angular velocity of connecting rod, s^{-1}
r_3	radius of gear fixed to connecting rod, in. or mm
r_4	radius of output gear, in. or mm
R	length of crank, in. or mm
S	displacement of the slider from its farthest away position (dead-center position), in. or mm

θ	angle between the direction of motion of the slider and the crank, degrees
θ_0	crank angle rotation during which the output gear reverses its motion, degrees
V_B	velocity of the slider, in./s or m/s

4

Geneva and Star-Wheel Mechanisms

CLASSIFICATION

Geneva mechanisms are characterized by a roller entering a slot (Fig. 4.1) and indexing the Geneva wheel, which, when the roller exits the slot, is locked in its position until the roller starts entering the next slot. Geneva mechanisms can be characterized further by the fact that the path of the driving roller can be circular or noncircular and that the slots can be straight (radial or offset) or curved. Geneva mechanisms can be arranged in series with other mechanisms, for instance, with other Geneva mechanisms, four-bar linkages, noncircular gears, and so forth.

Geneva mechanisms belong to a group of mechanisms having intermittent motion, i.e., when the driving member rotates continuously the output member has a prolonged dwell.

DESCRIPTION AND MOTION CHARACTERISTICS

Figures 4.1, 4.2, and 4.3 show three positions of a 4-station Geneva mechanism. In Fig. 4.1 is shown the position where the roller is just about to enter the slot in the Geneva wheel. The Geneva wheel itself consists of a disc with four radial slots, and the Geneva wheel is fastened to the output

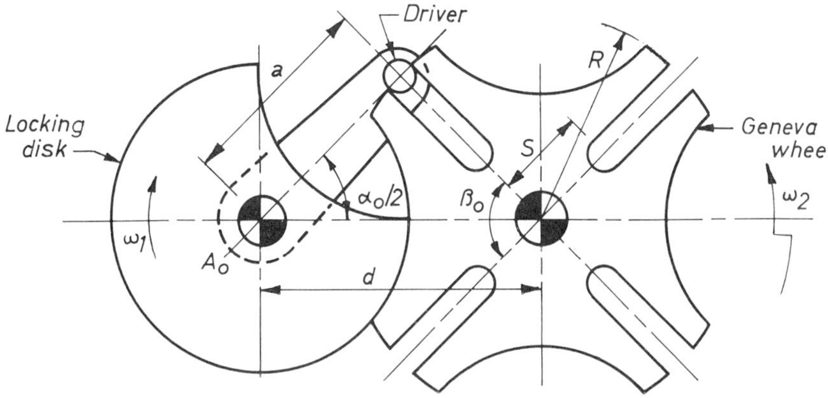

Figure 4.1 A 4-station external Geneva mechanism. The driving roller has just entered the slot, and the locking disc starts unlocking.

shaft. The driving roller, which is just about to enter the slot, is fixed on an arm of the input shaft which is assumed to rotate with a constant angular velocity, but formulas developed also take into account a driving arm rotating with a nonuniform velocity. To the input shaft is fastened a circular locking disc with a cutout. This locking disc fits into a circular cutout in the Geneva wheel.

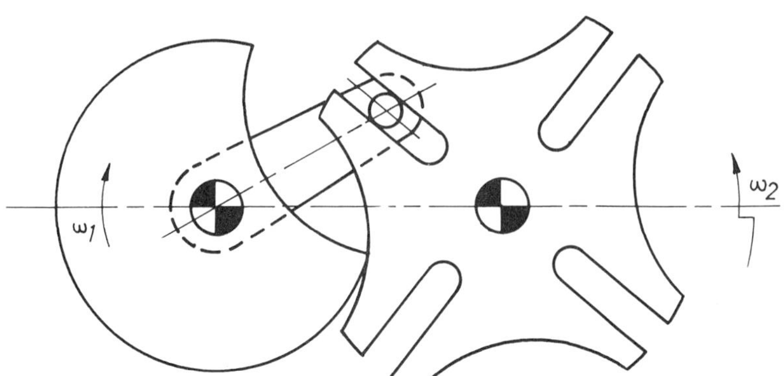

Figure 4.2 The Geneva mechanism in Fig. 4.1. The driving roller is rotating the Geneva wheel.

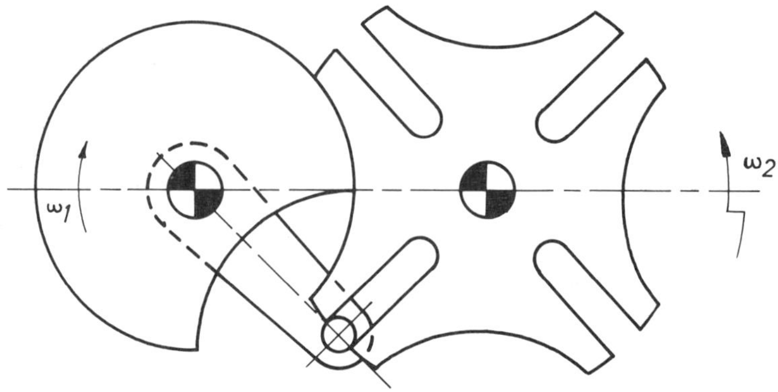

Figure 4.3 The Geneva mechanism in Fig. 4.1. The driving roller is just about to leave the slot. Locking disc starts unlocking.

Figure 4.2 shows a position where the roller has entered the slot and has turned the Geneva wheel an angle in a CCW direction (assuming that the input shaft turns CW). It is seen how the circular locking disc has moved away from the Geneva wheel so as to allow it to rotate.

Figure 4.3 shows the position where the roller is just about to leave the slot so that the Geneva wheel ends its indexing motion. It is also seen how the locking disc has just reached a position where it starts to lock the Geneva wheel in its present position.

For a 4-station Geneva mechanism, motion takes place as follows: For 90° rotation CW of the driving crank from the position shown in Fig. 4.1 to the position shown in Fig. 4.3, the Geneva wheel rotates 90° CCW. For the next 270° rotation of the input shaft, the Geneva wheel has a complete dwell, because the roller has left the slot, and the locking disc prevents any further motion of the Geneva wheel in this position.

The Geneva mechanism is called "4-station" because of the four radial slots. The 4-station Geneva mechanism described in Figs. 4.1–4.3 is called "external" because the roller penetrates the slot from the outside. Figure 4.4 shows a 4-station internal Geneva mechanism where the roller enters the slot from the inside. The latter type is not used very often, because a through-going shaft cannot be used.

In a Geneva mechanism, it is very important that the roller enters the slot tangentially because otherwise shocks are generated and this might lead to a destruction of the entire mechanism at high speed and/or high loads.

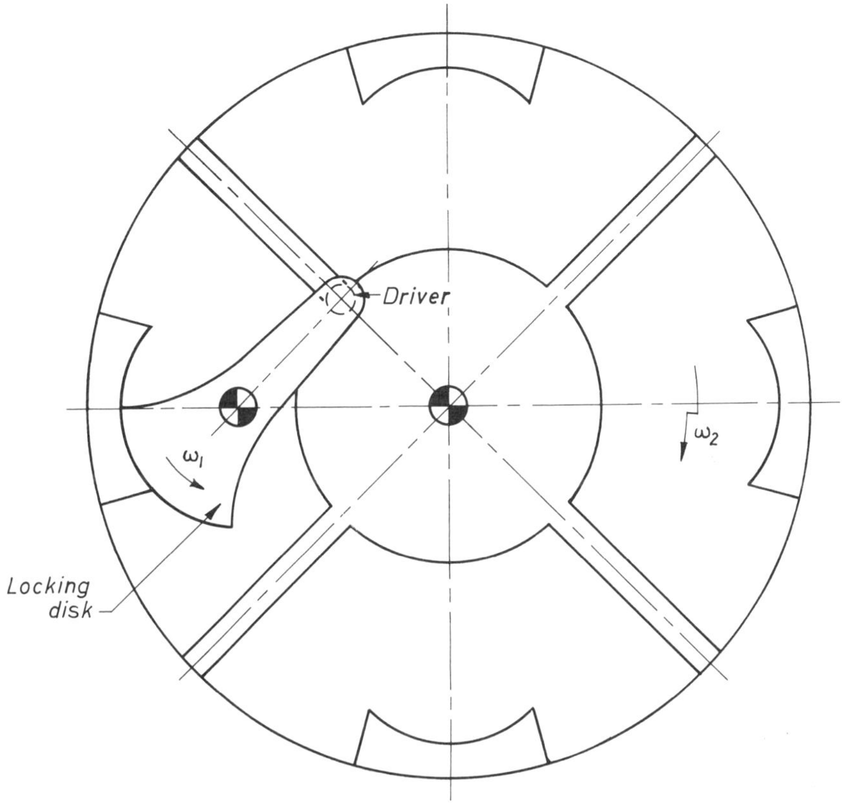

Figure 4.4 A 4-station internal Geneva mechanism. The roller is entering the slot and starts driving the Geneva wheel in a CW direction.

GEOMETRY OF THE GENEVA MECHANISM

Equations 4.1–4.10 are easily derived (see Fig. 4.1).

$$\beta_0 = \frac{360°}{n} \qquad \text{(external and internal)} \tag{4.1}$$

$$\alpha_0 + \beta_0 = 180° \qquad \text{(external)} \tag{4.2}$$

$$\alpha_0 - \beta_0 = 180° \qquad \text{(internal)} \tag{4.3}$$

$$a = d \sin (\beta_0/2) \tag{4.4}$$

$$R = d \cos \beta_0/2 \qquad \text{(external)} \tag{4.5}$$

$$R = d - a \qquad \text{(external)} \tag{4.6a}$$

$$R = d + a \qquad \text{(internal) (compare eq. 34)} \tag{4.6b}$$

$$S \le d - a = d(1 - \sin \beta_0/2) \qquad \text{(external)} \tag{4.7}$$

$$S \ge d + a = d(1 + \sin \beta_0/2) \qquad \text{(internal)} \tag{4.8}$$

$$\left(\frac{\omega_2}{\omega_1}\right)_{max} = \frac{a}{d - a} \qquad \text{(external)} \tag{4.9}$$

$$\left(\frac{\omega_2}{\omega_1}\right)_{max} = \frac{a}{d + a} \qquad \text{(internal)} \tag{4.10}$$

The results of eqs. (4.1–4.4) and (4.7–4.10) are listed in Table 4.1 (external Genevas) and Table 4.2 (internal Genevas).

Figure 4.5 shows a 4-station Geneva mechanism with its corresponding angular velocities and angular accelerations (ω_2/ω_1 and ω_2/ω_1^2, respectively) dependent on the position of the driving arm. Also shown is the relationship between the angular position of the Geneva wheel and the angular position of the driving arm. Only an external Geneva mechanism is shown, but the curves are valid for the corresponding internal Geneva mechanism, too. The formulas for angular velocities and angular accelerations are developed in the following section.

ANGULAR VELOCITIES AND ACCELERATIONS OF THE GENEVA WHEEL

Figure 4.6 shows the slot and the roller arm of a Geneva mechanism. The angles α and β are measured from the center line, connecting the centers of the input and output shafts. All angles and directions are measured positive if CW, negative if CCW.

To find the angular velocity of the slotted part, consider two points A and B. The two points are coincident but A is a point on the roller arm and B a point on the slotted part.

The velocity vector equation is written as usual:

$$V_A = V_B \leftrightarrow V_{A/B}$$

These vectors may now be written as complex vectors:

$$V_B = iV_Be^{i\beta} = ir\omega_2e^{i\beta}$$

$$V_A = ia\omega_1e^{i\alpha}$$

$$V_{A/B} = V_{A/B}e^{i\beta}$$

but

Table 4.1 External Geneva Mechanisms

	Eq. no.								
	4.1	4.2	4.4	4.7	4.9	4.15	4.16	4.17	4.29
n	β_0	α_0	a/d	s/d	$(\omega_2/\omega_1)_{max}$	$(\alpha_2/\omega_1^2)_{max}$	$\alpha(F_{12})_{max}$	$\alpha(F_{12}^t)_{max}$	m_{max}
3	120°	60°	0.8660	0.1340	6.46	31.40	175.7°	176.3°	≤6
4	90°	90°	0.7071	0.2929	2.41	5.41	169.4°	171.1°	≤4
5	72°	108°	0.5878	0.4122	1.43	2.30	163.6°	166.8°	3
6	60°	120°	0.5000	0.5000	1.00	1.35	158.8°	163.4°	≤3
7	51.43°	128.73°	0.4339	0.5661	0.77	0.93	154.6°	160.6°	2
8	45°	135°	0.3827	0.6173	0.62	0.70	152.4°	158.4°	2
9	40°	140°	0.3420	0.6580	0.52	0.56	148.2°	156.6°	2
10	36°	144°	0.3090	0.6910	0.45	0.46	145.6°	155.0°	2
11	32.73°	147.27°	0.2818	0.7182	0.39	0.40	143.4°	153.7°	2
12	30°	150°	0.2588	0.7412	0.35	0.35	141.4°	152.6°	2
13	27.69°	152.31°	0.2393	0.7607	0.31	0.31	139.7°	151.7°	2
14	25.71°	154.29°	0.2225	0.7775	0.29	0.28	138.1°	150.8°	2
15	24°	156°	0.2079	0.7921	0.26	0.25	136.8°	150.1°	2
∞	0	180°	—	d − a	0	0	90°	135°	≤2

Table 4.2 Internal Geneva Mechanisms

	Eq. no.							
	4.1	4.3	4.4	4.8	4.10	4.15	4.16	4.17
n	β_0	α_0	a/d	s/d	$(\omega_2/\omega_1)_{max}$	$(\alpha_2/\omega_1^2)_{max}$	$\alpha(F_{12})_{max}$	$\alpha(F^t_{12})_{max}$
3	120°	300°	0.8660	1.8660	0.46	1.73	150°	132.5°
4	90°	270°	0.7071	1.7071	0.41	1.00	135°	110°
5	72°	252°	0.5878	1.7878	0.37	0.73	126°	97.3°
6	60°	240°	0.5000	1.5000	0.33	0.58	120°	89.2°
7	51.43°	231.37°	0.4339	1.4339	0.30	0.48	115.7°	83.6°
8	45°	225°	0.3827	1.3827	0.28	0.414	112.5°	79.6°
9	40°	220°	0.3420	1.3420	0.25	0.364	110°	76.5°
10	36°	216°	0.3090	1.3090	0.24	0.325	108°	74.1°
11	32.73°	212.73°	0.2818	1.2817	0.22	0.294	106.4°	72.1°
12	30°	210°	0.2588	1.2588	0.21	0.268	105°	70.5°
13	27.69°	207.69°	0.2393	1.2393	0.19	0.246	103.3°	69.2°
14	25.71°	205.71°	0.2225	1.2225	0.18	0.228	102.9°	68.0°
15	24°	204°	0.2079	1.2079	0.17	0.213	102°	67.1°
∞	0°	180°	—	d + a	0	0	90°	45°

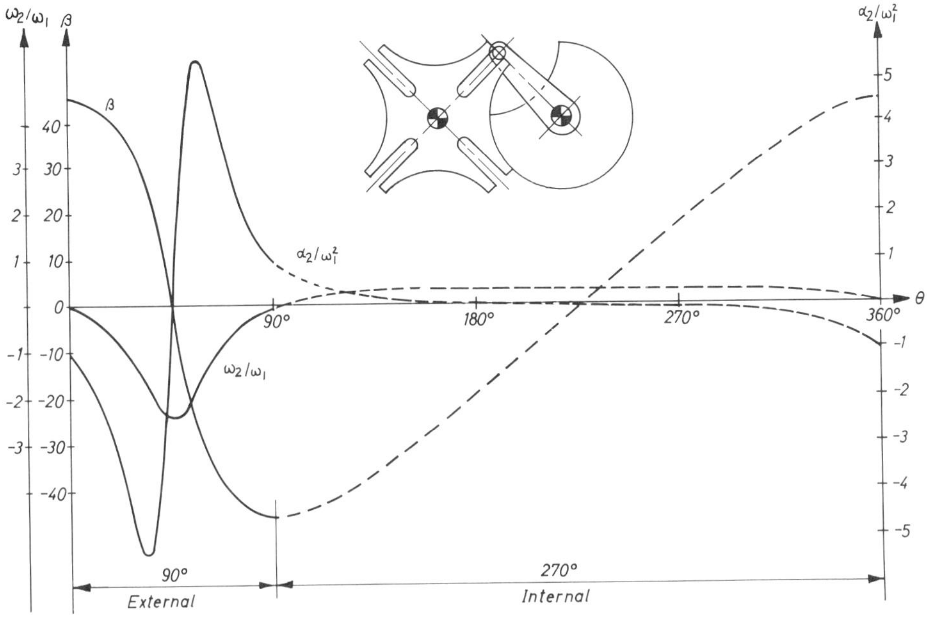

Figure 4.5 Angular displacement, angular velocity, and angular acceleration of a 4-station external and internal Geneva mechanism. (See also Fig. 4.4.)

$$e^{i\theta} = \cos\theta + i\sin\theta$$

and therefore

$$ia\omega_1(\cos\alpha + i\sin\alpha) = ir\omega_2(\cos\beta + i\sin\beta) + V_{A/B}(\cos\beta + i\sin\beta)$$

Solving the above equations the following results are obtained:

$$\frac{\omega_2}{\omega_1} = \frac{a}{r}\cos(\beta - \alpha) \tag{4.11}$$

$$V_{A/B} = a\omega_1 \sin(\beta - \alpha) \tag{4.12}$$

The distance $B_0B = r$ is found from

$$r = \sqrt{a^2 + d^2 - 2ad\cos(180^\circ - \alpha)} \tag{4.13}$$

and

$$r\sin\beta = a\sin(180^\circ - \alpha)$$

$$\beta = \sin^{-1}\left(\frac{a}{r}\sin\alpha\right) \tag{4.14}$$

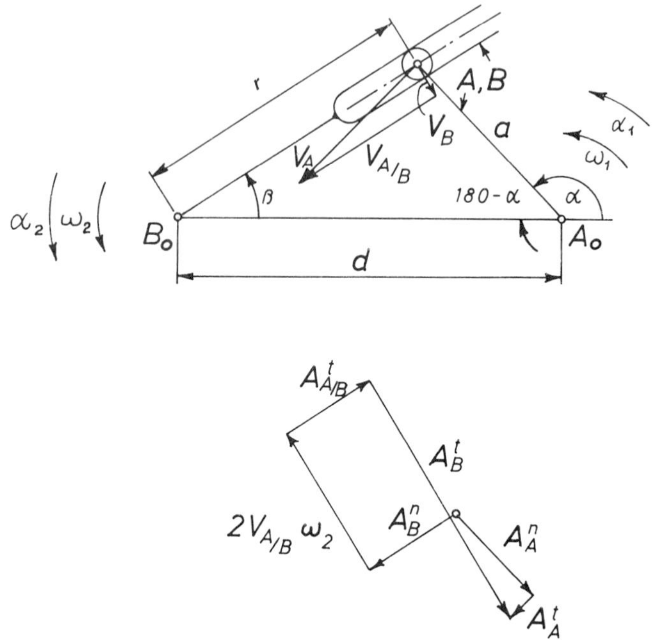

Figure 4.6 Velocity and acceleration vector diagrams.

The vector equation for the accelerations can be written

$$\underset{=}{A_A{}^n} \leftrightarrow \underset{=}{A_A{}^t} = \underset{=}{A_B{}^n} \leftrightarrow \underset{\sphericalangle}{A_B{}^t} \leftrightarrow \underset{=}{A^n_{A/B}} \leftrightarrow \underset{\sphericalangle}{A^t_{A/B}} \leftrightarrow \underset{=}{2V_{A/B}\omega_2}$$

where the last term is Corioli's acceleration*

$$A_A{}^n = -a\omega_1{}^2\, e^{i\alpha}$$
$$A_A{}^t = ia\alpha_1 e^{i\alpha}$$
$$A_B{}^n = -r\omega_2{}^2\, e^{i\beta}$$
$$A_B{}^t = i\, A_B{}^t\, e^{i\beta}$$

where α_1 is the angular acceleration of the driving arm.

$$A^n_{A/B} = 0$$
$$A^t_{A/B} = A^t_{A/B}\, e^{i\beta}$$
$$V_{A/B} = a\omega_1 \sin(\beta - \alpha) \tag{4.12}$$

---

*Gaspard G. de Corioli (1792–1843), Sur les Équations de Mouvements Relatif des Systèmes des Corps, *Journal de l'École Polytechnique*, 1825.

$$\omega_2 = \frac{a}{r}\,\omega_1 \cos(\beta - \alpha) \tag{4.11}$$

From the above equations the angular acceleration α_2 of the driven member is found:

$$\alpha_2 = \frac{a\omega_1^{\,2}\sin(\beta - \alpha) - a\alpha_1 \cos(\beta - \alpha)}{r} + \frac{a^2}{r^2}\,\omega_1^{\,2}\sin(2\beta - 2\alpha) \tag{4.15}$$

Example 1

A 4-station external Geneva mechanism is driven by an arm rotating with a constant angular velocity of N = 60 RPM CCW. Determine the angular velocity and angular acceleration of the Geneva wheel when the driving arm has rotated 25° from the position where the roller has just entered the slot. In symbols:

$n = 4$
$N = 60$ RPM (CCW)
$\alpha_1 = 0$ (constant angulur velocity of input member is given)

Solution

Assuming that $A_0B_0 = d = 3$ in (any other value may be assumed), the following is obtained:

$$\beta_0 = \frac{360°}{4} = 90° \tag{4.1}$$

$$\alpha_0 = 180° - 90° = 90° \tag{4.2}$$

$$a = 3 \sin 45° = 2.12 \text{ in.} \tag{4.4}$$

$$\alpha = 180° - \beta_0/2 + 25° = 160° \qquad \text{(see Fig. 4.6)}$$

$$r = \sqrt{2.12^2 + 3^2 - (2)(2.12)(3)\cos 20°} = 1.242 \text{ in.} \tag{4.13}$$

$$\beta = \sin^{-1}\left(\frac{2.12}{1.242}\sin 20°\right) = 35.72° \tag{4.14}$$

$$\omega_1 = -\frac{N}{60}2\pi = -2\pi \text{ rad/s}$$

$$\omega_2 = \frac{2.12}{1.242}(2\pi)\cos(35.72 - 160°) = -6.04 \text{ rad/s (CW)} \tag{4.11}$$

$$\alpha_2 = 49.3 \text{ rad/s}^2 \tag{4.15}$$

The equations developed in the foregoing allow the determination of the angular acceleration of the Geneva wheel even if the driving arm does not rotate with a uniform angular velocity. This means that the formulas may be used for Geneva mechanisms where the output of the first Geneva mechanism is input to the next Geneva mechanism, that is, for Geneva mechanisms in series. The formulas are also valid for all other types of mechanisms that drive the Geneva mechanism. Driving by other mechanisms may be done to obtain improved acceleration characteristics or to improve the angular motion characteristics of the Geneva mechanism.

FORCE ANALYSIS

In the following, a force analysis of the Geneva mechanism in a certain position is made, based partly on formulas from the foregoing section. A complete analysis for a whole revolution can be made with a pocket calculator or a home computer, based on these formulas.

The mass moment of inertia of the Geneva wheel is denoted I_G. The force with which the roller on the driving arm a presses against the Geneva wheel is F_{12} (Fig. 4.7). The force F_{12} is resolved into two components, namely F^t_{12}, which is perpendicular to the roller arm a, and F^n_{12}, which is perpendicular to F^t_{12}, that is, in the direction of the center line A_0A of the roller arm. The force F_{12} is the force that accelerates or decelerates the Geneva wheel and is determined from the equation

$$I_G\alpha_2 = F_{12}r$$

or

$$F_{12} = \frac{I_G\alpha_2}{r} \tag{4.16}$$

From the geometry of Fig. 4.7

$$F_{12} = F^t_{12} \leftrightarrow F^n_{12}$$

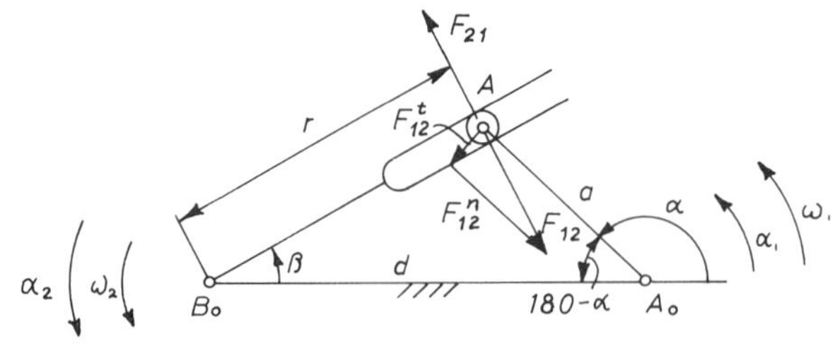

Figure 4.7 Determination of forces.

but

$$F_{12} = -F_{12}ie^{i\beta}$$
$$F^t_{12} = F^t_{12}ie^{i\alpha}$$
$$F^n_{12} = -F^n_{12}e^{i\alpha}$$

or

$$-F_{12}i(\cos\beta + i\sin\beta) = F^t_{12}i(\cos\alpha + i\sin\alpha) - F^n_{12}(\cos\alpha + i\sin\alpha)$$

from which

$$\frac{F^t_{12}}{F_{12}} = \cos(\alpha - \beta) \tag{4.17}$$

$$\frac{F^n_{12}}{F_{12}} = \sin(\alpha - \beta) \tag{4.18}$$

The force F_{12} varies depending on the motion of the Geneva wheel, the distance r, and the angular acceleration α_2 during motion. The maximum value of F_{12} determines the maximum pressure between the roller and the slot and is, therefore, part of the engineering calculations.

Force Analysis for a Single Position

Example 2

For a 6-station external Geneva mechanism (n = 6)

$$R = 3.0 \text{ in.}$$
$$r_f = .25 \text{ in.}$$
$$t = .4 \text{ in.}$$
$$I_G = .0165 \text{ lb-in.-s}^2$$

The roller arm (a) has from the position where it enters the slot (CW) turned an angle of 30°. N = 60 RPM CCW and is constant, i.e., $\alpha_1 = 0$.

Solution

Using the procedure from Example 1 the following results are obtained:

$$\omega_1 = 2\pi \text{ rad/s}$$
$$\beta_0 = 60° \tag{4.1}$$
$$d = 3.464 \text{ in.} \tag{4.5}$$
$$a = 1.732 \text{ in.} \tag{4.4}$$

The roller arm (a) has rotated 30° from the position where it started its motion into the slot, and, therefore, it has to rotate 30° further before it reaches the center line (see Fig. 4.6):

$$\alpha = 180° - 30° = 150°$$

$$r = 2.147 \text{ in.} \tag{4.13}$$

$$\beta = 23.8° \tag{4.14}$$

$$\alpha_2 = -50.21 \text{ rad/s}^2 \tag{4.15}$$

and now from the section Force Analysis:

$$F_{12} = \frac{I_G \alpha_2}{r} = 0.316 \text{ lb} \tag{4.16}$$

$$F^t_{12} = -0.316 \cos(150° - 23.8°) = -0.187 \text{ lb} \tag{4.17}$$

$$F^n_{12} = -0.316 \sin(150° - 23.8°) = 0.255 \text{ lb} \tag{4.18}$$

Having made the force analysis for one single position and for a 6-station Geneva wheel, one may now program all the equations developed. It is then possible to find the angle α for which $F^n_{12} = (F^n_{12})_{max}$. For a Geneva mechanism with a certain number of stations, angle α may be found, independent of the size of the Geneva wheel. The results are listed in Tables 4.1 and 4.2, where eqs. (4.16) and (4.18) are used.

DETERMINATION OF MASS MOMENT OF INERTIA OF GENEVA WHEELS

The Geneva wheel has a rather irregular shape, but the formulas for mass moment of inertia can be developed from the usual equations for the various sectors of the profile. Referring to Fig. 4.8 we have

$$I_G = I_1 - I_2 - I_3 - I_4 \tag{4.19}$$

$$I_1 = \frac{1}{2}\frac{\rho}{g} \cdot t\pi R^4 \tag{4.20}$$

$$I_2 = n\frac{\rho}{g} \cdot t \left\{ \frac{1}{6}[R - S] \cdot r_f [(R - S)^2 + 4r_f^2] + \left(\frac{R + S}{2}\right)^2 (R - S) \cdot 2r_f \right\} \tag{4.21}$$

$$S = \frac{R}{\cos(\beta_0/2)} - R\frac{\tan \beta_0}{2} \tag{4.22}$$

$$I_3 = n\frac{\rho}{g} t \cdot R^4 \left[\frac{\pi}{2}\frac{\alpha_3}{180} - \sin 2\alpha_3 \left(\frac{1}{4} - \frac{1}{6}\sin^2 \alpha_3 \right) \right] \tag{4.23}$$

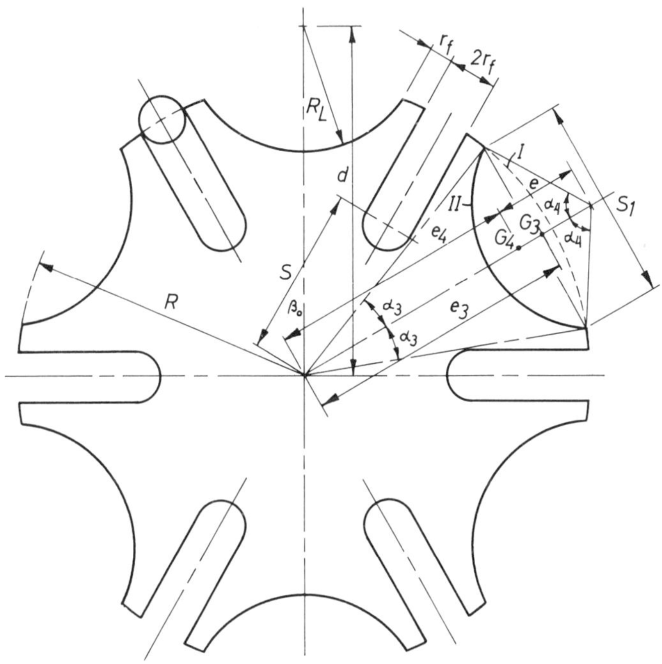

Figure 4.8 Geometry of the Geneva wheel.

$$\alpha_3 = \frac{\beta_0 \cdot \pi}{360} - \frac{2r_f}{R} \tag{4.24}$$

$$\begin{aligned} I_4 = {} & n\frac{\rho}{g} \cdot tR_L^4 \left[\frac{\pi}{2}\frac{\alpha_4}{180} - \sin 2\alpha_4 \left(\frac{1}{4} - \frac{1}{6}\sin^2 \alpha_4\right)\right] \\ & + n\frac{\rho}{g}t\left(\pi R_L^2 \frac{\alpha_4}{180} - R_L^2 \sin \alpha_4 \cos \alpha_4\right) \\ & \times \left(d^2 - \frac{4}{3} d\, R_L \cdot \frac{\sin^3 \alpha_4}{\pi\alpha_4/180 - \sin \alpha_4 \cos \alpha_4}\right) \end{aligned} \tag{4.25}$$

$$\alpha_4 = \tan^{-1}\left(\frac{R \sin \alpha_3}{d - R \cos \alpha_3}\right) \tag{4.26}$$

$$R_L = d \sin (\beta_0/2) - 2r_f \tag{4.27}$$

where

I_t = mass moment of inertia connected to geneva wheel, lb-m-s^2
I_G = mass moment of inertia of Geneva wheel, lb-m-s^2

I_1 = mass moment of inertia Geneva wheel with no cutouts, lb-m-s^2
I_2 = mass moment of inertia of n slots, lb-m-s^2
I_3 = mass moment of inertia of n circle segments nr. I, lb-m-s^2
I_4 = mass moment of inertia of n circle segments nr. II, lb-m-s^2
ρ = 0.284 lb/in.^{3} for steel
g = 386 in./s^2, gravitational constant
t = thickness of Geneva wheel, in.
R = radius of Geneva wheel, in.
S = minimum distance between center of Geneva wheel and roller, in.
r_f = roller radius, in.
α_3 = angle between corner of circular cutout and center line of cutout, measured from center of Geneva wheel
R_L = locking radius of Geneva wheel, in.
α_4 = angle between corner of circular cutout and center line of cutout, measured from center of driving arm

When calculating the mass moment of inertia it is assumed that the width at the tip of the Geneva wheel is r_f which is considered to establish safety against the tip breaking off.

The eqs. (4.19–4.27) have been used to calculate the curves in Fig. 4.9. It is assumed that ρ = .284 lb/in.^{3} (steel), R = 1 in., and t = 1 in. The mass moment of inertia of a Geneva wheel with dimensions R = 3 in. and t = .4 in. is found by multiplying the value found in Fig. 4.9 with

$$K = \left(\frac{.4}{10}\right)\left(\frac{3}{10}\right)^4$$

Notice that the values in Fig. 4.9 have been calculated assuming that the tip width is equal to the roller radius r_f.

OPTIMAL DESIGN

The Geneva mechanism can be used in rather heavy machines, for instance, round tables or packaging machines. A question in this connection arises: For a given mass to be indexed, how large should the dimensions of the Geneva wheel be in order that the maximum force between the roller and the slot is as small as possible? The equation

$$F_{12} = (I_G + I_t)\frac{\alpha_2}{r} \tag{4.28}$$

shows F_{12} as a function of some variables. I_t may be considered constant for a given case, and the remaining variables are radius R and thickness t

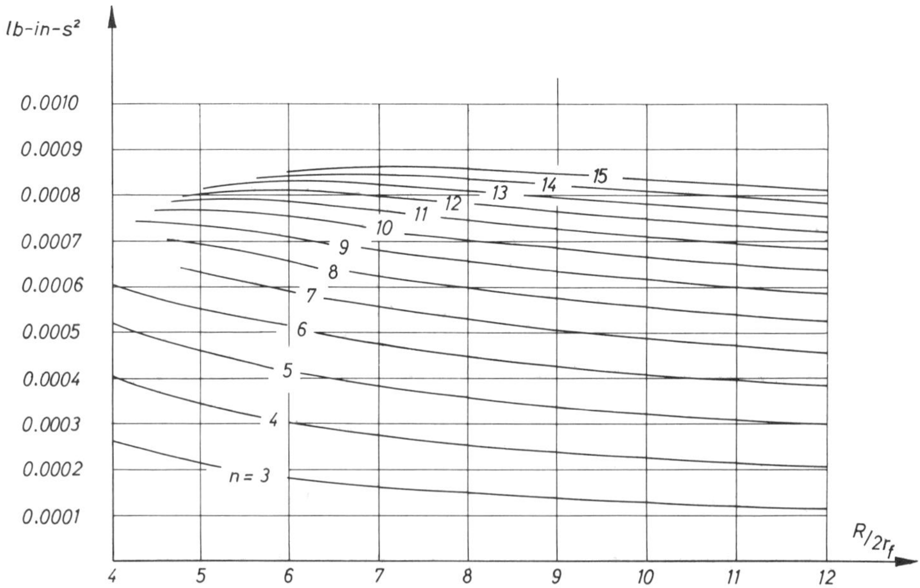

Figure 4.9 Chart to determine mass moment of inertia of Geneva wheels, dependent on the ratio $R/2r_f$ and the number of stations, n. The width of the tip equals r_f, thickness t equals 1, radius of Geneva wheel $R = 1$, and $\rho = .284$ lb/in.^{3} (steel).

of the Geneva wheel because the number of slots may be considered constant, too.

For any given size of the Geneva wheel, $(F_{12})_{max}$ may be found from Table 4.1 (eq. 4.16). By varying the size of the Geneva wheel, it is possible to find a value of R so that $(F_{12})_{max}$ is a minimum. The search is started assuming the thickness of the Geneva wheel. If the resulting Hertz pressure is too large or too small, then the thickness may be increased or decreased, respectively, or the roller radius and/or width may be changed, and the procedure repeated.

The results of an optimization are shown in the following:

Example 3

Given

$n = 4$
$r_f = .5$ in.
$t = .75$ in.

ρ = .284 lb/in.^{3}
N = 60 RPM
I_t = .208 lb-in.-s^2

The Geneva wheel is to drive a machine member with the above given mass moment of inertia (I_t). Find the minimum value of R.

Computer Solution

The eqs. (4.19–4.27) are programmed and the display shows the following values:

R (in.)	$(F_{12})_{max}$ (lb)
1.00	94.3128
3.75	32.6141
4.00	32.3621
4.25	32.1989

For R = 4.25 in., the corresponding compression stress is S_c = 21,295.7 lb/in.^{2}

GENEVA MECHANISMS IN SERIES

The output motion of the first Geneva mechanism is used as input for the second Geneva mechanism. Although there are many possibilities for connecting Geneva mechanisms in series, the most common case is probably the use of Geneva mechanisms having the same number of stations and same indexing during the same time interval, that is, the start of one roller entering the slot in the first mechanism coincides with the same phase of the second mechanism.

The formulas developed in the foregoing can be directly applied by using the output (angular position, angular velocity, and angular acceleration) of the first mechanism as the input value for the second mechanism. Connecting two Geneva mechanisms as suggested above results in the second Geneva mechanism having an output angular acceleration starting and ending with zero and may be advantageous for the dynamics of the combined mechanism.

SPECIAL TYPES OF GENEVA MECHANISMS

In the foregoing, Geneva mechanisms with one roller only were considered but there are a number of possibilities for design changes that make this type of mechanism more versatile.

Maximum Number of Rollers When Stations Are Evenly Distributed

If more than one driving roller is used, then the first condition is that a roller in mesh with the Geneva wheel must be out of mesh or just about to leave the slot before another roller may enter. The maximum number of rollers is obtained if they are placed as close to each other as possible. The maximum number of rollers (m_{max}) is determined by

$$m_{max} \leq \frac{360°}{\alpha_0} \tag{4.29}$$

Table 4.1 lists m_{max} as a function of n, the number of evenly distributed slots (or stations). The sign "$\leq$" in the column for m_{max} means that if the maximum number of rollers is used, then one roller leaves a slot at the instant another roller enters another slot.

One such example (Fig. 4.10) is a 4-station Geneva drive with four evenly spaced slots and four evenly spaced driving rollers. The maximum number of driving rollers is 4, and only instantaneous dwells are possible (Fig. 4.10a). Figure 4.10b shows an arrangement with three driving rollers; it follows that the three possible dwell angles are determined by

$$\alpha_{D1} + \alpha_{D2} + \alpha_{D3} = 90° \tag{4.30}$$

If the rollers are unevenly distributed, eq. (4.29) still holds, but the three dwell angles are not equal.

If two driving rollers are used, then Fig. 4.10c:

$$\alpha_{D1} + \alpha_{D2} = 180° \tag{4.31}$$

If one driving roller only is used, then

$$\alpha_{D1} = 270° \tag{4.32}$$

which corresponds to an ordinary Geneva Mechanism. In Fig. 4.10 one or more of the dwell angles may be zero, but eqs. (4.29–4.31) must be fulfilled.

It is also possible to vary the length of the driving arms when using two or more driving rollers. Figure 4.11a shows two unevenly distributed rollers with unequal arm lengths. In the position shown, the short arm rotates through 100° CW in order to turn the Geneva wheel 80° CCW. The long arm is now in the position when it enters a slot and indexes the Geneva wheel 100° in a CCW direction when the roller arm moves through 80° in a CW direction. The 180° dwell period indicated in Fig. 11b can be distributed over two dwell periods.

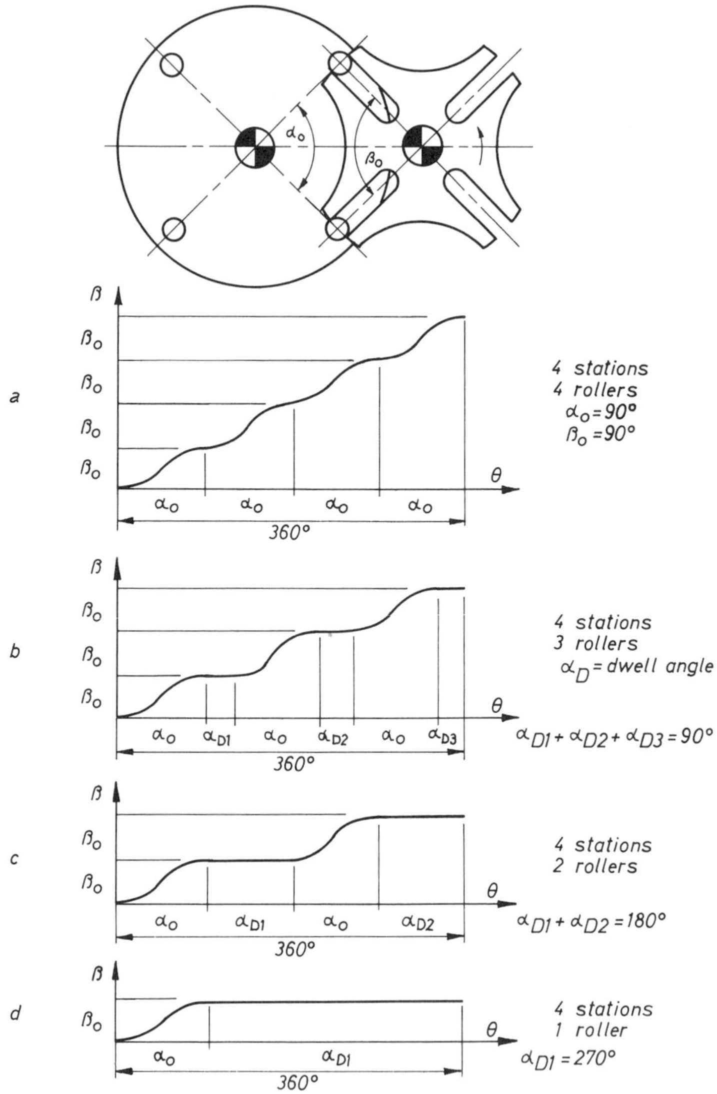

Figure 4.10 4-station Geneva mechanism. Correlation between number of driving rollers and dwell angles.

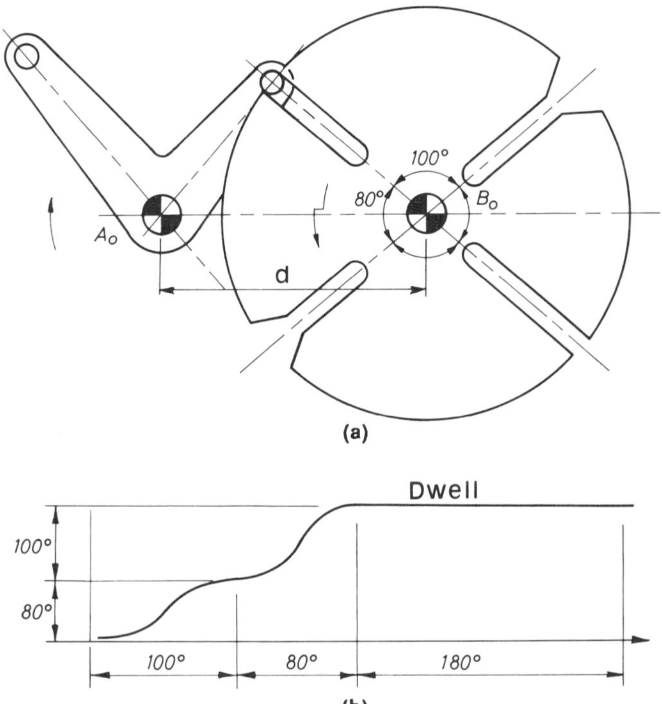

Figure 4.11 Geneva mechanism with two driving rollers. (a) The roller arms are of unequal length. (b) Time-displacement diagram for the Geneva mechanism.

Geneva Mechanisms with an Uneven Number of Slots Uniformly Distributed

Another possibility of a design variation is to make the number of stations uneven and then let the roller enter each second or third station instead of each station or slot. One example is shown in Fig. 4.12. The Geneva wheel shown has 9 slots and one roller. For each full revolution of the driving roller, the wheel is indexed 80°. The equation

$$n = \frac{360°}{\beta_0} \tag{4.33}$$

defines a nominal number of stations. For the mechanism shown in Fig. 4.12,

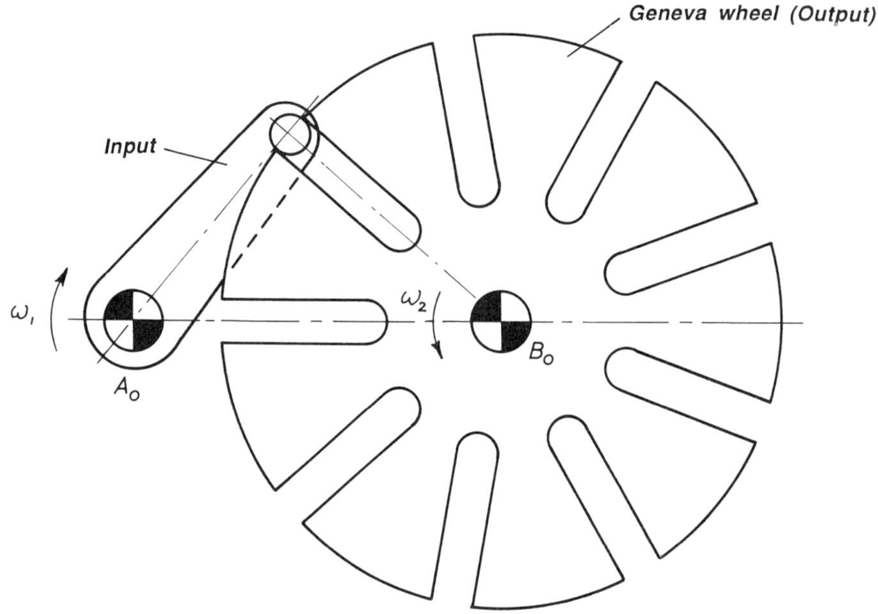

Figure 4.12 Geneva mechanism where the roller enters each second slot so that the Geneva wheel must complete two revolutions before it is back in its starting position.

$$n = \frac{360^\circ}{80^\circ} = 4.5 \qquad (4.34)$$

Table 4.3 shows some of the possibilities for external Geneva mechanisms.

Design Considerations

The actual design is, to a large extent, dependent upon the speed and/or loading of the Geneva wheel. In general two kinds of failure may take place, namely, too great a compressive stress between the roller and the wheel or direct fracture close to the tip of a slot or at a point where two neighboring slots are closest to each other.

As a rule the thickness of the wheel is made equal to the width of the roller. To ensure that the roller enters correctly into the slot, that is, tangential relative to the center line of the slot, the maximum diameter R_m of the wheel is calculated from

$$R_m = \sqrt{R^2 + r_f^2} \qquad (4.35)$$

Table 4.3 External Geneva Mechanisms

	Eq. no.								
	4.1		4.2	4.4	4.7	4.9	4.15	4.16	4.17
n	β_0	β_0'*	α_0	a/d	s/d	$(\omega_2/\omega_1)_{max}$	$(\alpha_2/\omega_1^2)_{max}$	$\alpha(F_{12})_{max}$	$\alpha(F_{12}^t)_{max}$
2-1/3	154.29°	51.43°	25.71°	0.9749	0.025	38.84	1007	179°	179.3°
2-1/2	144°	72°	36°	0.9510	0.049	19.41	258	178°	178.7°
2-2/3	135°	45°	45°	0.9239	0.0761	12.14	104	176.8°	178.0°
3-1/3	108°	36°	72°	0.8090	0.1910	4.24	14.5	171.4°	174.7°
3-1/2	102.86°	51.43°	77.14°	0.7818	0.2182	3.58	10.7	170.0°	173.9°
3-2/3	98.18°	32.73°	81.82°	0.7557	0.2443	3.09	8.3	168.7°	173.1°
4-1/3	83.08°	27.69°	96.92°	0.6631	0.3369	1.97	3.8	163.5°	170.0°
4-1/2	80°	40°	100°	0.6428	0.3572	1.80	3.3	162.2°	169.3°
4-2/3	77.14°	25.71°	102.86°	0.6235	0.3765	1.66	2.9	161°	168.7°

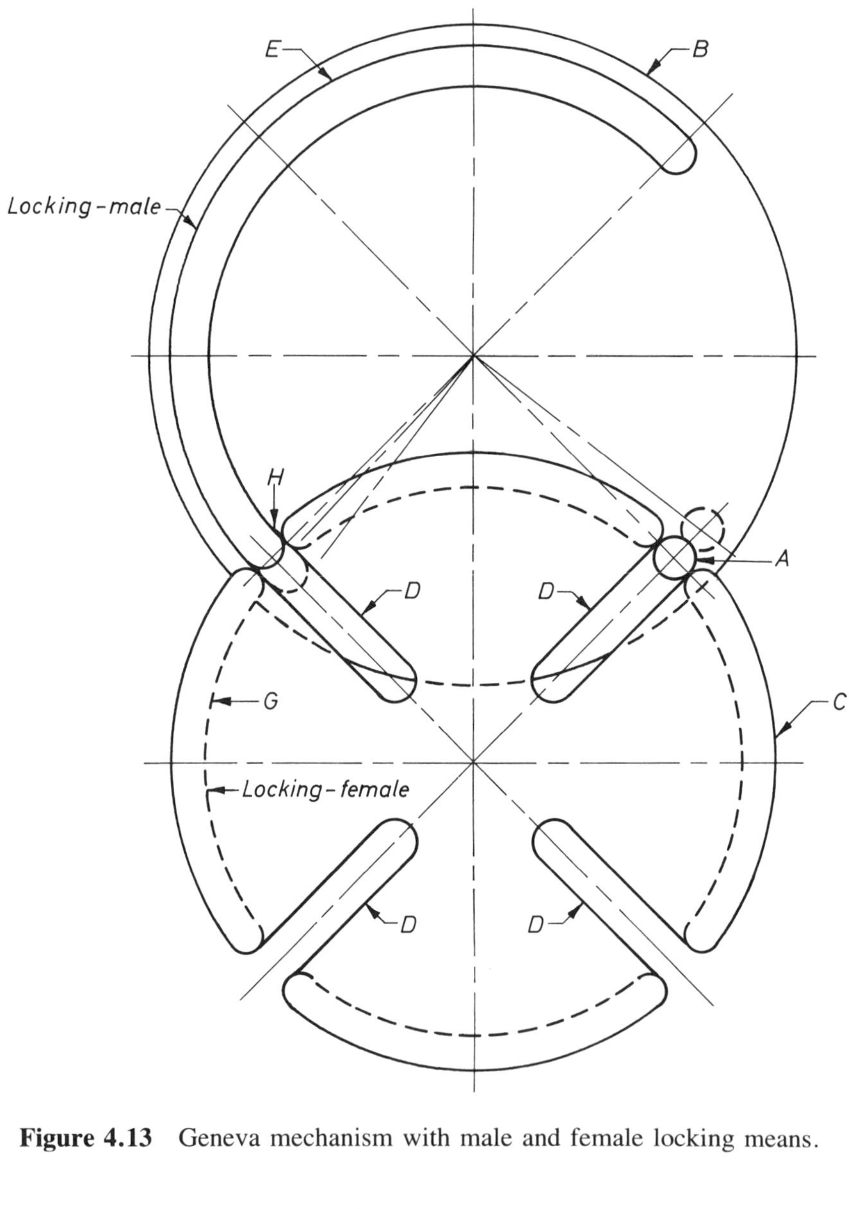

Figure 4.13 Geneva mechanism with male and female locking means.

To lock the position of the wheel when the roller is not moving it, a locking disc is employed as already described, but the radius of this disc must be very accurate, which necessitates that the distance between the center of the wheel and the center of the shaft for the driving arm be very accurate, too.

When using the maximum number of rollers of 4 for the 4-station Geneva mechanisms, no locking disc is required. It is also possible to lock the wheel as shown in Fig. 4.13. The roller A is just about to enter the slot D of a 4-station Geneva wheel. The locking male E is just about to unlock the Geneva wheel. A recommended design (Figure 4.14) shows how a cam, which is fixed to the driving arm, controls the motion of a locking roller, which is in mesh with the wheel when the driving roller is out of mesh. It can also be recommended to mount the rollers on an eccentric so that it locks the wheel in the correct position after being adjusted.

The 3-station external Geneva mechanism should be avoided because of its poor dynamic characteristic. Other factors of importance are choice of materials, surface treatment, tolerances, loads, and lubrication.

DESIGN VARIATIONS

The foregoing has treated regular (evenly spaced stations) and irregular Geneva drives having straight slots and one or more rollers driving the Geneva

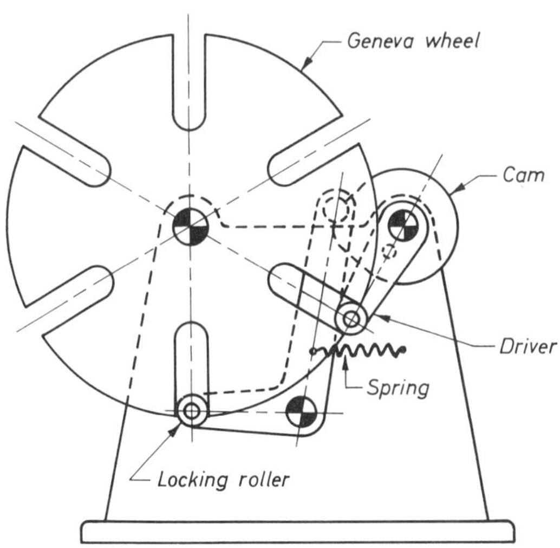

Figure 4.14 At high speeds it is recommended that the Geneva wheel be locked in position by means of a cam mechanism. The use of a locking disc is unsatisfactory at high speeds.

wheel. Following is a selection of Geneva mechanisms or Geneva-type mechanisms in which the driving member, usually a roller, enters a slot, indexes the Geneva wheel, and then leaves the slot.

Other design possibilities are to let the slots not be radial but offset, to use curved slots that will yield desired acceleration characteristics, and to use another mechanism as input to the driving arm, for instance a four-bar linkage, noncircular gears, chain-driven mechanisms and so forth. There are many possibilities. If the input mechanism that drives the Geneva mechanism is a four-bar linkage, refer to the worked-out examples on four-bar linkages in Chapter 2.

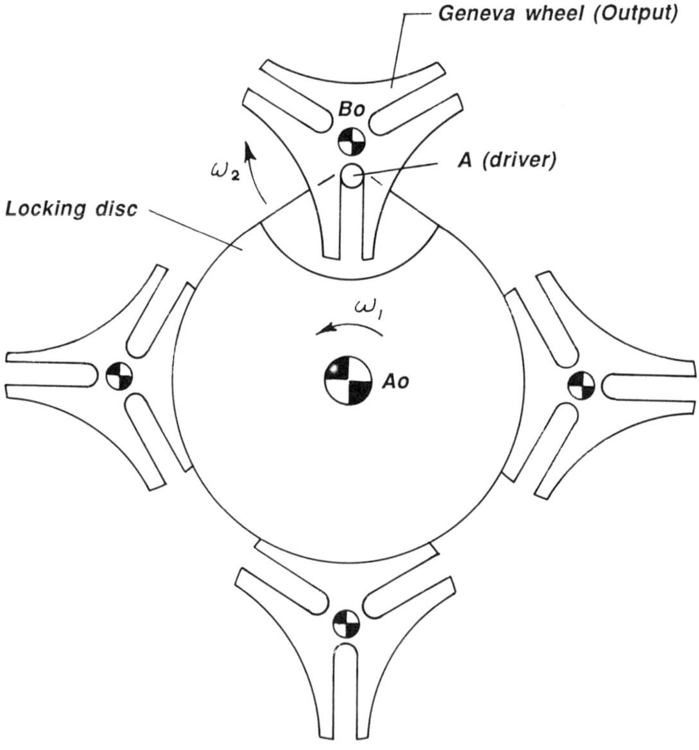

Figure 4.15 The roller A drives four 3-station Geneva mechanisms, one after the other. But remember to avoid 3-station Geneva drives because of the high acceleration values at the start and end of motion of the Geneva wheel, unless speed and masses are low.

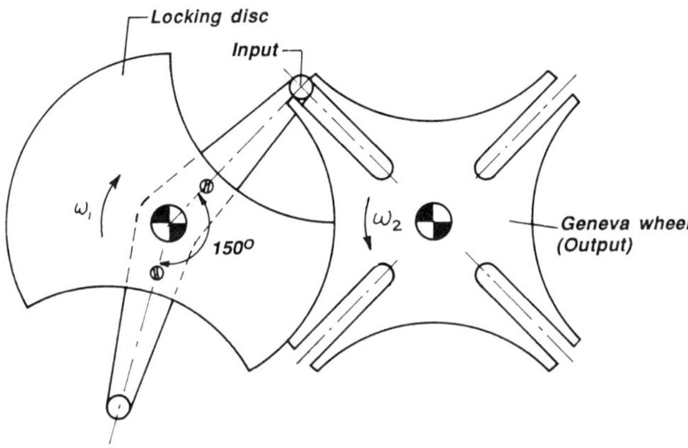

Figure 4.16 The input member has two driving arms spaced 150° apart. One of the rollers is just about to enter the slot, and for 90° rotation of the input member, the Geneva wheel is driven 90°. After a further rotation of 120° the second roller is about to enter a slot. (See also Fig. 4.10.)

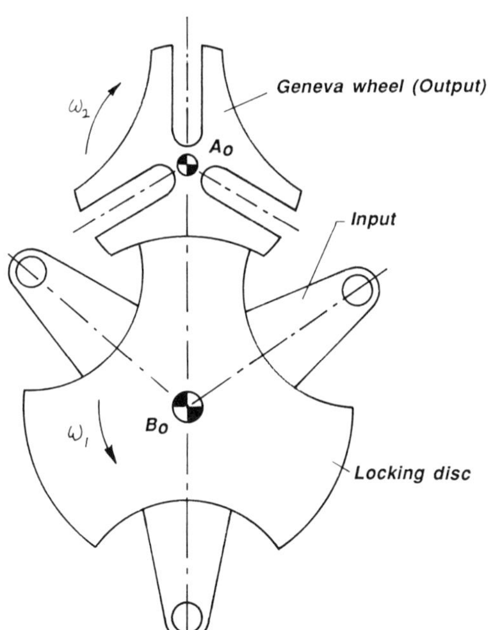

Figure 4.17 A 3-station Geneva wheel is driven by three rollers. Because of the evenly spaced rollers, each dwell of the Geneva wheel will be 60°.

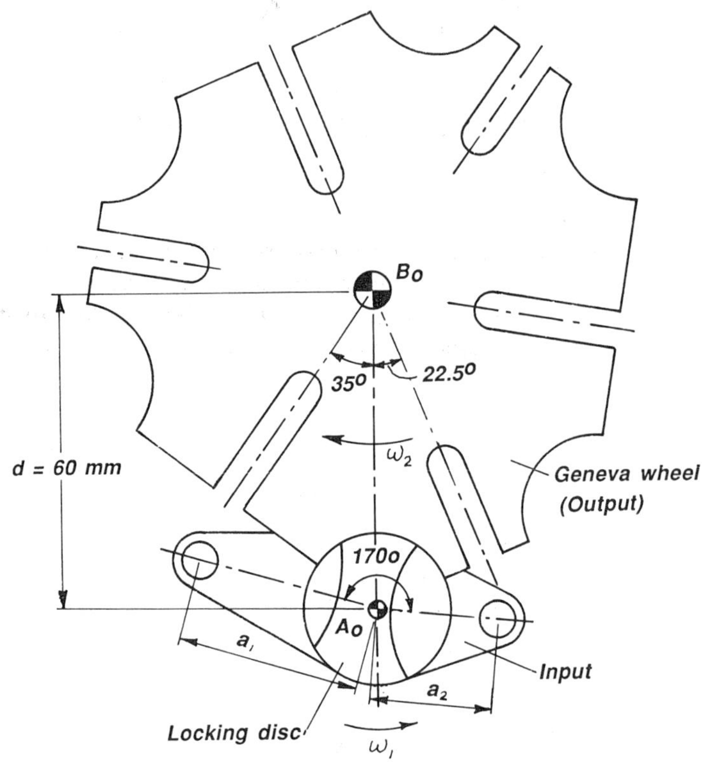

Figure 4.18 A Geneva mechanism with two driving arms unequally spaced and having unequal length. From the geometry of the Geneva wheel the two crank radii are calculated from $a_1 = 60 \sin 35° = 34.41$ mm and $a_2 = 60 \sin 22.5° = 22.96$ mm. For 110° rotation of arm a_1, the Geneva wheel indexes 70°. For 135° rotation of arm a_2 the Geneva wheel indexes 45°.

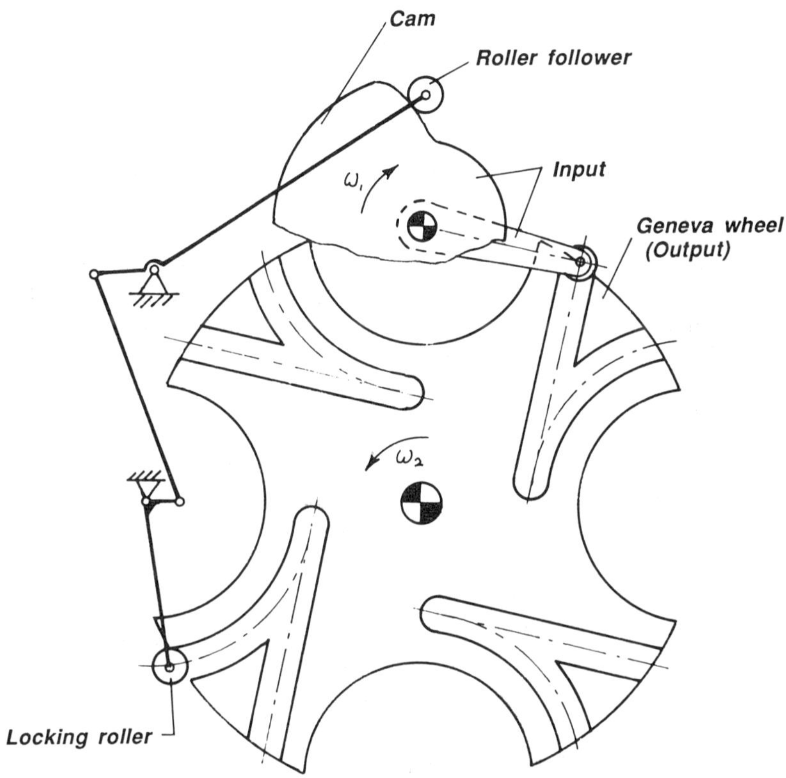

Figure 4.19 This 4-station Geneva wheel has offset slots (as opposed to radial slots). Indexing takes place for 90° rotation of the input arm, causing a 90° indexing of the Geneva wheel. The offset characteristic of the slots results in a changed displacement characteristic as well as in higher maximum values of the angular acceleration of the Geneva wheel. A rather elaborate locking means is provided. Because of the circular shape of the slot into which the locking roller enters, the locking action is speeded up.

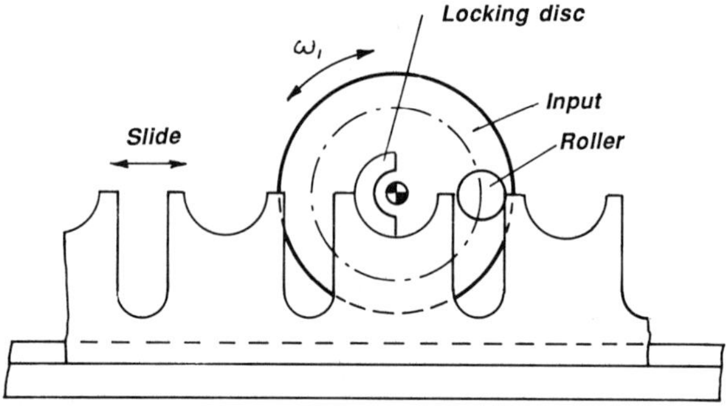

Figure 4.20 The rotary input is converted to the translating output of the slide (for each revolution of the driving roller the slide is indexed). However, motion of input must reverse at some time because otherwise the slide would have to be of infinite length.

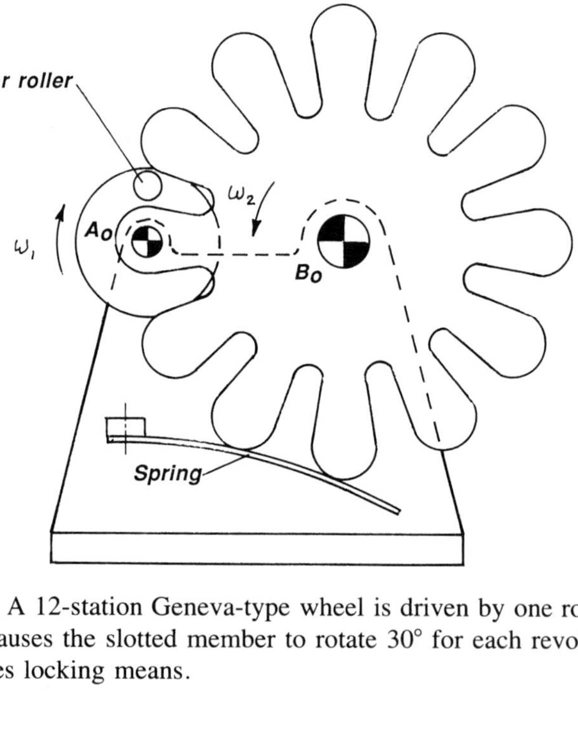

Figure 4.21 A 12-station Geneva-type wheel is driven by one roller. The rotary input causes the slotted member to rotate 30° for each revolution. A spring provides locking means.

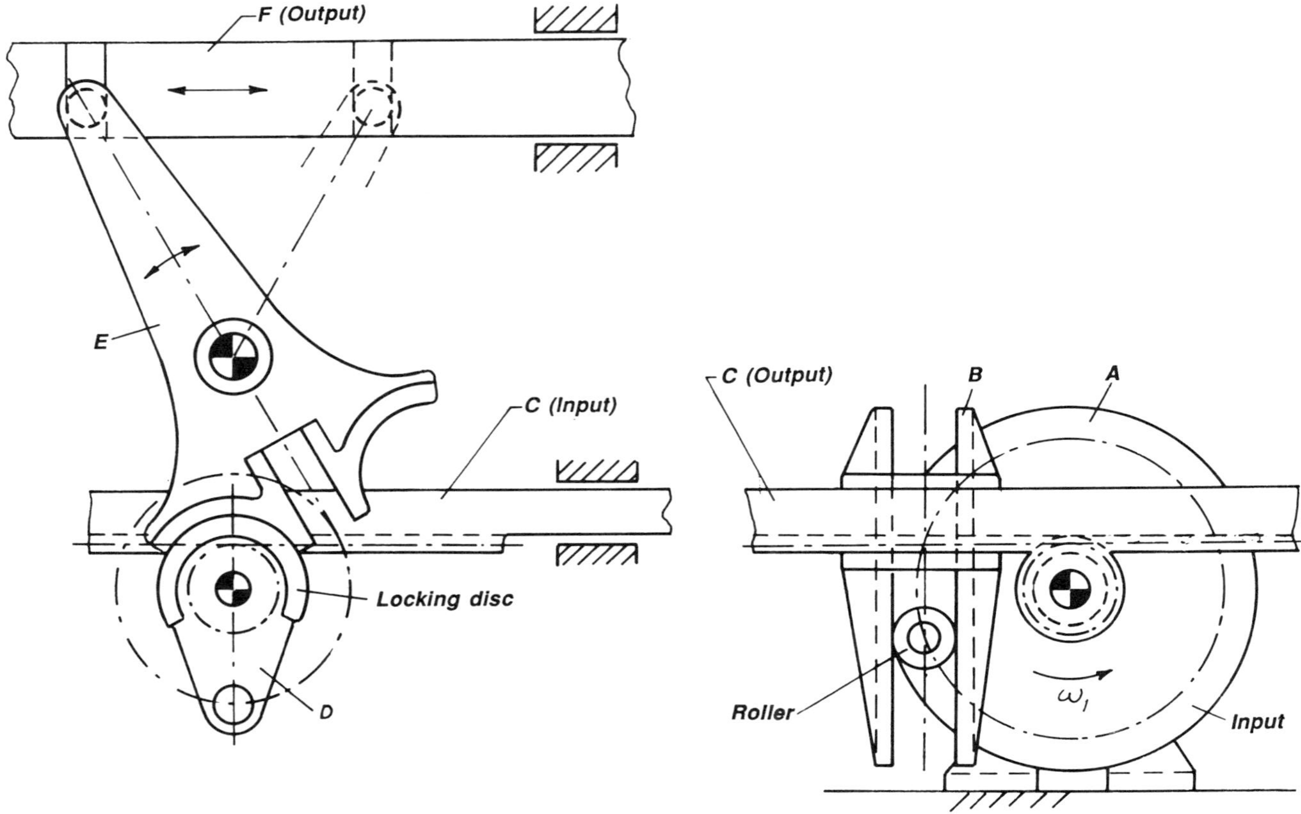

Figure 4.22 Input is to rotary member A on which is fastened a roller, which oscillates the slotted member B at right (also called a Scotch yoke). The yoke is integral with a rack C, which oscillates arm E, which in turn oscillates output member F.

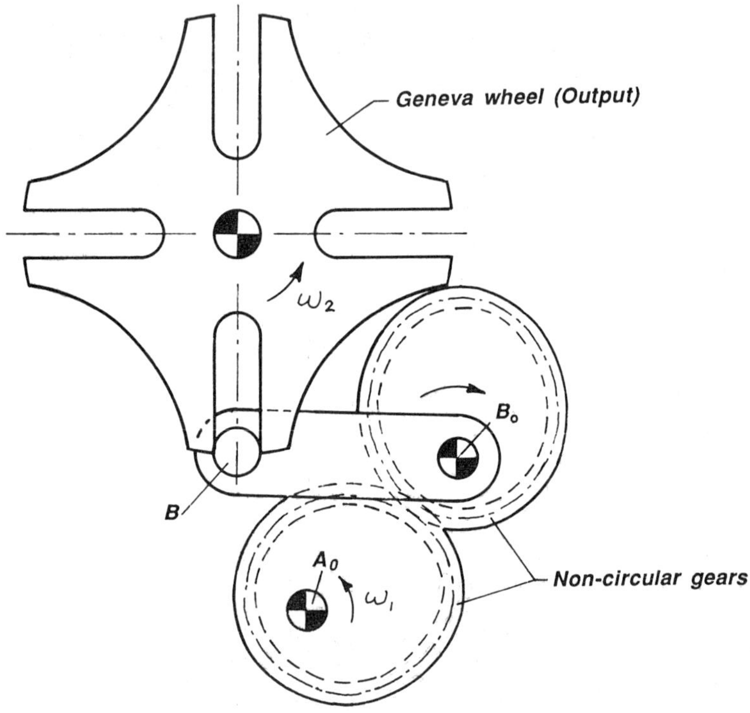

Figure 4.23 Using noncircular gears to convert the uniform rotation of the input gear at A_0 results in rotation of driving arm B_0B with nonuniform velocity, thereby changing the motion characteristics of the 4-station Geneva wheel. The driving roller has already penetrated the slot.

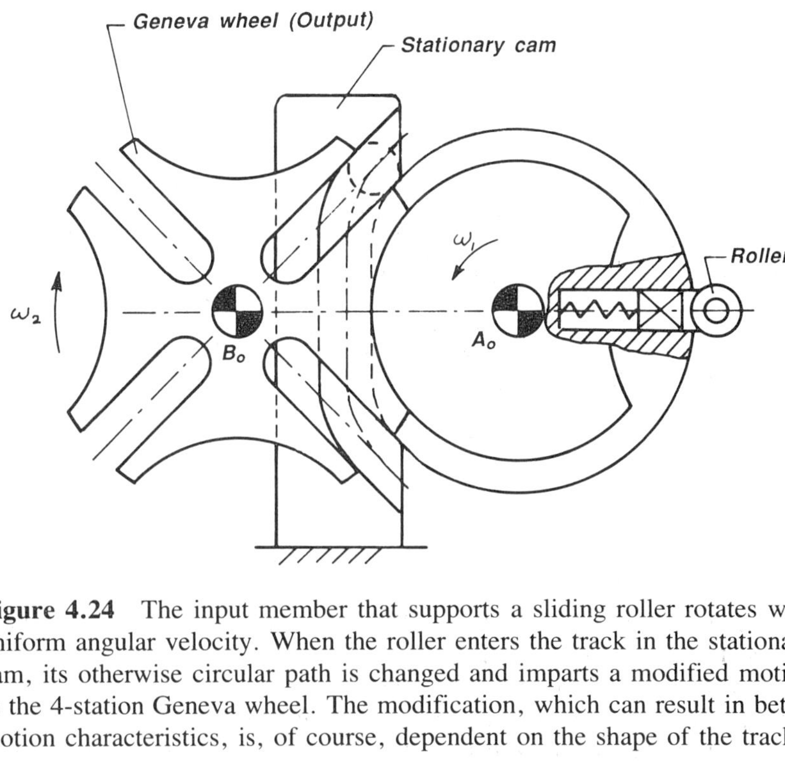

Figure 4.24 The input member that supports a sliding roller rotates with uniform angular velocity. When the roller enters the track in the stationary cam, its otherwise circular path is changed and imparts a modified motion to the 4-station Geneva wheel. The modification, which can result in better motion characteristics, is, of course, dependent on the shape of the track.

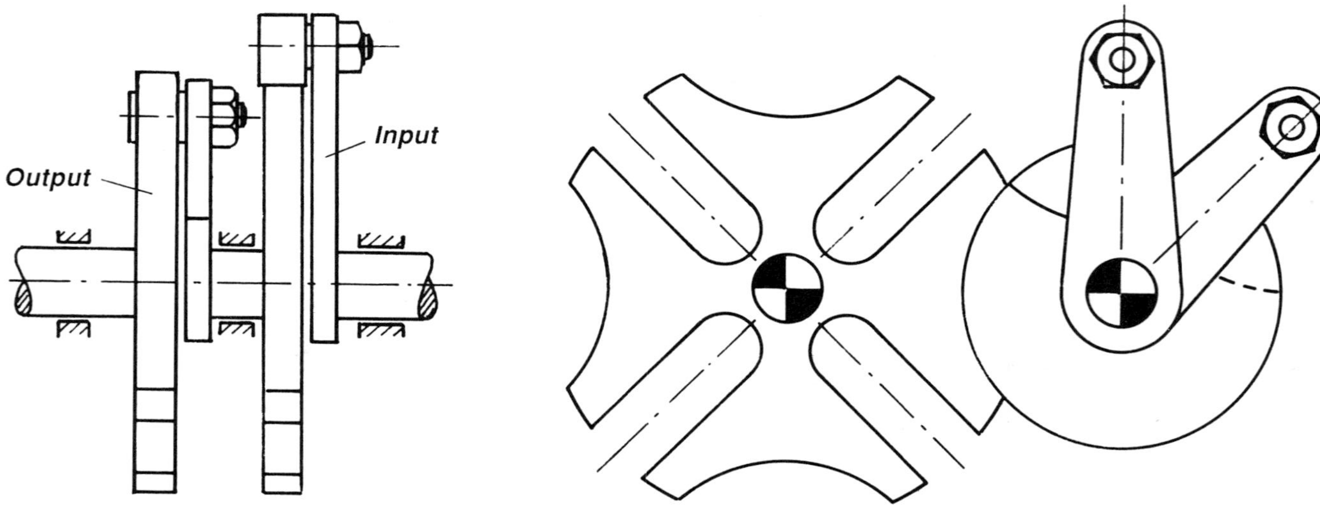

Figure 4.25 Geneva mechanisms can be arranged in series so that the output from one Geneva mechanism is the input to another Geneva mechanism.

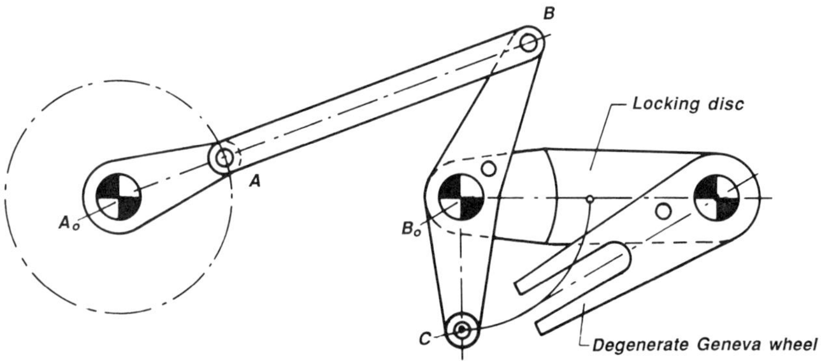

Figure 4.26 A four-bar linkage A_0ABB_0 drives a degenerate Geneva wheel through the roller C.

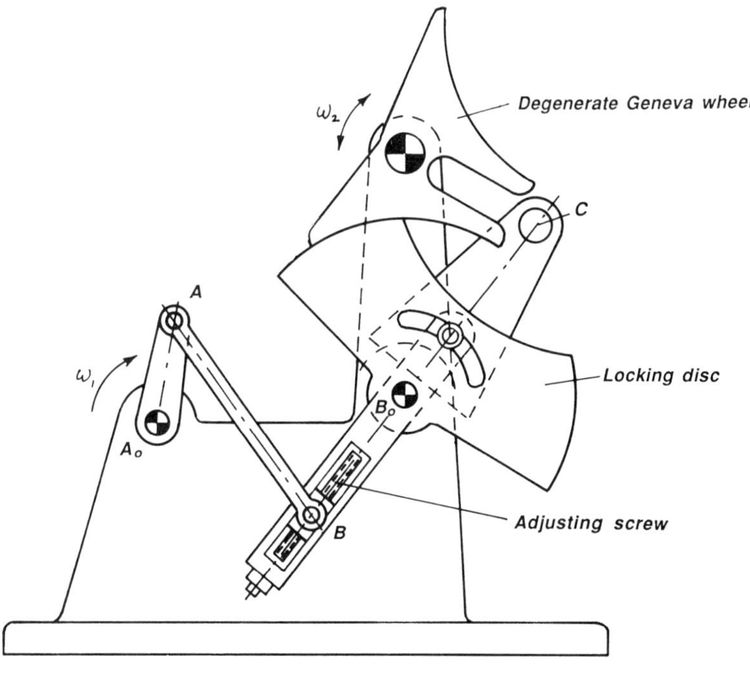

Figure 4.27 A four-bar linkage A_0ABB_0 drives a degenerate Geneva wheel through the roller C. The length of the rocker is made adjustable by means of the adjusting screw, and the angular position of the roller C can be changed relative to the rocker B_0B.

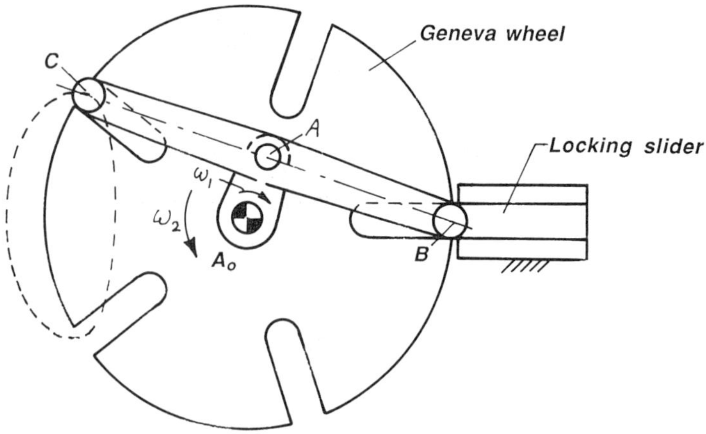

Figure 4.28 The driving roller at C is fastened to the connecting rod AB of a slider crank A_0AB. The slider at B provides locking means for the 5-station Geneva wheel.

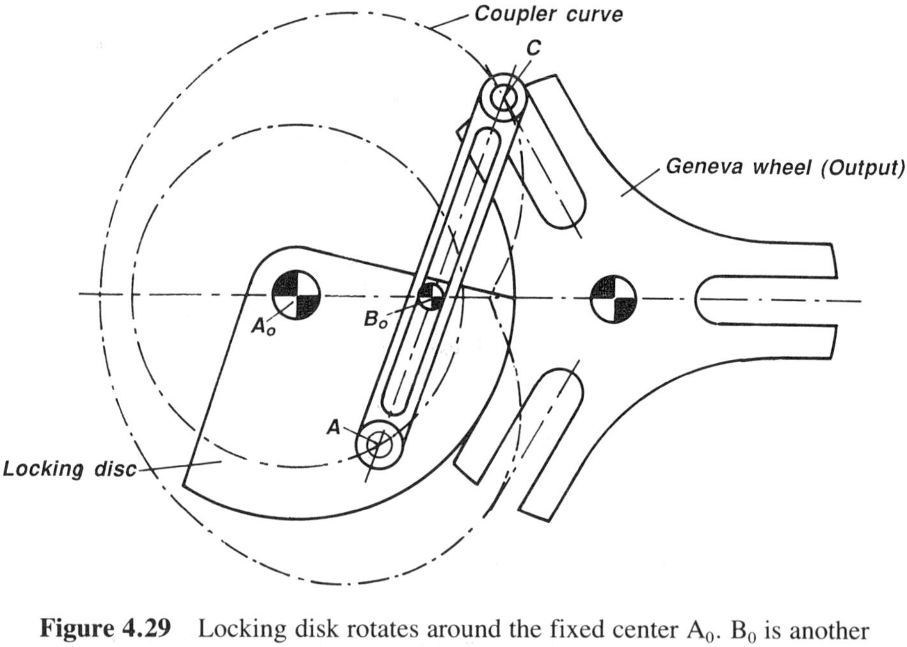

Figure 4.29 Locking disk rotates around the fixed center A_0. B_0 is another fixed axis which guides the slotted link AC with pin connection to the locking disc. When the locking disk rotates, it causes point C of the slotted link to trace the coupler curve shown. The roller at C drives a 3-station Geneva wheel.

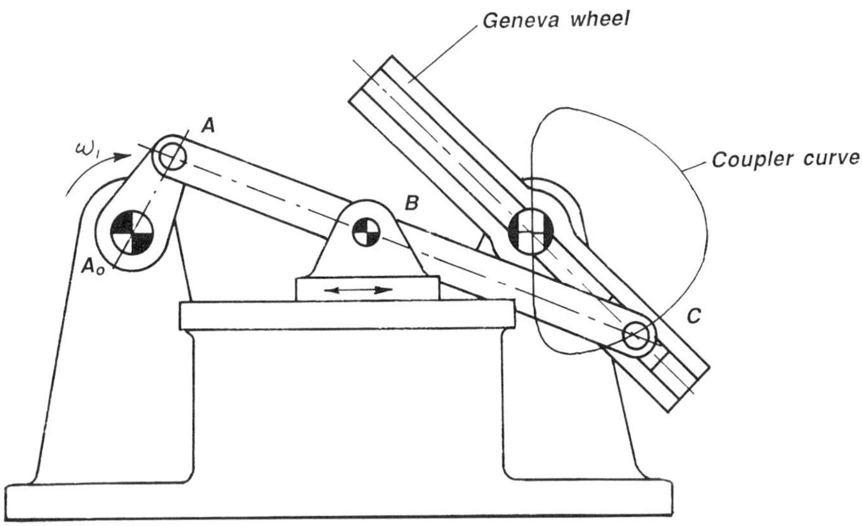

Figure 4.30 The slider crank A_0AB with the coupler point C traces the coupler curve shown. The roller at C drives the 2-station Geneva wheel, and as long as the roller traces the approximate linear path, the Geneva wheel (indexed 180° for each revolution of the crank A_0A) remains in its (vertical) dwell position.

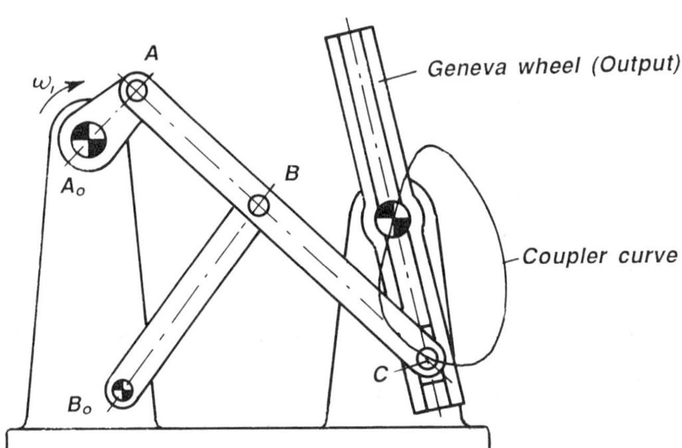

Figure 4.31 The four-bar linkage A_0ABB_0 with the coupler point C traces the coupler curve shown. The coupler curve has an approximate straight line portion passing through the center of rotation of the 2-station Geneva wheel, and as long as the coupler curve traces the linear portion, the Geneva wheel (indexed 180° for each revolution of the crank A_0A) remains in its dwell position.

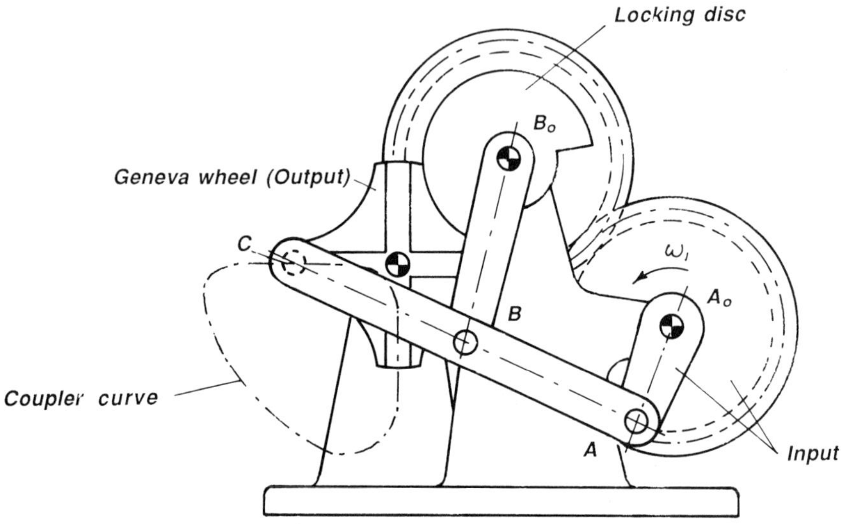

Figure 4.32 The coupler curve of the four-bar linkage A_0ABB_0 traces a coupler curve with two approximate straight lines perpendicular to each other. The roller at C is just about to enter a slot of a 4-station Geneva wheel. As long as it follows the approximate straight-line portion, the Geneva wheel remains in its dwell position, but then it is indexed 90°. At that point the roller reaches the other approximate straight-line portion, and the Geneva wheel remains in its dwell position. The dynamic motion characteristics of this mechanism is very good because the angular acceleration at start and end of indexing can be made zero, and the straight-line portions provide locking means over an extended period and the locking disc then takes over the locking. (P. W. Jensen, Synthesis of Four-Bar Linkages with a Coupler Point Passing Through 12 Points, *Mechanisms and Machine Theory* 19, no. 1, 1984, pp. 149–156.)

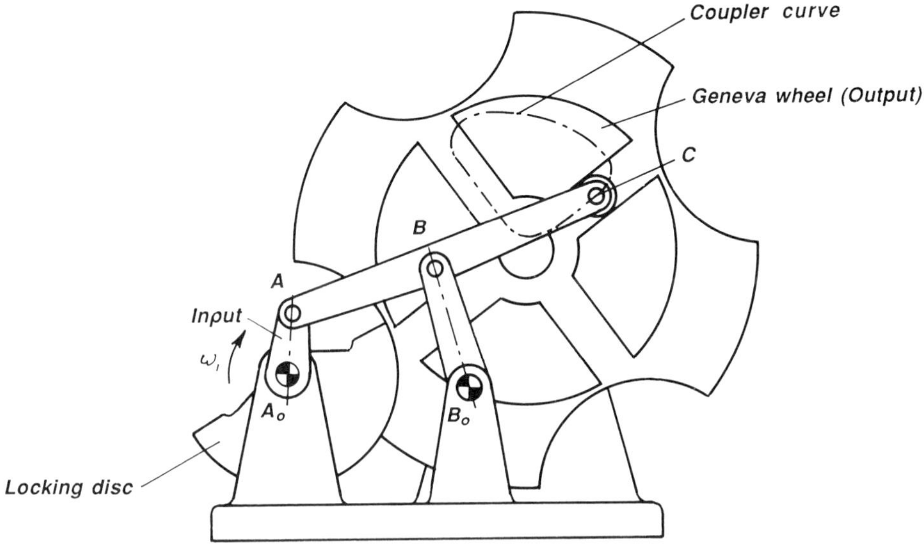

Figure 4.33 The coupler curve of the four-bar linkage A_0ABB_0 traces a coupler curve with two approximate straight lines perpendicular to each other. The roller at C is just about to enter a slot of a 4-station Geneva wheel. As long as it follows the approximate straight-line portion, the Geneva wheel remains in its dwell position, but then it is indexed 90°. At that point the roller now reaches the other approximate straight-line portion, and the Geneva wheel remains in its dwell position. The dynamic motion characteristics of this mechanism is also very good. The straight line portions of the coupler curve provide locking means over an extended period and the locking disc then takes over the locking. The nonsymmetrical coupler curve provides another motion characteristic as compared with Fig. 4.32.

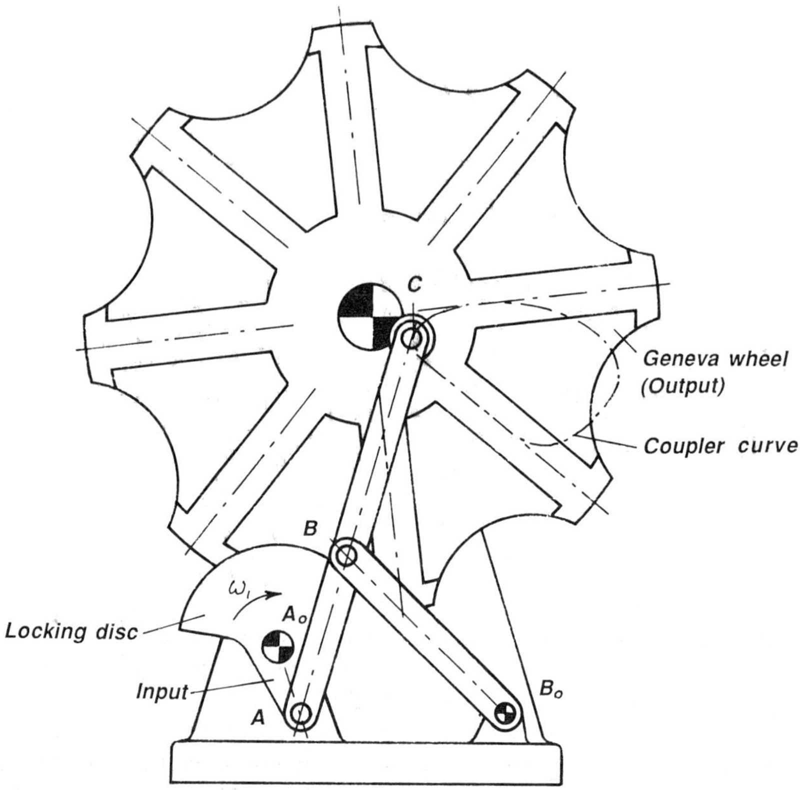

Figure 4.34 The four-bar linkage A_0ABB_0 traces the coupler curve shown. The coupler curve has an approximate straight-line portion that is radially directed relative to the center of the 8-station Geneva wheel. For a prolonged period of time the Geneva wheel remains in its dwell position and then indexes rather fast.

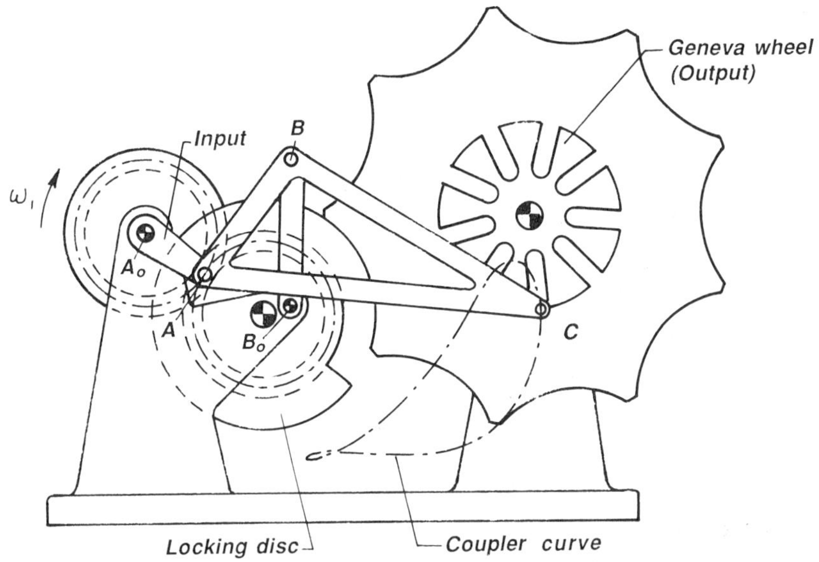

Figure 4.35 The four-bar linkage A_0ABB_0 traces the coupler curve shown. The coupler curve has portions that are radially directed relative to the center of the 9-station Geneva wheel. For a prolonged period of time the Geneva wheel remains in its dwell position and then indexes very fast.

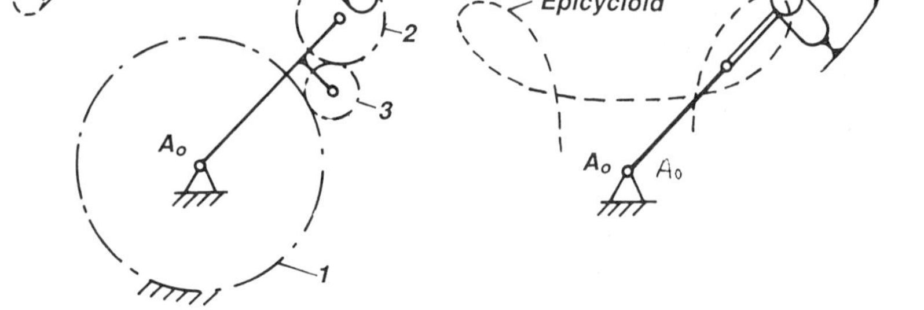

Figure 4.36 The three gears 1, 2, and 3 are in mesh. Gear 1 is fixed and when the carrier C, which drives gear 2, is rotated, a point on gear 2 traces a so-called epicycloid, which is shown to the right. The driving pin can be used to drive an internal Geneva wheel, as shown. (See also Chapter 6 on cycloidal mechanisms.)

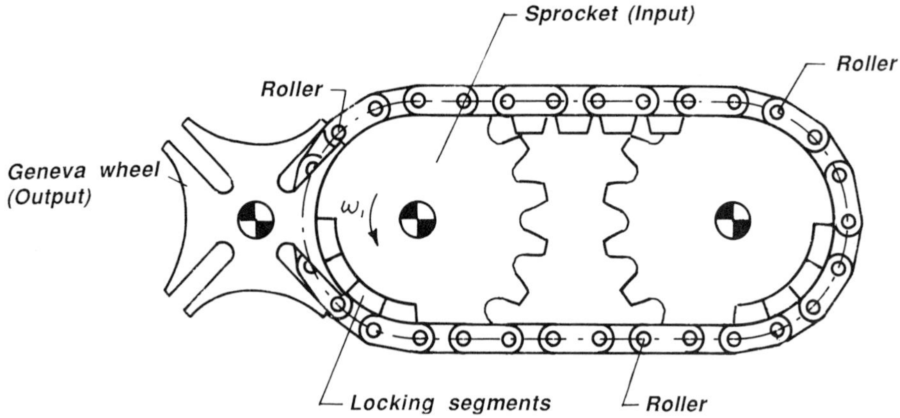

Figure 4.37 Prolonged dwell of this Geneva-type mechanism can be achieved using a chain, as shown. The roller on the chain is just about to enter the slot of the Geneva wheel. The locking segment is just about to unlock. For 90° rotation of the sprockets the Geneva wheel is indexed 90°. This design is very flexible. (See also Chapter 7 on chain-driven mechanisms.)

The normal Geneva drive in which the roller is moved around a circle has limitations that make the mechanism useless for certain purposes. The most severe limitation is that the sum of the time consumed in indexing and the time of dwell corresponds to the time that it takes the roller to move one revolution around the axis of the driving component of the device.

Instead of moving the roller around a circle, it is possible to move it along a different path and thereby obtain more freedom of design. Figure 4.38 shows one possible divergence from conventional design. Sprockets A, B, C, and D are driven by a chain E on which there is fastened roller F. Indexing disc G has two slots spaced 180° apart. The roller is shown in a position where it is just about to enter a slot of the indexing disc. With proportions shown in Figure 4.38 the disc G is accelerated during 30° of movement, then moved with a constant angular velocity through 120°, and decelerated during 30°. The disc then remains in its dwell position until roller F enters the next slot. Flexibility is a feature of this design because the number of slots, the number of rollers, and the length of the chain can be changed to suit various purposes.

Figure 4.39a shows a modified Geneva drive of which the time for motion is dependent upon gear ratios. In this drive, input shaft A rotates with uniform velocity and drives gear B, which in turn drives sun gear C. The

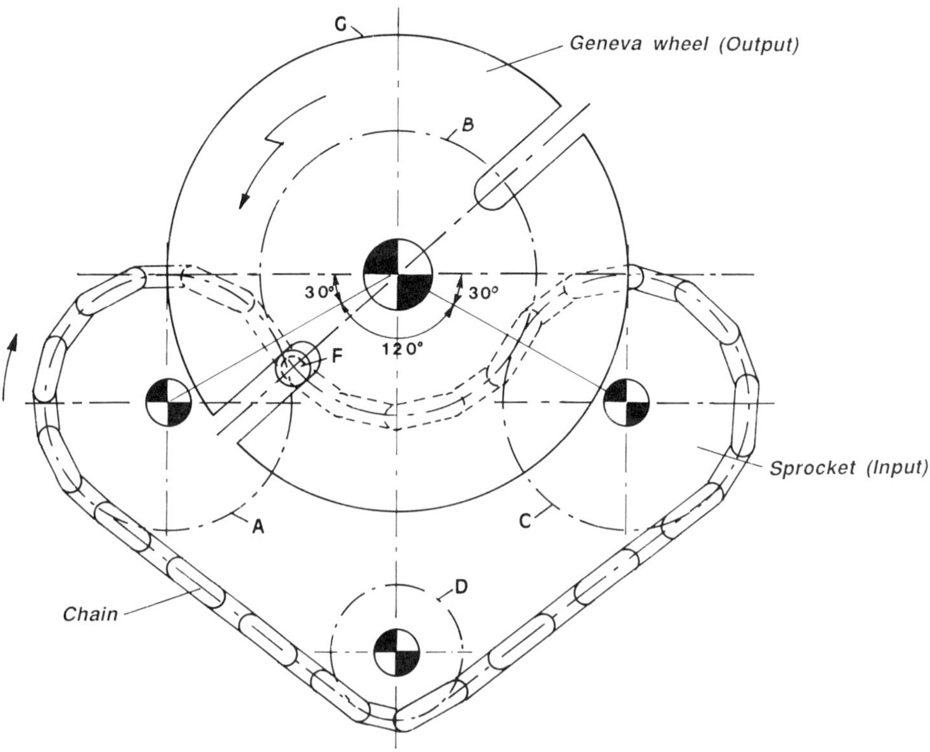

Figure 4.38 Driving pin F is fastened to a triple-strand chain and guided on the path shown. The indexing motion of the Geneva wheel is 180°.

latter is free to rotate on the shaft D. Shafts A and D are supported in the gear housing. Sun gear C drives planet gear E, and as long as the roller H is outside of the slotted member F, gear E rotates with uniform velocity because the planet block G is detended. The detent device is not shown. Roller H is just starting to enter a slot in Fig. 4.39b, while Fig. 4.39c shows the mechanism some time after the roller has entered the slot.

At the moment that the roller entered the slot, the planet carrier was unlocked. The roller, however, is now in the slot, and because of the angular motion of link J, which is driven by planet gear E, the roller will penetrate deep into the slot and cause shaft K to rotate CCW around shaft D. In general, the size of an idler gear has no influence on gear ratio, but in this case the size of gear C is important because planet gear E rolls on gear C during motion.

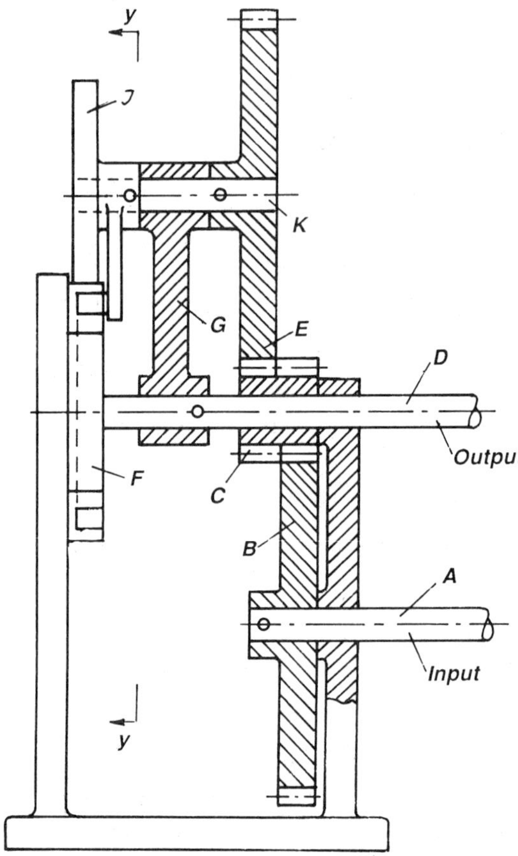

SIDE VIEW
(a)

Figure 4.39 Inverted Geneva mechanism. (a) Side view of mechanism. (b) Front view of mechanism. Roller is just about to enter slot. (c) Front view of mechanism. Indexing has started.

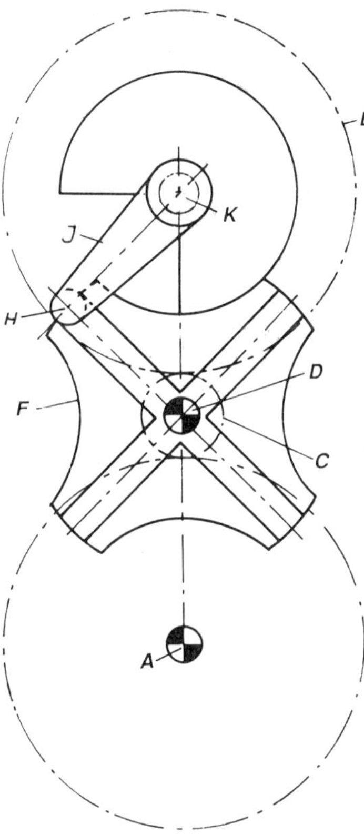

FRONT VIEW
(b)

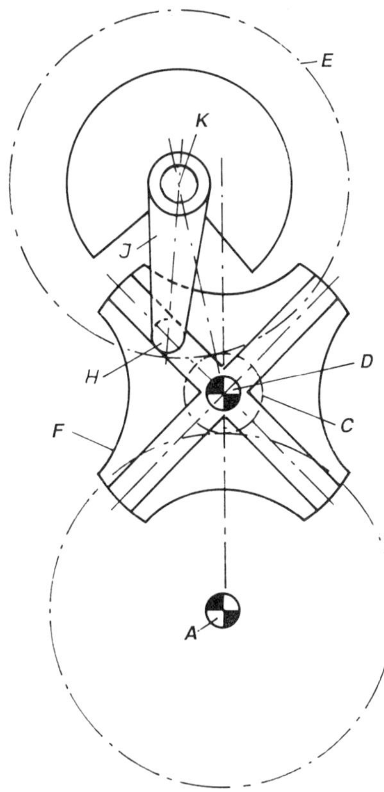

FRONT VIEW
(c)

By superimposition of the different motions it can be shown that time for indexing by using a 4-slotted member is

$$T = 90^\circ \frac{D_4 - D_3}{D_2}$$

For the mechanism described, $4D_3 = D_2 = D_4$, where D_2, D_3, and D_4 are the pitch diameters of gears B, C, and E, respectively.

$$T = 90^\circ \frac{D_2 - .25D_2}{D_2} = 90^\circ \times .75 = 67.5^\circ$$

and the time for dwell is $360^\circ - 67.5^\circ = 292.5^\circ$. The output motion of the planet carrier by each indexing is 90°.

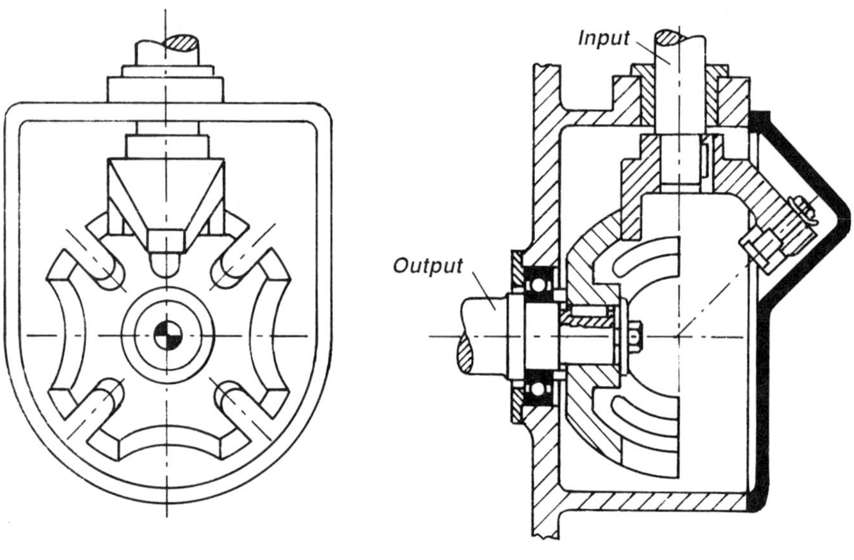

Figure 4.40 A spatial 4-station Geneva mechanism, where the input and output shaft are perpendicular to each other. For 180° rotation of the input shaft the output shaft indexes 90°. The axis of the shafts intersect each other under an angle of 90°.

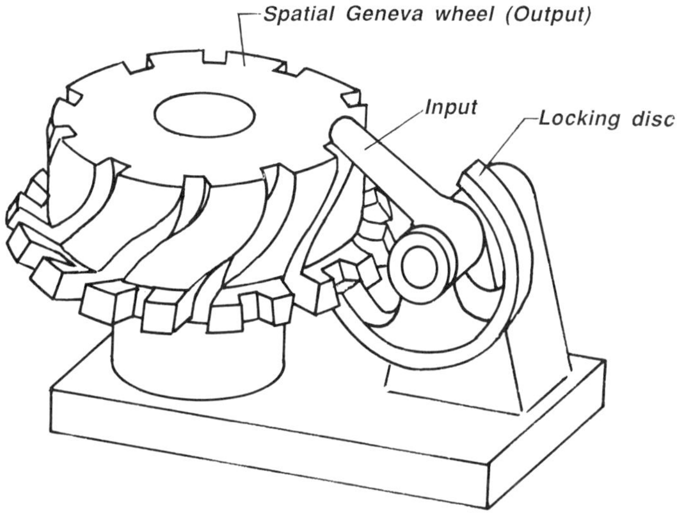

Figure 4.41 A spatial 10-station Geneva mechanism, where the input and output shaft are perpendicular to each other. For 70° rotation of the input shaft the output shaft indexes 36°. The axis of the shafts are perpendicular to each other, but they do not intersect.

STAR WHEELS

Star wheels are a combination of gears and modified Geneva mechanisms. The gears may be made with pin gearing or with teeth as in ordinary gears. Because of their complex geometry and accuracy required in manufacture, they are not used very often. Their dynamic characteristics may theoretically be better than ordinary Geneva mechanisms, but whether the improvement is such that they really operate better is questionable. They have interesting motion characteristics that cannot in general be achieved with other mechanisms. With star wheels it is important to have an angular acceleration that is continuous, at least for relatively high speeds.

A star wheel is shown in Fig. 4.42 with accompanying angular displacement, angular velocity, and angular acceleration diagram. The mechanism is made of four parts: A, B, C, and D. Part A is a locking disc and has the same function as the locking disc of a conventional Geneva mechanism, namely, to lock the mechanism in its dwell position. Part B is driven by the rollers T_1 and T_2. The rollers R_1, R_2, R_3, and R_4 follow the contour of the locking disc, thereby preventing rotation of member B during the dwell period. C is a gear segment that, when turned, gets into mesh with gear D for a limited rotation.

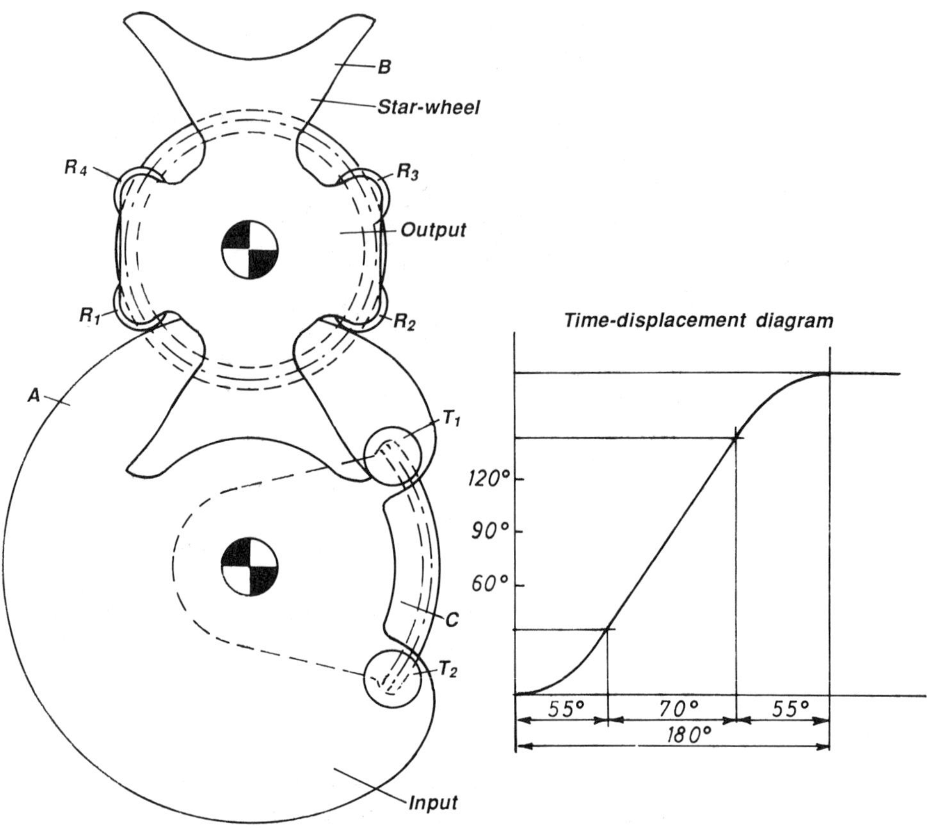

Figure 4.42 A 2-station star wheel.

The roller R_1 is just about to drive the star wheel in a CW direction. The shape of the cam surface on the star wheel is determined from the desired displacement diagram and can be constructed by knowing the relative angular displacements of the input and output shafts, and then tracking the position of roller R_1 relative to the star wheel.

The shape of the locking disc is determined by the motion of roller R_2 relative to the locking disc and can also be determined from the displacement diagram.

When gear segment C has rotated 55° from its initial position, it enters into mesh with gear D, and for a rotation of 70° the gear segment is in mesh with gear D, which now rotates with uniform velocity. The two gears subsequently get out of mesh, and B is slowed down all the way to a complete stop. At this point B is locked in its position by rollers R_3 and R_4 over the remaining portion of 180°, and then the whole cycle starts over again.

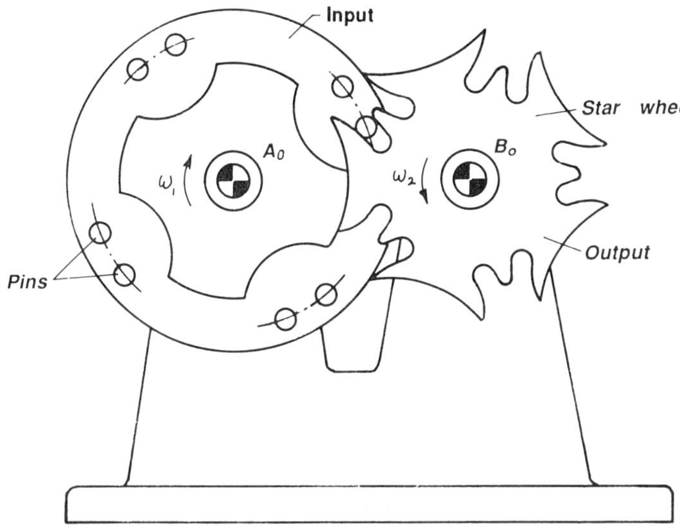

Figure 4.43 A 4-station star wheel (notice its similarity to a 4-station external Geneva mechanism). For each 90° revolution of member 1, member 2 is driven 72°. After five revolutions of member 1, member 2 has rotated four revolutions. Member 1 has a circular-shaped locking part.

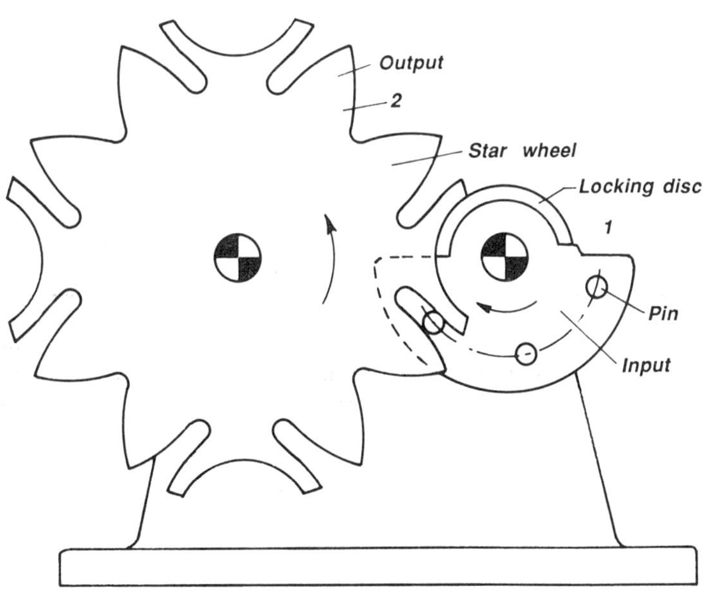

Figure 4.44 The three pins on member 1 drive the star wheel with uniform motion over part of its rotation. For 180° rotation of member 1, member 2 is rotated through 90°. For the next 180° rotation of member 1, the star wheel remains in its dwell position.

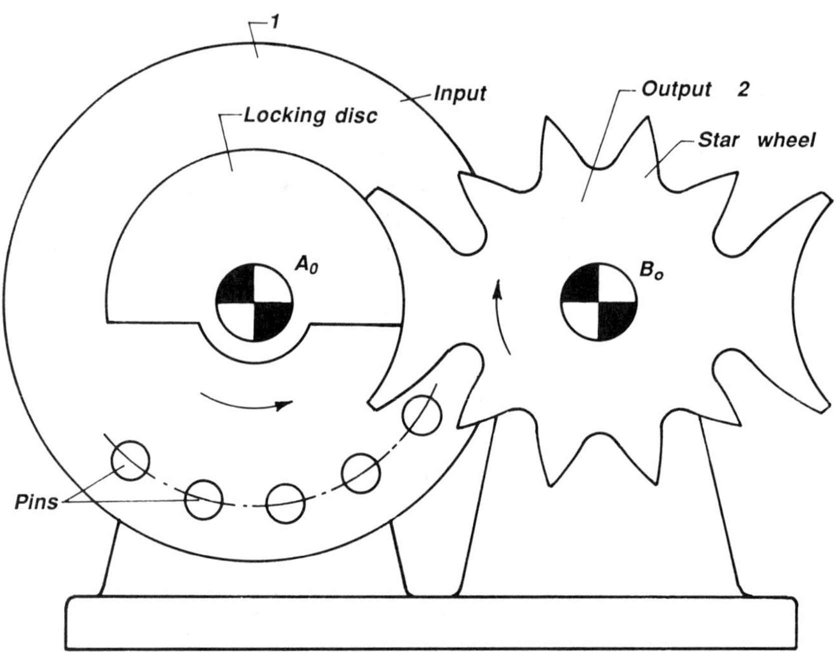

Figure 4.45 The five pins on member 1 drive the star wheel through 180°. During this motion member 1 is rotating through 180°. Member 1 is fitted with a circular-shaped locking disc that locks the star wheel over 180° of rotation of member 1.

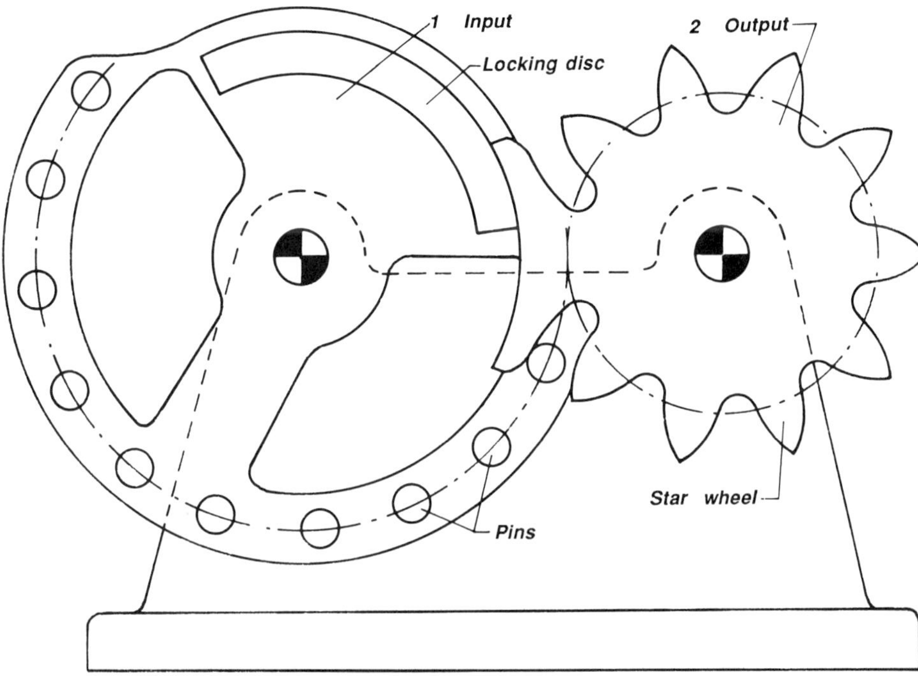

Figure 4.46 The input member 1 drives the star wheel 2 through 360°, while it rotates through 192.5°. For the remaining revolution of 167.5° the star wheel remains in its dwell position.

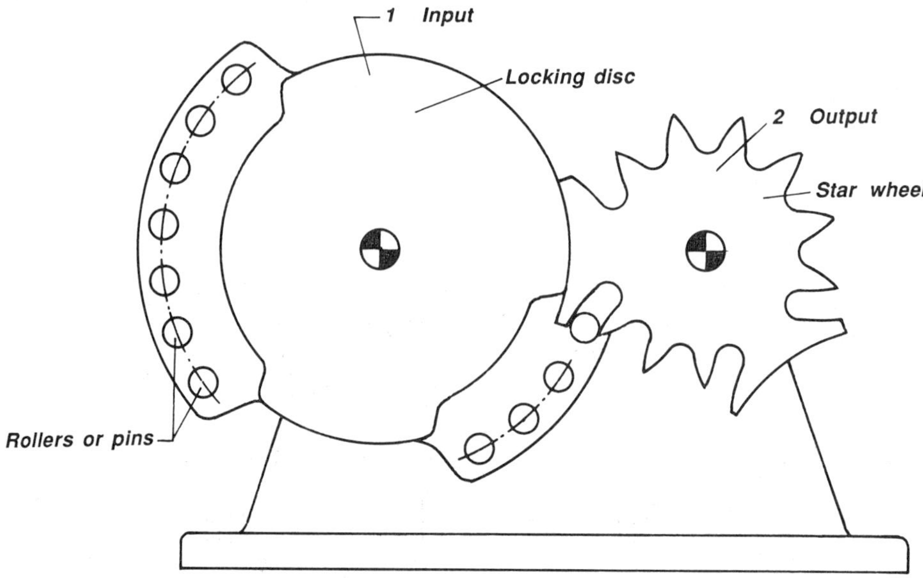

Figure 4.47 An even greater variation of rotations and dwells is achieved with this irregular star wheel mechanism, where two different motions and angles of indexing are obtained for each revolution of the input member (Compare with Fig. 4.45.)

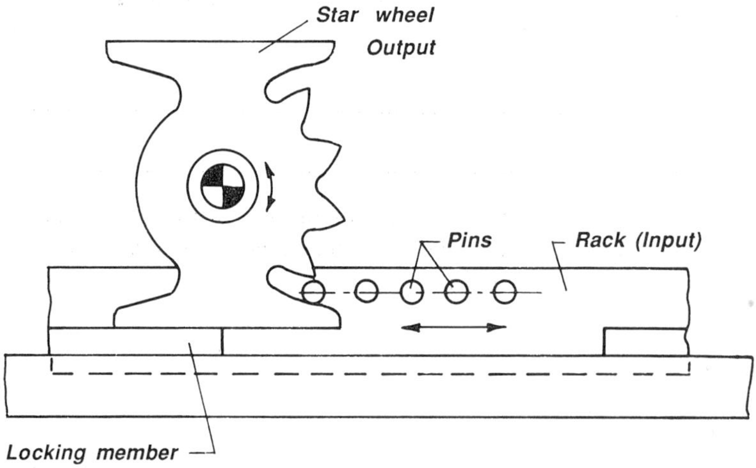

Figure 4.48 The star wheel is driven from a rack-type input member. The motion of the rack causes the star wheel to rotate 180°. The rack must make an oscillating motion. Notice that the locking member is straight.

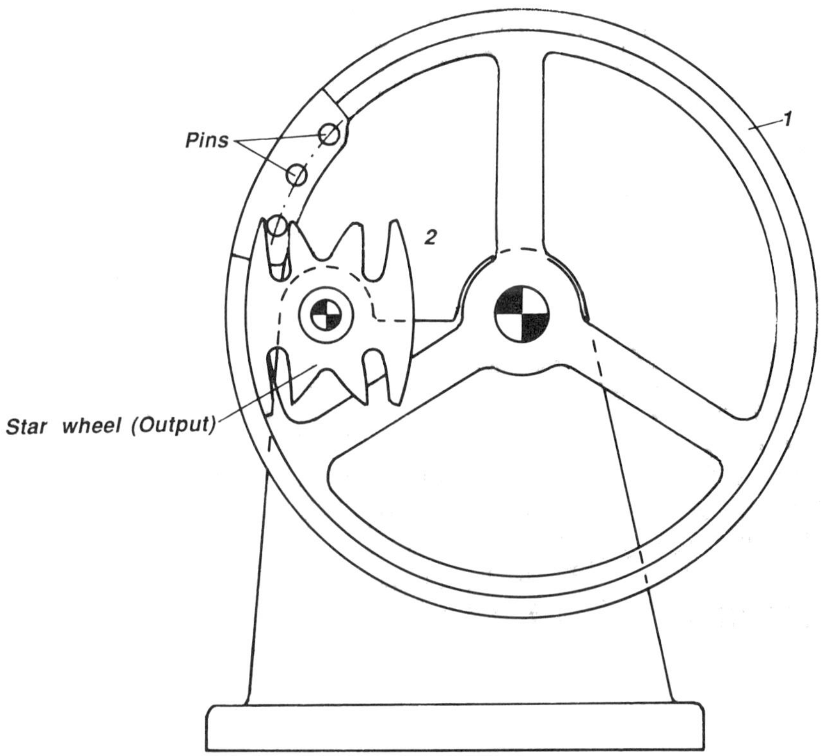

Figure 4.49 A very long dwell is achieved with this mechanism. As a matter of fact, 22.5° rotation of the input member causes the star wheel to rotate through 180°. For the remaining 337.5° of the complete revolution of the input shaft, the star wheel remains in its dwell position. Notice that the star wheel shaft is supported "in front of" driving member.

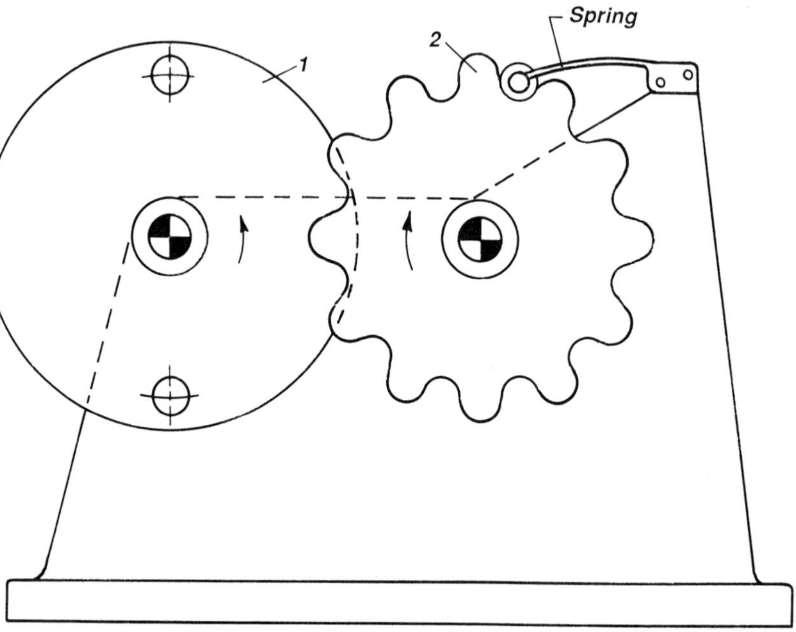

Figure 4.50 This is a very crude form of star wheel. For each revolution of the input shaft 1, wheel 2 is driven two teeth or 30°. The locking of the output member in its dwell position is done with a spring.

SYMBOLS

n	number of stations
d	distance between centers, in.
a	length of driver arm, in.
ω_1	angular velocity of driver arm, rad/s
ω_2	angular velocity of Geneva wheel, rad/s
α_1	angular acceleration of driver arm, rad/s^2
α_2	angular acceleration of Geneva wheel, rad/s^2
β_0	indexing angle of Geneva wheel, degrees
α_0	angle of rotation of the driver arm, corresponding to β_0
N	angular speed of driver arm, RPM
I_t	mass moment of inertia of machine member, driven by Geneva mechanism, lb-in.-s^2

5

Planetary Gear Systems

INTRODUCTION

Planetary gears have enjoyed an increasing application over a period of years partly because of the design improvements but also, and especially, because of the advantages associated with planetary gears in comparison with ordinary gear trains.

APPLICATIONS

Applications of planetary gears range from hoisting units in cranes to use in helicopters, tanks, airplane engines, turbines, pumps, textile machines, paper and cardboard-fabricating machinery, printing presses, and controls for a variety of instruments. They can also be used as subunits in variable speed changers.

The advantages are

Colinear input and output shafts
Small space requirements
Good efficiency
Distribution of forces on several planet gears and decrease in space requirement for same output power

The disadvantages are

Large space requirements for certain applications; can be improved by selecting the right system
Poor efficiency; can be improved by selecting the right system
Difficulty distributing forces evenly on several planet gears; can be taken care of by good design

The apparent contradictions above are due to the variety of planetary gear systems from which to choose. Also good design engineering can improve the system. For instance, an improvement of the efficiency can be made by making the tooth flanks smooth (by grinding), by choosing a low pressure angle, using internal gearing, increasing the number of teeth, take care of sufficient lubrication, and so on.

Fundamental to any calculations of planetary gear trains is the following ratio:

$$\frac{r_1}{r_2} = \frac{Z_1}{Z_2}$$

or, the ratio of the pitch radii of two gears is equal to the ratio of the numbers of teeth of the two gears.

CALCULATING THE TRANSMISSION RATIO

The transmission ratio for all types of gear trains is defined by

$$R = \frac{n_{in}}{n_{out}} = \frac{\omega_{in}}{\omega_{out}} \tag{5.1}$$

where n_{in} is the speed of the input shaft in RPM and n_{out} is the speed of the output shaft, also in RPM. If both shafts rotate in the same direction, then $R > 0$, and if they rotate in opposite directions, then $R < 0$. ω_{in} and ω_{out} are the respective angular velocities.

There are several methods available to calculate the transmission ratio for planetary gears. To start with, the graphical and the tabular methods will be explained.

Graphical Method

Consider first the ordinary gear train (Fig. 5.1a) that consists of two gears 1 and 2 with fixed centers of rotation A_0 and B_0. Gears 1 and 2 are in mesh, and the two gears must therefore have the same tangential velocity at the pitch point P. The tangential velocity of the two gears at P is represented

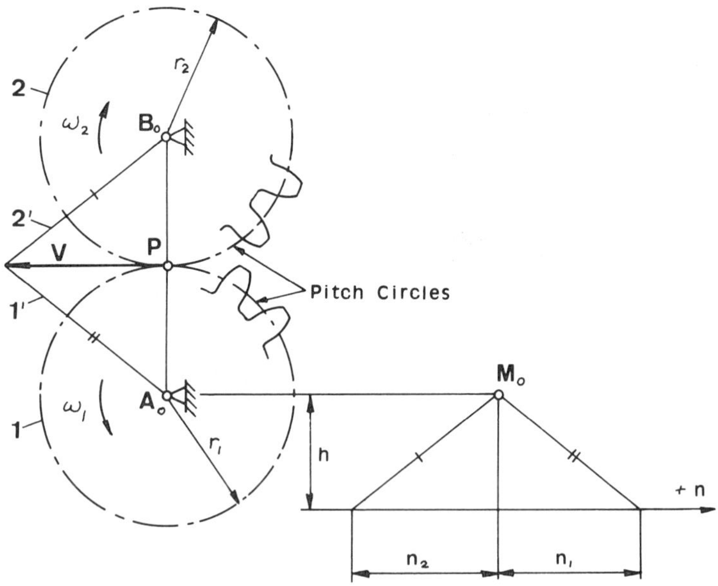

Figure 5.1 Ordinary gear train (with fixed centers).

by the vector V. The head of V is connected to A_0 and B_0 whereby the velocity lines 1′ and 2′ are determined. These velocity lines are a measure for the velocities of points on the gears along line A_0B_0.

The direction of the velocity lines (also called gauge lines) is laid out from M_0, and a parallel to A_0B_0 is drawn through M_0. Next, a horizontal line with an arbitrary distance h from M_0 is drawn. The relative length of the distances n_1 and n_2 is equal to the relative velocity ratio of n_1 and n_2, respectively. The distance h can be chosen so that a scale for the +n axis corresponds to the RPM of the various gears. Because n_1 is positive and n_2 is negative, we have

$$r_1 n_1 = -r_2 n_2$$

$$R_{12} = \frac{n_1}{n_2} = -\frac{r_2}{r_1} \tag{5.2}$$

The method described above is normally not used on ordinary gear trains but is most useful when applied to planetary gear trains, as will be shown in the following.

Consider the simple planetary gear train in Fig. 5.2 with the central gear or sun gear 1 being fixed. The arm or planet carrier C carries the planet

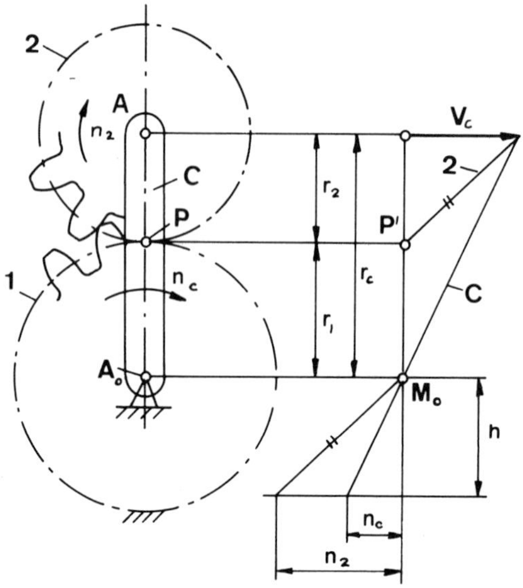

Figure 5.2 Velocity diagram for planet gear 2 and carrier C.

gear 2, which is in mesh with gear 1. It is most convenient to draw the carrier in a vertical position as shown, and through all points where two bodies are in contact and have the same velocity, horizontal lines are drawn. The carrier and gear 1 have the same velocity at A_0—namely, zero—so one horizontal line is drawn through that point. The carrier C and gear 2 have the same velocity at B_0 because whether point B_0 is considered a point on carrier A_0B_0 or the center of gear 2, these two points follow each other at all times, so another horizontal is drawn through B_0. Point P, where the pitch circles of the gears contact each other, has the same velocity whether it is a point on gear 1 or a point on gear 2, and a third horizontal line is drawn as shown.

Through point M_0 a vertical is drawn. The carrier C is assumed to rotate CW with n_c, and, therefore, the velocity of C is $V_c = r_cn_c$. Connecting the head of V_c with M_0 determines the velocity line for the carrier C. However, V_c is also the velocity of the center A of gear 2. The velocity of P, whether considered as a point on gear 1 or 2, is the same in both cases and must be zero because P is a point on the stationary gear 1. Thus the velocity line for gear 2 can be drawn, and through M_0 a parallel to 2 is drawn. As before, a horizontal line at a distance h from M_0 is drawn. The values for n_c and n_2 are now determined.

Because n_2 and n_c are on the same side of the vertical line through M_0, both members rotate in the same direction. The following formula, based on the velocity diagram in Fig. 5.2, is now applied:

$$n_2 = \frac{V_c}{r_2} = \frac{(r_1 + r_2)n_c}{r_2}$$

$$n_2 = \left(\frac{r_1}{r_2} + 1\right)n_c = \left(\frac{Z_1}{Z_2} + 1\right)n_c \tag{5.3}$$

The foregoing examples are meant as an introduction to the more complicated cases that follow.

Simple Planetary Gear Trains

Fig. 5.3 shows a simple planetary gear train consisting of the sun gear 1, which is fixed, and the planet gear 3 in mesh with gears 1 and the ring gear 2. Planet gear 3 is carried by the planet carrier C. The carrier C is rotating clockwise with n_c. n and ω are used interchangeably because the transmission ratio R is expressed also by a ratio of angular velocities or angular speed in RPM. The velocity V_c is $V_c = r_c n_c$, and a line from the head of V_c through M_0 determines the velocity line for the carrier C. The center of gear 3 has the same velocity V_c. Gears 3 and 1 have the same velocity at

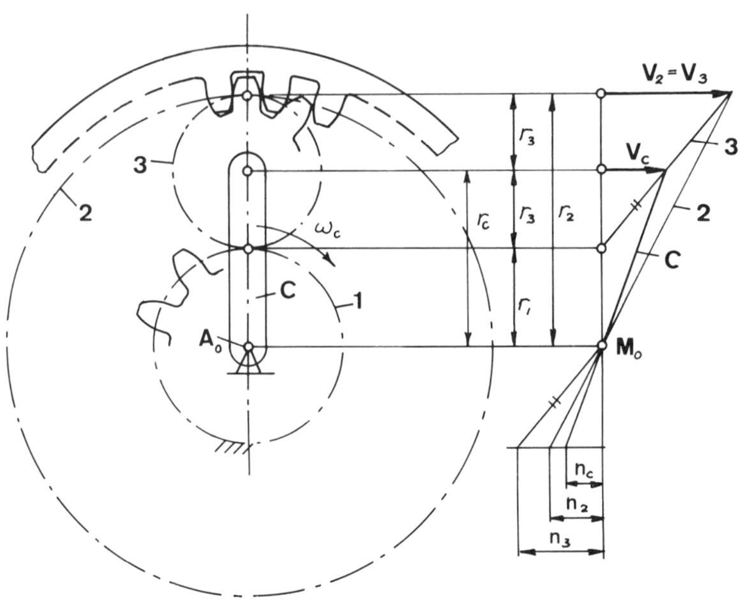

Figure 5.3 Simple planetary gear train: types F and D, Table 5.3.

their point of contact, and therefore gear 3 has zero velocity at that point corresponding to point A in the velocity diagram. The velocity line for gear 3 is now drawn through A and the head of V_c. Gears 2 and 3 have the same velocity at their point of contact, so the velocity line for gear 3 intersects the horizontal through B at the head of the velocity vector $V_3 = V_2$. A line through the head of V_2 and M_0 determines the velocity line for ring gear 2.

Because n_2, n_3, and n_c all lie to the left of the vertical through M_0, all members turn in the same direction. Based on the diagram in Fig. 5.3, the following formulas can be developed:

$$n_2 = \frac{V_2}{r_2}$$

$$V_2 = 2V_c$$

$$V_c = \left(r_1 + \frac{r_2 - r_1}{2}\right)n_c = \left(\frac{r_1 + r_2}{2}\right)n_c$$

or

$$n_2 = \left(\frac{r_1 + r_2}{r_2}\right)n_c = \left(\frac{Z_1 + Z_2}{Z_2}\right)n_c \tag{5.4}$$

In Fig. 5.4 the ring gear 2 is kept stationary, but essentially the same approach is used again. The velocity V_c of B_0 is $V_c = r_c n_c$, and a line through the head of V_c and M_0 determines the velocity line for the carrier C. The head of V_c connected to B determines the velocity line for gear 3, and extending this line to intersect the horizontal through A determines the head of the velocity vectors $V_1 = V_3$. The velocity line for gear 2 is found by connecting M_0 with the head of B_3. As before, a relationship between n_1 and n_c can be found:

$$n_1 = \frac{V_3}{r_1}$$

$$V_3 = 2V_c$$

$$V_c = \left(r_1 + \frac{r_2 - r_1}{2}\right)n_c$$

or

$$n_1 = \left(\frac{r_1 + r_2}{r_1}\right)n_c = \left(\frac{Z_1 + Z_2}{Z_1}\right)n_c \tag{5.5}$$

The more general case in which both gears 1 and 2 rotate will now be considered (Fig. 5.5). It is given that gears 1 and 2 rotate CW with angular

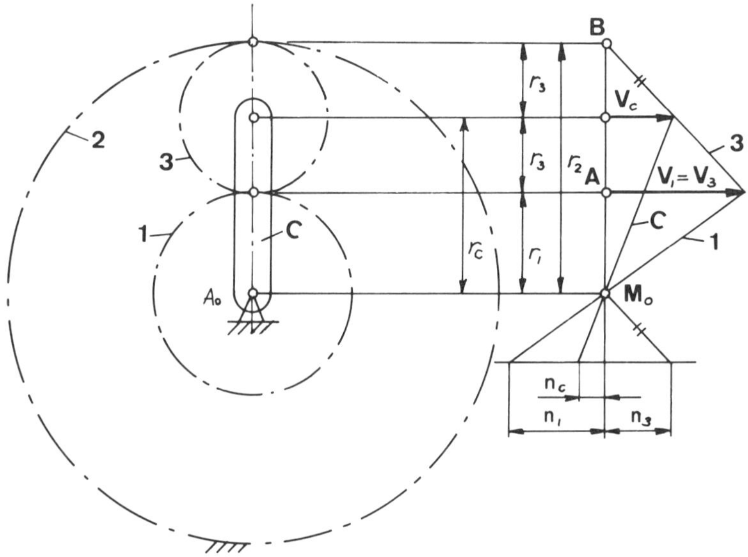

Figure 5.4 Simple planetary gear train: types B and C, Table 5.3.

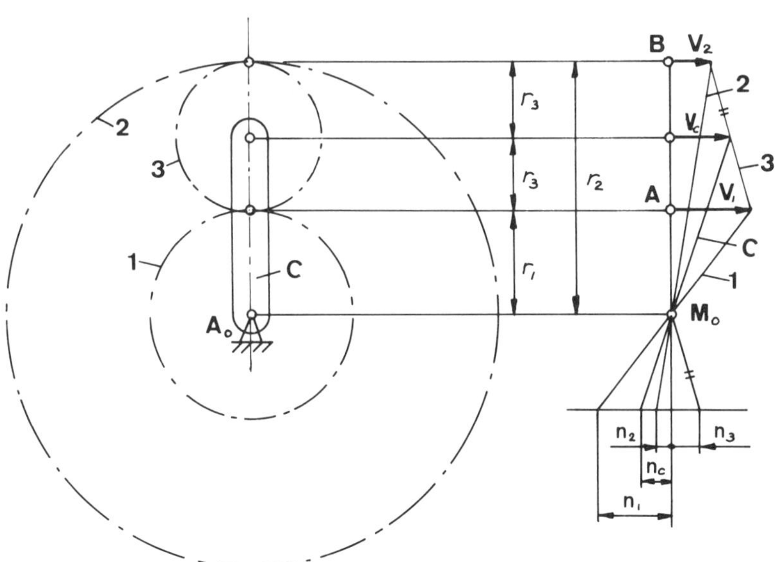

Figure 5.5 Simple planetary gear train with all members rotating.

RPM of n_1 and n_2. The velocity vectors $V_1 = r_1n_1$ and $V_2 = r_2n_2$ are drawn in the diagram in Fig. 5.5. A line through the heads of V_1 and V_2 determines the velocity line for gear 3 and, therefore, also the head of V_c.

A relationship among n_1, n_2, and n_c can be found as follows:

$$2V_c = V_1 + V_2$$

$$V_1 = r_1n_1 \quad V_2 = r_2n_2$$

$$V_c = \left(r_2 - \frac{r_2 - r_1}{2}\right)n_2$$

$$(2r_2 - r_2 + r_1)n_c = r_1n_1 + r_2n_2$$

$$n_c = \frac{r_1n_1 + r_2n_2}{r_1 + r_2} = \frac{Z_1n_1 + Z_2n_2}{Z_1 + Z_2} \tag{5.6}$$

Letting n_1 or n_2 equal zero results in eqs. (5.4) and (5.3), respectively.

The Tabular Method

This method to calculate the transmission ratio of planetary gears uses the principle of superimposition, that is, a combined motion is broken down into separate motion and then the separate motions are added together to yield the resultant motion.

Fig. 5.6 shows a compound planetary gear system. Gears 2 and 2′ are fastened to each other so that they act as one double gear. 2-2′ turns around the central shaft A_0 guided by the carrier C. Gears 1 and 3 are in mesh with gears 2 and 2′, respectively. It is now assumed that the carrier rotates with n_c RPM and gear 1 with n_1 RPM.

First, the carrier C is rotated n_c revolutions (Table 5.1), and all parts are considered as being locked together. This is shown in the first row in Table 5.1. Next, the carrier C is kept fixed, and gear 1 is first rotated through $-n_c$ revolutions to get back to its starting position and then rotated n_1 revolutions to get to the position corresponding to when the carrier C has rotated n_c. Because the carrier is fixed, the planetary gear train is now an ordinary gear train with fixed axis. The second row in Table 5.1 shows the corresponding rotations of the various members. When gear 1 rotates $-n_c + n_1$, then gear 2 = 2′ rotates $(-n_c + n_1)(-r_1/r_2)$. The minus sign in front of r_1/r_2 designates that the direction of rotation is changed between gears 1 and 2. Gear 2′ will now rotate gear 3 through $(n_c + n_1)(-r_1/r_2)(-r_2'/r_3)$, and again the motion is reversed between gears 2′ and 3. Adding the second and third rows in Table 5.1 yields the resultant motion of the various members.

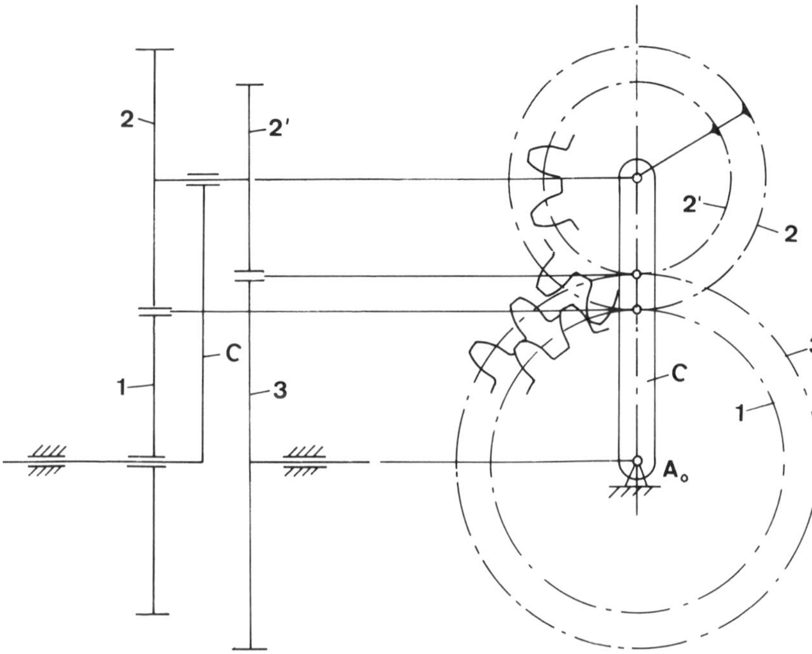

Figure 5.6 Compound planetary gear train with external gearing.

Table 5.1[a]

	1	2 = 2′	3	C
All parts locked together	n_c	n_c	n_c	n_c
Carries C stationary; gear 1 rotated	$-n_c + n_1$	$-\frac{r_1}{r_2}(-n_c + n_1)$	$\frac{r_1}{r_2}\frac{r_2'}{r_3}(-n_c + n_1)$	0
Resultant motion	n_1	$-\frac{r_1}{r_2}(-n_c + n_1) + n_c$	$\frac{r_1}{r_2}\frac{r_2'}{r_3}(-n_c + n_1) + n_c$	n_c

[a]See also Fig. 5.6.

It follows from Table 5.1 that

$$n_3 = \frac{r_1}{r_2}\frac{r_2'}{r_3}(n_1 - n_c) + n_c \qquad (5.7)$$

or that

$$n_3 = \frac{Z_1}{Z_2}\frac{Z_2'}{Z_3}(n_1 - n_c) + n_c \qquad (5.8)$$

where Z is the number of teeth but where the pitch of each meshing pair of gears must be the same.

Example

Given (Fig. 5.6) $Z_1 = 74$, $Z_2 = 36$, $Z_2' = 35$, $Z_3 = 75$, $n_1 = 500$ RPM CW, and $n_c = 750$ RPM CCW, Find n_3.

Solution

A CW direction is chosen to be positive so that $n_1 = 500$ RPM and $n_c = -750$ RPM. From eq. (5.8)

$$n_3 = \frac{74}{36}\frac{35}{75}[500 - (-750)] - 750$$

$$n_3 = (+)470 \text{ RPM (CW)}$$

Example

Find the value of n_3, if $n_1 = 0$, or gear 1 is fixed. The rest of the values of the preceding are assumed unchanged.

Solution

From eq. (5.8):

$$n_3 = \frac{74}{36}\frac{36}{75}(750) - 750$$

$$n_3 = -28.5 \text{ RPM (CCW)}$$

Fig. 5.7 shows another compound planetary gear train where the planet gears 2-2′ are fastened together. Input members are any two of members 1, 3, and C, and output is to the third member. The same procedure is used as explained in the foregoing and the results are shown in Table 5.2.

Space Requirements

A planetary gear train has certain space requirement for a given transmission ratio. In the following it is shown that for values of the transmission ratio lying within a certain interval the train can be made to a reasonable size.

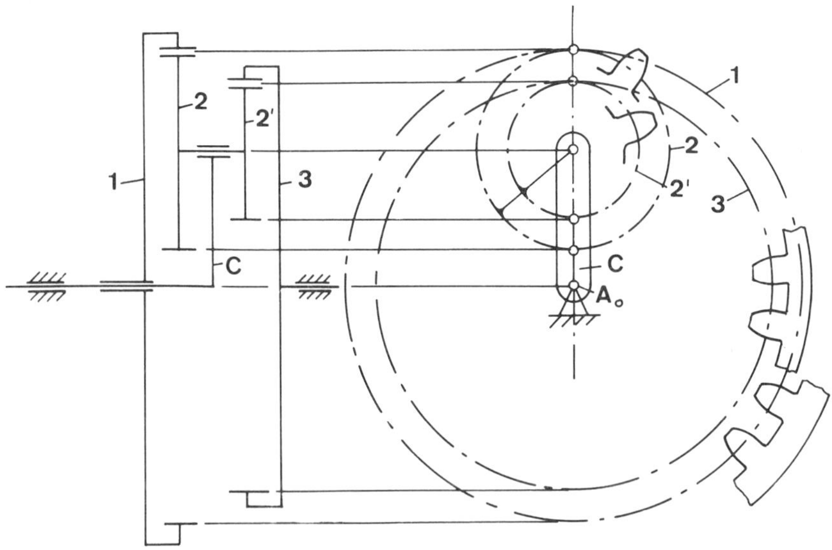

Figure 5.7 Compound planetary gear train with internal gearing.

Outside this interval it either builds too large to be of practical value or it does not exist at all.

Simple Planetary Gear Trains

A simple planetary gear train was shown in Fig. 5.3 in which the diameter D of the ring gear determines the maximum space requirement in radial direction. The smallest gear with radius r_{min} can be gear 1 and/or gear 3.

Table 5.2[a]

	1	2 = 2′	3	C
All parts locked together	n_c	n_c	n_c	n_c
Carries C stationary; gear 1 rotated	$-n_c + n_1$	$\frac{r_1}{r_2}(-n_c + n_1)$	$\frac{r_1}{r_2}\frac{r_2'}{r_3}(-n_c + n_1)$	0
Resultant motion	n_1	$\frac{r_1}{r_2}(-n_c + n_1) + n_c$	$\frac{r_1}{r_2}\frac{r_{2'}}{r_3}(-n_c + n_1) + n_c$	n_c

[a] See also Fig. 5.7.

Table 5.3 shows the various combinations of input, output, and fixed gears that can be applied to the simple planet gear train of Fig. 5.4.

Space Requirements for Train F

The transmission ratio was already found to be

$$R = \frac{n_2}{n_c} = \frac{r_1 + r_2}{r_2} \tag{5.3}$$

There are two possibilities concerning the sizes of gears 1 and 3, namely, that either 1 and/or 3 is the smallest possible gear.

Case 1

$$r_1 = r_{min}$$

$$R = \frac{r_1 + r_2}{r_2} = \frac{r_1 + D/2}{D/2}$$

$$\frac{D}{r_{min}} = \frac{2}{R - 1}$$

Case 2

$$r_3 = r_{min}$$

$$R = \frac{r_1 + r_2}{r_2} = \frac{D/2 - 2r_3 + D/2}{D/2}$$

$$\frac{D}{r_{min}} = \frac{4}{2 - R}$$

The above equation for transmission ratio and space requirements are not valid for all values of D/r_{min} because the lower value of $D/r_{min} = 6$, namely, when both gears 1 and 3 are as small as possible.

On the other hand, there is an upper value because there is a limit to how large gear 2 can be made. Here, the maximum value of D/r_{min} is assumed to be 20. The interval limit for system F can now be found.

Case 1

$$r_1 = r_{min}$$

$$6 \leq \frac{2}{R - 1} \leq 20$$

$$1.1 \leq R \leq 1.33$$

Case 2

$$r_3 = r_{min}$$

$$6 \leq \frac{4}{2 - R} \leq 20$$

$$1.33 \leq R \leq 1.8$$

Table 5.3 Simple Planetary Gear trains (Fig. 5.4, Chart 5.1)

	Gear train					
	A	E	B	C	F	D
Input	1	2	1	C	2	C
Output	2	1	C	1	C	2
Fixed	C[a]	C[a]	2	2	1	1
Transmission ratio	$R = -\frac{r_2}{r_1}$	$R = -\frac{r_1}{r_2}$	$R = \frac{r_1 + r_2}{r_1}$	$R = \frac{r_1}{r_1 + r_2}$	$R = \frac{r_1 + r_2}{r_2}$	$R = \frac{r_1 + r_2}{r_1}$
Space requirements	$-10 \leqq R \leqq -3$	$-0.333 \leqq R \leqq -0.1$	$4 \leqq R \leqq 11$	$0.09 \leqq R \leqq 0.25$	$1.1 \leqq R \leqq 1.33$	$0.75 \leqq R \leqq 0.91$
	$\frac{D}{r_{min}} = -2R$		$\frac{D}{r_{min}} = 2(R - 1)$		$\frac{D}{r_{min}} = \frac{2}{R - 1}$	
	$r_1 = r_{min}$		$r_1 = r_{min}$		$r_1 = {}_{min}$	
	$-3 \leqq R \leqq -1.25$	$-0.8 \leqq R \leqq -0.333$	$\frac{9}{4} \leqq R \leqq 4$	$0.25 \leqq R \leqq 0.445$	$1.33 \leqq R \leqq 1.80$	$0.55 \leqq R \leqq 0.75$
	$\frac{D}{r_{min}} = \frac{4R}{R + 1}$		$\frac{D}{r_{min}} = \frac{4(R - 1)}{R - 2}$		$\frac{D}{r_{min}} = \frac{4}{2 - R}$	
	$r_3 = r_{min}$		$r_3 = r_{min}$		$r_3 = r_{min}$	
Shown in	Fig. 5.1		Fig. 5.4		Fig. 5.3	

[a]The carrier C is here the frame link A_0B_0.

The space requirement curves are shown in Chart 5.1, and the formulas in Table 5.3.

Compound Planetary Gear Trains

Compound planetary gear train are shown in Figs. 5.6–5.13. The space requirements for these systems are listed in Table 5.4 and shown as curves in Chart 5.2.

It should be noticed that if the double gear 3-3′ protrudes from ring gear 2′, then the greatest diameter D is calculated as shown in Fig. 5.8.

Simple Planetary Gear Trains in Series

New planetary gear trains can be developed by coupling simple gear trains in series, where output members of the first unit connect to input members

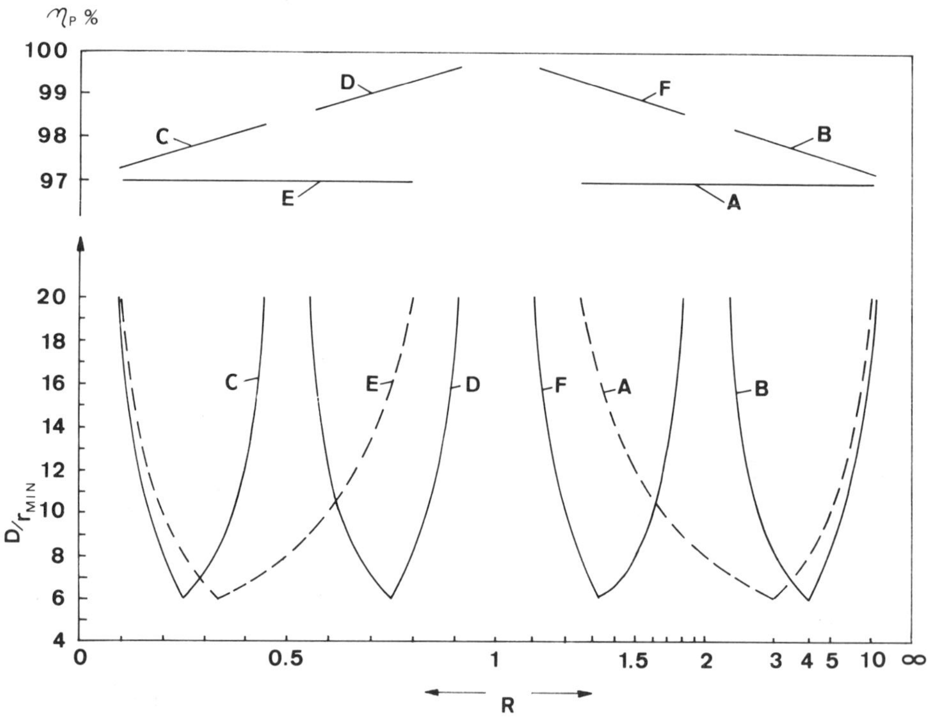

Chart 5.1 Design chart for simple planetary gear trains: R = transmission ratio, D/r_{min} = ratio between the outside diameter of the ring gear and the radius of the pitch circle of the smallest gear, η_p = Efficiency of planetary gear train. (Dashed curves indicate a negative transmission ratio.)

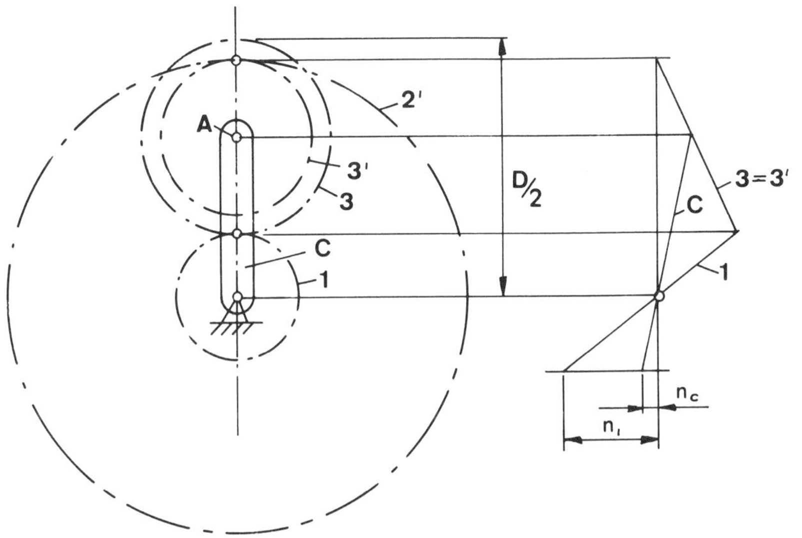

Figure 5.8 Compound planetary gear train. (See Table 5.4 for this figure and for Figs. 5.9–5.13.)

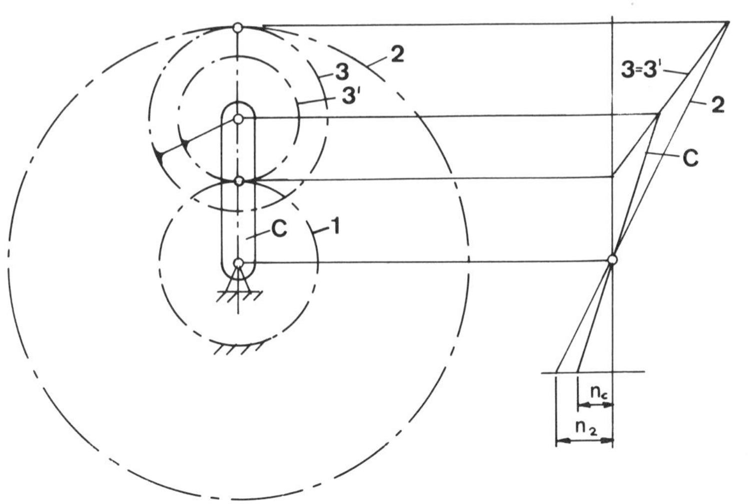

Figure 5.9

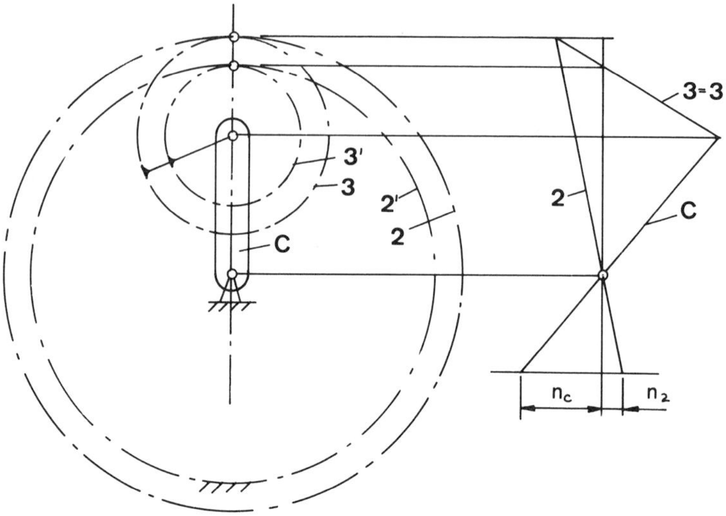

Figure 5.10

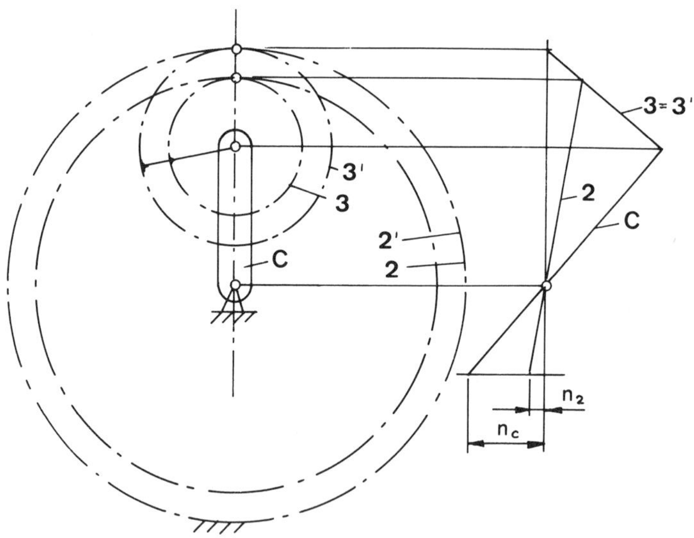

Figure 5.11

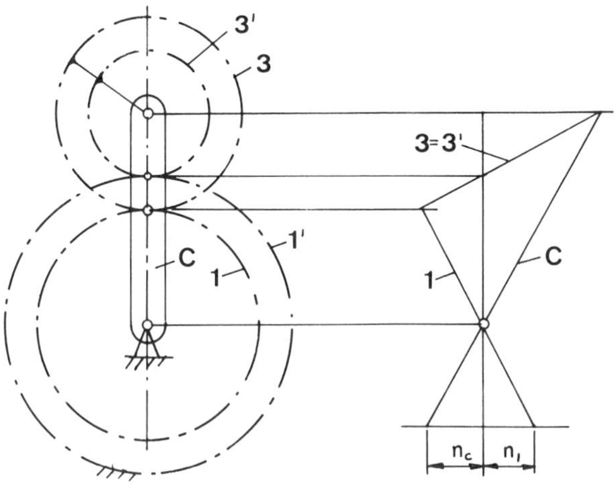

Figure 5.12

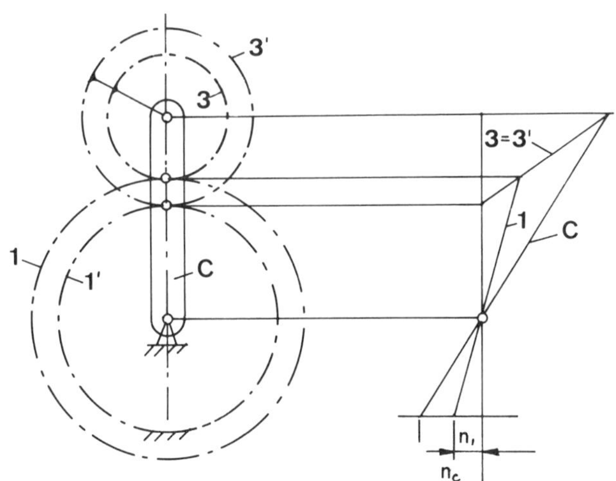

Figure 5.13

Table 5.4 Compound Planetary Gear Trains (Figs. 5.6–5.13, Chart 5.2)

	Gear train							
	a-4	4-a	a-5	5-a	a-6	6-a	a-7	7-a
Input	1	C	2	C	2	C	1	C
Output	C	1	C	2	C	2	C	1
Fixed	2′	2′	1′	1′	2′	2′	1′	1′
Transmission ratio	$R = 1 + \frac{r_3}{r_3'}\frac{r_2'}{r_1}$	$\frac{1}{R}$	$R = 1 + \frac{r_3}{r_3'}\frac{r_1'}{r_2}$	$\frac{1}{R}$	$R = 1 - \frac{r_3}{r_3'}\frac{r_2'}{r_2}$	$\frac{1}{R}$	$R = 1 - \frac{r_3}{r_3'}\frac{r_1'}{r_1}$	$\frac{1}{R}$
Space requirements	$4 \leqq R \leqq 30.25$	$0.333 \leqq R \leqq 0.25$	$1.035 \leqq R \leqq 1.33$	$0.75 \leqq R \leqq 0.967$	$-2.025 \leqq R \leqq 0$	$-\infty \leqq R \leqq -0.494$	$-19.25 \leqq R \leqq 0$	$-\infty \leqq R \leqq -0.052$
	$\frac{D}{r_{min}} = -2 + 4\sqrt{R}$		$\frac{D}{r_{min}} = -2 + 4\sqrt{\frac{R}{R-1}}$		$\frac{D}{r_{min}} = \frac{-2}{2\sqrt{R_2 - R} + 2R - 1}$		$\frac{D}{r_{min}} = 2 + 4\sqrt{1-R}$	
	$r_1 = r_3' = r_{min}$		$r_1' = r_3 = r_{min}$		$r_3' = r_{min}$		$r_1 = r_3' = r_{min}$	
	$1.49 \leqq R \leqq 4$	$0.25 \leqq R \leqq 0.67$	$1.33 \leqq R \leqq 3.025$	$0.331 \leqq R \leqq 0.75$	$0 \leqq R \leqq 0.67$	$1.49 \leqq R \leqq \infty$	$0 \leqq R \leqq 0.95$	$1.05 \leqq R \leqq \infty$
	$\frac{D}{r_{min}} = \frac{2\sqrt{R}(R-1)}{R\sqrt{R} + \sqrt{R} - 2R}$		$\frac{D}{r_{min}} = \frac{2}{2R - 2\sqrt{R^2 - R} - 1}$		$\frac{D}{r_{min}} = \frac{2(1-R)}{(1-\sqrt{R})^2}$		$\frac{R}{r_{min}} = 2 + \frac{4}{\sqrt{1-R}}$	
	$r_3 = r_{min}$		$r_1' = r_3' = r_{min}$		$r_3 = r_{min}$		$r_1' = r_3 = r_{min}$	
Shown in	Fig. 5.8		Fig. 5.9		Figs. 5.10 and 5.11		Figs. 5.12 and 5.13	

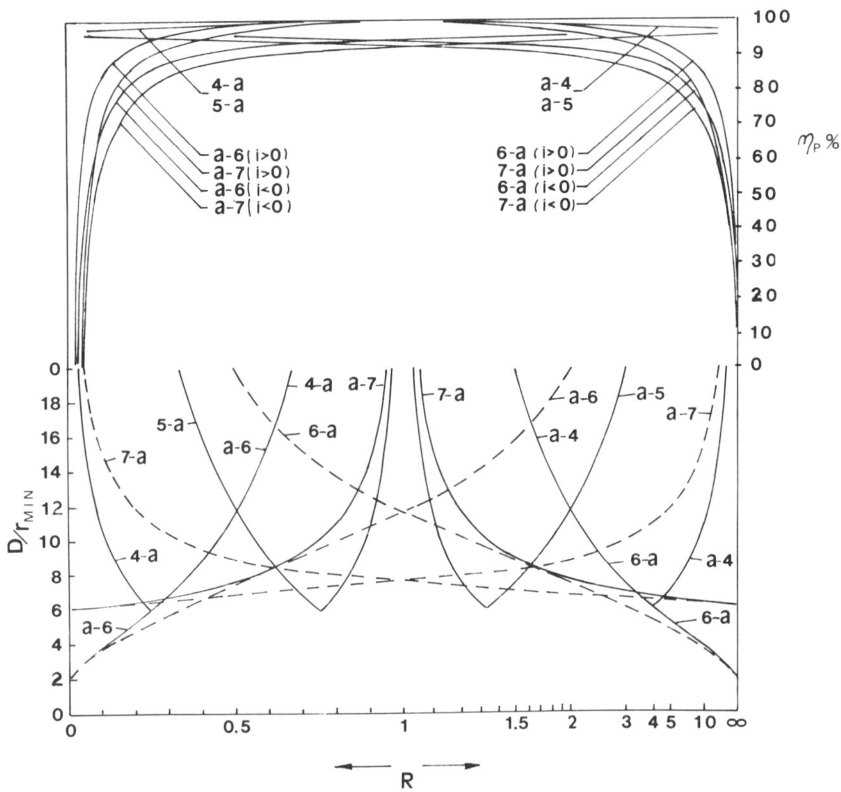

Chart 5.2 Design chart for compound planetary gear trains: R = transmission ratio, D/r_{min} = ratio between the outside diameter of the ring gear and the radius of the pitch circle of the smallest gear, η_p = efficiency of planetary gear train. (Dashed curves indicate a negative transmission ratio.)

of the second unit. Table 5.5 shows the possible combinations and their characteristics, and the corresponding curves are drawn in Chart 5.3. For all simple trains in series, in order to build them as compactly as possible, it is required that $D = 2r_2 = 2r_2'$.

EFFICIENCY

Ordinary Gear Trains

When two gears mesh, a friction loss occurs when opposing teeth roll and slide on each other, and the sliding is proportional to the distance between

Table 5.5 Simple Planetary Gear

	Gear A − B	Gear B − A	Gear A − C	Gear C − A
Input	1	C′	1	1′
Output	C′	1	1′	1
Transmission ratio	$R = -\frac{r_2(r_1' + r_2')}{r_1 r_1'}$	$R = -\frac{r_1 r_1'}{r_2(r_1' + r_2')}$	$R = -\frac{r_2 r_1'}{r_1(r_1' + r_2')}$	$R = -\frac{r_1(r_1' + r_2')}{r_2 r_1'}$
Space requirements	$-110 \leqq R \leqq -12$	$-\frac{1}{12} \leqq R \leqq -\frac{1}{110}$	$-4.45 \leqq R \leqq -0.75$	$-1.33 \leqq R \leqq -0.225$
	$\frac{D}{r_{min}} = -1 + \sqrt{1 - 4R}$; $r_1 = r_1' = r_{min}$		$\frac{D}{r_{min}} = -2R + 2 + 2\sqrt{R^2 + 1}$; $r_1 = r_3' = r_{min}$	
	$-12 \leqq R \leqq -2.81$	$-0.356 \leqq R \leqq -0.0833$	$-0.75 \leqq R \leqq -0.114$	$-8.77 \leqq R \leqq -1.33$
	$\frac{D}{r_{min}} = \frac{4R + 2 - 2\sqrt{-4R + 1}}{R + 2}$; $r_3 = r_3' = r_{min}$		$\frac{D}{r_{min}} = \frac{R - 1 - \sqrt{9R^2 - 2R + 1}}{R}$; $r_3 = r_1' = r_{min}$	

	Gear B − C	Gear C − B	Gear B − D	Gear D − B
Input	1	1′	1	2′
Output	1′	1	2′	1
Transmission ratio	$R = \frac{r_1 + r_2}{r_1} \frac{r_1'}{r_1' + r_2'}$	$R = \frac{r_1}{r_1 + r_2} \frac{r_1' + r_2'}{r_1'}$	$R = \frac{r_1 + r_2}{r_1} \frac{r_2'}{r_1' + r_2'}$	$R = \frac{r_1}{r_1 + r_2} \frac{r_1' + r_2'}{r_2'}$
Space requirements	$1 \leqq R \leqq 4.89$	$0.204 \leqq R \leqq 1$	$3 \leqq R \leqq 10$	$0.1 \leqq R \leqq 0.333$
	$\frac{D}{r_{min}} = 2R + 1 + \sqrt{4R^2 - 4R + 9}$; $r_2 = r_3' = r_{min}$		$\frac{D}{r_{min}} = 2R$; $r_1 = r_1' = r_{min}$	
	$0.204 \leqq R \leqq 1$	$1 \leqq R \leqq 4.89$	$1.25 \leqq R \leqq 3$	$0.333 \leqq R \leqq 0.8$
	$\frac{D}{r_{min}} = \frac{R + 2 + \sqrt{9R^2 - 4R + 4}}{R}$; $r_3 = r_1' = r_{min}$		$\frac{D}{r_{min}} = \frac{4R}{R - 1}$; $r_3 = r_3' = r_{min}$	

the point of contact between opposing teeth and the pitch point. The friction loss is also dependent on other factors such as pressure angle, height of teeth (addendum), surface finish, greasing and number of teeth. Practical tests have shown that with a good approximation the efficiency for two external gears in mesh is 0.98 and by internal gearing the efficiency is 0.99.

Trains in Series (Chart 5.3)

train	A − D	D − A	A − F	F − A	B − B	C − C
	1	2′	1	C′	1	C′
	2′	1	C′	1	C′	1
	$R = -\dfrac{r_2 r_2'}{r_1(r_1' + r_2')}$	$R = -\dfrac{r_1(r_1' + r_2')}{r_2 r_2'}$	$R = -\dfrac{r_2(r_1' + r_2')}{r_1 r_2'}$	$R = -\dfrac{r_1 r_2'}{r_2(r_1' + r_2')}$	$R = \dfrac{(r_1 + r_2)(r_1' + r_2')}{r_1 r_1'}$	$R = \dfrac{r_1 r_1'}{(r_1 + r_2)(r_1' + r_2')}$
	$-9.1 \leqq R \leqq -2.25$	$-0.445 \leqq R \leqq -0.11$	$-18 \leqq R \leqq -4$	$-0.25 \leqq R \leqq -0.056$	$16 \leqq R \leqq 121$	$0.0083 \leqq R \leqq 0.0625$
	$\dfrac{D}{r_{min}} = -R + \sqrt{R^2 - 4R}$		$\dfrac{D}{r_{min}} = 2 - R$		$\dfrac{D}{r_{min}} = -2 + 2\sqrt{R}$	
	$r_1 = r_1' = r_{min}$		$r_1 = r_3' = r_{min}$		$r_1 = r_1' = r_{min}$	
	$-2.25 \leqq R \leqq -0.695$	$-1.44 \leqq R \leqq -0.445$	$-4 \leqq R \leqq -1.375$	$-0.728 \leqq R \leqq -0.25$	$5.06 \leqq R \leqq 16$	$0.0625 \leqq R \leqq 0.198$
	$\dfrac{D}{r_{min}} = \dfrac{6R - 2\sqrt{R^2 - 4R}}{2R + 1}$		$\dfrac{D}{r_{min}} = \dfrac{4R - 2}{R + 1}$		$\dfrac{D}{r_{min}} = \dfrac{4R - 8 + 4\sqrt{R}}{R - 4}$	
	$r_3 = r_3' = r_{min}$		$r_3 = r_1' = r_{min}$		$r_3 = r_3' = r_{min}$	
train	**B − F**	**F − B**	**D − D**	**F − F**	**D − F**	**F − D**
	1	C′	C	2′	C	C′
	C′	1	2′	C	C′	C
	$R = \dfrac{r_1 + r_2}{r_1}\dfrac{r_1' + r_2'}{r_2'}$	$R = \dfrac{r_1}{r_1 + r_2}\dfrac{r_2'}{r_1' + r_2'}$	$R = \dfrac{r_2}{r_1 + r_2}\dfrac{r_2'}{r_1' + r_2'}$	$R = \dfrac{r_1 + r_2}{r_1}\dfrac{r_1' + r_2'}{r_2'}$	$R = \dfrac{r_2}{r_1 + r_2}\dfrac{r_1' + r_2'}{r_2'}$	$R = \dfrac{r_1 + r_2}{r_2}\dfrac{r_2'}{r_1' + r_2'}$
	$5.33 \leqq R \leqq 19.8$	$0.051 \leqq R \leqq 0.188$	$0.5625 \leqq R \leqq 0.825$	$1.21 \leqq R \leqq 1.78$	$1 \leqq R \leqq 1.64$	$0.61 \leqq R \leqq 1$
	$\dfrac{D}{r_{min}} = \dfrac{R + \sqrt{R^2 + 16}}{2}$		$\dfrac{D}{r_{min}} = \dfrac{-2R - 2\sqrt{R}}{R - 1}$		$\dfrac{D}{r_{min}} = \dfrac{-2R - 4}{R - 2}$	
	$r_1 = r_2' = r_{min}$		$r_1 = r_1' = r_{min}$		$r_1 = r_3' = r_{min}$	
	$2.47 \leqq R \leqq 5.33$	$0.188 \leqq R \leqq 0.405$	$0.308 \leqq R \leqq 0.5625$	$1.78 \leqq R \leqq 3.25$	$0.611 \leqq R \leqq 1$	$1 \leqq R \leqq 1.64$
	$\dfrac{D}{r_{min}} = \dfrac{2R + 2\sqrt{R^2 - 2R + 4}}{R - 2}$		$\dfrac{D}{r_{min}} = \dfrac{8R + 4\sqrt{R}}{4R - 1}$		$\dfrac{D}{r_{min}} = \dfrac{4R + 2}{2R - 1}$	
	$r_3 = r_1' = r_{min}$		$r_3 = r_3' = r_{min}$		$r_3 = r_1 = r_{min}$	

Planetary Gear Trains

The efficiency of a planetary gear train is found by comparing it with an ordinary gear train having the same losses.

The efficiency of a planetary gear train (or any other gear train) is defined by

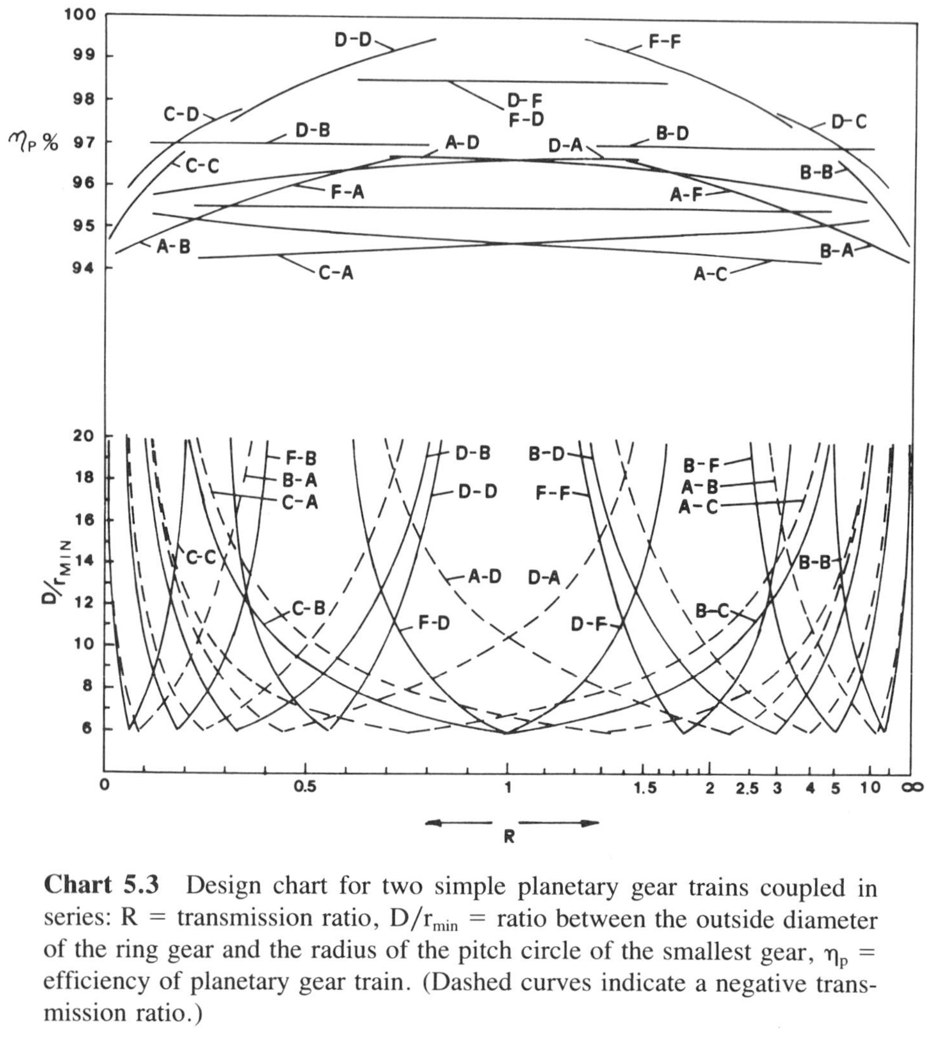

Chart 5.3 Design chart for two simple planetary gear trains coupled in series: R = transmission ratio, D/r_{min} = ratio between the outside diameter of the ring gear and the radius of the pitch circle of the smallest gear, η_p = efficiency of planetary gear train. (Dashed curves indicate a negative transmission ratio.)

$$\eta_p = \frac{P_{out}}{P_{in}}$$

where P_{out} is the effect leaving and P_{in} is the effect entering the gear train.

Designating η_0 as the efficiency of the ordinary gear train derived from the planetary gear train by letting the carrier be stationary, but letting the relative velocities of the gears remain the same, the following is obtained: For gear trains A and E the efficiency is

$$\eta_p = \eta_0$$

For gear trains B, F, a-4, a-5, a-6, and a-7 the efficiency is

$$\eta_p = 1 - \left(1 - \frac{1}{R}\right)(1 - \eta_0)$$

For C, D, 4-a, 5-a, 6-a, and 7-a the efficiency is

$$\eta_p = \frac{1}{1 + \left(1 - \frac{1}{R}\right)\left(\frac{1}{\eta_0} - 1\right)}$$

The efficiency of the planetary gear train in series shown in Table 5.5 is found as the product of the efficiency of the individual trains. The efficiency curves of the charts are based on the assumption that

$\eta_0 = 0.97$	for the trains in Table 5.3 and for a-4, 4-a, a-5, and 5-a
$\eta_0 = 0.98$	for a-6 and 6-a
$\eta_0 = 0.96$	for a-7 and 7-a

based on the previously mentioned efficiencies for external and internal gearing. The efficiency of planetary gear systems may be lower than indicated above because of bearing friction and losses due to the gears moving through an oil bath.

Use of Several Planet Gears on One Carrier

The use of n planet gears equally spaced on one carrier requires that

$$\frac{Z_1 + Z_2}{n} = \text{integer}$$

Coupled Planetary Gear Trains

Consider the simple planetary gear train of Fig. 5.14. If none of the members are fixed but can rotate freely on this axis, then this mechanism has two degrees of freedom because the motion of two of the three members, gear 1, gear 2, and planet carrier 3, determine the motion of the third member.

The planetary gear train of Fig. 5.14 with one fixed member is shown symbolically in Fig. 5.15, where *a* designates a frame member, *b* the input, and *c* or *d* the output. Table 5.6 shows the six possible combinations A-F of a simple planetary gear train.

The planetary gear train of Fig. 5.14 is shown symbolically in Fig. 5.16, and Table 5.7 shows the six possible combinations I–VI.

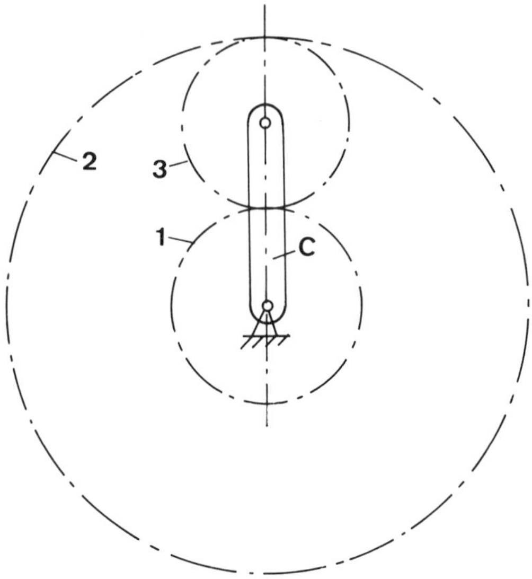

Figure 5.14 Planetary gear train with two degrees of freedom.

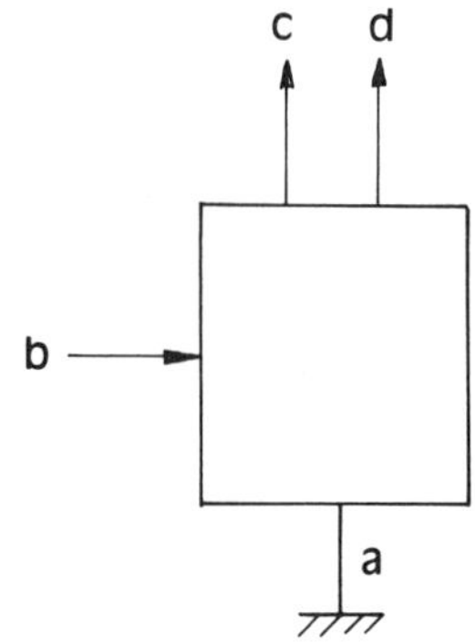

Figure 5.15 Symbolic representation of Fig. 5.14 with one member fixed.

Table 5.6

	A	B	C	D	E	F
a	C	2	2	1	C	1
b	1	1	C	C	2	2
c	1	1	C	C	2	2
d	2	C	1	2	1	C

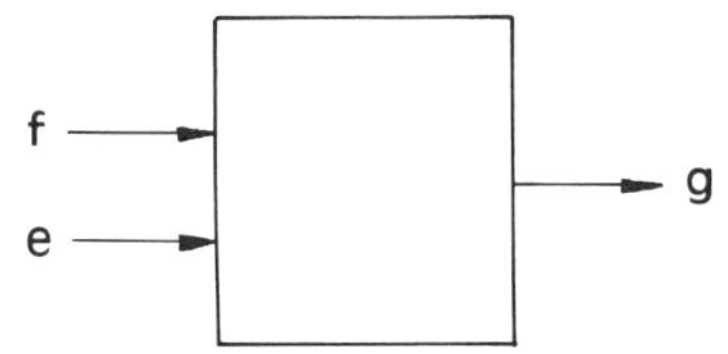

Figure 5.16 Symbolic representation of Fig. 5.14.

Table 5.7

	I	II	III	IV	V	VI
e	1	1	C	C	2	2
f	C	2	2	1	C	1
g	2	C	1	2	1	C

It is now possible to combine A–F with each of I to VI by letting member *a* be fixed and where *b* is the input, *c* is combined with *e*, and *d* with *f*. Member *g* is output. In this way, 36 planetary gear trains are developed (upper half of Table 5.8), and if input and output are interchanged, the 36 planetary gear trains of the lower half of Table 5.8 are developed.

All 36 coupled planetary gear trains A-I to F-VI are shown in Fig. 5.25. Unless otherwise indicated, input is from the left and output to the right.

Table 5.8

A-I	A-II	A-III	A-IV	A-V	A-VI
B-I	B-II	B-III	B-IV	B-V	B-VI
C-I	C-II	C-III	C-IV	C-V	C-VI
D-I	D-II	D-III	D-IV	D-V	D-VI
E-I	E-II	E-III	E-IV	E-V	E-VI
F-I	F-II	F-III	F-IV	F-V	F-VI
I-A	II-A	III-A	IV-A	V-A	VI-A
I-B	II-B	III-B	IV-B	V-B	VI-B
I-C	II-C	III-C	IV-C	V-C	VI-C
I-D	II-D	III-D	IV-D	V-D	VI-D
I-E	II-E	III-E	IV-E	V-E	VI-E
I-F	II-F	III-F	IV-F	V-F	VI-F

Transmission Ratio

Disregarding, for the time being, frictional losses, then (Fig. 5.14)

$$\frac{M_{r2}}{M_{r1}} = \frac{r_2}{r_1} = m$$

$$M_c = -(M_{r1} + M_{r2})$$

and the result is Table 5.9.

Applying Table 5.9 to the planetary gear train A-III, Fig. 5.17, the following equations are found:

$$M_{in} = M_{r1} + M'_{st}$$

$$M_{out} = M'_{r1}$$

$$\frac{M'_{r2}}{M'_{r1}} = m'$$

$$\frac{M'_c}{M'_{r1}} = -(m' + 1)$$

$$M_{r2} = -M'_{r2}$$

$$\frac{M_{r2}}{M_{r1}} = m$$

$$M_{in} = -\frac{1}{m} \cdot m' \cdot M'_{r1} - (m' + 1) \cdot M'_{r1}$$

$$\frac{M_{out}}{M_{in}} = -\frac{m}{m' + m(m' + 1)}$$

$$R = -\frac{M_{out}}{M_{in}} = \frac{m}{m + mm' + m'}$$

The above equation can also be written as

Table 5.9

$\frac{M_{r2}}{M_{r1}} = m$	$\frac{M_{r1}}{M_{r2}} = \frac{1}{m}$
$\frac{M_{r2}}{M_c} = -\frac{m}{m + 1}$	$\frac{M_c}{M_{r1}} = -\frac{m + 1}{m}$
$\frac{M_{r1}}{M_c} = -\frac{1}{m + 1}$	$\frac{M_c}{M_{r1}} = -(m + 1)$

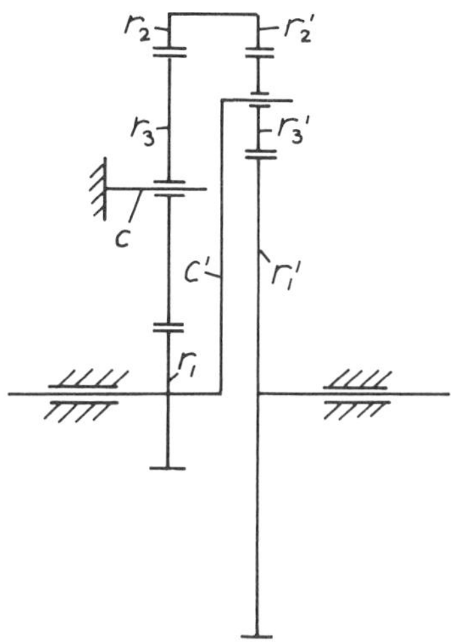

Figure 5.17 Compound planetary gear train: type A-III.

$$R = \frac{\dfrac{r_2}{r_1}}{\dfrac{r_2}{r_1} + \dfrac{r_2}{r_1}\dfrac{r_2'}{r_1'} + \dfrac{r_2'}{r_1'}}$$

The planetary gear train III-A has a transmission ratio, which is the inverse of gear train A-III, namely,

$$R = \frac{m + mm' + m'}{m}$$

To show the applicability of the method of Table 5.9, the method is applied to the planetary gear train A-I-V of Fig. 5.18:

$$M_{out} = M''_{r1}$$
$$M_{in} = M_{r1} + M'_{r1} + M''_{r2}$$

and from Table 5.9:

$$\frac{M''_{r2}}{M''_{r1}} = m''$$

$$\frac{M''_c}{M''_{r1}} = -(m'' + 1)$$

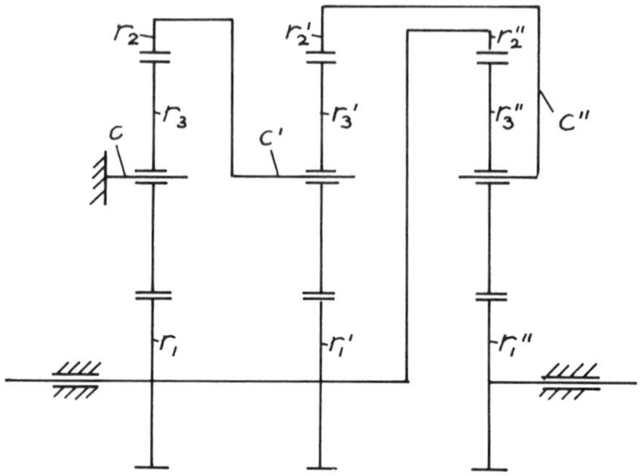

Figure 5.18 Three simple planetary gear trains coupled together: type A-I-V.

$$M_c'' = -M_{r2}'$$

$$\frac{M_{r1}'}{M_{r2}'} = \frac{1}{m'}$$

$$\frac{M_c'}{M_{r2}'} = -\frac{m' + 1}{m'}$$

$$M_c' = -M_{r2}$$

$$\frac{M_{r1}}{M_{r2}} = \frac{1}{m}$$

$$M_{in} = \frac{1}{m}\frac{m' + 1}{m'}(m'' + 1)M_{r1}'' + \frac{1}{m'}(m'' + 1)M_{r1}'' + m''M_{r1}''$$

$$M_{in} = \frac{m' + m'm'' + m' + 1 + mm'' + m + mm'm''}{mm'}M_{r1}''$$

$$R = -\frac{M_{out}}{M_{in}} = -\frac{mm'}{m + mm'' + mm'm'' + m' + m'm'' + m' + 1}$$

POWER BRANCHING AND CIRCULATING POWER

One of the characteristics of planetary gear trains is that the input power at some point in the gear train is divided (Fig. 5.19a), and later the power flow emerges. If there are no losses, then $N = N_1 + N_2$ (power branching). How-

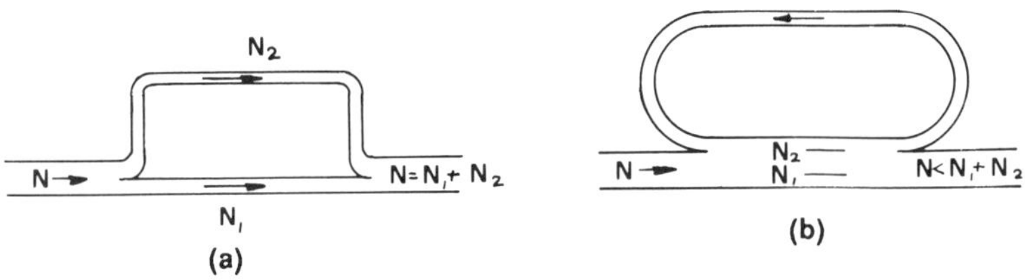

Figure 5.19 (a) Power branching. (b) Circulating power.

ever, there are planetary gear trains (Fig. 5.19b) where $N_1 + N_2 > N$ (circulating power).

The determination of whether there is branching or circulation is carried out as follows. Fig. 5.20 shows a simplified diagram of the forces on the planet gear. The forces F_1 and F_2 are directed opposite to the carrier force F_c. Therefore, the moments M_{r2} and M_{r1} are directed opposite to M_c.

This is applied to A-III, Fig. 5.21 the following way: Between 1′ and 3′ a "+" sign is assumed, whereby the "+" determines the direction of the moment on 1′. The numbers in parentheses give the sequence of the way in which the "+" and "−" signs are assumed or found. A "+" sign must be placed between 3′ and 2′ and a "−" sign by C′. A sign is placed between 3 and 2 because a "−" sign is placed between 3′ and 2′ (the ring gear 2-2′ must be in equilibrium). Finally, a "−" sign can be placed between 1 and 3. The driving member 1-C′ has two "−" signs, namely, (5) and (3). The result is, therefore, a power branching. If the two signs had been opposite, then the result would have been power circulation. Applying the same method to A-I-V (Fig. 5.22) shows that this planetary gear train has power branching.

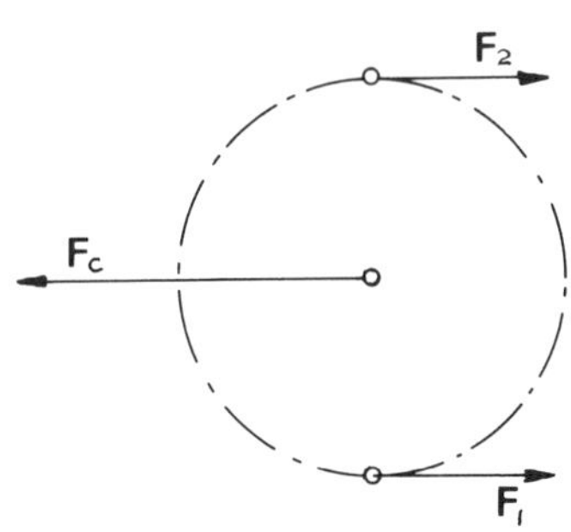

Figure 5.20 Free-body diagram of planet gear.

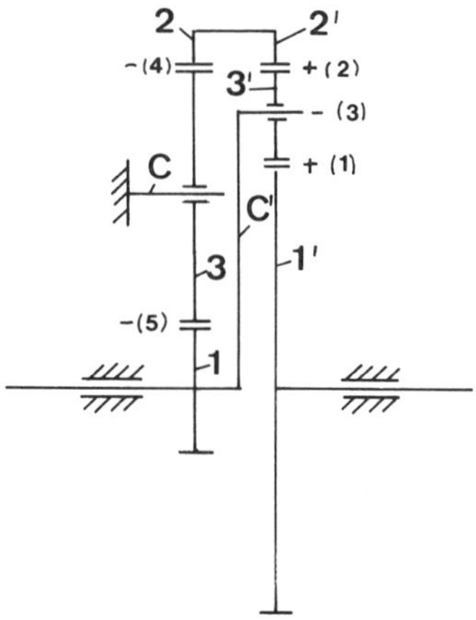

Figure 5.21 Determination of power flow in coupled planetary gear train: type A-III.

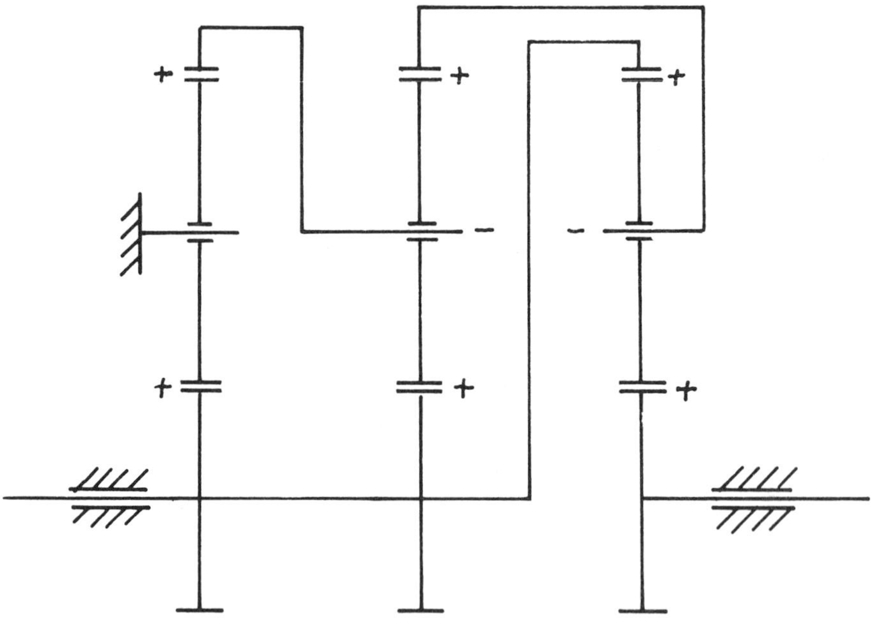

Figure 5.22 Determination of power flow in coupled planetary gear train: type A-I-V.

Space Requirements

Considering Fig. 5.14, it is clear that the smallest gear is either 1 and/or 3. The diameter D of the ring gear 2 is dependent on the size of gears 1 and 3. A reasonable assumption is that the maximum value of D is 20 r_{min}. This means that if the smallest gear has 18–20 teeth, then the ring gear has at least 180–200 teeth. The minimum size of D is 6 r_{min}, namely, when r_1 and r_2 are as small as possible or when the number of teeth Z_1 and Z_2 of the gears 1 and 3, respectively, is as small as possible.

D designates the greatest possible diameter of the ring gear 2. When considering A-III (Fig. 5.17), then D designates either $2r_2$ or $2r_2'$ or both. It is first assumed that sun gear 1 is the smallest gear. The size of the gears 1′ and 3′ is expressed by the variables x and y, and it is required that $0 < x \leq 1$ and $0 < y \leq 1$.

For the planetary gear train A-III (Fig. 5.17), the following equations can be developed:

Case 1

$$
\begin{aligned}
D &= 2r_2 \\
r_1 &= r_{min} \\
r_1 &= xr_1' \qquad 0 < x \leq 1 \\
r_1 &= yr_3' \qquad 0 < y \leq 1
\end{aligned}
\tag{5.9}
$$

The transmission ratio is

$$R = \frac{r_1'}{r_1' + r_2' + \dfrac{r_1 r_2'}{r_2}}$$

$$R = \frac{\dfrac{r_{min}}{x}}{\dfrac{2r_{min}}{x} + \dfrac{2r_{min}}{y} + \dfrac{r_{min}}{D/2}\left(\dfrac{r_{min}}{x} + \dfrac{2r_{min}}{y}\right)}$$

or

$$\frac{D}{r_{min}} = \frac{-\dfrac{2R}{x} - \dfrac{4R}{y}}{\dfrac{2R}{x} - \dfrac{1}{x} + \dfrac{2R}{y}}$$

Differentiating with respect to x and then to y:

$$\frac{\delta(D/r_{min})}{\delta x} = \frac{-\dfrac{4R}{x^2 y}(R-1)}{\left[\dfrac{2R}{x} - \dfrac{1}{x} + \dfrac{2R}{y}\right]^2}$$

$$\frac{\delta(D/r_{min})}{\delta y} = \frac{\dfrac{4R}{xy^2}(R-1)}{\left[\dfrac{2R}{x} - \dfrac{1}{x} + \dfrac{2R}{y}\right]^2}$$

Case 1a $R < 0$; no solution
Case 1b $0 < R < 1$; from Case 1

$$r_1 = r_{min}$$
$$r'_3 = r_{min}$$
$$r_2 = r'_2$$

Substituting these values into eq. (5.9):

$$\frac{D}{r_{min}} = 2\frac{R-2}{2R-1}$$

It was already stated that practical limits imposed on D are

$$6 \le \frac{D}{r_{min}} \le 20$$

Combining the above-two equations:

$$6 \le 2\frac{R-2}{2R-1} \le 20$$

or

$$0.2 \le R \le 0.421$$

Case 1c $1 < R$; no solution

Next, when r_3 is assumed to be the smallest gear, the following equations can be developed:

Case 2

$$r_3 = r_{min}$$
$$r_3 = xr'_1 \qquad 0 < x \le 1$$
$$r_3 = yr'_3 \qquad 0 < y \le 1$$

$$R = \frac{\dfrac{r_{min}}{x}}{\dfrac{2r_{min}}{x} + \dfrac{2r_{min}}{y} + \left(\dfrac{r_{min}}{x} + \dfrac{2r_{min}}{y}\right)\dfrac{D/2 - 2r_{min}}{D/2}}$$

or

$$\frac{D}{r_{min}} = \frac{\dfrac{4R}{x} + \dfrac{8R}{y}}{\dfrac{2R}{x} - \dfrac{1}{x} + \dfrac{4R}{y}}$$

Differentiating with respect to x and then to y:

$$\frac{\delta(D/r_{min)})}{\delta x} = \frac{\dfrac{8R}{xy^2}(R - 1)}{\left[\dfrac{3R}{x} - \dfrac{1}{x} + \dfrac{4R}{y}\right]^2}$$

$$\frac{\delta(D/r_{min})}{\delta y} = \frac{-\dfrac{8R}{xy^2}(R - 1)}{\left[\dfrac{3R}{x} - \dfrac{1}{x} + \dfrac{4R}{y}\right]^2}$$

Case 2a $R < 0$; no solution
Case 2b $0 < R < 1$

$$r_3 = r_{min}$$
$$r_1' = r_{min}$$
$$r_2 = r_2'$$

$$R = \frac{r_{min}}{r_{min} + D/2 + D/2 - 2r_{min}}$$

or

$$\frac{D}{r_{min}} = \frac{R + 1}{R}$$

$$6 \le \frac{R + 1}{R} \le 20$$

$$0.0526 \le R \le 0.2$$

Case 2c $1 < R$; no solution

Assuming $D = 2r_2'$ and then repeating the procedure from Cases 1 and 2 yields the same results.

The space requirement curves for all 72 planetary gear trains are shown in Charts 5.4–5.7. A dashed curve means that the input and output shaft rotate in opposite directions (negative transmission ratio).

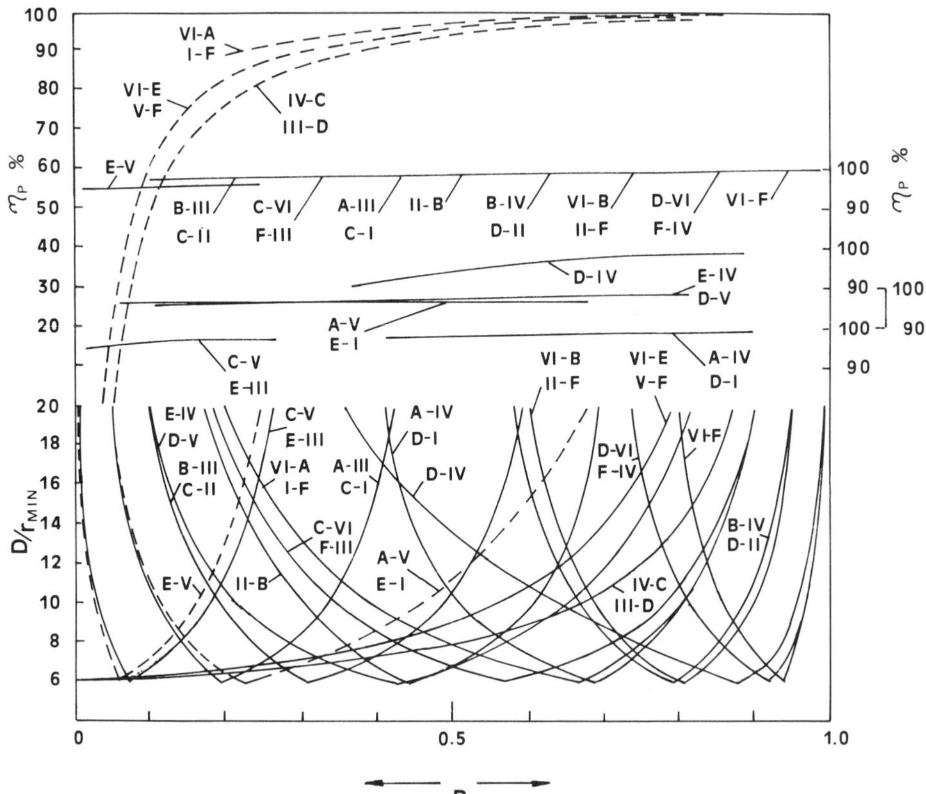

Chart 5.4 Design chart for coupled planetary gear trains: R = transmission ratio, D/r_{min} = ratio between the outside diameter of the ring gear and the radius of the pitch circle of the smallest gear, η_p = efficiency of planetary gear train. (Dashed curves indicate a negative transmission ratio.)

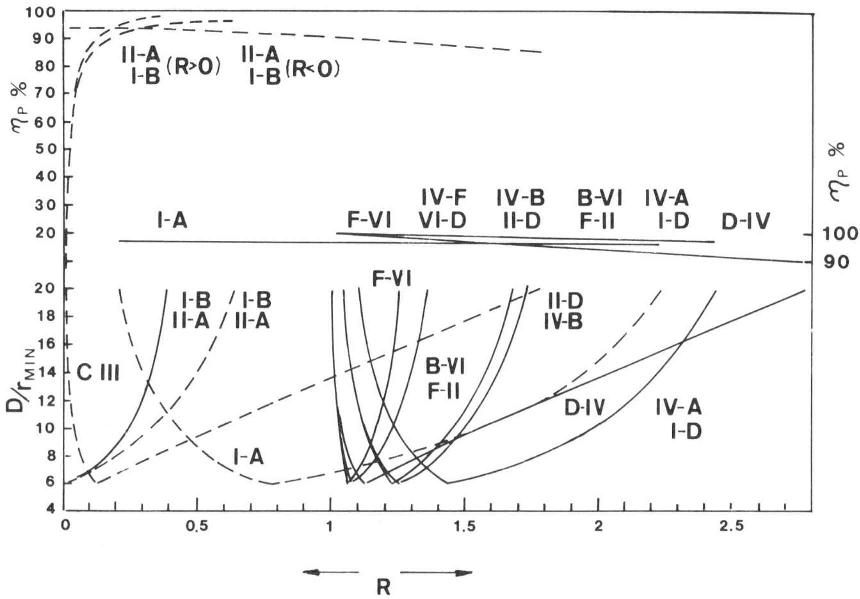

Chart 5.5 Design chart for coupled planetary gear trains: R = transmission ratio, D/r_{min} = ratio between the outside diameter of the ring gear and the radius of the pitch circle of the smallest gear, η_p = efficiency of planetary gear train. (Dashed curves indicate a negative transmission ratio.)

Efficiency of Simple Planetary Gear Train with Two Degrees of Freedom

For the mechanism in Fig. 5.23a, it is assumed that gear 1 is input and turns CW with an angular velocity ω_1. The ring gear 2 turns in a CW direction with the angular velocity ω_2.

First, a simplified free-body diagram of gear 3 is drawn (Fig. 5.23b). F_1 is directed to the right because gear 1 is the input. The other forces are determined from equilibrium conditions. Assume now that $\omega_1 > \omega_2$ results in the velocity diagram of Fig. 5.23c. Friction losses by gear mesh are dependent not on the absolute angular velocities, but only on the relative rolling velocities. To neutralize the absolute angular velocities, one has to think of oneself as standing on the carrier C. The velocity diagram will then look like Fig. 5.23d.

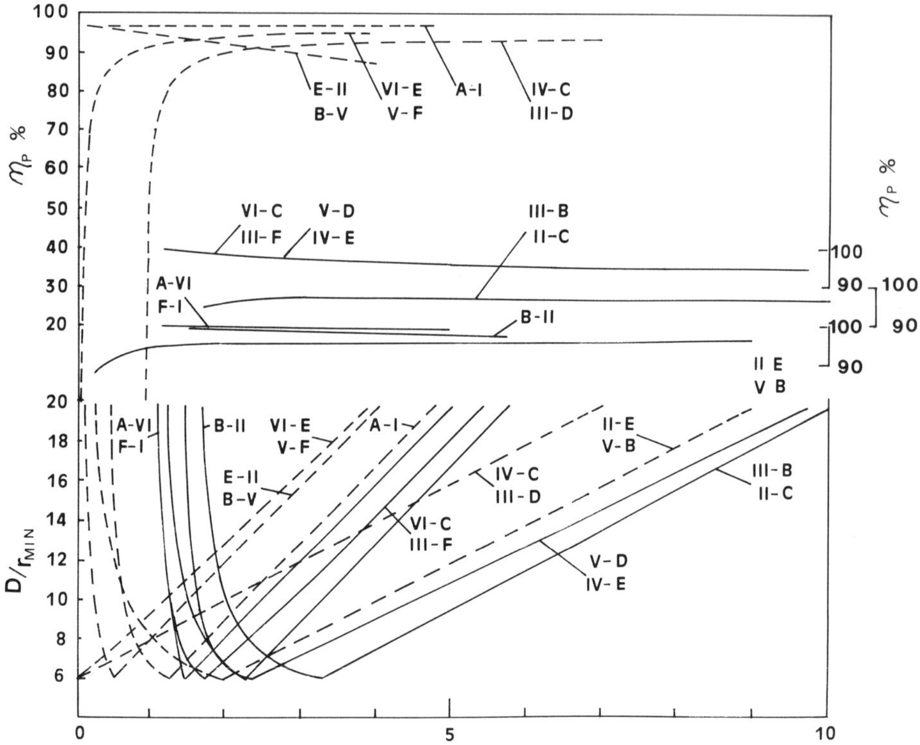

Chart 5.6 Design chart for coupled planetary gear trains: R = transmission ratio, D/r_{min} = ratio between the outside diameter of the ring gear and the radius of the pitch circle of the smallest gear, η_p = efficiency of planetary gear train. (Dashed curves indicate a negative transmission ratio.)

F_1 and V_1' have the same direction. Therefore, moments are calculated as if gear 1 is the input and gear 2 the output, although both gear 1 and gear 2 are input. The following equations can now be derived:

$$M_{r1} = M_{r2} \frac{r_1}{r_2} \frac{1}{\eta}$$

$$M_{r1} = M_{r2} \frac{1}{m\eta}$$

$$M_c = -(M_{r1} + M_{r2}) = -M_{r2}\left(1 + \frac{1}{m\eta}\right)$$

designate the other links. All four links are connected by turning joints E_t. Each joint is numbered as shown in Fig. 17.12(a).

The four-bar linkage can be written symbolically as

$$\begin{matrix} 1 & 2 & 3 & 4 \\ E_t & E_t & E_t & E_t \end{matrix}$$

A symbolic representation is shown in Fig. 17.12(b). Fig. 17.12(c) explains the symbols. Because input and output are not defined at this point, and input and output members are assumed to be links connected to the frame link, it follows that the arrangement in Fig. 17.12(b) can be mirrored about its vertical axis and therefore the following numbers can be written:

$$\begin{matrix} 1 & 2 & 3 & 4 \\ 4 & 3 & 2 & 1 \end{matrix}$$

This is done to decide whether two solutions are identical. For instance, it is seen that

$$\begin{matrix} 1 & 2 & 3 & 4 & = & 1 & 2 & 3 & 4 \\ E_g & E_t & E_t & E_t & = & E_t & E_t & E_t & E_g \end{matrix}$$

The following formula can be written:

$$\Sigma E_{rs}, E_t, E_r, E_s, Z_{rs} = 4$$

This means that the summation of the degrees of freedom of the various joints is 4. Continuing systematically,

$$\begin{matrix}
4E_{rs}: & \begin{matrix} 1 & 2 & 3 & 4 \\ E_{rs} & E_{rs} & E_{rs} & E_{rs} \end{matrix} \\
3E_{rs},\ 1E_t: & \begin{matrix} 1 & 2 & 3 & 4 \\ E_{rs} & E_{rs} & E_{rs} & E_t \end{matrix} \\
 & \begin{matrix} 1 & 2 & 3 & 4 \\ E_{rs} & E_{rs} & E_t & E_{rs} \end{matrix} \\
3E_{rs},\ 1E_r: & \begin{matrix} 1 & 2 & 3 & 4 \\ E_{rs} & E_{rs} & E_{rs} & E_r \end{matrix} \\
 & \begin{matrix} 1 & 2 & 3 & 4 \\ E_{rs} & E_{rs} & E_r & E_{rs} \end{matrix} \\
3E_{rs},\ 1E_s: & \begin{matrix} 1 & 2 & 3 & 4 \\ E_{rs} & E_{rs} & E_{rs} & E_s \end{matrix} \\
 & \begin{matrix} 1 & 2 & 3 & 4 \\ E_{rs} & E_{rs} & E_s & E_{rs} \end{matrix} \\
2E_{rs},\ 1Z_{rs}: & \text{No solution}
\end{matrix}$$

etc.

The above solutions are shown in Table 17.11. The number of solutions

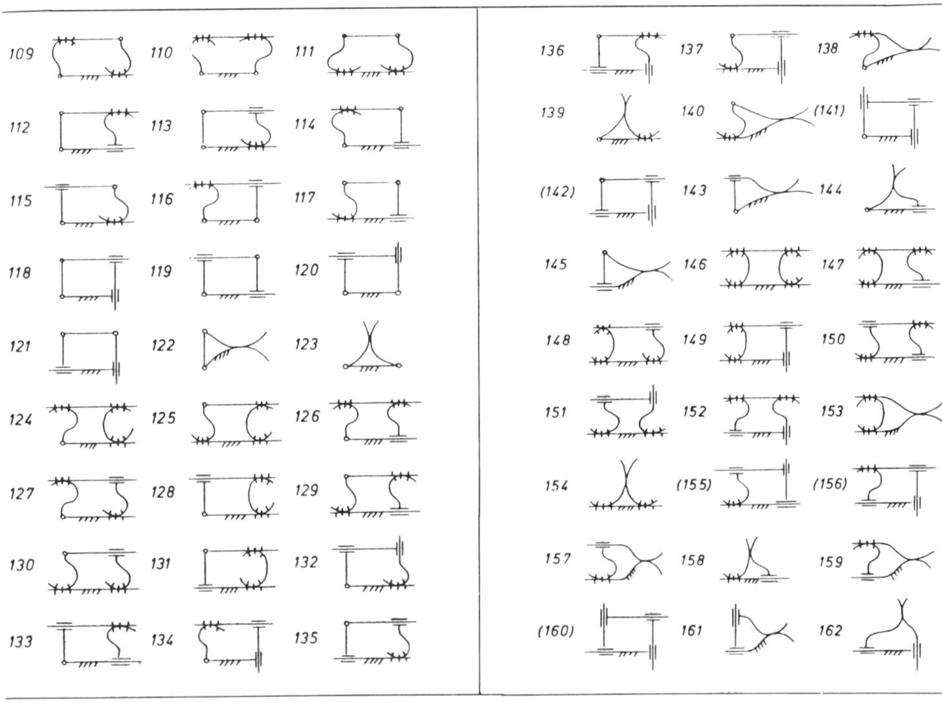

Table 17.12 (Continued)

1 2-3 5-6 7 4 $E_t Z_g\ Z_g E_s\ E_t$ 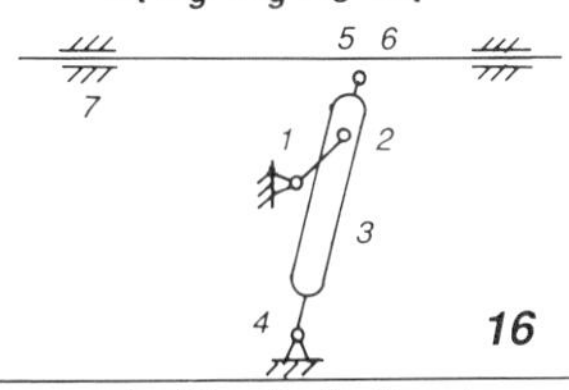 16	1 2-3 5-6 7 4 $E_t Z_g'\ Z_g E_s\ E_t$ 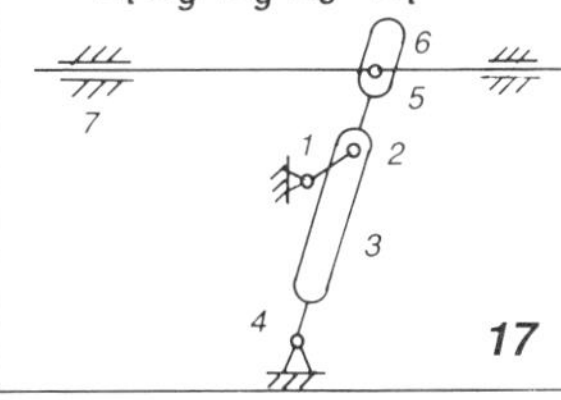 17	1 2-3 5-6 7 4 $E_t Z_g\ Z_w E_s\ E_t$ 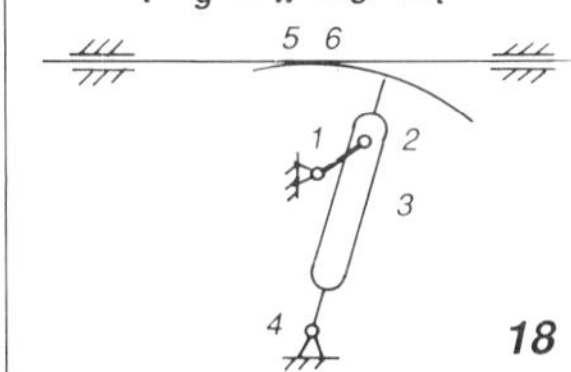 18
1 2-3 5 6 7 4 $E_t Z_g'\ E_t E_t E_s\ E_t$ 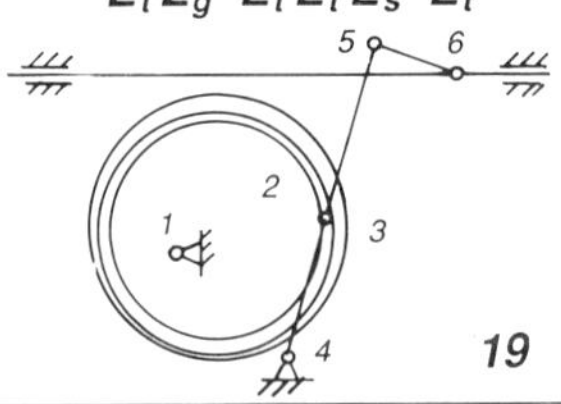 19	1 2-3 5 6 7 4 $E_t Z_g'\ E_t E_s E_s\ E_t$ 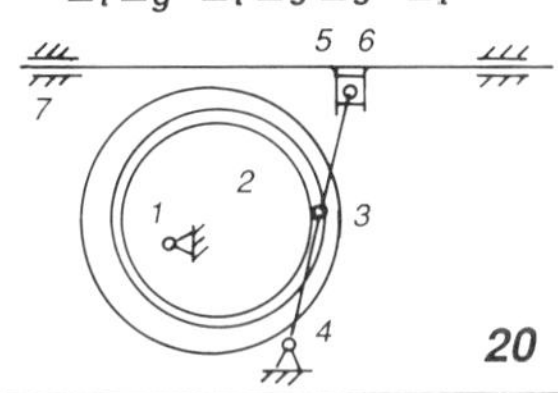 20	1 2-3 5 6 7 4 $E_t Z_g'\ E_s E_t E_s\ E_t$ 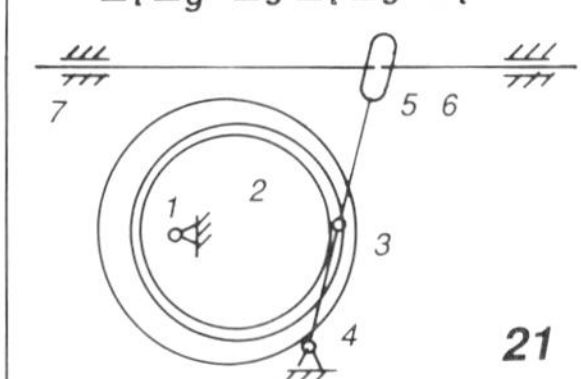 21
1 2-3 5-6 7 4 $E_t Z_g'\ Z_g E_s\ E_t$ 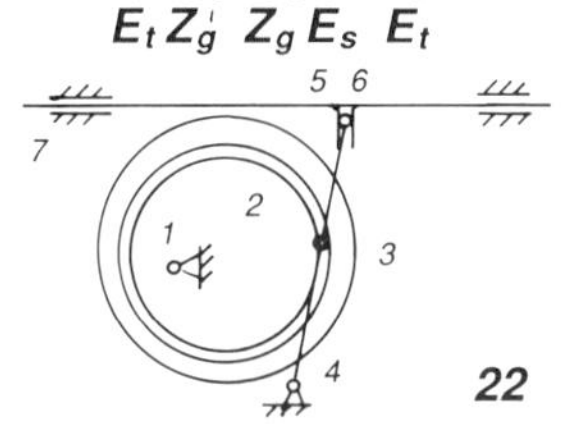 22	1 2-3 5-6 7 4 $E_t Z_g'\ Z_g' E_s\ E_t$ 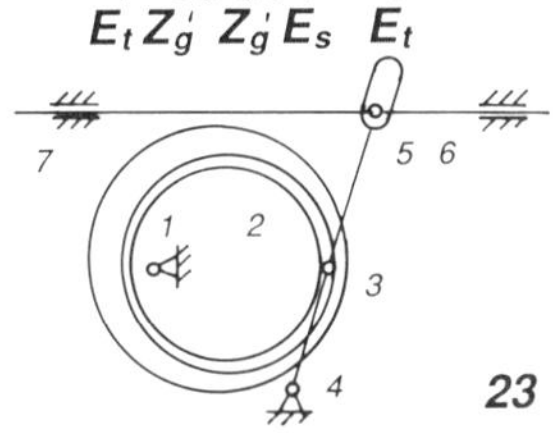 23	1 2-3 5-6 7 4 $E_t Z_g'\ Z_w E_s\ E_t$ 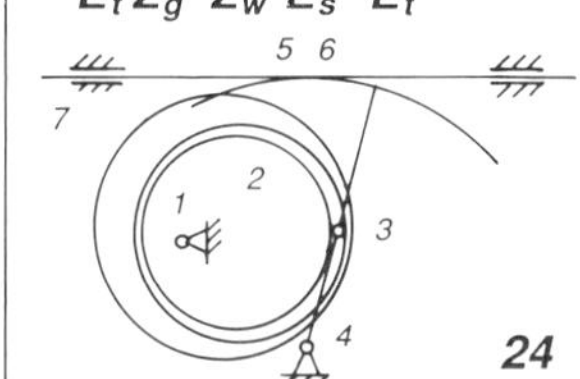 24
1 2-3 5 6 7 4 $E_t Z_w\ E_t E_t E_s\ E_t$ 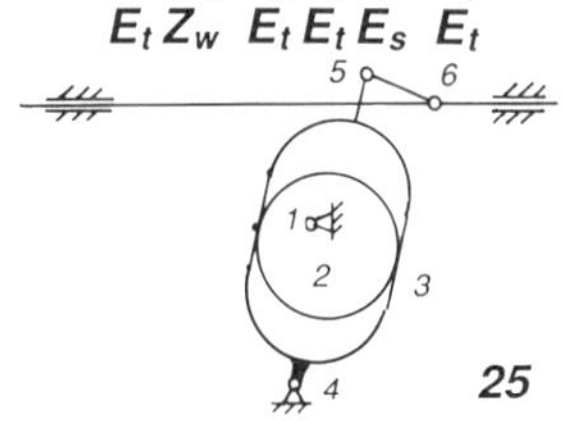 25	1 2-3 5 6 7 4 $E_t Z_w\ E_t E_s E_s\ E_t$ 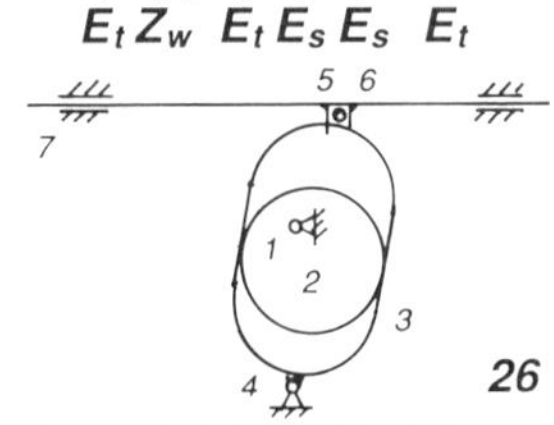 26	1 2-3 5 6 7 4 $E_t Z_w\ E_s E_t E_s\ E_t$ 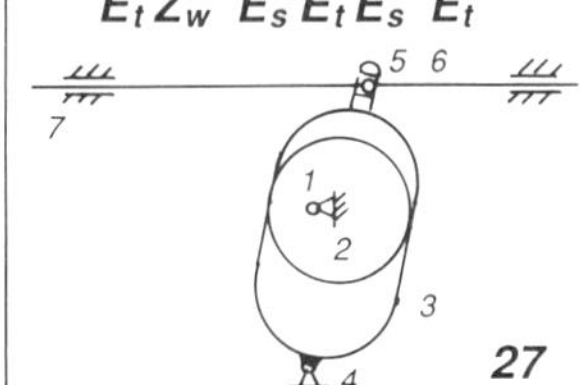 27
1 2-3 5-6 7 4 $E_t Z_w\ Z_g E_s\ E_t$ 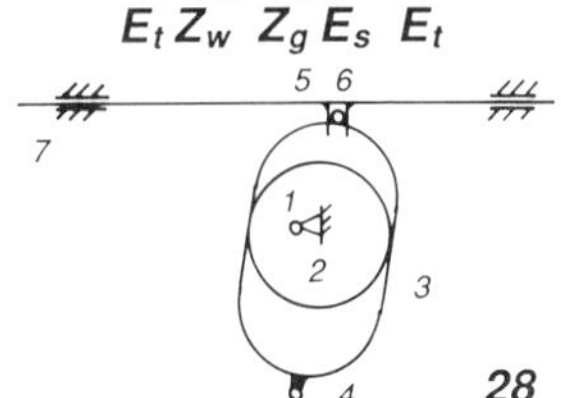 28	1 2-3 5-6 7 4 $E_t Z_w\ Z_g' E_s\ E_t$ 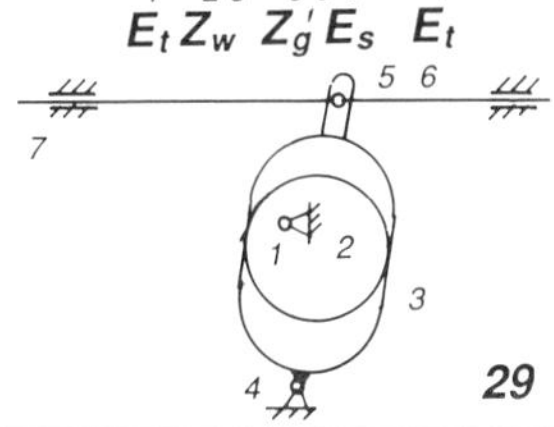 29	1 2-3 5-6 7 4 $E_t Z_w\ Z_w E_s\ E_t$ 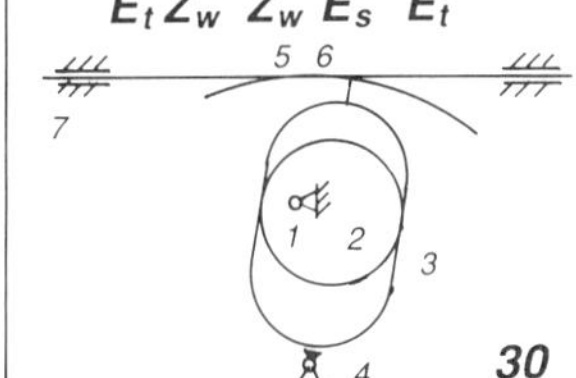 30

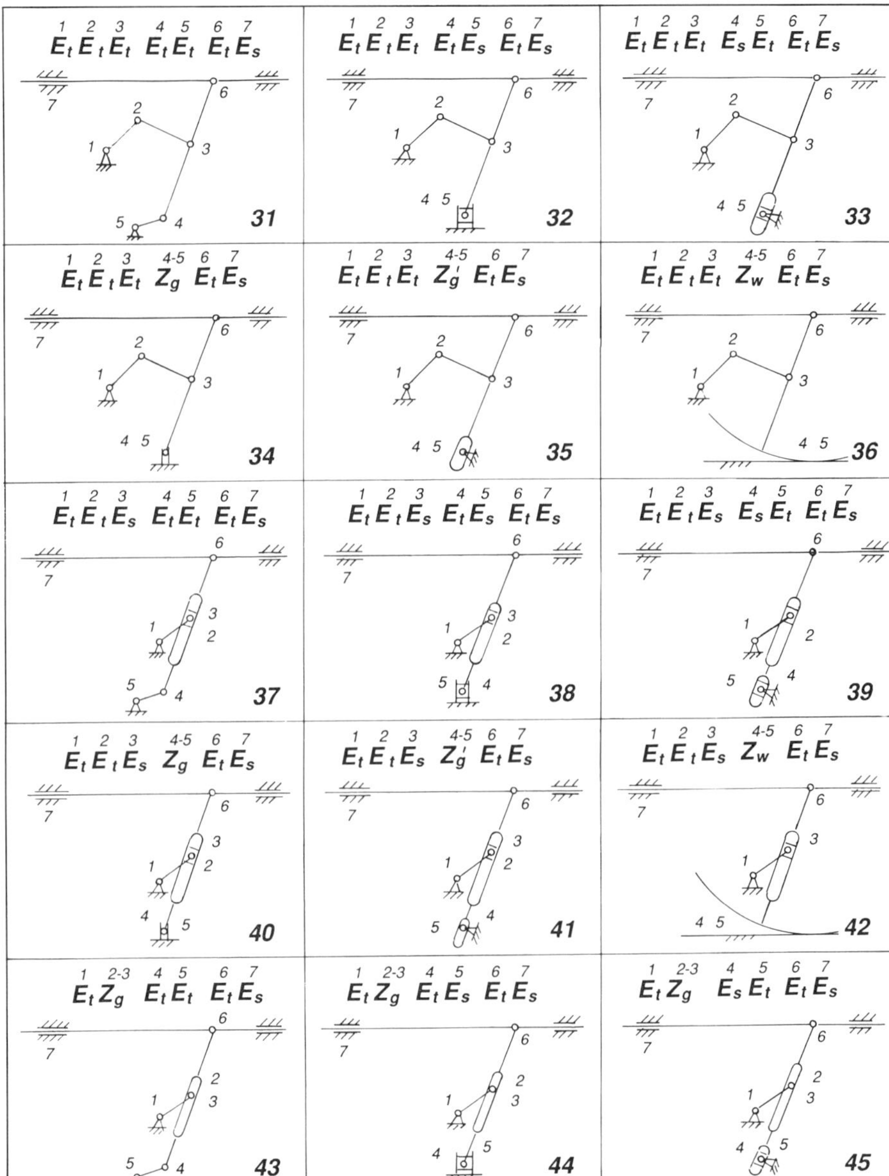

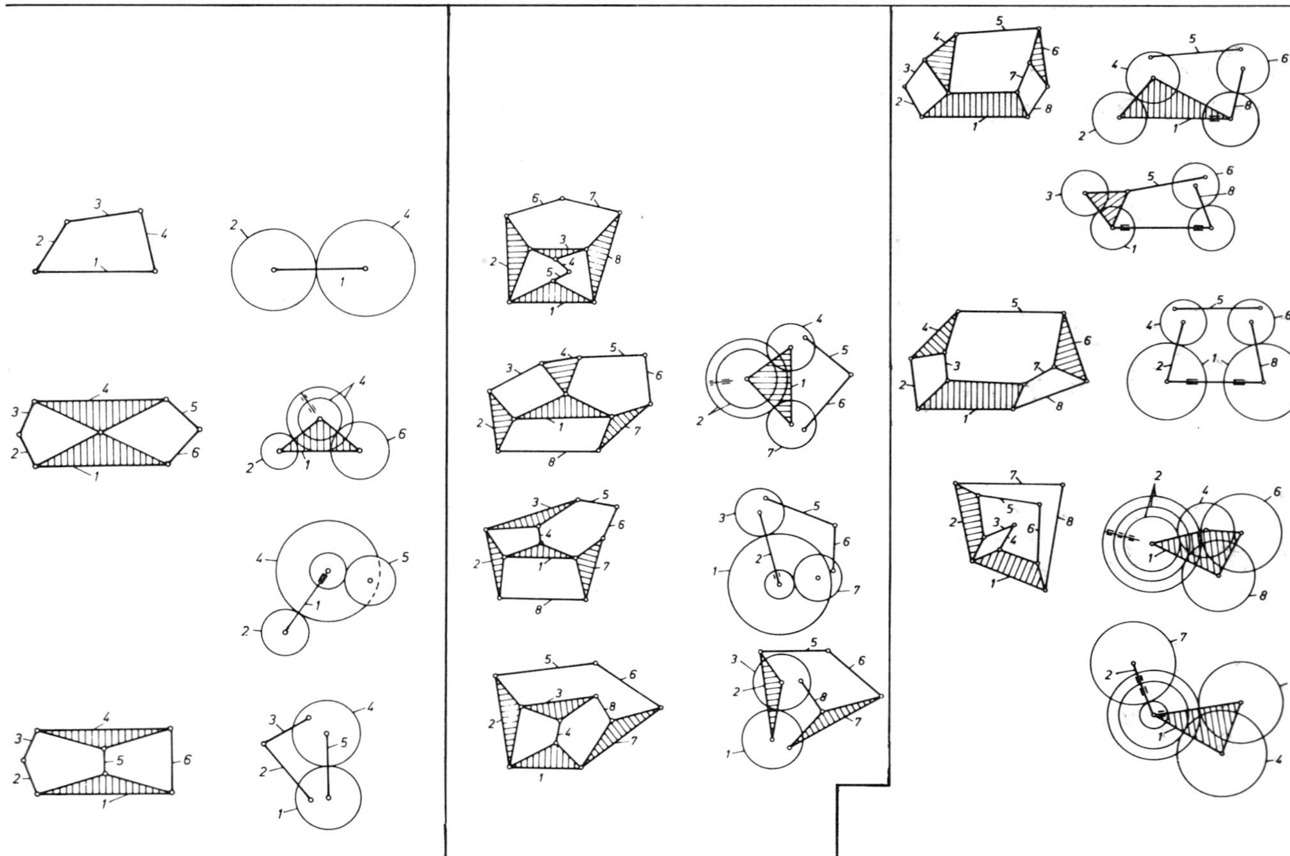

Table 17.15 (Continued)

$n_4 = 1 \quad g_3 = 2$

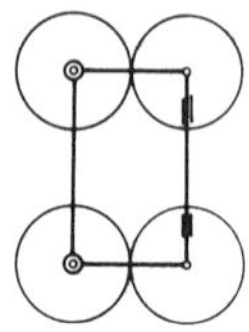

$n_3 = 2 \quad g_4 = 1$

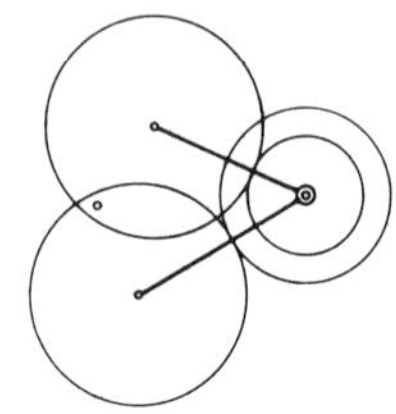

$n_3 = 1 \quad g_3 = 1 \quad g_4 = 1$

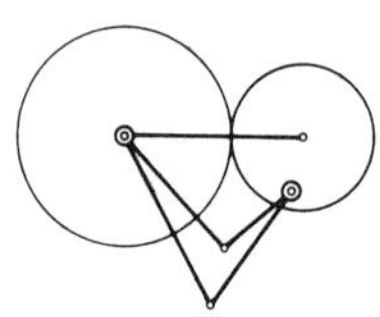

$n_4 = 1 \quad g_4 = 1$

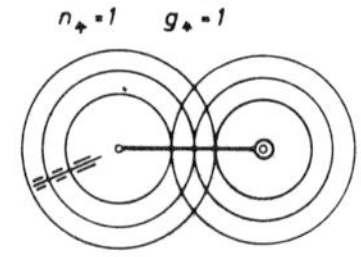

$g_4 = 2$

no solution

Development of Planetary Gears from Chains with Multiple Joints

Planetary gears can be developed from kinematic chains with multiple joints. Two planetary gear systems are found in Table 17.15 and are shown in Fig. 17.16.

A planetary gear system is characterized by colinearity of the axes of the input and output shafts. In many cases they are also characterized by the fact that one or more gears can be blocked to provide different output speeds (for instance in automatic automobile transmissions). The latter distinction kinematically characterizes planetary gear systems with two or more degrees of freedom. They can be developed from kinematic chains with two or more degrees of freedom.

Complex Planetary Gear Systems

Complex planetary gear systems are developed from kinematic chains with multiple joints, and the combination $n_3 = 4$, $g_4 = 1$, and $f = 1$ is chosen. The investigation is started by findig the corresponding kinematic chains and then reducing the chains to multiple-joint chains.

Four chains I-IV are found (Figs. 17.17–17.20). From each of these

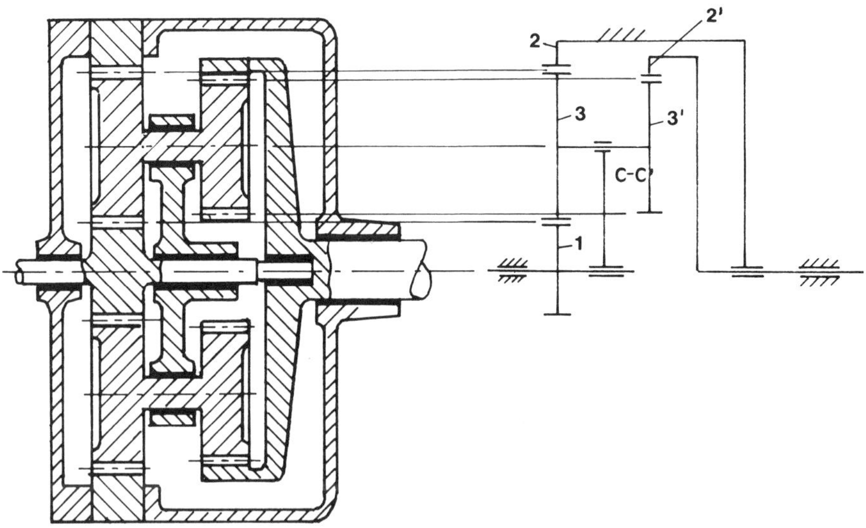

Figure 5.56 There is no sun gear to the right and the carrier carries two integral gears. If input is to the left, it is of type a-4, otherwise of type 4-a.

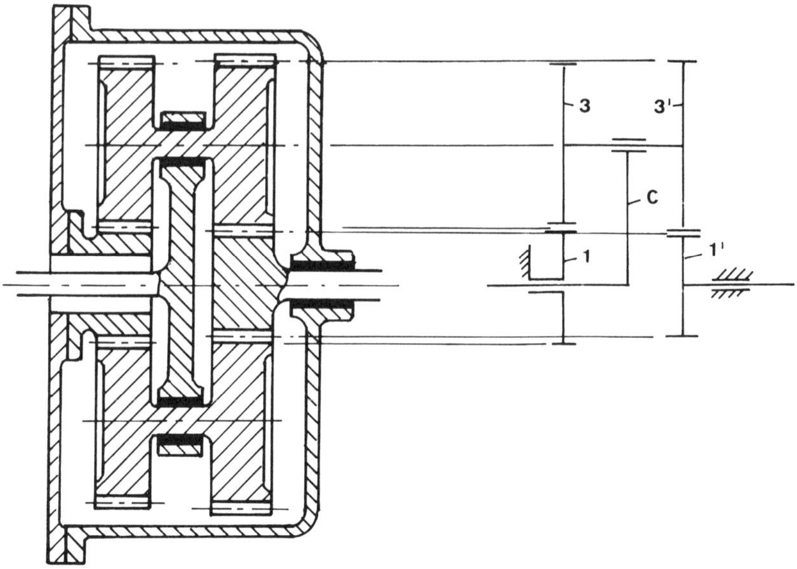

Figure 5.57 This system is of type a-7, Table 5.4, if input is to gear 1.

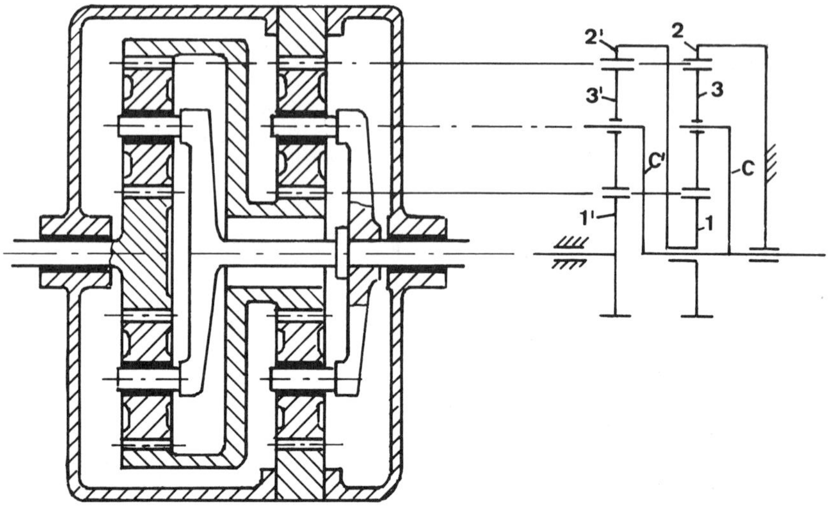

Figure 5.58 The ring gear to the right is fixed and the members are, therefore, designated 1, 2, 3, and C. The members to the left are designated 1′, 2′, 3′, and C′. Refer to Fig. 5.25. The system must be of type B- or C-. Because input is to gear 1′, it is a III-C system.

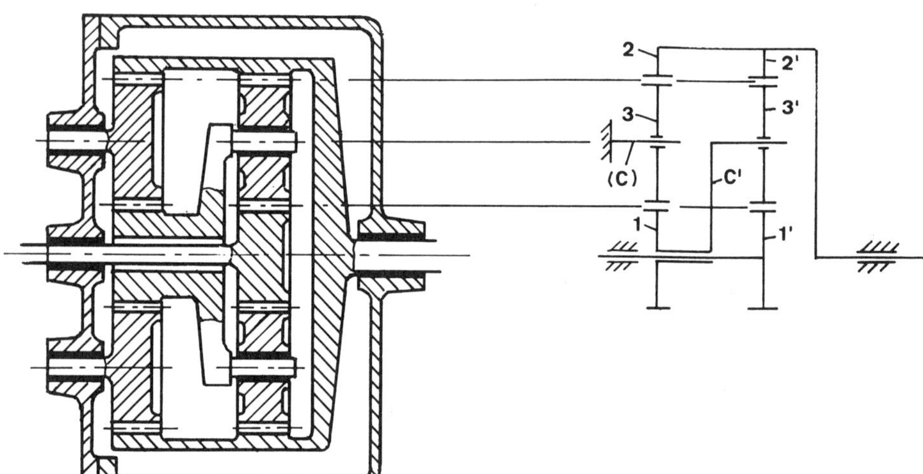

Figure 5.59 This system is identified as a C-III system if input is to gear 1. (See fig. 5.25.)

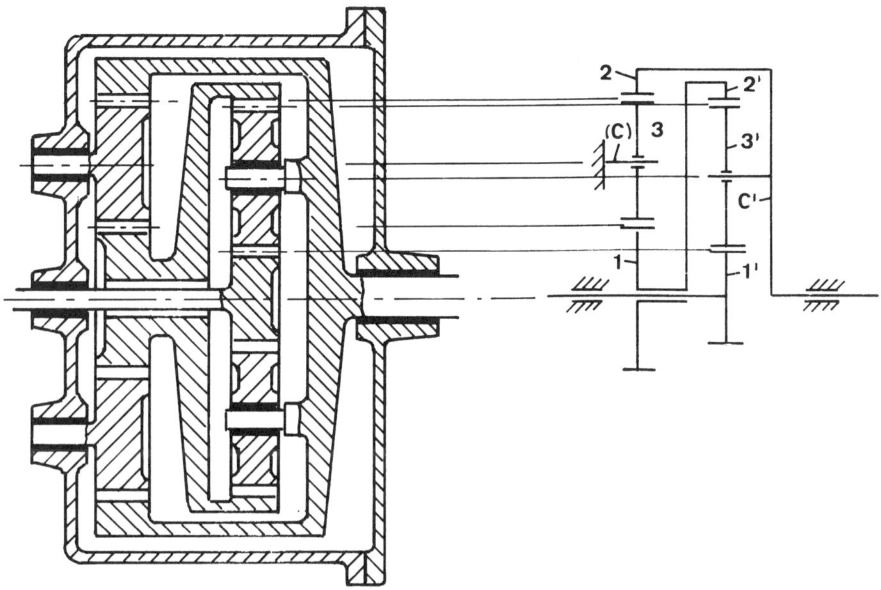

Figure 5.60 This is an E-IV system. (See Fig. 5.25.)

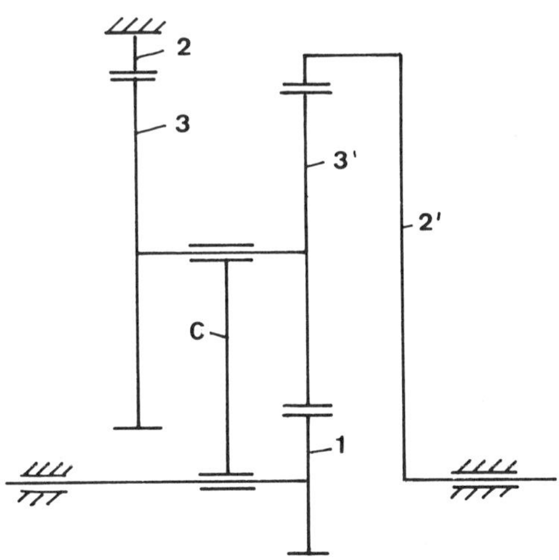

Figure 5.61 The Minuteman cover drive. This is a so-called Wolfram planetary gear system. It is characterized by the carrier C-C′ acting only to support the planetary gear 3-3′.

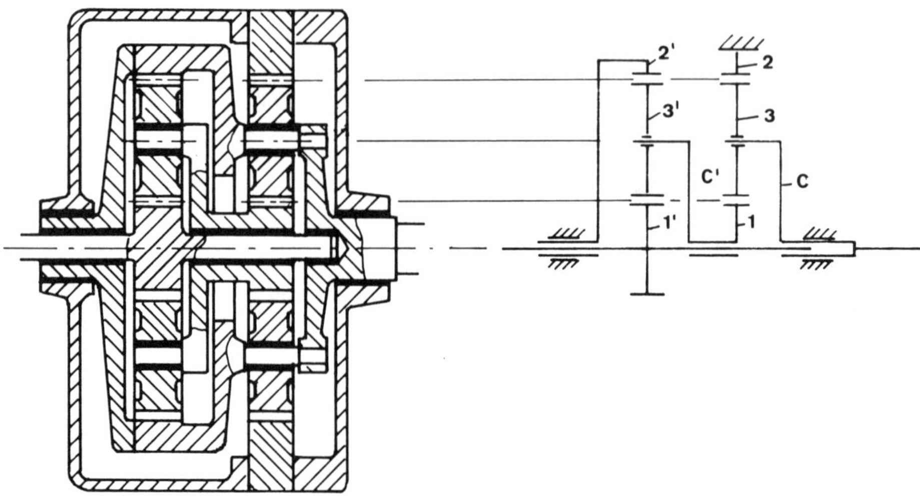

Figure 5.62 This is of type V-C if input is to gear 1′. (See Fig. 5.25.)

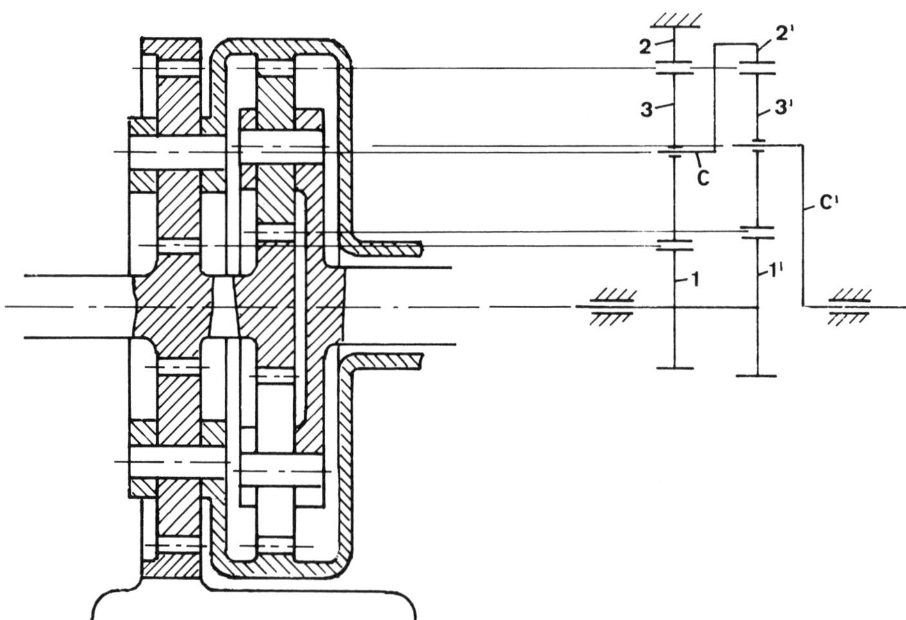

Figure 5.63 This is of type B-II. (See Fig. 5.25.)

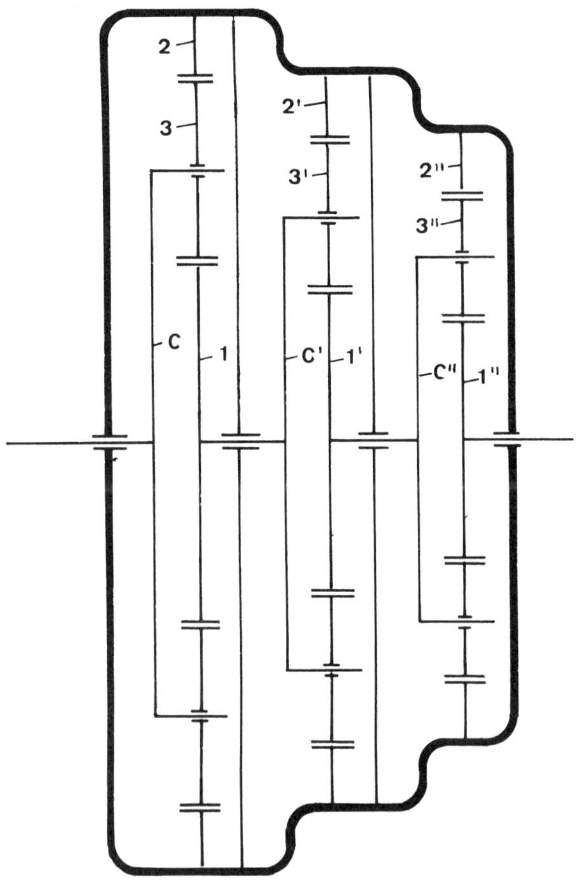

Figure 5.64 The first unit in this planetary gear drive is to type B, and because all three units are identical, the designation is B-B-B.

$$\frac{n_1''}{n_c} = \left(1 + \frac{z_2''}{z_1''}\right)\left(1 + \frac{z_2'}{z_1'}\right)\left(1 + \frac{z_2}{z_1}\right)$$

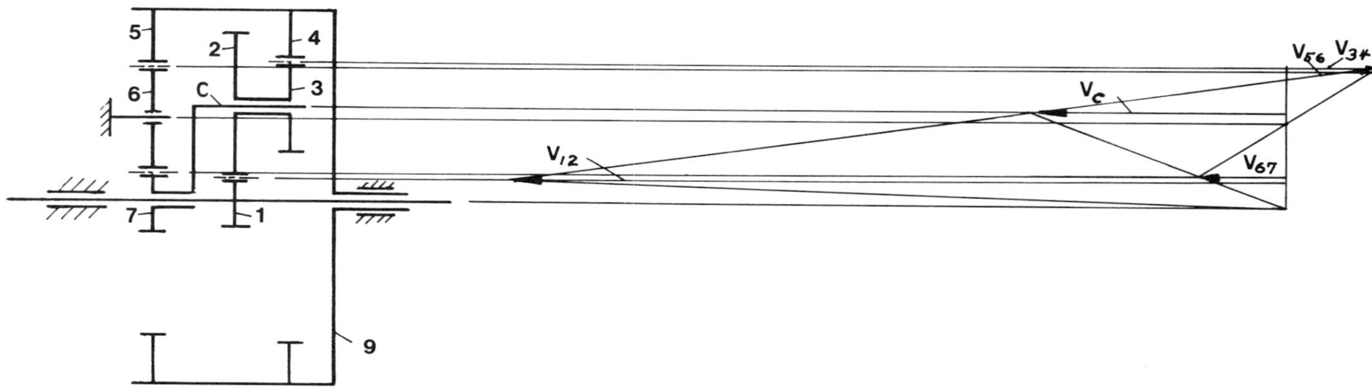

Figure 5.65 This system is a combination of a type a-4 system with members 1, 3, 3′, 2′, and C, where members C and 2′ are coupled to gears 4 and 5, respectively.

$$\frac{n_a}{n_1} = \frac{Z_1Z_3Z_7}{Z_2Z_4Z_5 - Z_2Z_4Z_7 + Z_1Z_3Z_5}$$

PLANETARY GEAR SYSTEMS WITH BEVEL GEARS

In the foregoing, planetary gear systems with speed gears were investigated. Planetary gears systems with bevel gears are employed extensively in gear differentials in automobiles but they also have many other uses.

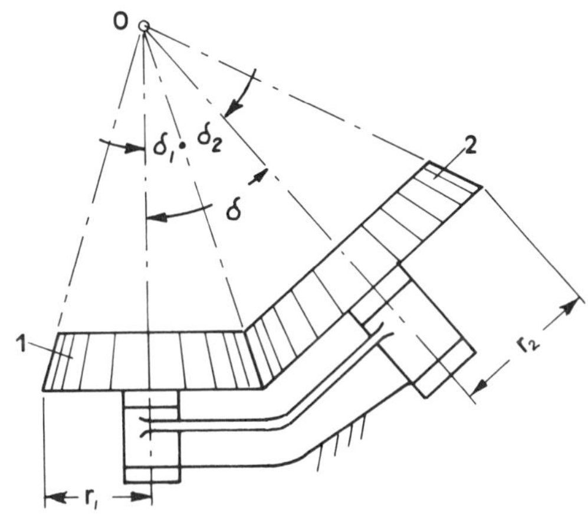

Figure 5.66 Two bevel gears 1 and 2 in mesh. The speed reduction ratio is

$$i_{1\cdot 2} = \frac{\omega_1}{\omega_2} = -\frac{r_2}{r_1} = -\frac{N_2}{N_1} = -\frac{\sin\delta_2}{\sin\delta_1}$$

The above formula is correct when using the number of teeth, but it must be remembered that $2\delta_1$ is determined by the pitch cone, where the base of the cone corresponds to the pitch circle.

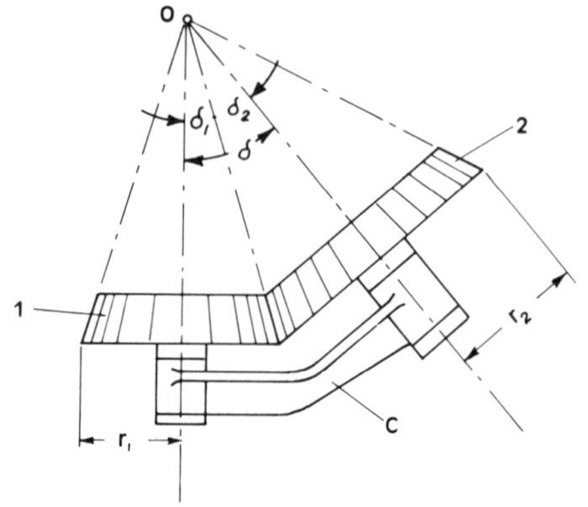

Figure 5.67 Two bevel gears 1 and 2 in mesh, but the axis of 1 remains stationary whereas the axis of 2 rotates around the axis of 1; in other words, C is the carrier of a simple planetary gear system.

$$i_{2C} = \frac{\omega_2}{\omega_C} = 1 + \frac{r_1}{r_2} = 1 + \frac{Z_1}{Z_2} = 1 + \frac{\sin\delta_1}{\sin\delta_2}$$

The same conditions hold for this formula as for the formula developed in Fig. 5.66.

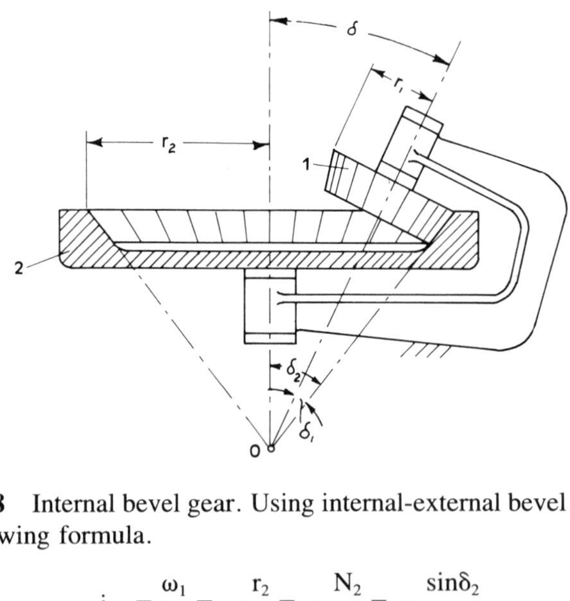

Figure 5.68 Internal bevel gear. Using internal-external bevel gears results in the following formula.

$$i_{i\cdot 2} = \frac{\omega_1}{\omega_2} = -\frac{r_2}{r_1} = -\frac{N_2}{N_1} = -\frac{\sin\delta_2}{\sin\delta_1}$$

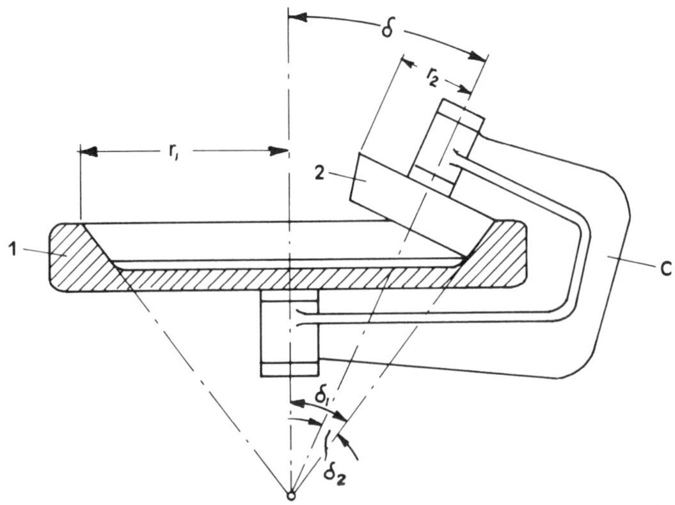

Figure 5.69 This bevel gear, planetary gear system has an internal bevel gear 1 in mesh with an external bevel gear 2. The speed reduction ratio is

$$i_{2C} = \frac{\omega_2}{\omega_C} = 1 - \frac{r_1}{r_2} = 1 - \frac{N_1}{N_2} = 1 - \frac{\sin\delta_1}{\sin\delta_2}$$

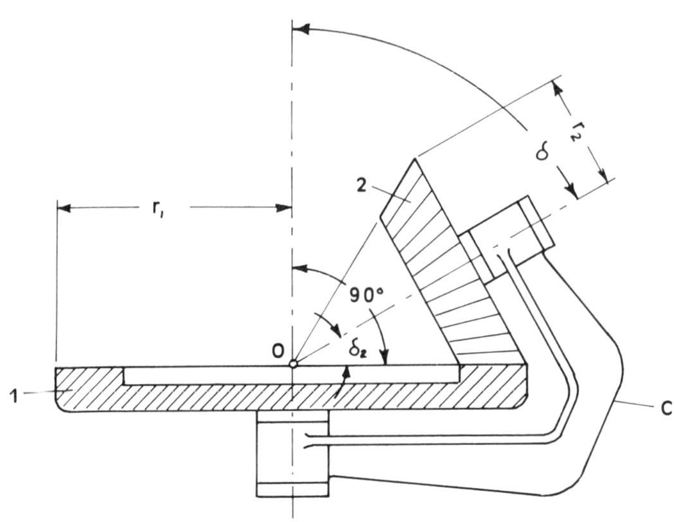

Figure 5.70 Making angle $\delta_1 = 90°$ in Fig. 5.77 results in the planetary gear system shown. The speed reduction ratio formula follows that of Fig. 5.69.

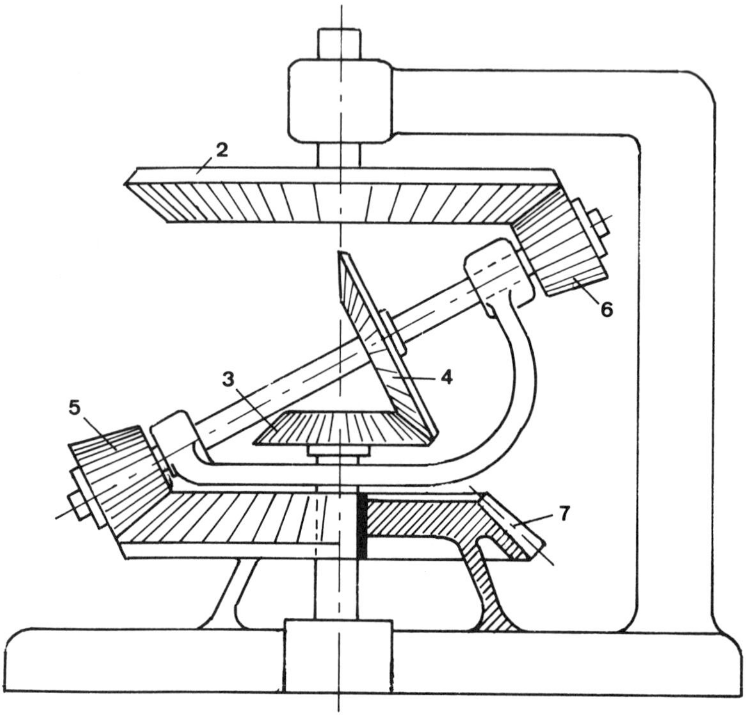

Figure 5.71 Input gear 1 meshes with gear 2, which is fixed to a shaft that carries two more gears, 3 and 5. Gear 3 meshes with 4, and gear 5 with 6, which is fixed to the stationary frame. The speed reduction ratio is

$$R = \frac{n_2}{n_1} = \frac{Z_3}{Z_2}\,\frac{Z_2 Z_5 + Z_6 Z_7}{Z_3 Z_5 - Z_4 Z_7}$$

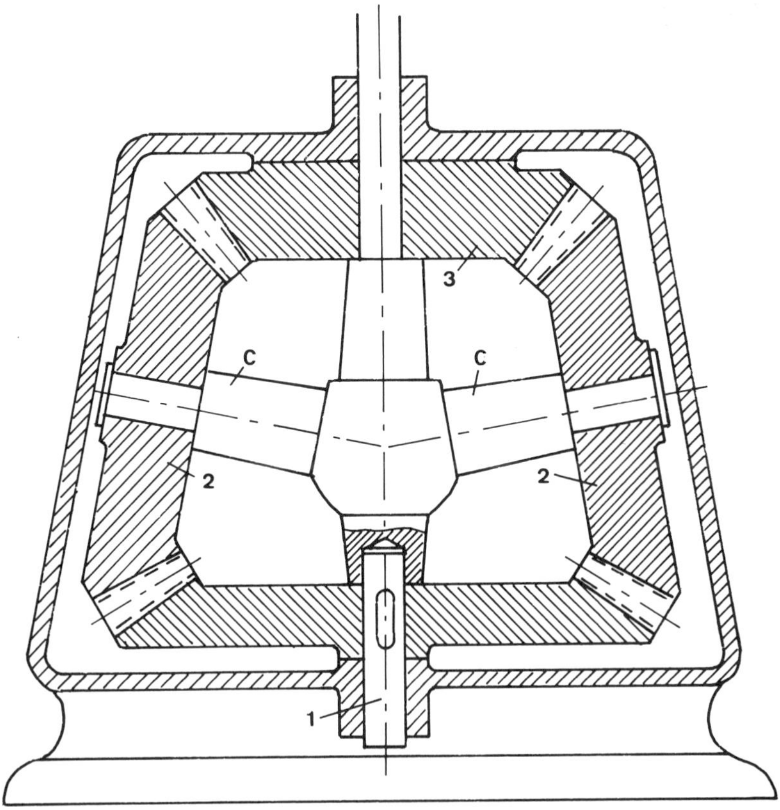

Figure 5.72 Gear differential. Input is to member 4. Carrier has two arms, C, on which the two gears, 2, are free to rotate. Gears 2 mesh with gear 3, which is fixed to the housing, and with gear 1, which is keyed to the output shaft. The reduction ratio is

$$\frac{n_c}{n_1} = \frac{Z_1}{Z_1 - Z_3}$$

CIRCULATING POWER

In a gear system where the energy flow follows a path from the input member to the output member, the flow may be divided into two flow branches, the sum of which is equal to the input (apart from friction losses), or within a system there may be created a circulating power that can be many times the input power.

Example of Circulating Power

A mixing unit is to be designed. The speed of the motor is 800 RPM and the mixing blades are to rotate at 2 RPM. The speed reduction ratio, therefore, is

$$R = \frac{n_{in}}{n_{out}} = 400 = 20 \times 20$$

A fixed gear train with a double reduction could be used. If the small gears have 18 teeth, then the large gears have 360 teeth, resulting in a rather large gear box. However, it is immaterial for mixing purposes whether the mixing container is stationary and the mixing blades rotate at 2 RPM or the container and the blades rotate as long as their relative angular velocity is 2 RPM. A unit that makes use of the latter solution is shown in Fig. 5.73, where the number of teeth for the four gears is $N_A = 27$, $N_B = 28$, $N_C = 30$, and $N_D = 29$. The input gear A turns at 800 RPM and drives the container at the same time. The mixing arms turn $800\ (-Z_A/Z_B)\ (-Z_C/Z_D) = 800\ 27/28\ 30/29 = 798.03$ RPM.

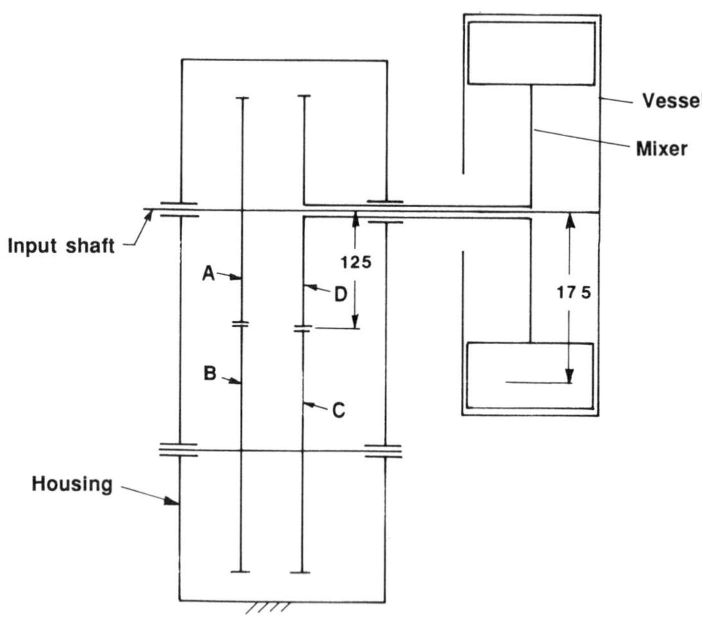

Figure 5.73 Planetary gear system with high circulating power.

For the unit that was actually designed in the above example, it was assumed that the horsepower of the system would be 1 HP, and the dimensioning of it was done accordingly, but after a short while the unit broke down. A new and stronger unit was built which lasted longer but broke down again. It was noticed that the unit generated much heat. The explanation follows.

It is assumed that the active radius of the blades is 175 mm. With a relative speed between the container and the blades of 2 RPM, the relative circumferential speed between the blades and the container is

$$V_R = 2\frac{2\pi}{60} \times \frac{175}{1000} = 0.0367 \text{ m/s}$$

and the force in the blades is

$$F_R = \frac{75}{0.0367} = 2050 \text{ kp}$$

where 1 kp $\approx$ 2.2 lb force. The force between the gears A and B and between C and D is

$$F_t = \frac{175}{125} \times F_R = 2861 \text{ kp}$$

The pitch velocity between A-B and C-D is

$$\approx 800\frac{2\pi}{60} \times \frac{125}{1000} = 10.47 \text{ m/s}$$

The rolling power to be transmitted between the gears is a remarkable

$$\frac{2861 \times 10.47}{75} = 400 \text{ HP}$$

If it is assumed that the friction loss between each pair of gears in mesh is 1%, then the total loss is $400 \times 2 \times 0.01 = 8$ HP of the circulating power that is converted to heat and has to be dissipated by the gear box itself.

The efficiency of the unit is approximately

$$\frac{1}{1 + 8} = 11\%$$

The unit has to be dimensioned for 400 HP. A new design based on the above data was built and operated satisfactorily.

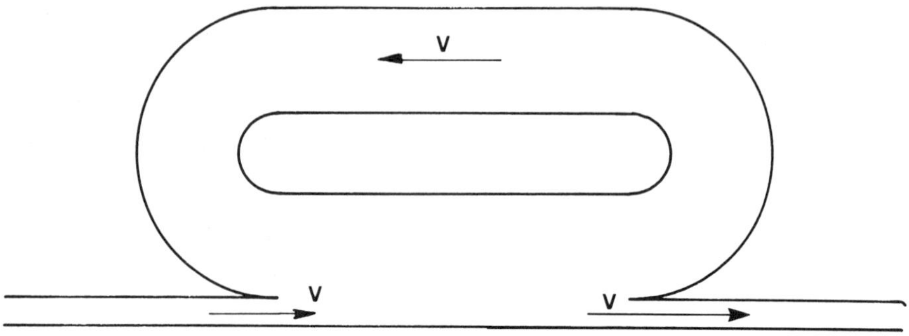

Figure 5.74 Illustration of circulating power.

An Explanation of the Explanation

As an illustration of the above phenomenon, visualize a lot through which a small brook flows. The owner is of the type who could sell the Brooklyn Bridge to a New Yorker. He wants to sell his lot and he could get more money if the brook were capble of driving a saw mill. However, measurements of the water flow reveal that the brook has a power of only 1 HP. This value, he has been told, is proportional to the velocity of the water and the cross-sectional area of the brook.

So, after a few days, he has figured it out. He builds a circuit, as shown in Fig. 5.74, where the velocity of the water in the branch is the same as the velocity of the brook but the cross-sectional area in the branch is 400 times greater. It does take some time before the small brook gets all the water circulating, but finally it does.

When he invites a prospective buyer to measure the circulating power in the branch, the horsepower measured is now 400 times greater than before because the cross-sectional area of the branch is 400 times that of the brook and the velocity is the same for both.

The lot is sold and the buyer builds a saw mill, which, when finally activated, stops immediately.

SYMBOLS

M_{r1}; M'_{r1}; M''_{r1}	moments acting on the sun gear; first, second and third unit, respectively
M_{r2}; M'_{r2}; M''_{r2}	moments acting on the ring gear; first, second and third unit, respectively
M_c; M'_c; M''_c	moments acting on planet carrier; first, second and third unit, respectively

M_{in}; M_{out}	input and output moment, respectively
C; C'; C''	planet carrier; first, second and third unit, respectively
1; $1'$; $1''$	sun gear; first, second and third unit, respectively
2; $2'$; $2''$	ring gear; first, second and third unit, respectively
3; $3'$; $3''$	planet gear; first, second and third unit, respectively
D	maximum diameter of ring gear
r_{min}	minimum radius of smallest gear
r_1; r_1'; r_1''	radius of sun gear; first, second and third unit, respectively
r_2; r_2'; r_2''	radius of ring gear; first, second and third unit, respectively
r_3; r_3'; r_3''	radius of planet gear; first, second and third unit, respectively
m; m'; m''	ratio between the radii of the ring gear and the sun gear; first, second, and third unit, respectively
η; η'; η''	efficiency of first, second and third unit; the planet carrier considered stationary
η_p	overall efficiency of planetary gear train
ω_{in}; ω_{out}	angular velocity of input and output shaft, respectively s^{-1}
n_{in}; n_{out}	angular speed of input and output shaft, respectively, RPM
Z_1; Z_2; Z_1'; Z_2''	number of teeth of the various members
$R = \omega_{in}/\omega_{out} = n_{in}/n_{out}$	transmission ratio

6

Cycloidal Mechanisms

CLASSIFICATION

Cycloidal mechanisms can easely be tailored to provide one of three common motion requirements: (1) *intermittent motion*, with either short or long dwells; (2) *rotary motion with progressive oscillations*, where the output undergoes a cycloidal motion during which the forward motion is greater than the return motion; or (3) *rotary-to-linear motion* with a dwell period. All cycloidal mechanisms are geared. This results in compact positive devices capable of operating at relatively high speeds with little backlash.

This type of mechanism can also be classified into three groups: (1) *epicycloid*, where the tracing points are on an external gear which rolls on an external (stationary) gear; (2) *hypocycloid*, where the points tracing the cycloidal curves are located on an external gear rolling inside a fixed internal gear; and (3) *pericycloid*, where the tracing points are located on an internal gear which rolls on an external (stationary) gear.

APPLICATIONS

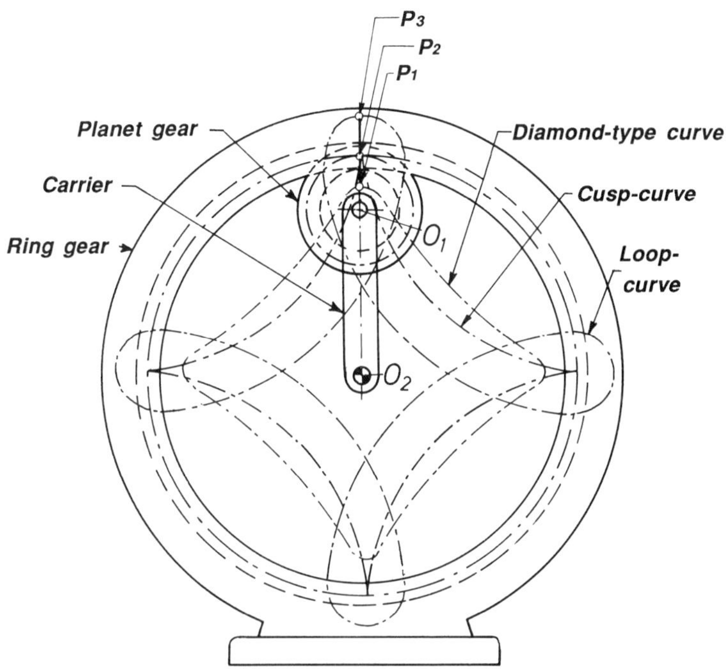

Figure 6.1 Basic hypocycloid curves. Input crank or carrier drives a planet gear in mesh with a stationary ring gear. Point P_1 on the planet gear traces a diamond-shaped curve, point P_2 on the pitch line of the planet traces the familiar cusp curve, and point P_3, which is on the extension rod fixed to the planet gear, traces a loop-type curve. In one known application, an end miller located at P_1 is employed in production for machining a diamond-shaped profile.

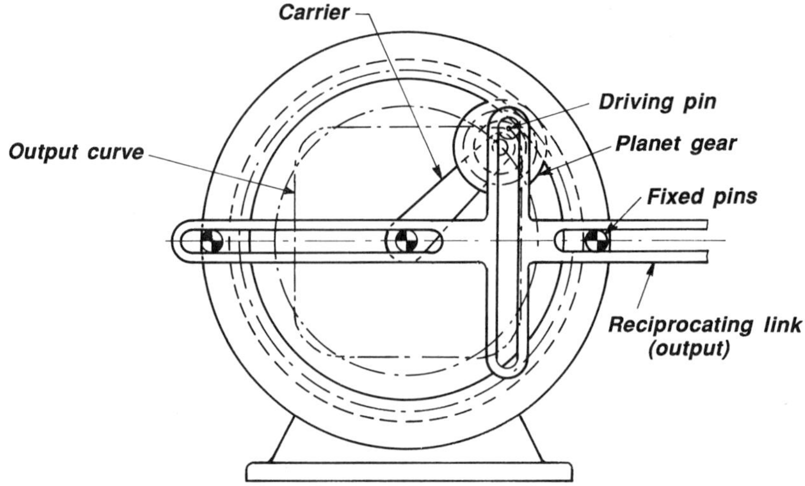

Figure 6.2 Double-dwell mechanism. Coupling the output pin to a slotted member produces a prolonged dwell in each of the extreme positions. This is another application of the diamond-type hypocycloidal curve.

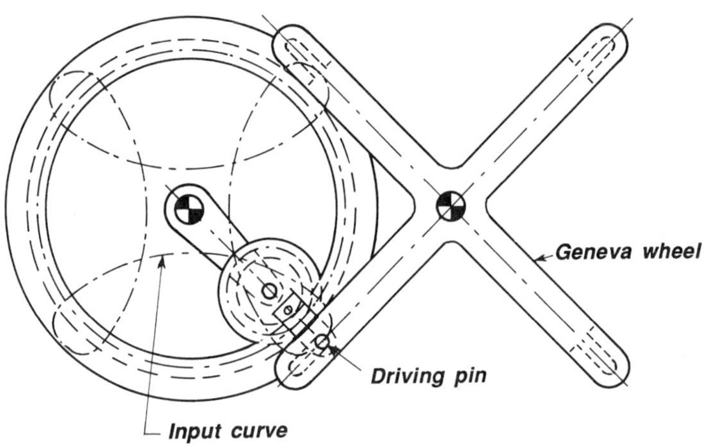

Figure 6.3 Long-dwell Geneva drive. As with standard 4-station Geneva wheels, each rotation of the input indexes the Geneva wheel 90°. By employing a pin fastened to the planet gear to obtain a rectangular-shaped cycloidal curve, a smoother indexing motion is obtained.

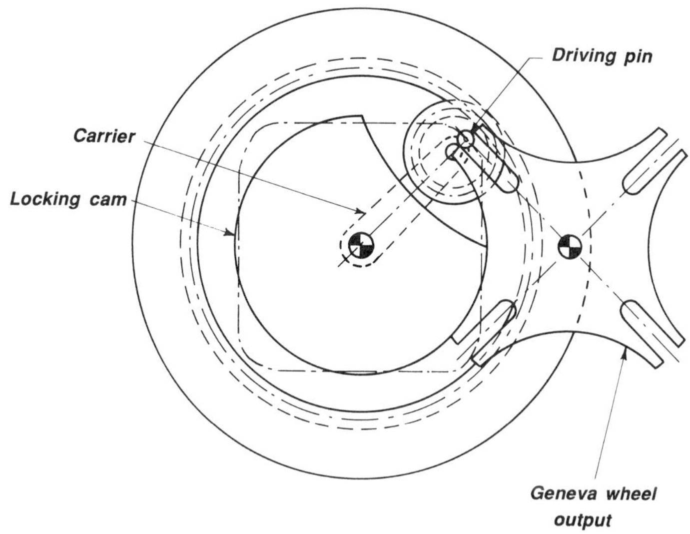

Figure 6.4 External Geneva drive. Loop-type curve permits driving pin to enter slot in a direction that is radially outward from the center, and then loop over to rapidly index the Geneva wheel. As with the design in Fig. 6.3, the output rotates 90°, then goes into a dwell period during each 270° rotation of the input.

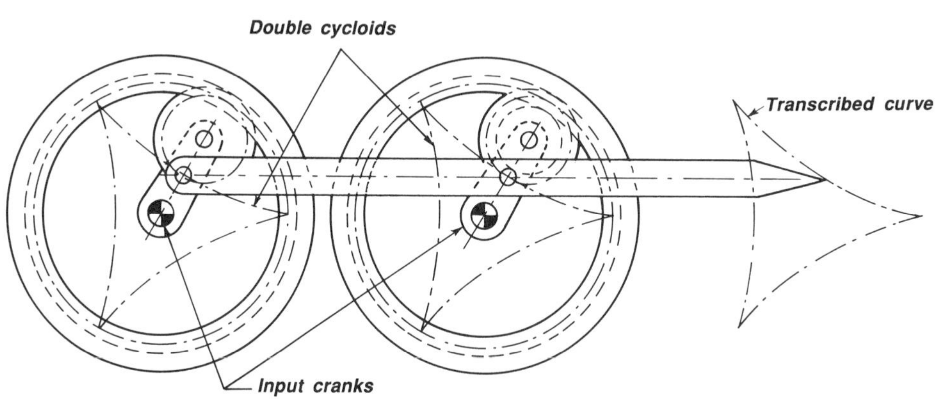

Figure 6.5 Cycloidal parallelogram. Two identical hypocycloid mechanisms guide the point of the bar along the triangular-shaped path. Mechanisms of this sort are useful in cases where there is limited space in the area where the curve must be traced. Such double-cycloidal mechanisms can be designed to produce other types of curves.

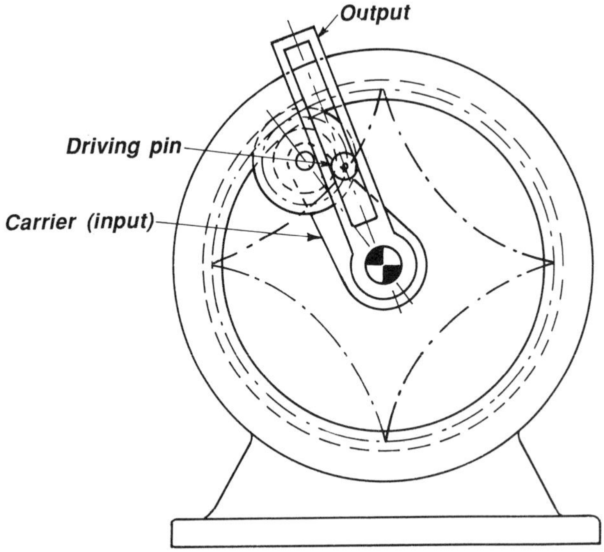

Figure 6.6 Short-dwell rotary. The pitch circle of the planet gear is exactly one-quarter that of the ring gear. A pin on the pitch circle of the planet gear will cause the slotted output member to have four instantaneous dwells for each revolution of the input shaft.

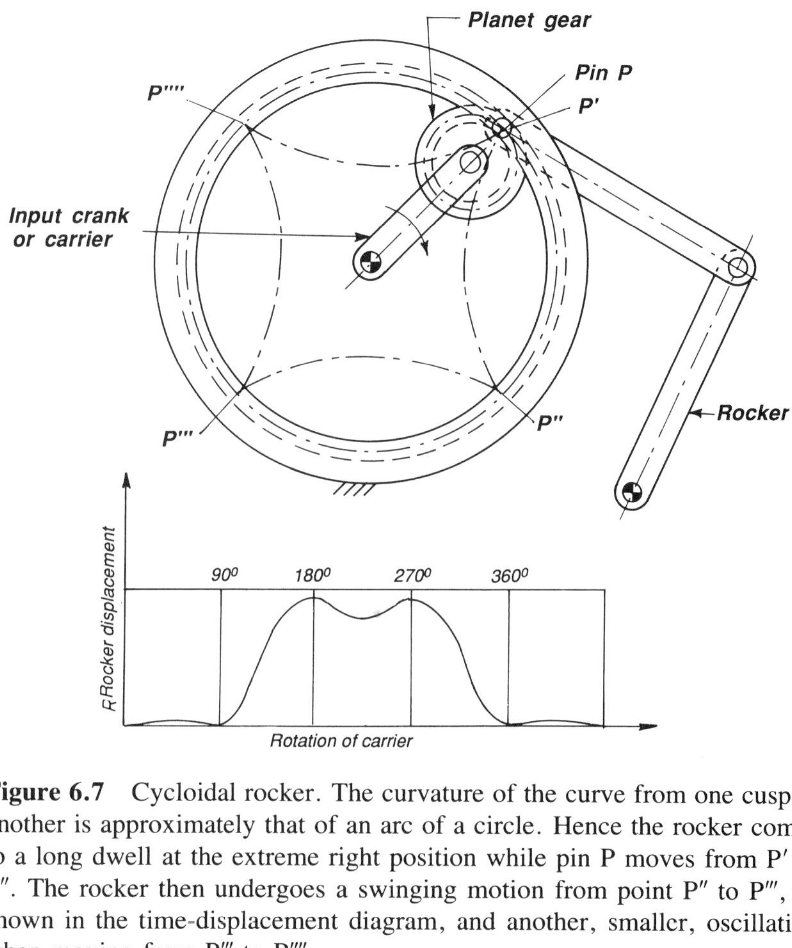

Figure 6.7 Cycloidal rocker. The curvature of the curve from one cusp to another is approximately that of an arc of a circle. Hence the rocker comes to a long dwell at the extreme right position while pin P moves from P′ to P″. The rocker then undergoes a swinging motion from point P″ to P‴, as shown in the time-displacement diagram, and another, smaller, oscillation when moving from P‴ to P⁗.

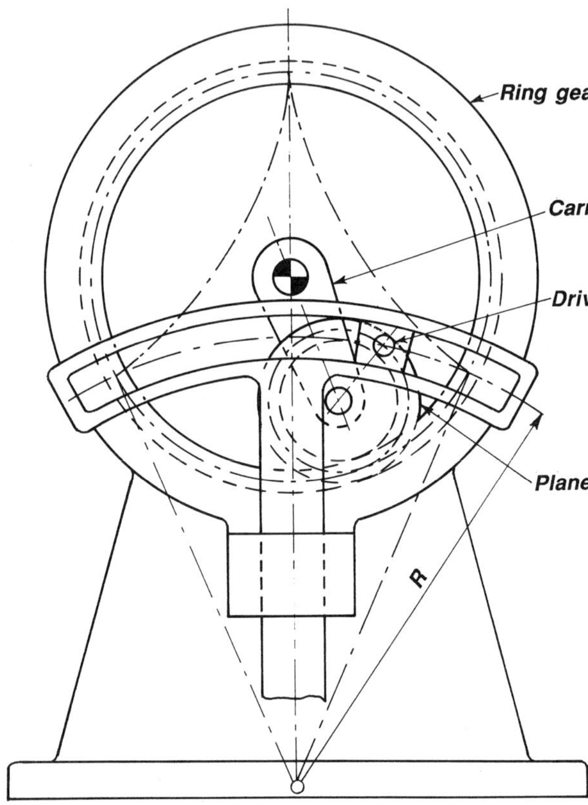

Figure 6.8 Modified Scotch yoke cycloidal mechanism. The curved slot of the Scotch yoke causes a prolonged, almost exact, dwell of the Scotch yoke for 120° rotation of the input shaft.

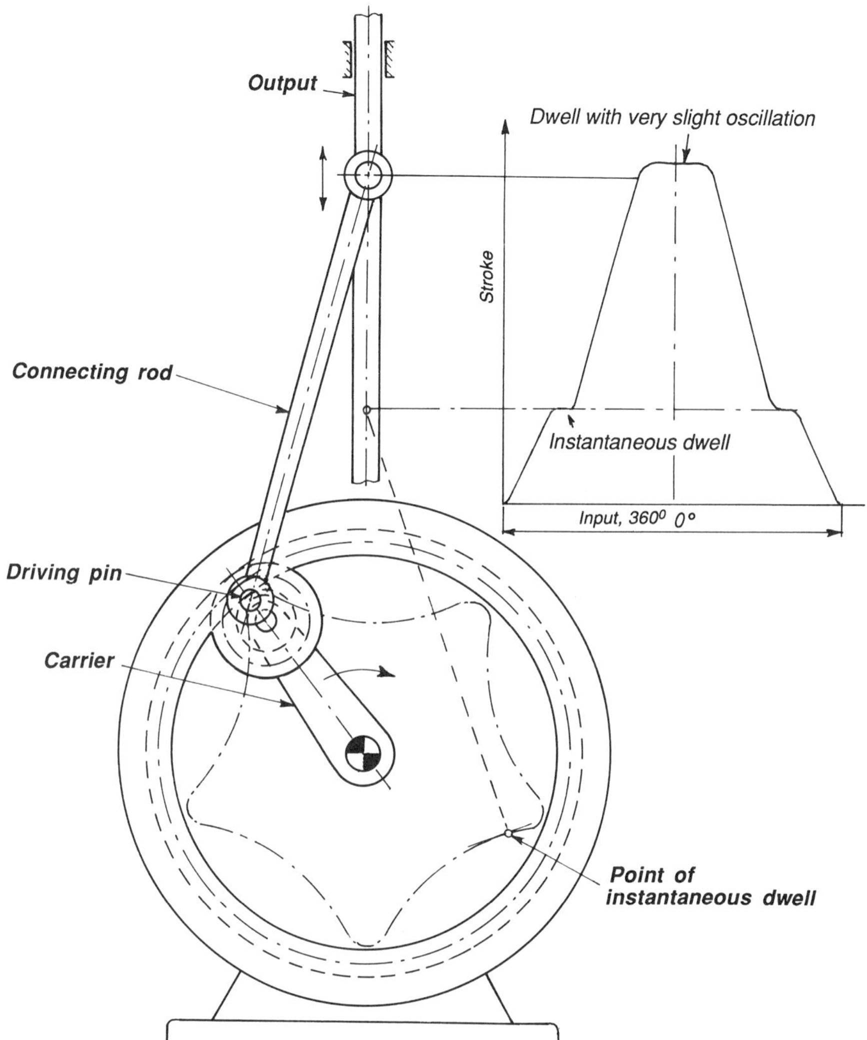

Figure 6.9 Cycloidal reciprocator. Portion of the curve traced by the driving pin produces a prolonged dwell (as in the previous mechanism). There are also two points of instantaneous dwell where the curve is perpendicular to the connecting rod, and a dwell at the lower extreme position.

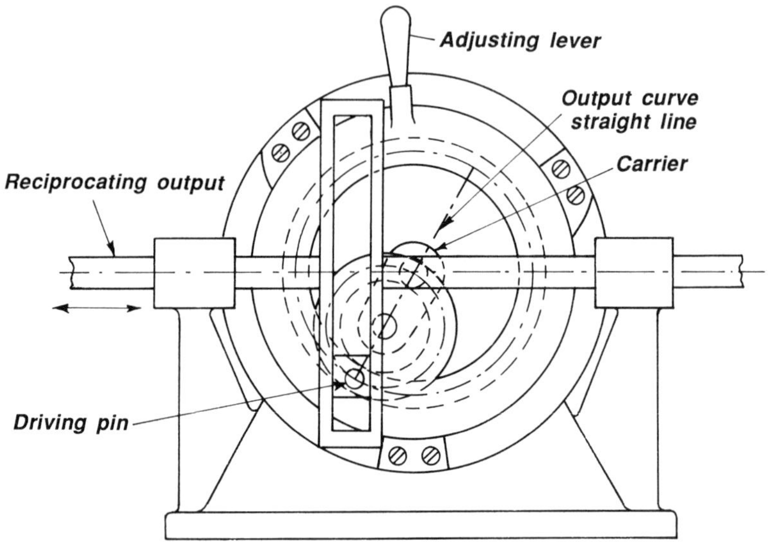

Figure 6.10 Adjustable Scotch yoke cycloidal mechanism. By making the size of the planet gear half that of the internal gear, a straight-line output curve is traced by the driving pin, which is fastened to the planet gear. The pin engages the Scotch yoke to cause the output to reciprocate back and forth with harmonic (sinusoidal) motion. The position of the fixed ring gear can be changed by adjusting the lever, which in turn rotates the straight-line output curve. When the curve is horizontal, the stroke is at a maximum; when the curve is vertical, the stroke is zero.

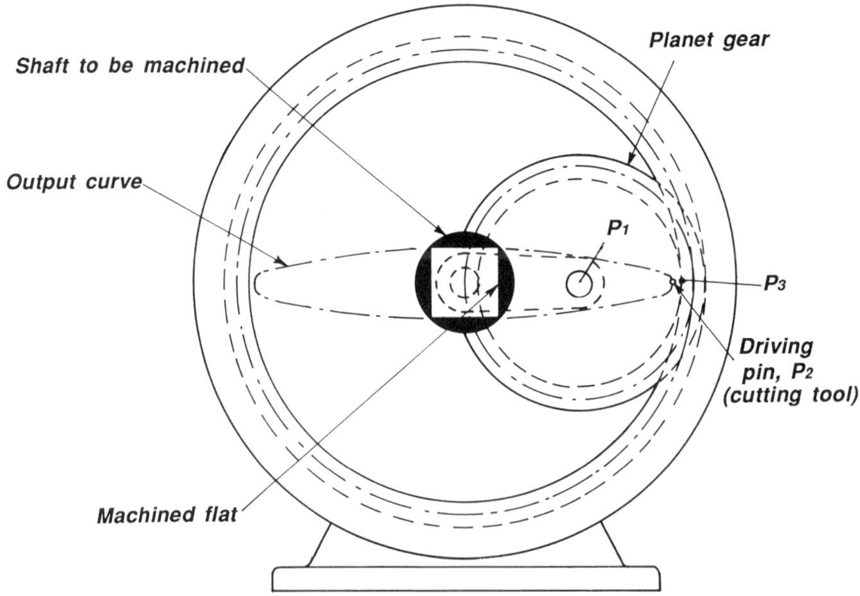

Figure 6.11 Elliptical-motion drive. When the pitch diameter of the planet gear is made equal to half that of the ring gear, every point on the planet gear, such as point P_2, will describe elliptical curves which get flatter as the points are selected closer to the pitch circle. Point P_1, at the center of the planet, traces a circle. Point P_3, at the pitch circle, traces a straight line. When a cutting tool is placed at P_2, it will cut almost-flat sections from round stock, as when machining a square-headed bolt. The other two sides of the bolt can be cut by rotating the bolt, or the cutting device, 90°.

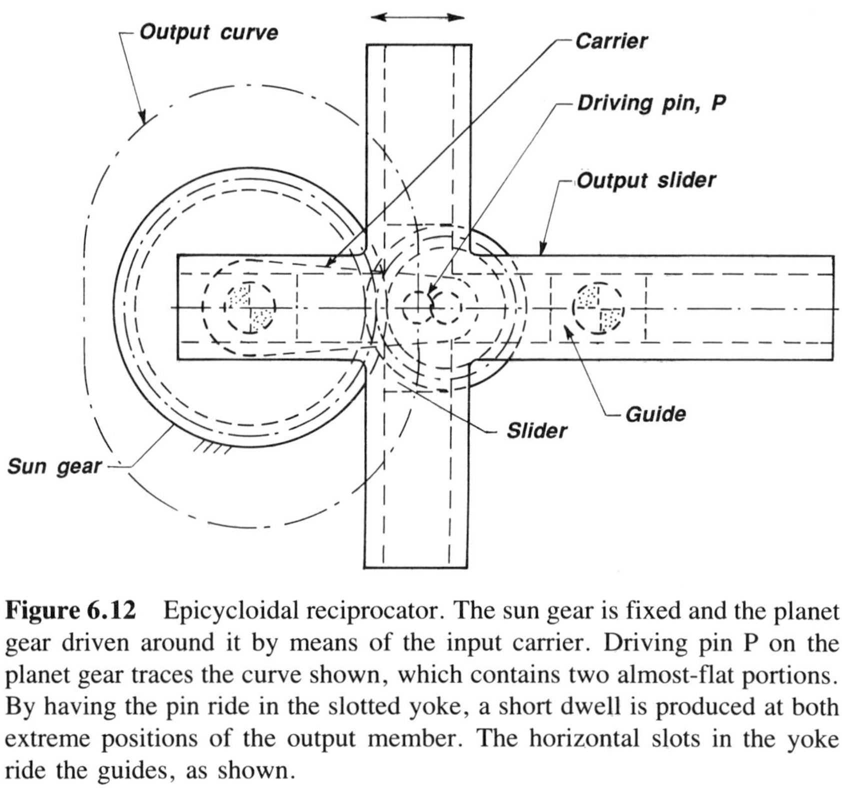

Figure 6.12 Epicycloidal reciprocator. The sun gear is fixed and the planet gear driven around it by means of the input carrier. Driving pin P on the planet gear traces the curve shown, which contains two almost-flat portions. By having the pin ride in the slotted yoke, a short dwell is produced at both extreme positions of the output member. The horizontal slots in the yoke ride the guides, as shown.

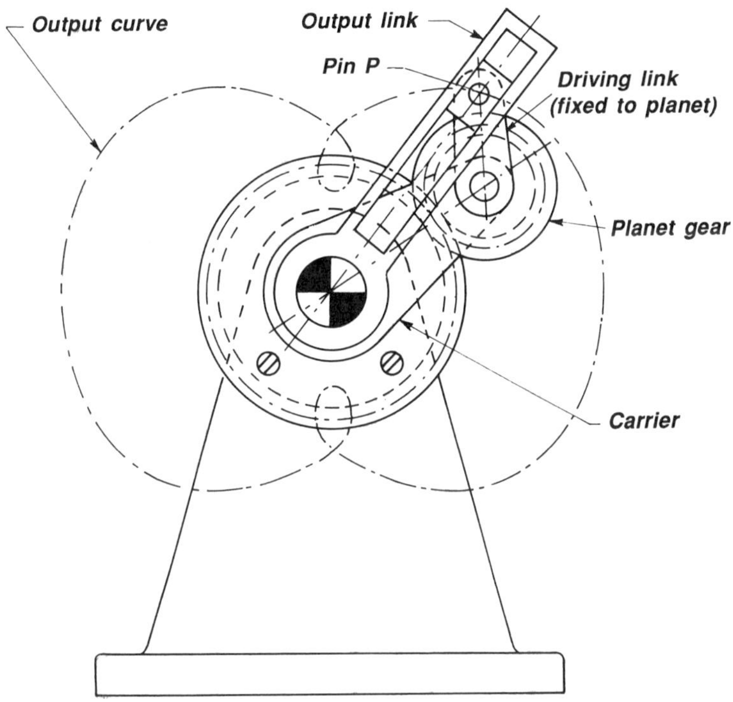

Figure 6.13 Progressive oscillating drive. By fixing a crank to the planet gear, pin P can be made to trace the looped curve illustrated. The slotted output crank oscillates briefly at the vertical looped portions of the curve.

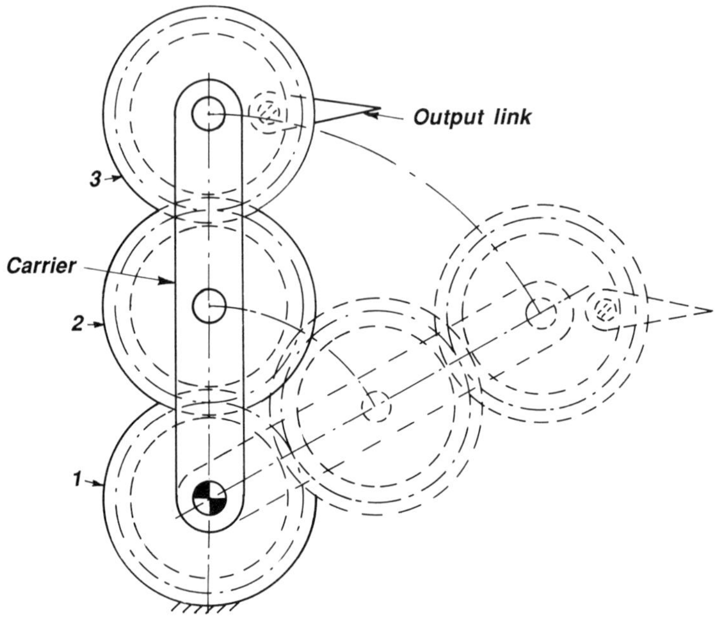

Figure 6.14 Parallel guidance mechanism. The input crank or carrier contains two planet gears. The center sun gear is fixed as in the epicycloid mechanism in Fig. 6.12. By making the three gears equal in diameter and having gear 2 serve as an idler, any member fixed to gear 3 will remain parallel to its previous positions throughout the rotation of the input crank.

Motion equations for an epicycloid drive (Fig. 6.15) are derived based on the designations shown:*
Angular displacement:

$$\tan\beta = \frac{(R + r)\sin\theta - b\sin(\theta + \gamma)}{(R + r)\cos\theta - b\cos(\theta + \gamma)} \tag{6.1}$$

Angular velocity:

$$\omega_0 = \omega_i \frac{1 + \dfrac{b^2}{r(R + r)} - \left(\dfrac{2r + R}{r}\right)\left(\dfrac{b}{R + r}\right)\left(\cos\dfrac{R}{r}\theta\right)}{1 + \left(\dfrac{b}{R + r}\right)^2 - \left(\dfrac{2b}{R + r}\right)\left(\cos\dfrac{R}{r}\theta\right)} \tag{6.2}$$

---

*E. H. Schmidt, Cycloidal Cranks, *Transactions of the 5th Conference on Mechanisms*, 1958, pp. 164–180.

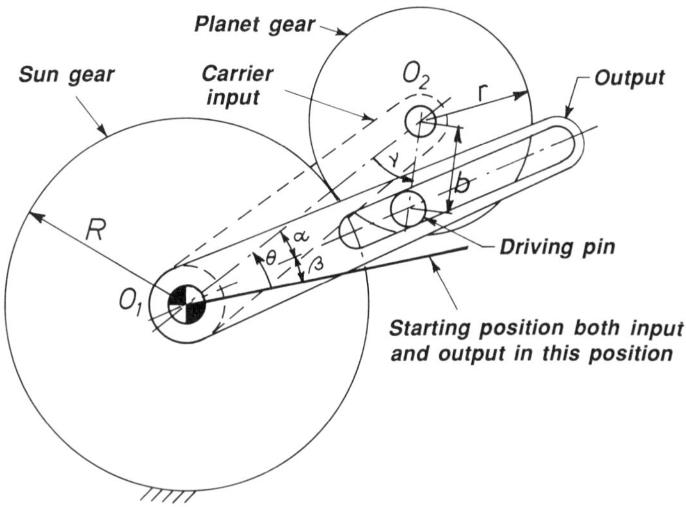

Figure 6.15 Symbols for deriving equations for epicycloidal drives.

Angular acceleration:

$$\alpha = \omega_i^2 \frac{\left(1 - \frac{b^2}{(R+r)^2}\right)\left(\frac{R^2}{r^2}\right)\left(\frac{b}{R+r}\right)\left(\sin\frac{R}{r}\theta\right)}{\left[1 + \frac{b^2}{(R+r)^2} - \left(\frac{2b}{R+r}\right)\left(\cos\frac{R}{r}\theta\right)\right]^2} \tag{6.3}$$

Motion equations for a hypocycloid drive (Fig. 6.16) are derived based on the designations shown:

$$\tan\beta = \frac{\sin\theta - \left(\frac{b}{R-r}\right)\left(\sin\frac{R-r}{r}\theta\right)}{\cos\theta + \left(\frac{b}{R-r}\right)\left(\cos\frac{R-r}{r}\theta\right)} \tag{6.4}$$

$$\omega_0 = \omega_i \frac{1 - \left(\frac{R-r}{r}\right)\left(\frac{b^2}{(R+r)^2}\right) + \left(\frac{2r-R}{r}\right)\left(\cos\frac{R}{r}\theta\right)}{1 + \frac{b^2}{(R+r)^2} + \left(\frac{2b}{R+r}\right)\left(\cos\frac{R}{r}\theta\right)} \tag{6.5}$$

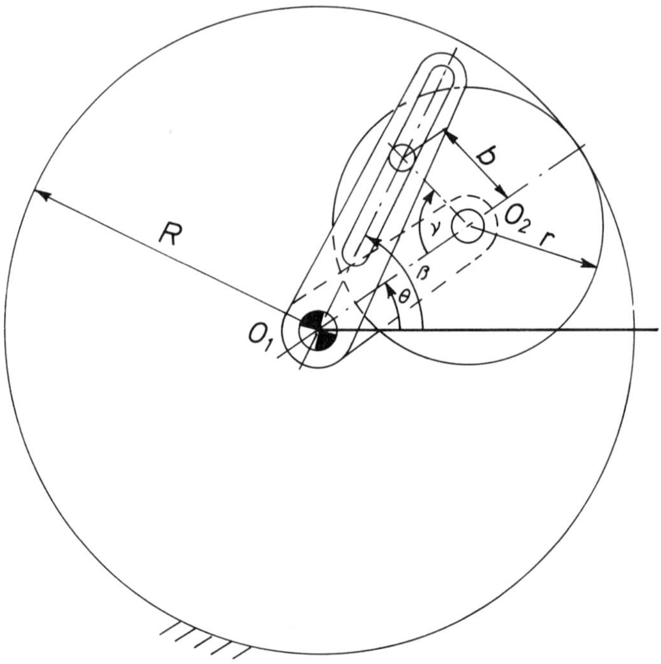

Figure 6.16 Symbols for deriving equations for hypocycloidal drives.

$$\alpha = \omega_i^2 \frac{\left(1 - \dfrac{b^2}{(R + r)^2}\right)\left(\dfrac{b}{R + r}\right)\left(\dfrac{R^2}{r^2}\right)\left(\sin \dfrac{R}{r} \theta\right)}{\left[1 + \dfrac{b^2}{(R + r)^2} + \left(\dfrac{2b}{R + r}\right)\left(\cos \dfrac{R}{r} \theta\right)\right]^2} \tag{6.6}$$

Figure 6.17 shows a hypocycloidal mechanism with curved slot for a prolonged dwell of the output member. Part of the slot is made identical to part of the curve traced by the driving roller on the planet gear.

GENERATING CYCLOIDAL CURVES CONTAINING APPROXIMATE STRAIGHT LINES

It is frequently desirable to find points on the planet gear that will trace approximately straight lines for portions of the output curve. Points lying on the so-called inflection circle have their center of curvature at infinity. Such points will yield dwell mechanisms as shown in the following.

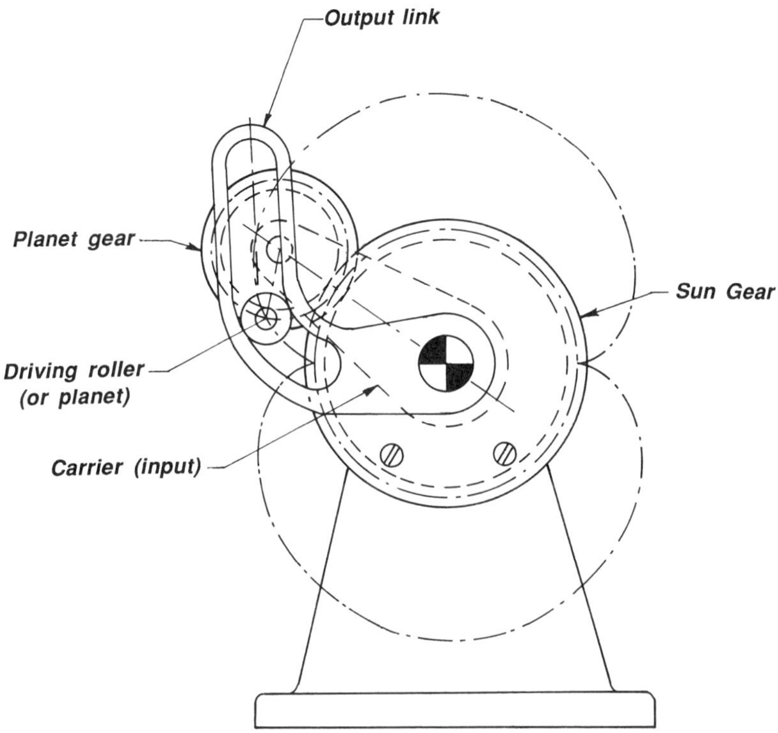

Figure 6.17 A prolonged-dwell epicycloidal drive.

Figure 6.18 shows how to find the inflection circle for an external gear rolling on another external gear:

1. Draw arbitrary line from the pitch point P subtending the angle ψ with the common tangent.
2. Draw a parallel through O_2.
3. Draw a perpendicular from P. Point of intersection is A.
4. Draw line AO_1 through the center of the rolling gear. This locates W_1.
5. Draw a perpendicular to PW_1 at W_1. Obtain W on O_1O_2.

PW is the diameter of the inflection circle. Point W will trace part of an approximate straight line when the planet gear rolls on the sun gear.

Figure 6.19 shows how to find the inflection circle for a gear rolling on a rack. Note that the diameter of the inflection circle is PO_1. All points on

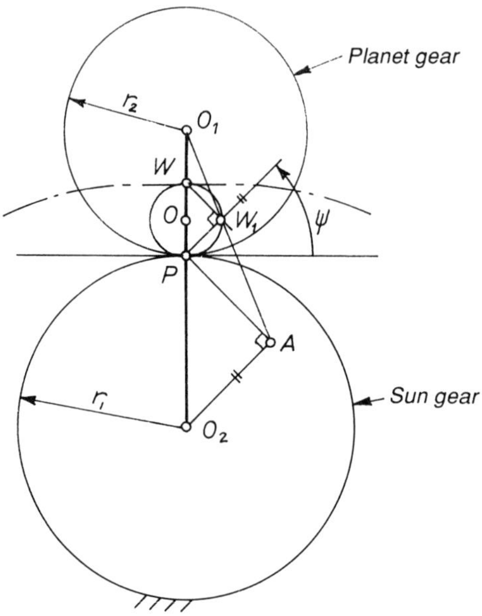

Figure 6.18 Gear rolling on gear. Flat portion of curve.

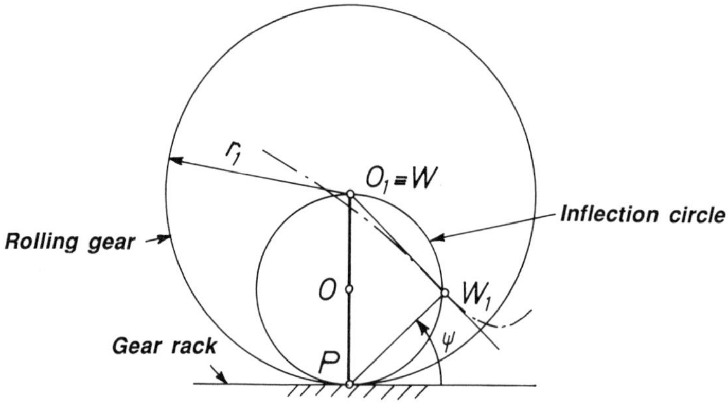

Figure 6.19 Gear rolling on rack: vee curves.

the inflection circle trace curves that are approximate straight lines for an interval of time, as shown by the curve traced by W_1.

Figure 6.20 shows how to find the inflection circle for a gear rolling inside an internal, fixed gear:

1. Draw arbitrary line from the pitch point P subtending the angle ψ with the common tangent.
2. Draw a parallel through O_2.
3. Draw a perpendicular from P. Point of intersection is A.
4. Draw line AO_1 through the center of the rolling gear. This locates W_1.
5. Draw a perpendicular to PW_1 at W_1. Obtain W on O_1O_2.

Point W_1, which is an arbitrary point on the circle, will trace a curve of repeated almost straight lines.

FINDING RADIUS OF CURVATURE OF CURVES TRACED BY POINTS ON A ROLLING GEAR (BOBILLIER'S CONSTRUCTION)

Figure 6.21 shows the case of a gear rolling on another gear. The geometrical method for locating the center of curvature C_0 for the path traced by C is given by these steps:

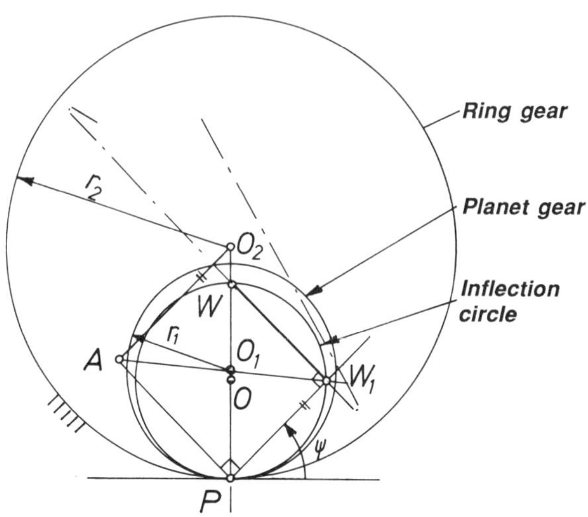

Figure 6.20 Gear rolling inside gear: zig-zag curves.

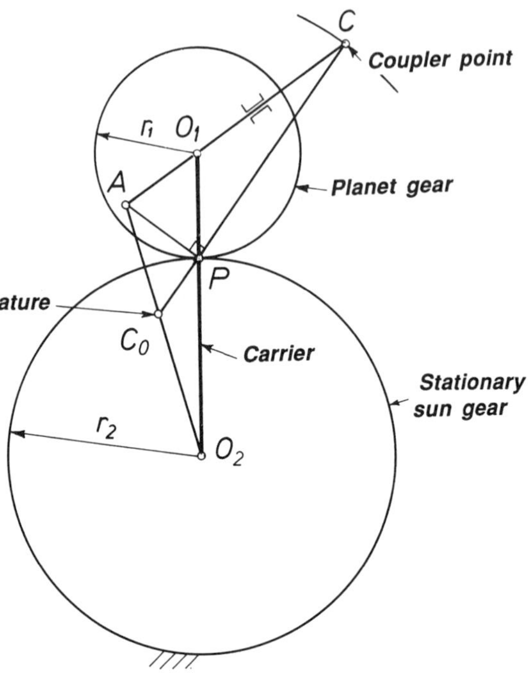

Figure 6.21 Finding center of curvature for epicycloidal motion.

1. Draw a line through points C and P.
2. Draw a line through points C and O_1.
3. Draw a perpendicular to CP at P. This locates point A.
4. Draw line AO_2, to locate C_0, the center of curvature.

If a gear rolls on a rack (Fig. 6.22), the construction is similar to that of the previous case.

1. Draw an extensions of line CP.
2. Draw a perpendicular at P to locate A.
3. Draw a perpendicular from A to the straight surface to locate C_0, the center of curvature.

To find radius of curvature of a point on a gear rolling inside another gear (Fig. 6.23):

1. Draw extensions of CP and CO_1.
2. Draw a perpendicular to PC at P to locate A.
3. Draw AO_2 to locate C_0, the center of curvature.

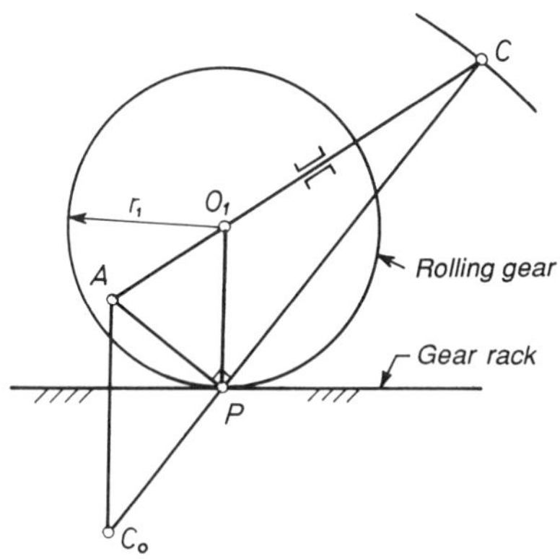

Figure 6.22 Finding center of curvature where gear rolls on rack.

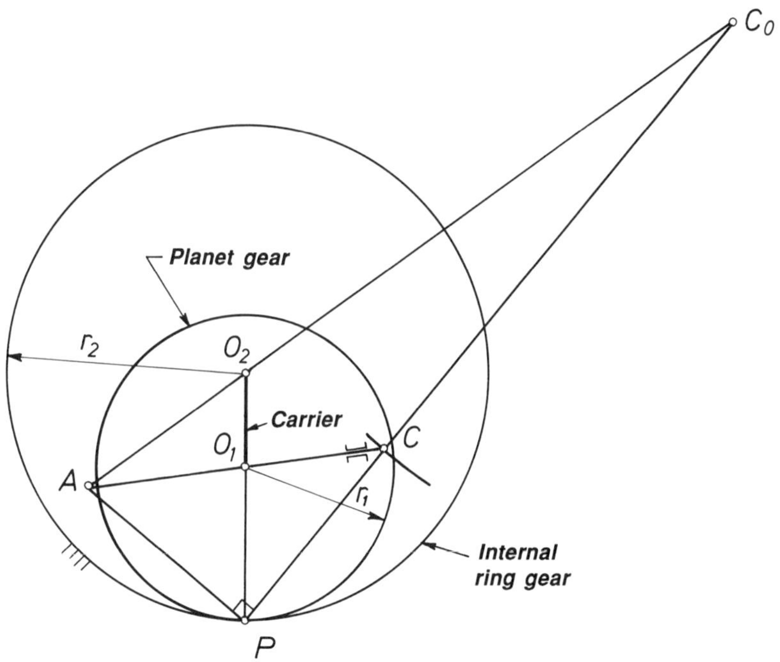

Figure 6.23 Finding center of curvature for hypocycloidal motion.

By locating the centers of curvature at various points, one can determine the proper length of the rocking or reciprocating arm to provide long dwells or proper entry conditions.

In Fig. 6.24 the center of curvature of a gear rolling on an external gear can be computed directly from the Euler-Savary equation:

$$\left(\frac{1}{r} - \frac{1}{r_c}\right) \sin\psi = \text{constant} \tag{6.7}$$

By appling this equation twice, specifically to points O_1 and O_2, where O_2 rotates on a circle around O_1, the following equation is obtained:

$$\left(\frac{1}{r_2} + \frac{1}{r_1}\right) \sin 90^\circ = \left(\frac{1}{r} + \frac{1}{r_c}\right) \sin\psi$$

or

$$\frac{1}{r_2} + \frac{1}{r_1} = \left(\frac{1}{r} + \frac{1}{r_c}\right) \sin\psi \tag{6.8}$$

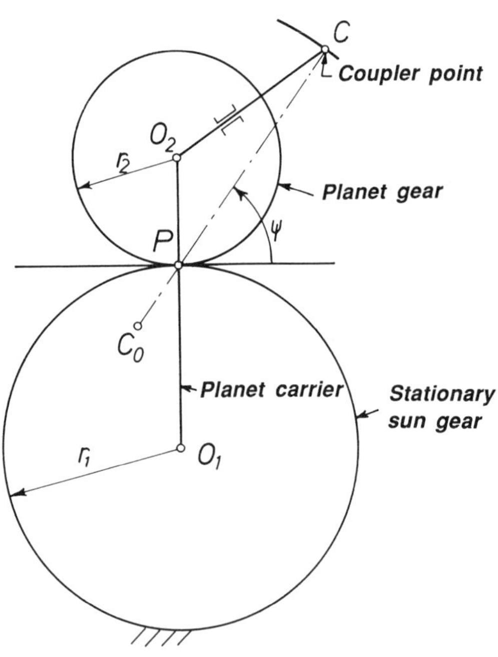

Figure 6.24 Finding center of curvature: analytical solution.

This is the final design equation. All factors except r_c are known; hence, solving for r_c leads to the location of C_0.

For a gear rolling inside an external gear, the Euler-Savary equation is

$$\left(\frac{1}{r} + \frac{1}{r_c}\right) \sin\psi = \text{constant}$$

which leads to

$$\frac{1}{r_2} - \frac{1}{r_1} = \left(\frac{1}{r} - \frac{1}{r_c}\right) \sin\psi$$

COGNATE CYCLOIDS

It is not always realized that cycloidal mechanisms can be replaced by other types of cycloidal mechanisms that produce exactly the same motion and yet are more compact.

Figure 6.25 shows a typical hypocycloidal mechanism. Gear 1 rolls inside gear 2 while point C traces a hypocycloidal curve. To find the substitute mechanism, draw parallels O_3O_2 and O_3C to locate point P_2. Then select O_2P_2 as the new radius of the large (internal) gear. Line P_2O_3 becomes the radius of the small gear. Point C has the same relative position and can be obtained by completing the triangles. The new mechanism is about two-thirds the size of the original.

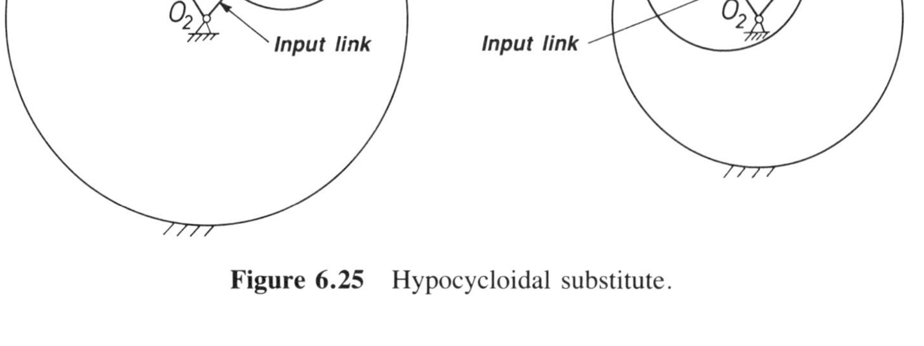

Figure 6.25 Hypocycloidal substitute.

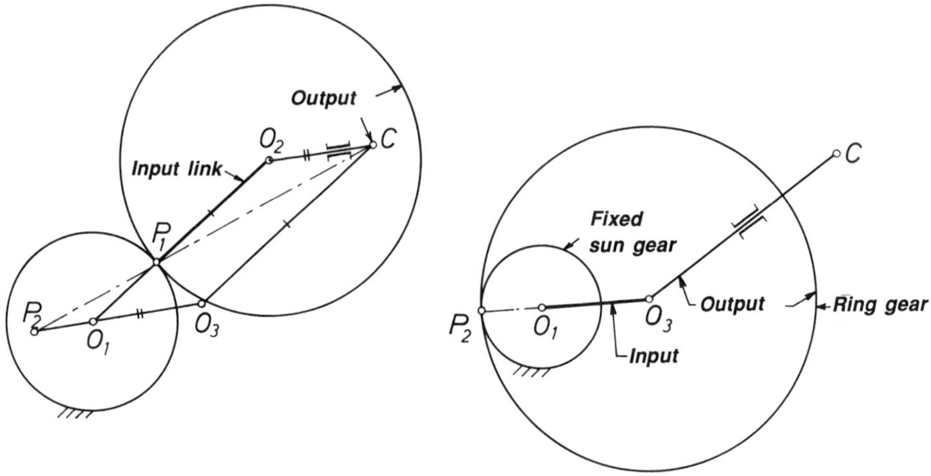

Figure 6.26 Epicycloidal substitute.

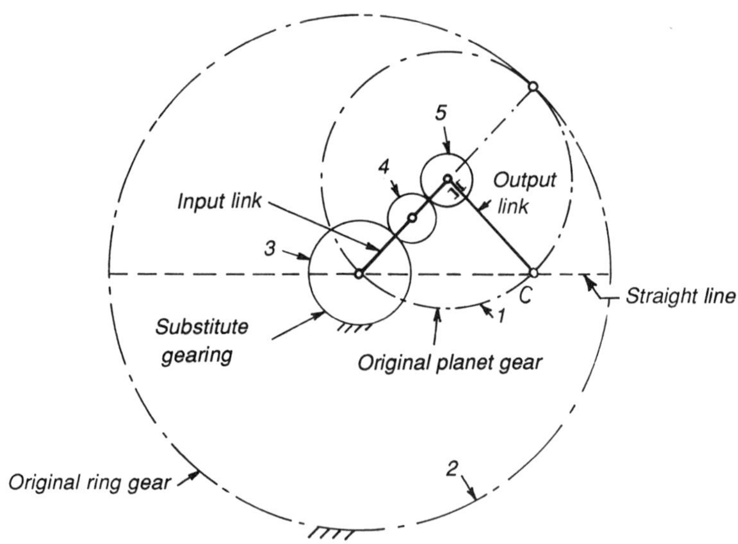

Figure 6.27 Multigear substitute.

The kinematic equivalent mechanisms of epicycloids are pericycloids, in which the planetary gear is stationary and the output is taken from the ring gear. Such arrangement usually lead to a more compact design.

In Fig. 6.26 Point C traces an epicycloid curve. Draw the proper parallels to find P_2, then use P_2O_3 to construct the compact substitute mechanism shown to the right of the original.

Figure 6.27 shows another way of producing a compact substitute for a hypocycloidal mechanism. The original mechanism is shown in dashed lines. Gear 1 rolls inside gear 2 and point C traces the curve. The three external gears, 3, 4, and 5, replace gears 1 and 2 with a remarkable savings in space. The only criterion is that gear 5 must be one-half the size of gear 3. Gear 4 is only an idler. The new mechanism thus has been reduced to approximately one-half that of the original in size. Point C traces an exact straight line.

SYMBOLS

α	angular acceleration, rad/s^2
b	radius of driving pin from center of planet gear
r	pitch radius of planet gear
R	pitch radius of fixed sun gear
ω_0	angular velocity of output, rad/s
β	angular displacement of output, deg
γ	$\theta R/r$
θ	input displacement, deg
ω_i	angular velocity of input, rad/s

7
Chain-Driven Mechanisms

Chains are, in general, used to transmit positive rotary motion from one shaft to another. There are many types of chains available, but here we are interested in using the chain as part of a mechanism so that a nonuniform motion can be obtained. From a design point of view, chains are low in cost, reliable as machine elements, and readily adaptable to a variety of tasks.

The trick when using chains to obtain a nonuniform motion is to fasten a roller, a pin, or a slider to a chain link and then arrange the chain sprockets in such a way that a desired output motion is obtained. When driving heavy loads, it might be necessary to use multiple-strand chains or parallel chains displaced a distance from each other, so that in combination they can support the pin or roller.

DESIGN CONSIDERATIONS

If an exact dwell is required, it is often necessary to support the motion of the chain-driven pin by steel or nylon rails, because it should be borne in mind that a chain is flexible when moving from one sprocket to another. Another way to obtain the desired motion is through the use of special guides.

KINDS OF MOTION POSSIBLE

Chain-driven mechanisms are very flexible, and motions can be obtained that are not possible with any other type of mechanisms, as will be evident from the following. Although many cases are described, still other solutions are possible.

Figure 7.1 shows two sprockets connected with a chain. When used as driving means for a mechanism, the length of the chain determines how many revolutions the driving sprocket must make in order for the motion to start all over again. If we let

L = length of chain, in.
r_1 = pitch radius of driving sprocket, in.
r_2 = pitch radius of driven sprocket, in.
N = number of revolutions for one complete cycle

then the length of the chain in Fig. 7.1 is calculated from the following equations:

$$\alpha = \tan^{-1}\frac{r_2 - r_1}{\overline{A_0B_0}} \text{ rad} \tag{7.1}$$

$$L = 2\pi r_1 \frac{180 - 2\alpha}{360} + 2\pi r_2 \frac{180 + 2\alpha}{360} + 2\overline{A_0B_0}\cos\alpha \tag{7.2}$$

If the length L of the chain is known, then the distance C between the two sprockets can be calculated from

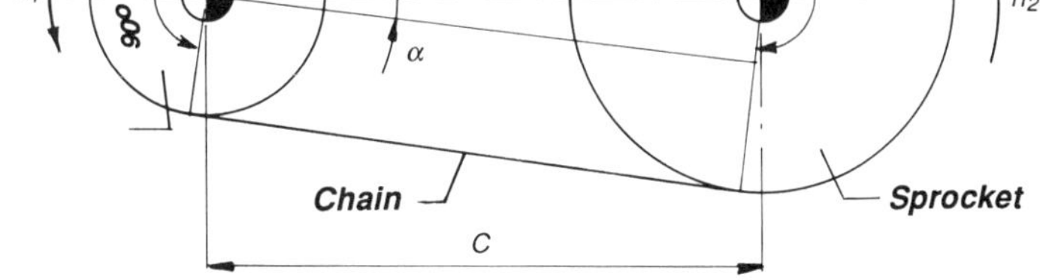

Figure 7.1 Chain drive with two sprockets.

$$C = \sqrt{0.25[L - \pi(r_1 + r_2)]^2 - (r_2 - r_1)^2} \quad (7.3)$$

The transmission ratio R is defined as the output speed over input speed of the output and input shafts:

$$r = \frac{n_1}{n_2} = \frac{r_2}{r_1}$$

APPLICATIONS

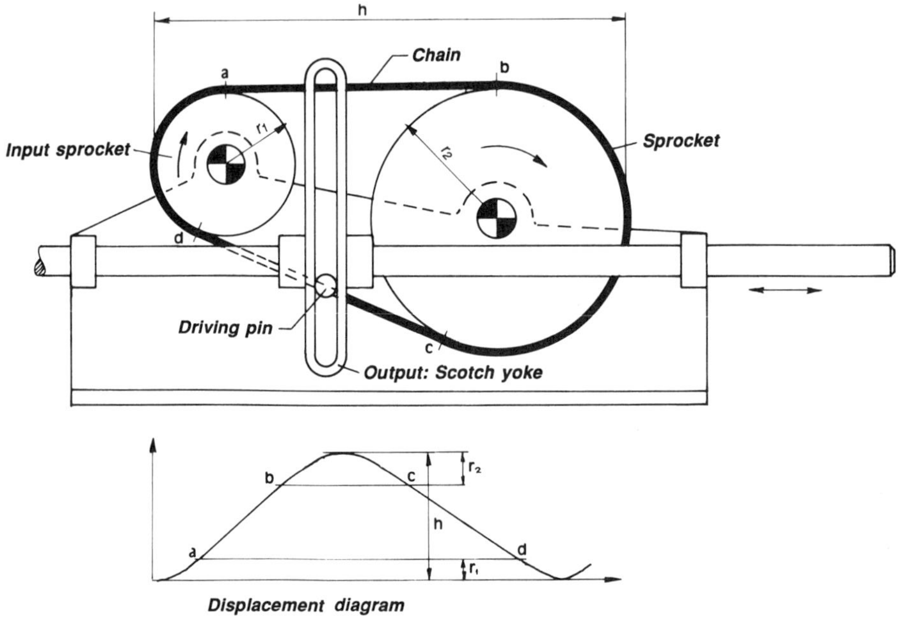

Figure 7.2 Chain-driven Scotch yoke, that is, a translating slotted member (the yoke) is driven by a roller or pin fastened to a chain. The chain is guided by two chain sprockets. The velocity of the yoke is uniform when the pin moves from *a* to *b* and from *c* to *d*. The motion from *b* to *c* and *d* to *a* is sinusoidal. The corresponding displacement diagram is shown below. It is possible to vary the size of the sprockets for this mechanism and in the cases that follow, so that different constant velocities can be obtained.

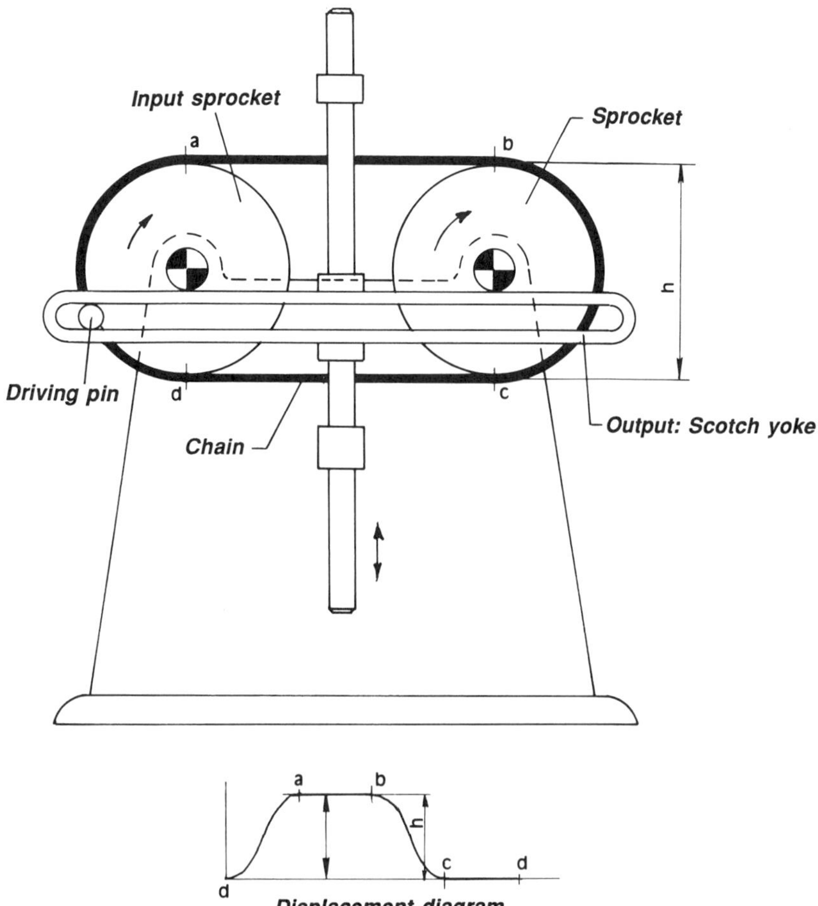

Figure 7.3 Here is an arrangement where the center line of the slot is parallel to the straight-line portions of the chain. The result is prolonged dwells in the extreme (upper and lower) positions. The corresponding time-displacement diagram is shown below.

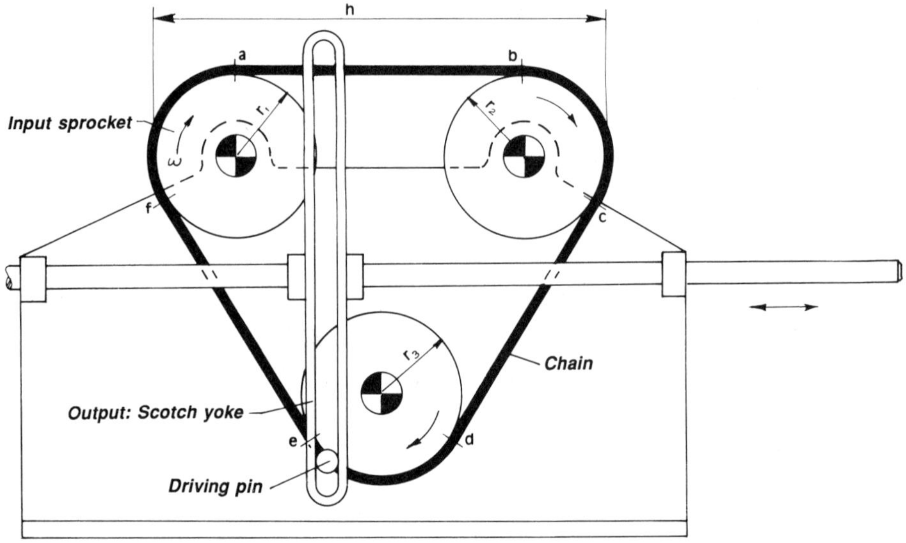

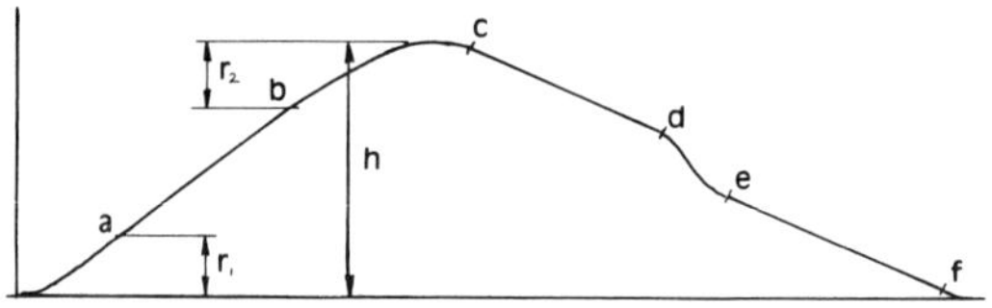

Displacement diagram

Figure 7.4 Three sprockets provide motion to the slotted member, with instantaneous dwell in the two extreme positions. The output motion from *a* to *b* provides constant velocity motion of the slotted link. From *b* to *c* it is part of a sinusoidal motion; from *c* to *d* it is a constant velocity motion, but the velocity is less than when moving from *a* to *b*. The motion starts repeating itself because of symmetry when the pin is immediately below the center of sprocket 3. From *d* to *e* the motion is part of a sinusoidal motion. The corresponding displacement diagram is shown below.

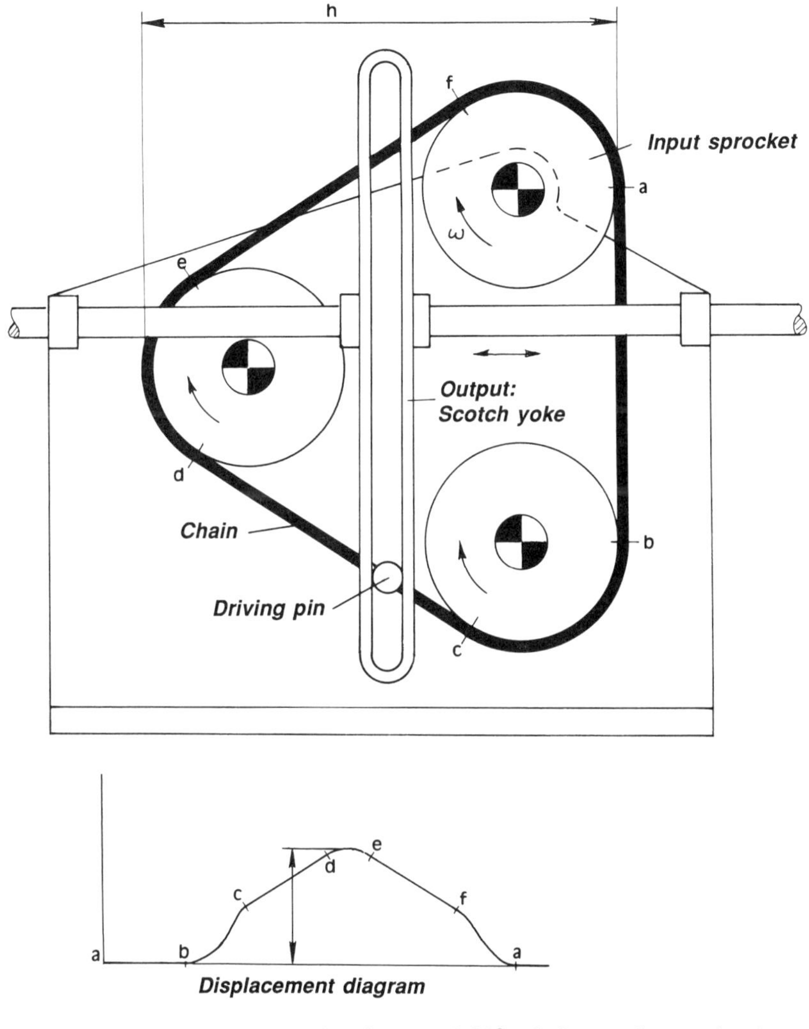

Figure 7.5 The slotted member is turned 90° relative to the mechanism of Fig. 7.4: Zero velocity from a to b, sinusoidal motion from b to c, constant velocity from c to d sinusoidal motion from d to e, constant velocity from e to f, and sinusoidal motion from f to a. The displacement diagram is symmetrical and is shown below.

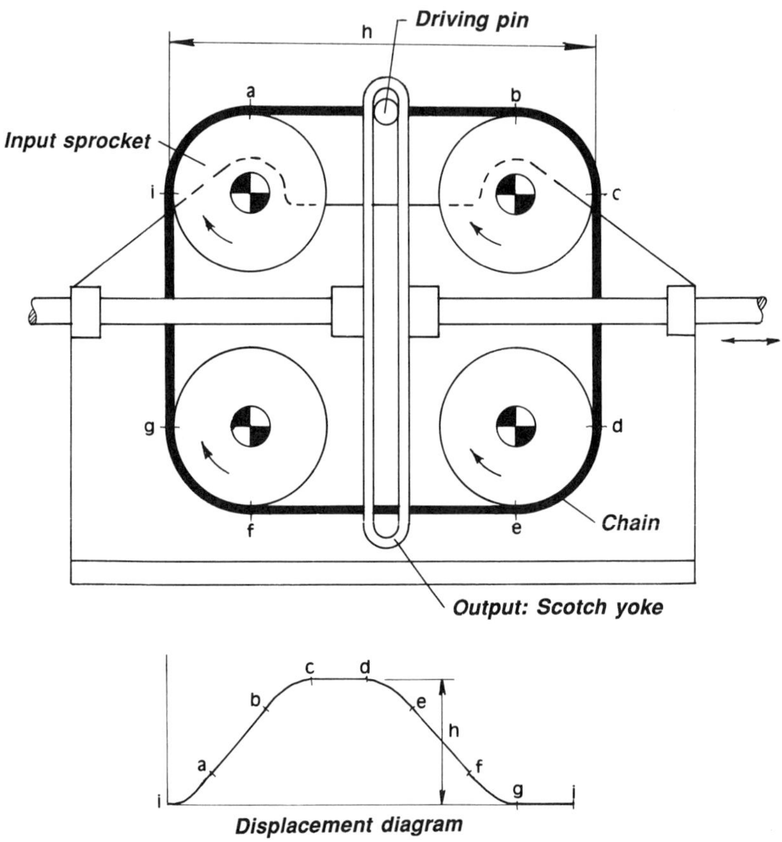

Figure 7.6 Four sprockets are used to drive the slotted member. There are two prolonged dwells and two constant-velocity portions (in the extreme left and right positions) for the motion of the slotted member. There are four partly sinusoidal motions for one complete revolution of the chain. The displacement diagram is shown below.

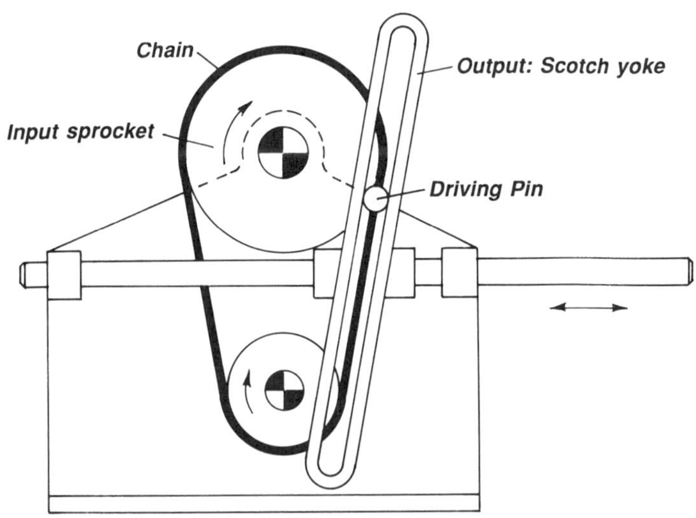

Figure 7.7 The center line of the slotted member subtends an angle with the shaft different from 90°. A prolonged dwell is obtained for the portion of travel of the pin when in the same direction as the slot. The same effect can also be achieved by making the slot perpendicular to the shaft and then letting one of the straight-line portions of the chain be parallel with the slot.

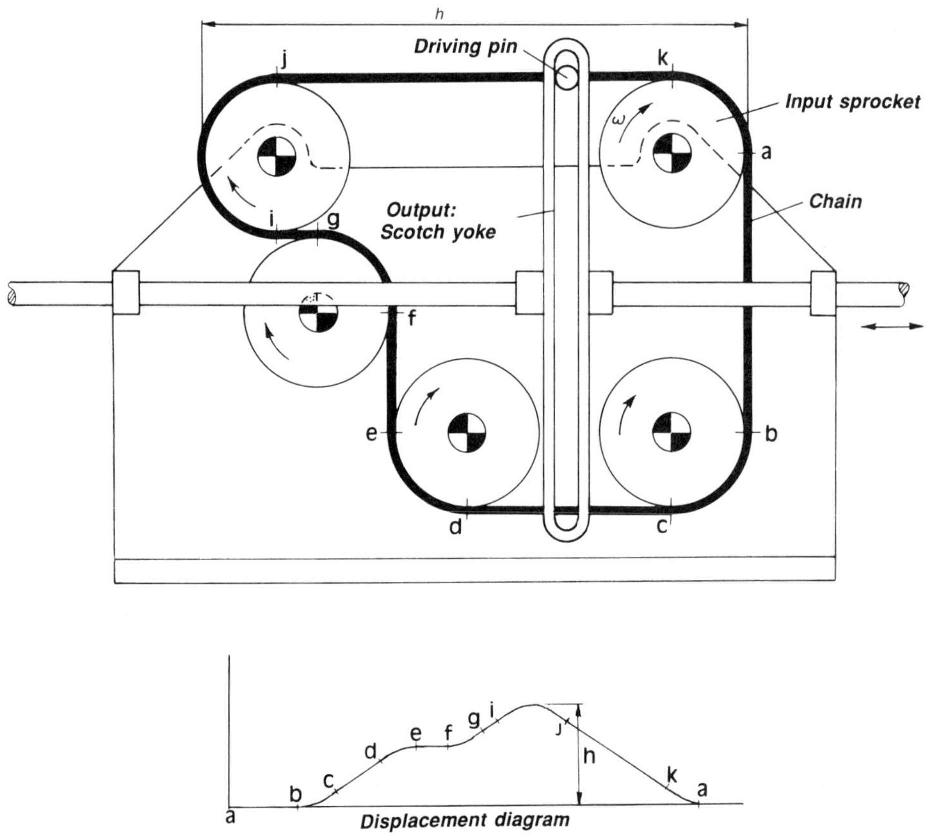

Figure 7.8 Arrangement with four sprockets, with a prolonged dwell in the extreme right position corresponding to *a* to *b*, a smaller prolonged dwell at an intermediate position corresponding to *e* to *f*, and an instantaneous dwell at the extreme left position. The corresponding time-displacement diagram is shown below.

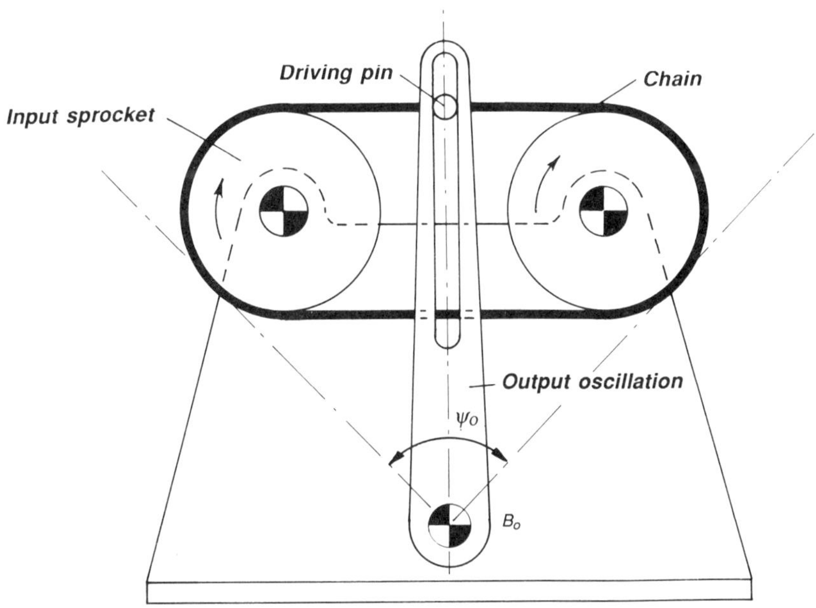

Figure 7.9 A chain and two equal-size sprockets impart an oscillating motion to the slotted member. The motion has an instantaneous dwell at the two extreme positions and oscillates through a total angle ψ_0.

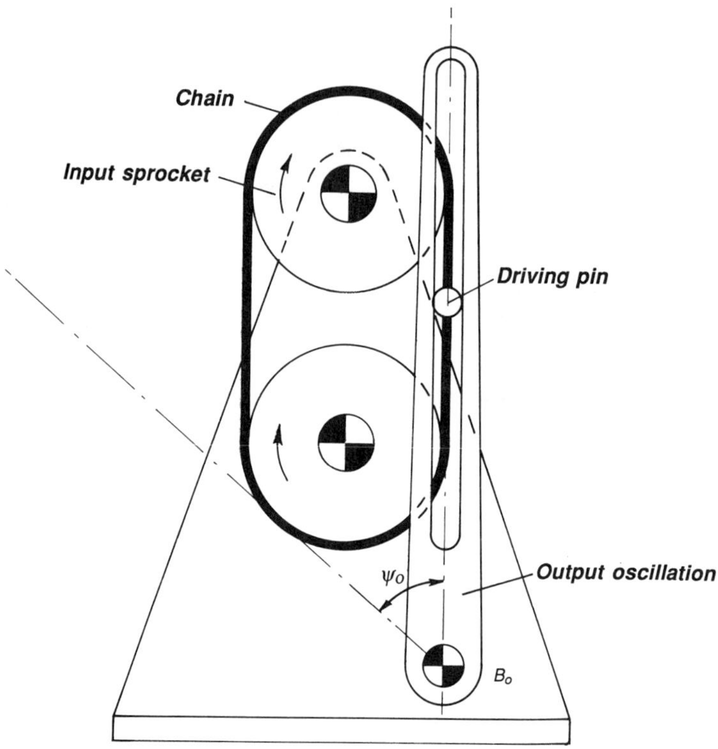

Figure 7.10 Chain-driven mechanism with slotted member oscillating. To obtain a prolonged dwell (in the extreme right position), the center line of the slotted member must be coincident with the straight-line portion of the path of the pin.

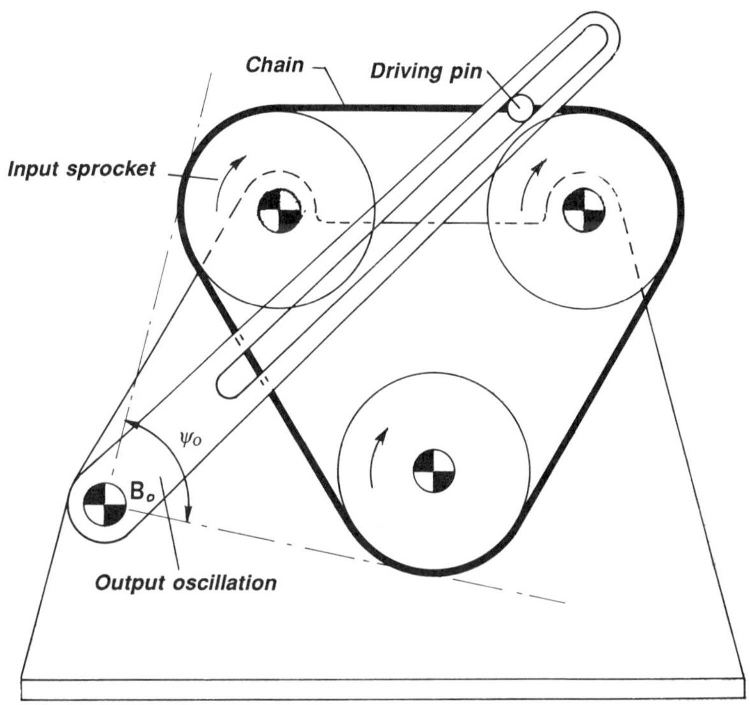

Figure 7.11 A general arrangement with three sprockets and an oscillating slotted member. There is an instantaneous dwell at the two extreme positions and the slotted member oscillates through a total angle ψ_0.

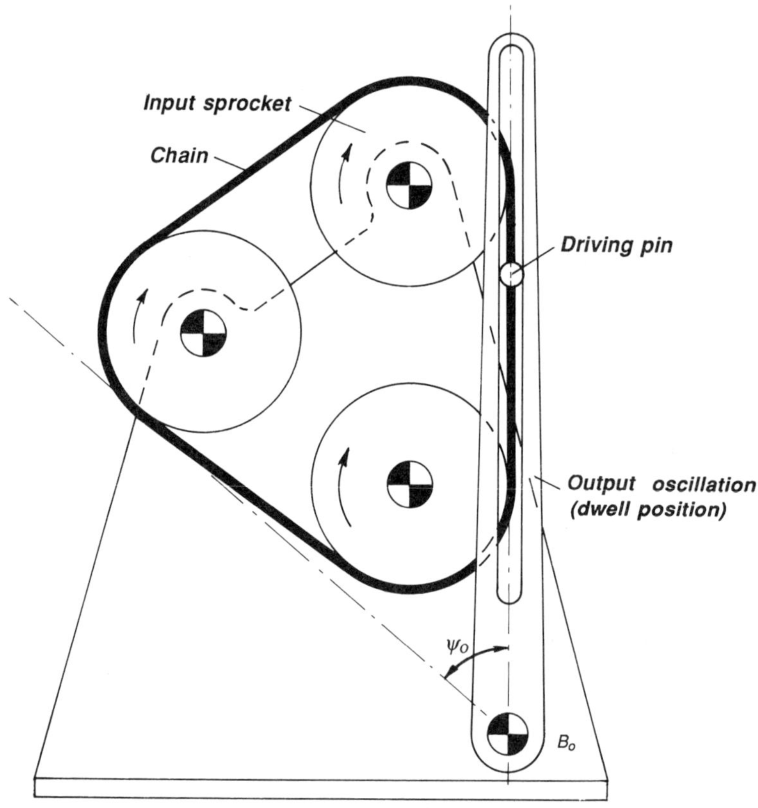

Figure 7.12 An arrangement with three sprockets and an oscillating slotted member. There is an instantaneous dwell at the extreme left position; a prolonged dwell at the extreme right position. The slotted member oscillates through a total angle ψ_0.

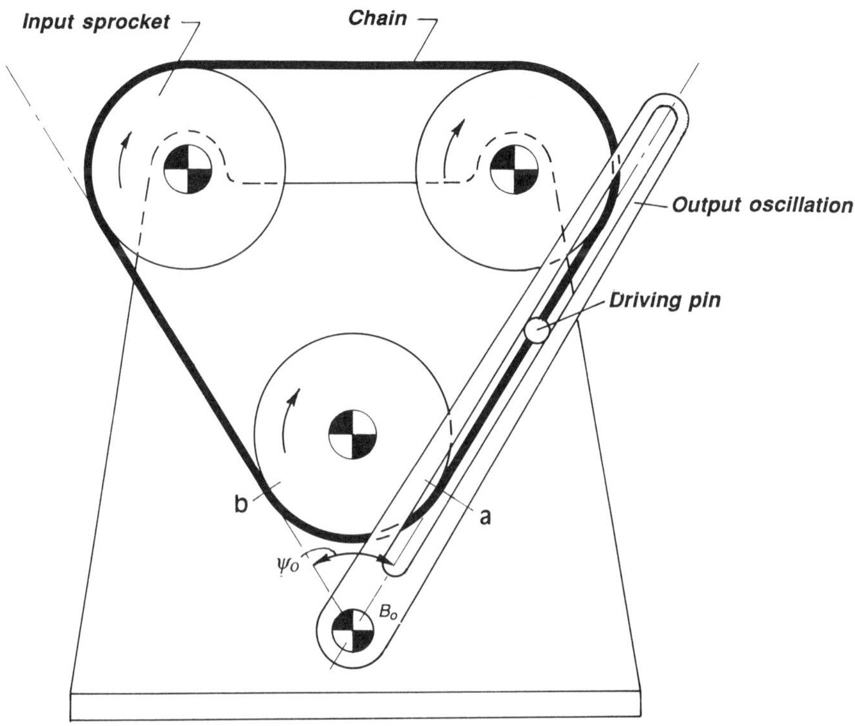

Figure 7.13 A three-sprocket arrangement similar to that in Figs. 7.11–7.13 but arranged so that two prolonged dwells are obtained in the extreme positions. The slotted member, which has high angular velocities and acceleration when the driving pin or roller is in its lowest position, oscillates through a total angle ψ_0.

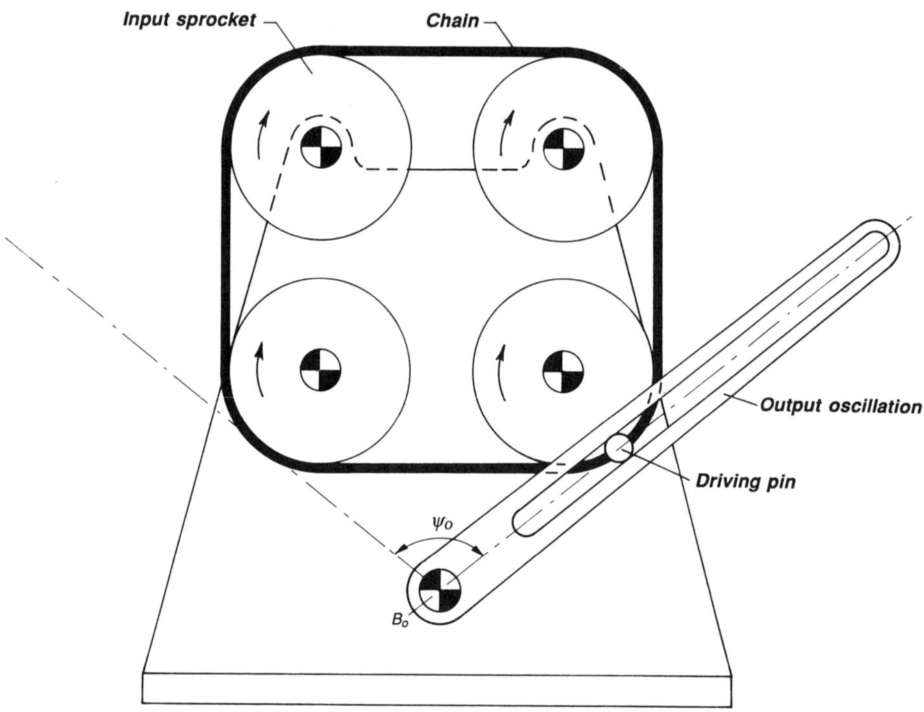

Figure 7.14 A general arrangement with four sprockets and a slotted output member. The slotted member oscillates through a total angle ψ_0.

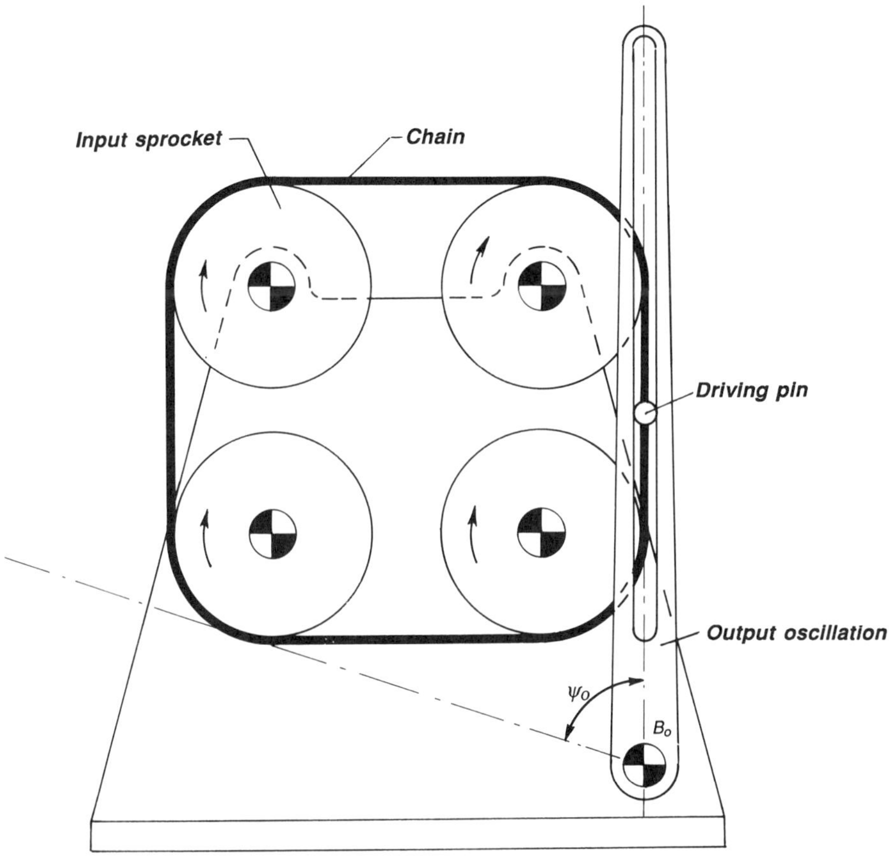

Figure 7.15 An arrangement with four sprockets. There is a prolonged dwell of the slotted output member in its extreme right position. The slotted member oscillates through a total angle ψ_0.

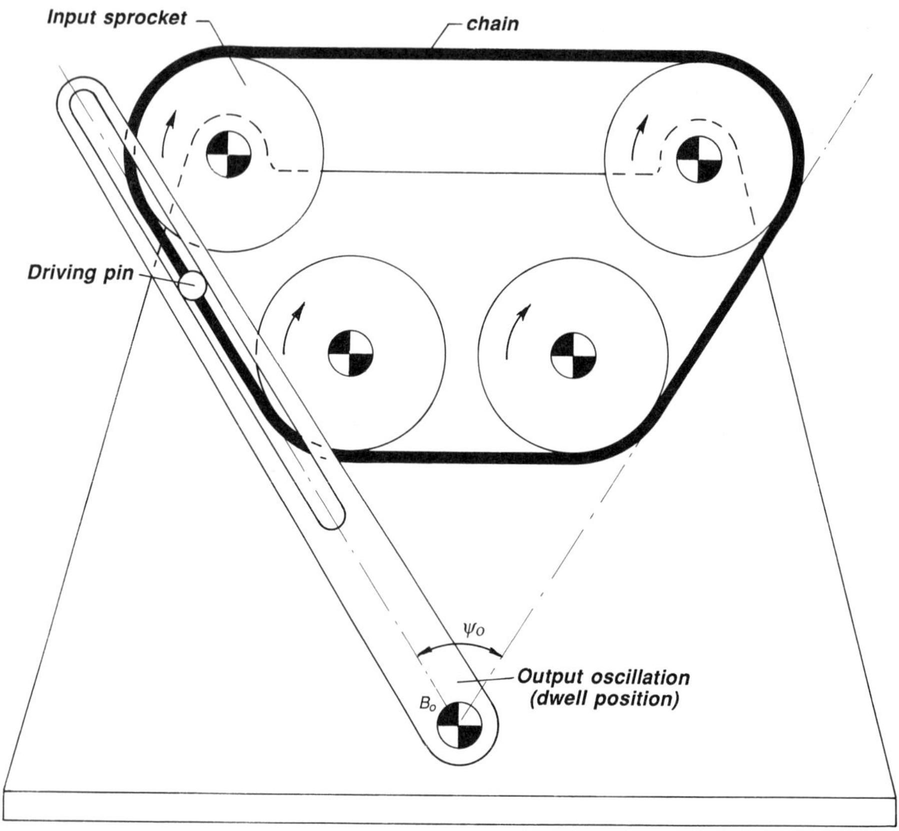

Figure 7.16 An arrangement with four sprockets. Two prolonged dwells at the two extreme positions. The slotted member oscillates through a total angle ψ_0. This arrangement is an improvement over that in Fig. 7.13.

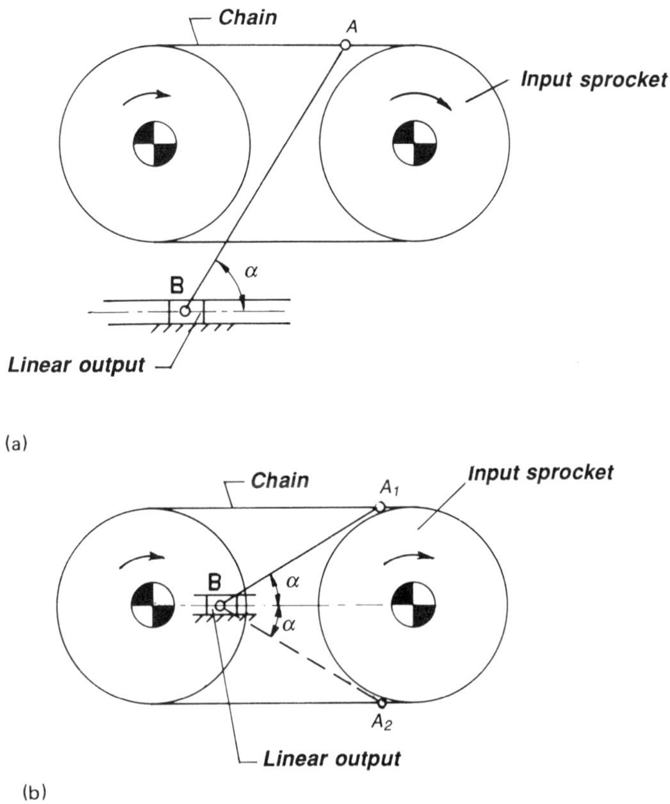

Figure 7.17 (a) Two-sprocket arrangement with linear output. The output motion of the slider B is rather similar to the Scotch-yoke motion, but it does not have the disadvantage of the yoke motion, where the length of the guides of the slotted member must be greater than the total stroke. However, the pressure angle α comes into the picture. A rule of thumb says that at all times the maximum pressure angle α_m by linear output must satisfy the condition $\alpha_m \leq 30°$, although this value may be increased when care is taken in reducing friction by using sliding elements with a low coefficient of friction. The best arrangement with respect to pressure angles is obtained when the maximum pressure angle is the same for the two extreme positions of A relative to the path of B. (b) The best arrangement of the motion of B would be for the path of B to pass through the centers of the two sprockets, because then α_m is the same, regardless of whether the rod AB is in position A_1B or A_2B. The same kind of reasoning holds for some of the following mechanisms.

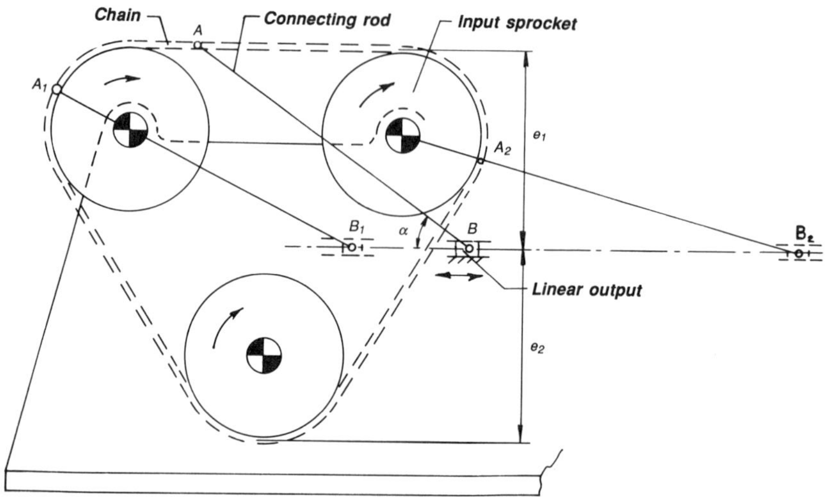

Figure 7.18 A chain-driven mechanism with three sprockets and linear output. Instantaneous dwells are at B_1 and B_2. To obtain the best motion characteristic with respect to transmission of forces, the slider B has been placed midway between the extreme positions of A, so that $e_1 = e_2$.

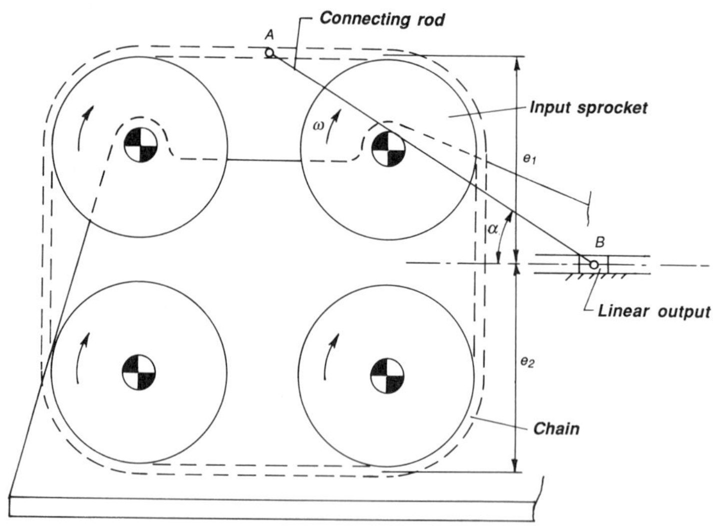

Figure 7.19 A chain-driven mechanism with four sprockets. The distances e_1 and e_2 should be approximately equal to obtain the best force transmission.

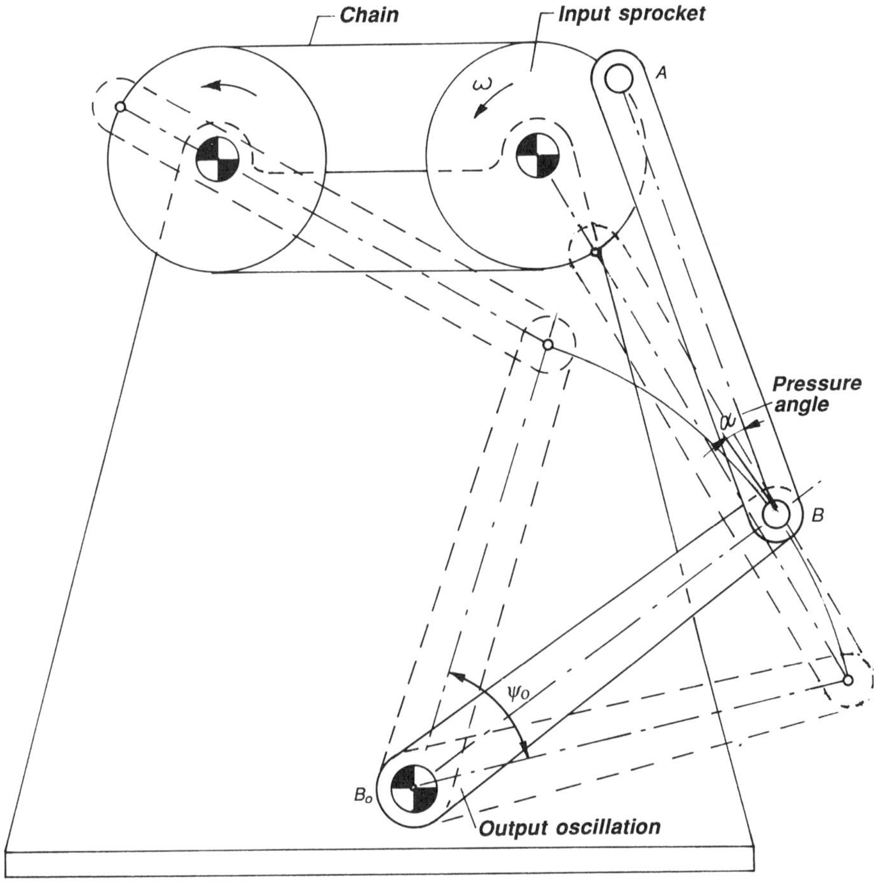

Figure 7.20 A two-sprocket mechanism with oscillating output. Here the pressure angle α = angle $B_0BA - 90°$. To get a good force transmission for all positions of A, $-45° \leq \alpha \leq 45°$.

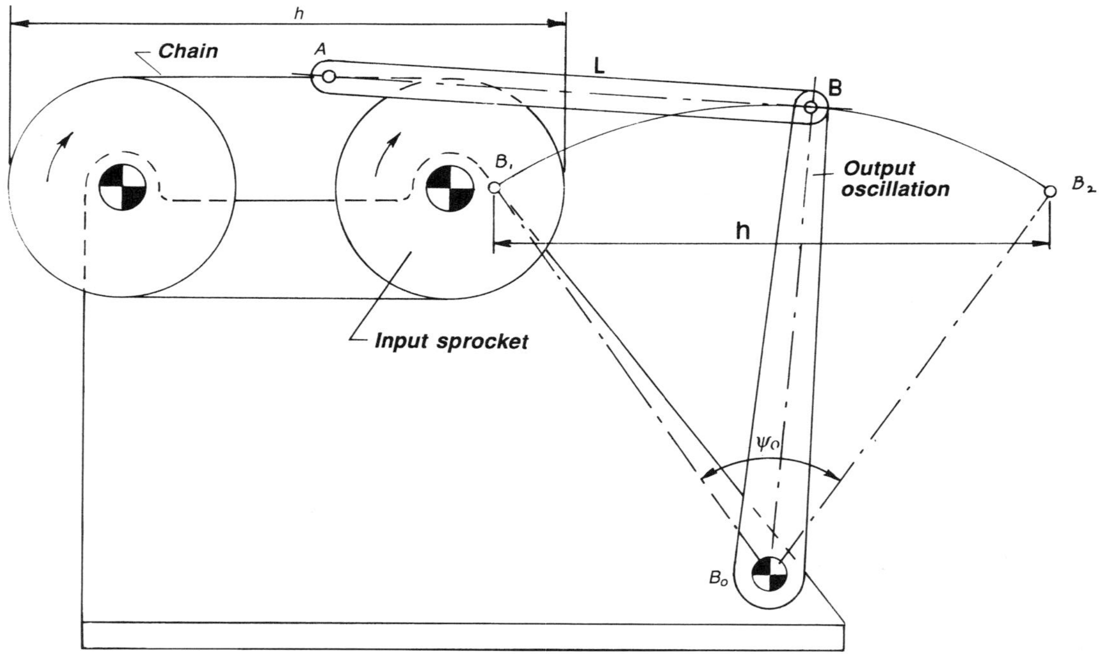

Figure 7.21 A good force transmission is obtained by this arrangement. The motion of pin A, which is carried by the chain, oscillates the output member B_0B through the angle ψ_0. The length of rod AB can be chosen arbitrarily. A good kinematic design is obtained by letting the distance $B_1B_2 = h$ and then choosing B_0 on the midnormal to B_1B_2.

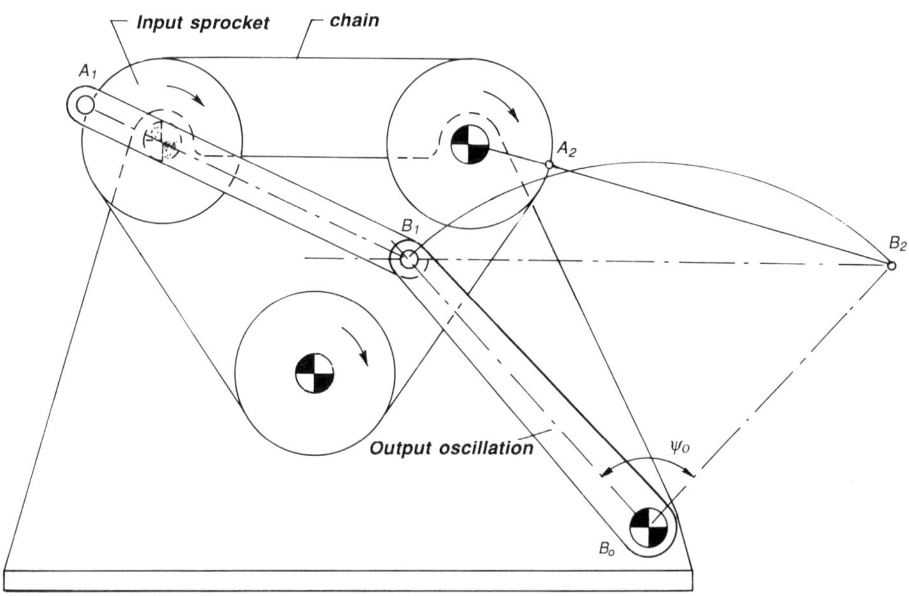

Figure 7.22 A three-sprocket mechanism with oscillating output. Instantaneous dwells are at B_1 and B_2. A good force transmission is obtained by this arrangement. The motion of pin A, which is carried by the chain, oscillates the output member B_0B through the angle ψ_0. The length of rod AB can be chosen arbitrarily. A good design is obtained by letting B_1B_2 lie between the extreme positions of the chain and then choosing B_0 on the midnormal to B_1B_2.

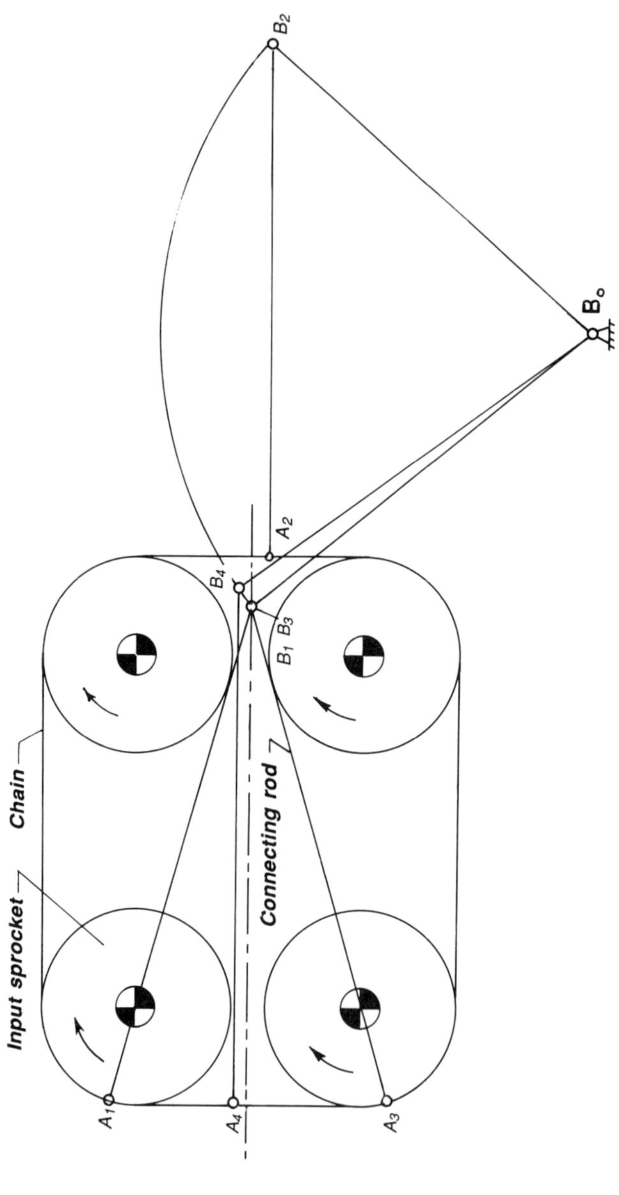

Figure 7.23 A four-sprocket arrangement with oscillating output and four instantaneous dwells. These are at positions B_1, B_2, B_3, and B_4.

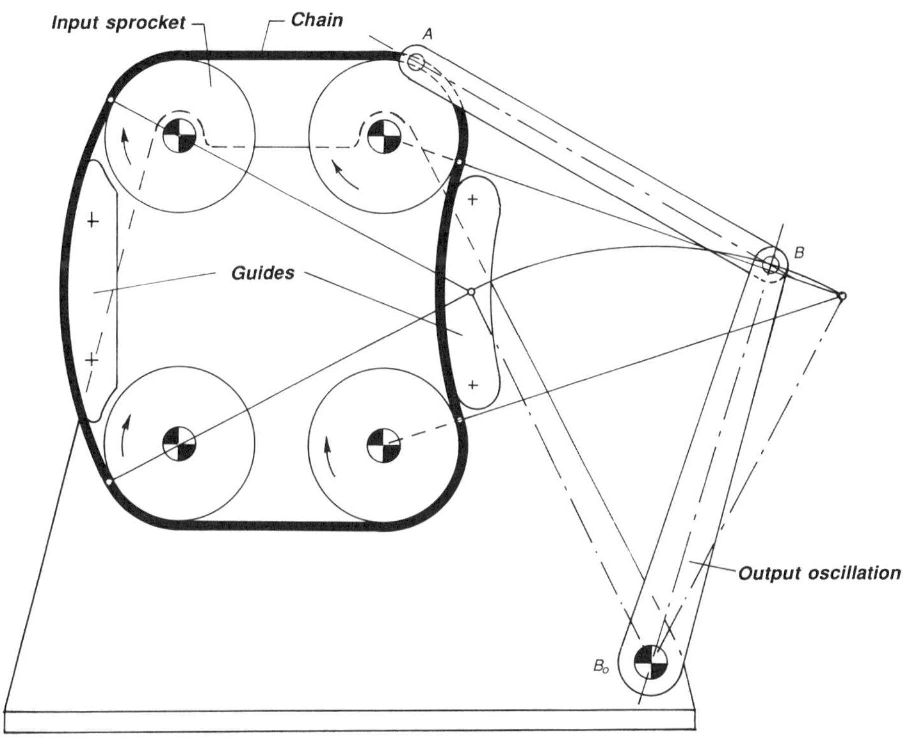

Figure 7.24 Changing the path of the driving pin between rollers results in this design, where a prolonged dwell is obtained in the extreme positions of the oscillating output member. The chain is guided on a circular path by nylon, bronze, or steel guides.

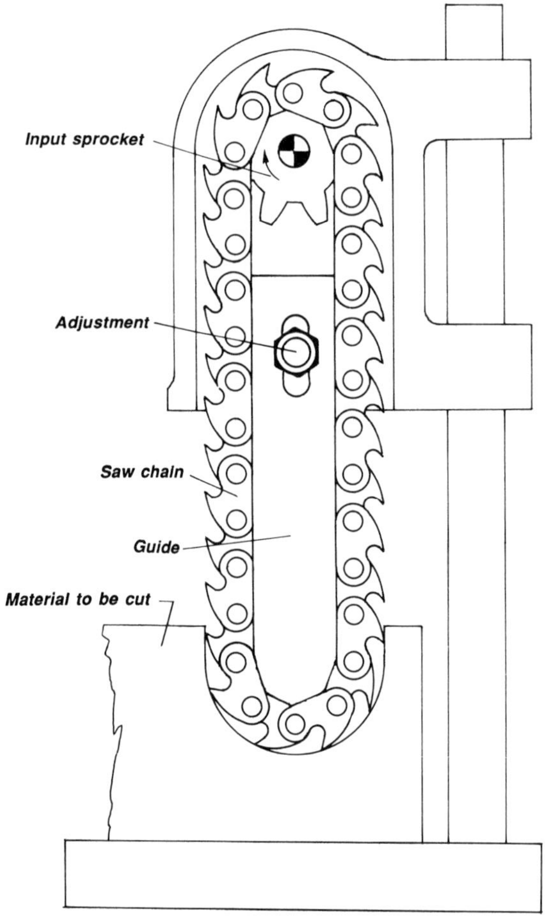

Figure 7.25 Application of chain-driven mechanism as a chain saw. Each link of the chain is formed as a cutting tooth.

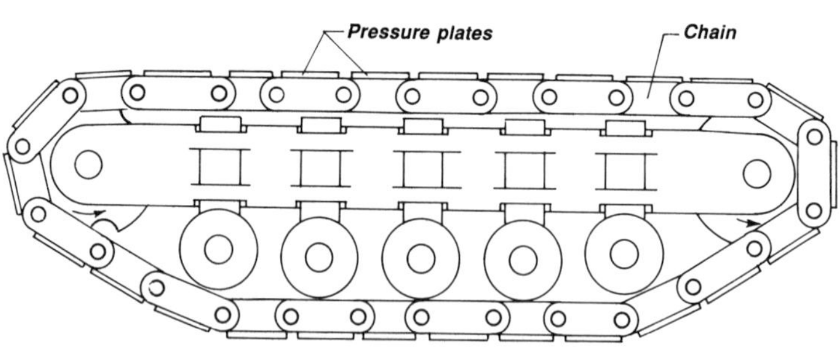

Figure 7.26 A chain used for tanks. One can actually define this chain as an endless transportable road.

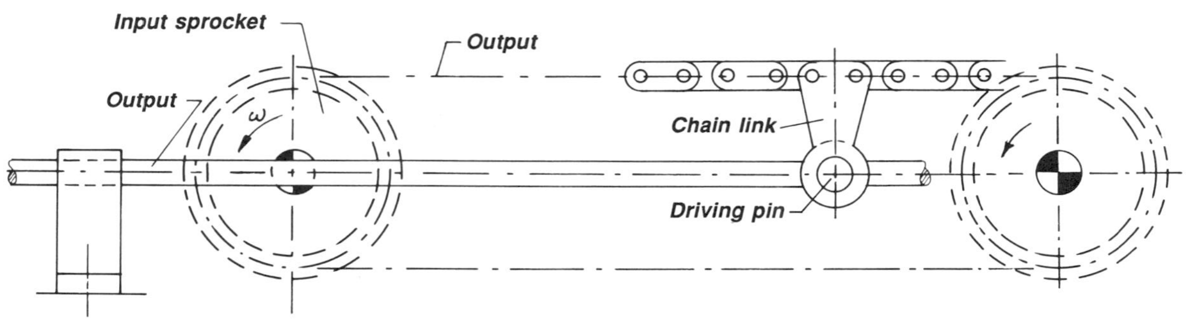

Figure 7.27 The driving pin, which is fastened to a link of the chain, reciprocates with a constant velocity over most of the rather long stroke. The driving link has a length equal to the pitch radius of the two gears, so that the pin will remain in a dwell position when the link reaches a gear. This arrangement is suited only for low speed because of the sudden change in velocity near the extreme positions.

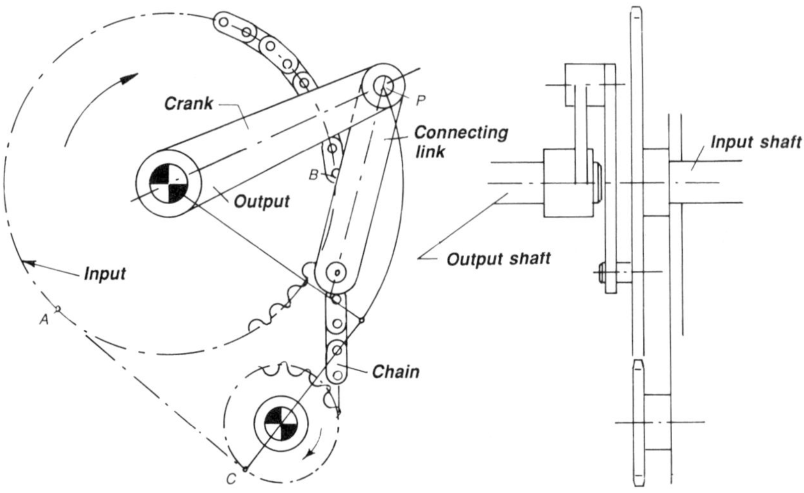

Figure 7.28 This mechanism can be adapted to produce a stop, a variable speed without stop, or a variable speed with momentary reverse motion. A uniformly rotating input sprocket drives the chain, which in turn, through the connecting link, drives the crank around. The two gears have fixed centers of rotation. When point P travels around the sprocket from point A to B, the crank rotates uniformly. Between B and C, P decelerates; between C and A, it accelerates. By changing the size and position of the small sprocket, a variety of motions can be obtained.

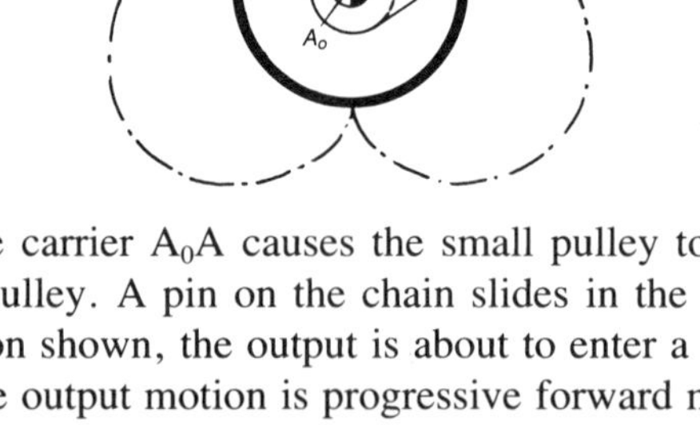

Figure 7.29 The carrier A_0A causes the small pulley to orbit around the stationary larger pulley. A pin on the chain slides in the slot of the output link. In the position shown, the output is about to enter a long dwell period of about 200°. The output motion is progressive forward motion interrupted by a prolonged dwell.

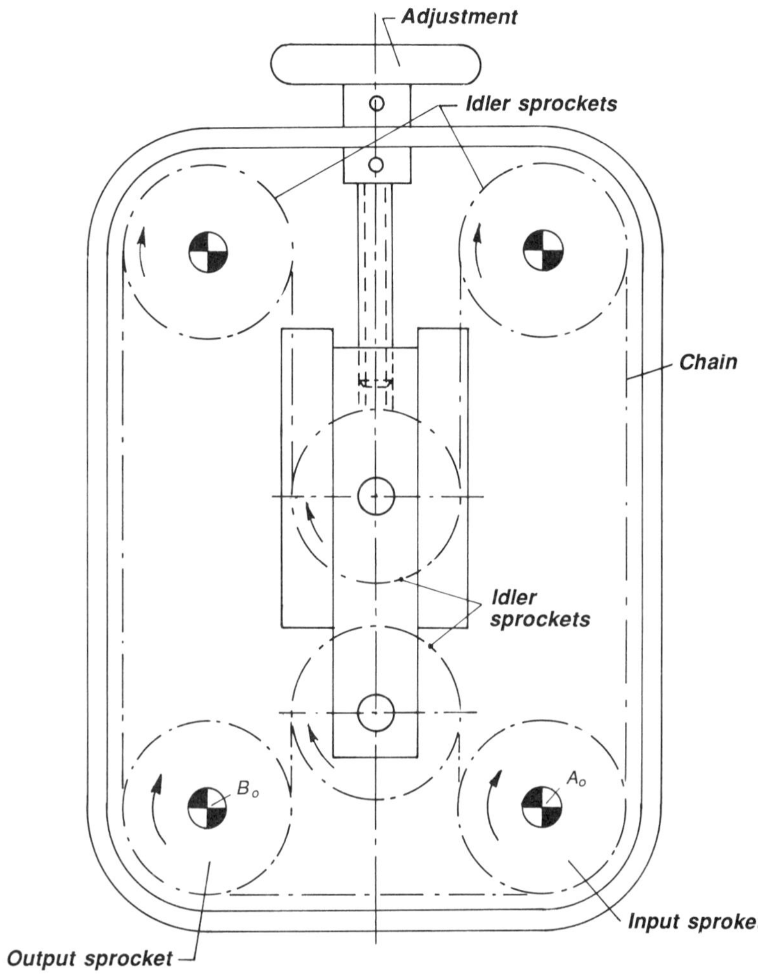

Figure 7.30 Synchronization between input and output shafts A_0 and B_0 is varied by moving the two idler sprockets by means of the adjusting screw.

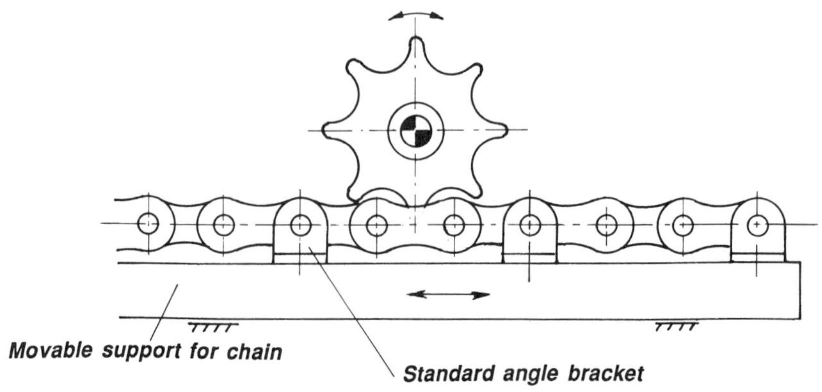

Figure 7.31 A low-cost rack-and-pinion device is easely assembled from a chain sprocket and a chain, fastened to the slider.

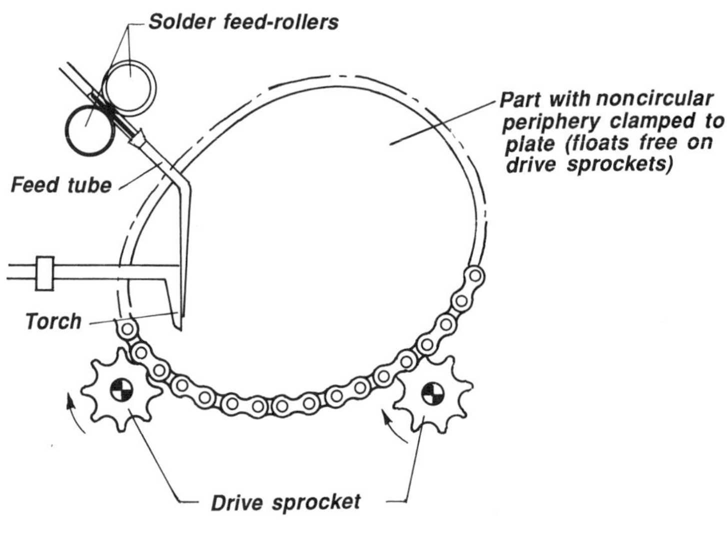

Figure 7.32 An extension of the rack-and-pinion principle of Fig. 7.31. Positive-action cams can be similarly designed. Standard angle brackets attach chain to cam. An irregular-shaped part can be made to rotate with constant circumferential speed for soldering purposes.

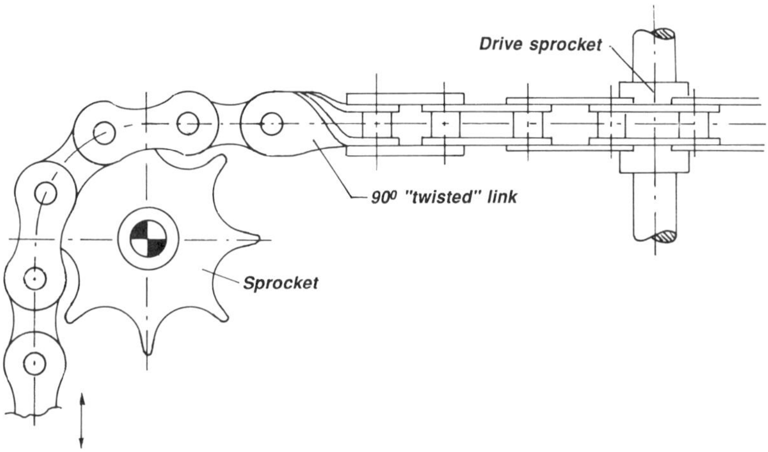

Figure 7.33 Chain with link bent 90° provides means for transmitting rotary motion between nonparallel (here, perpendicular) shafts.

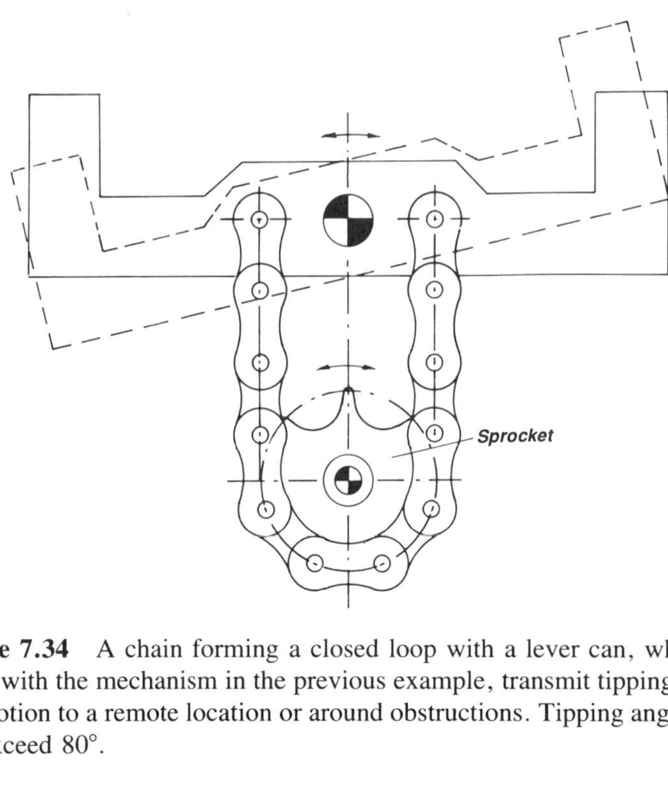

Figure 7.34 A chain forming a closed loop with a lever can, when combined with the mechanism in the previous example, transmit tipping or rocking motion to a remote location or around obstructions. Tipping angle should not exceed 80°.

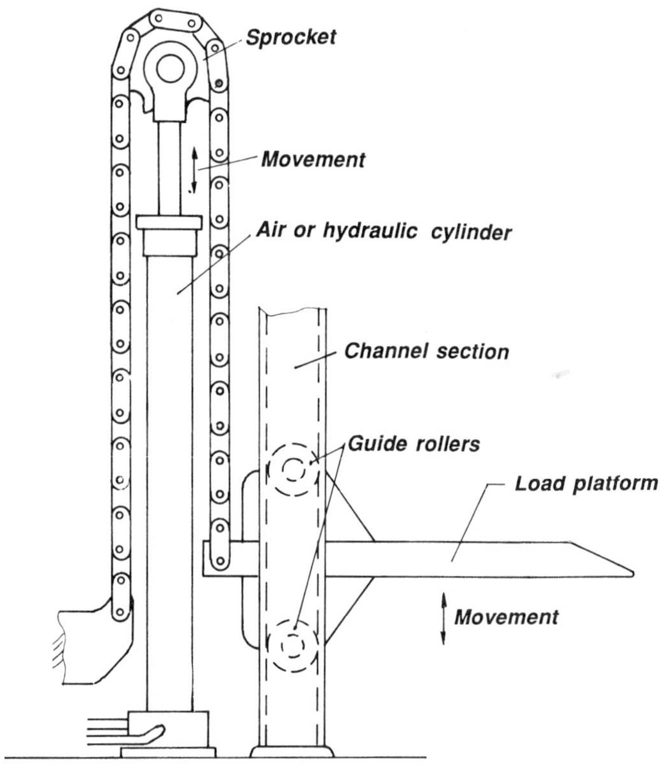

Figure 7.35 A hydraulic cylinder moves a chain sprocket. The output motion to the platform is twice the stroke of the hydraulic cylinder. The platform is guided in a horizontal position.

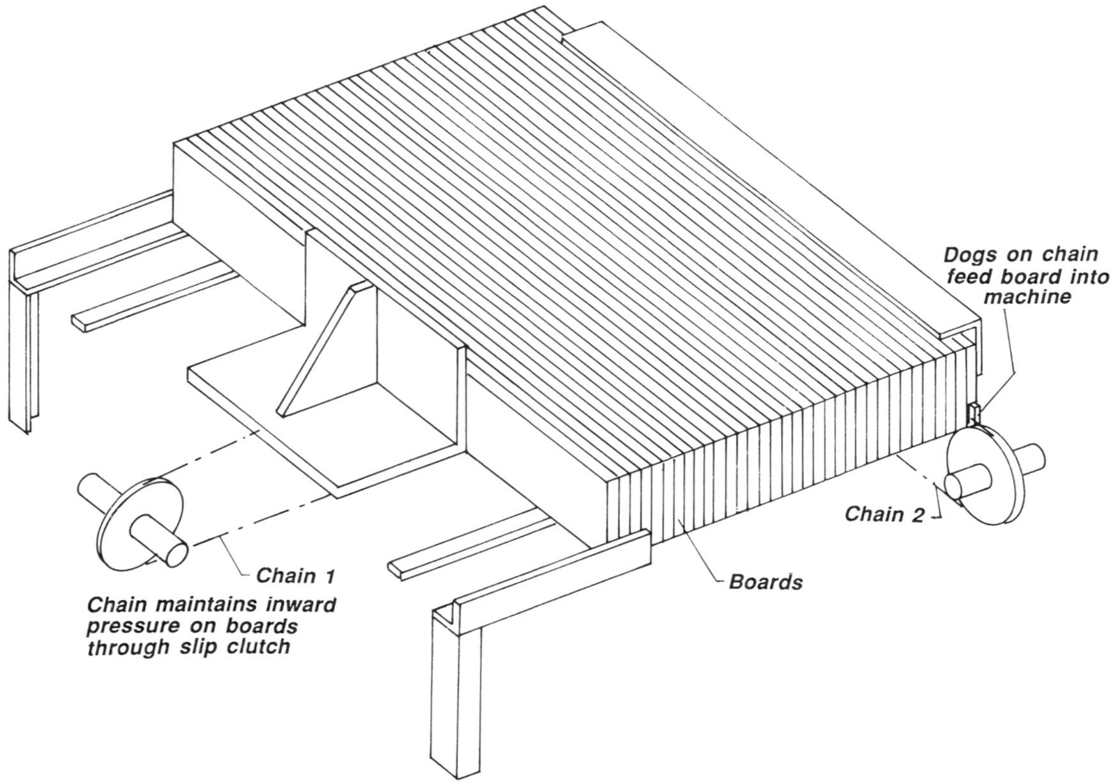

Figure 7.36 Two examples of indexing and feeding uses of roller chains are shown here in a setup that feeds plywood strips into a brush-making machine. Chain drive 1 (left) presses the boards towards the back through a slip clutch. Chain drive 2 (right) with dogs feeds board into machine.

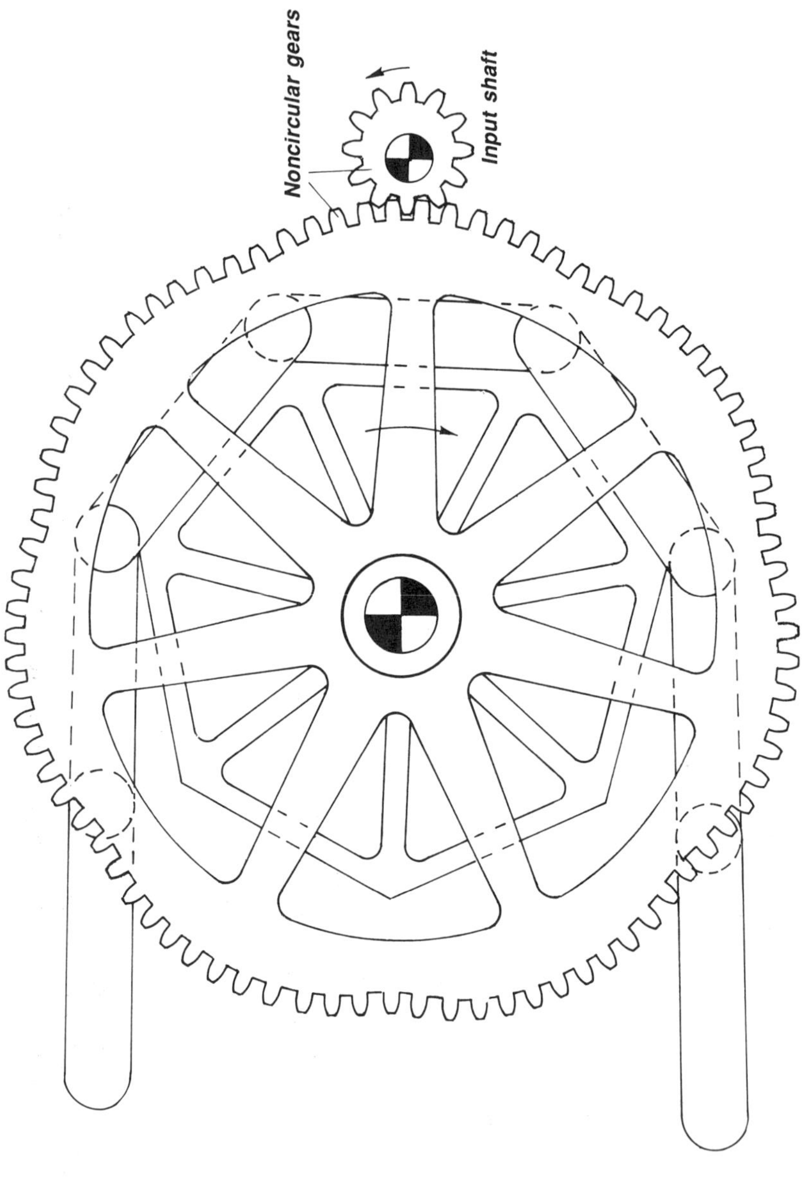

Figure 7.37 The large noncircular gear, mounted on the chain sprocket shaft, is given a nonuniform rotation, when driven from the noncircular pinion. This drive completely equalizes chain pulsations due to polygonal chordan actions of the chain and sprocket.

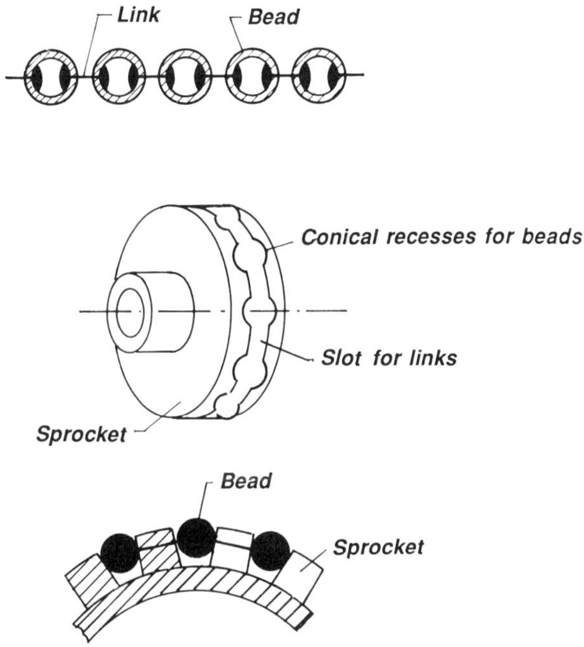

Figure 7.38 Details of bead chain and sprocket. Beads of chain seat themselves on conical recesses in the face of sprocket. Links ride freely in slots between recesses in sprocket.

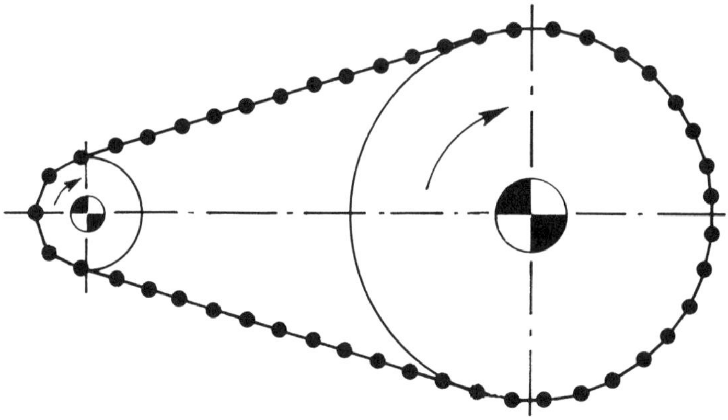

Figure 7.39 Bead chains allow high-ratio drives and are less expensive than gear trains. Bead chains and sprockets will transmit power without slippage.

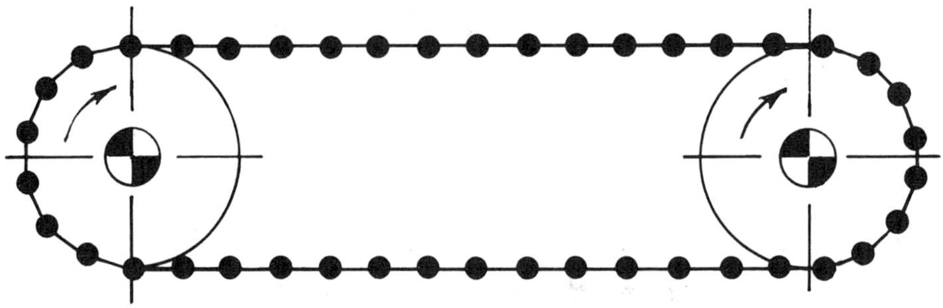

Figure 7.40 The drive possesses overload protection. Shallow sprocket gives positive drive for low loads; slips one bead at a time when overloaded.

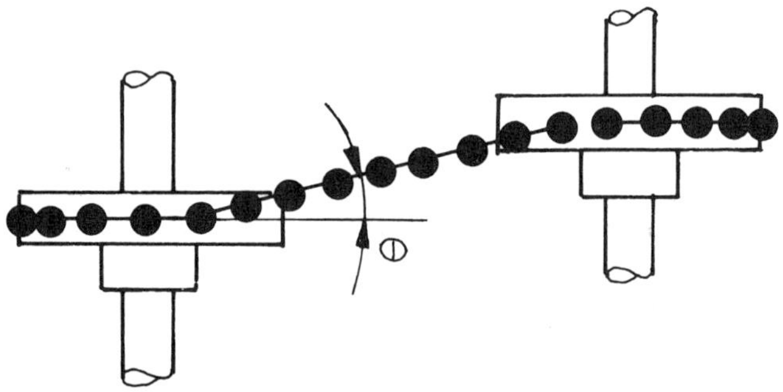

Figure 7.41 Misaligned sprockets can be displaced. Bead chains can operate at angles of up to $\theta = 20°$.

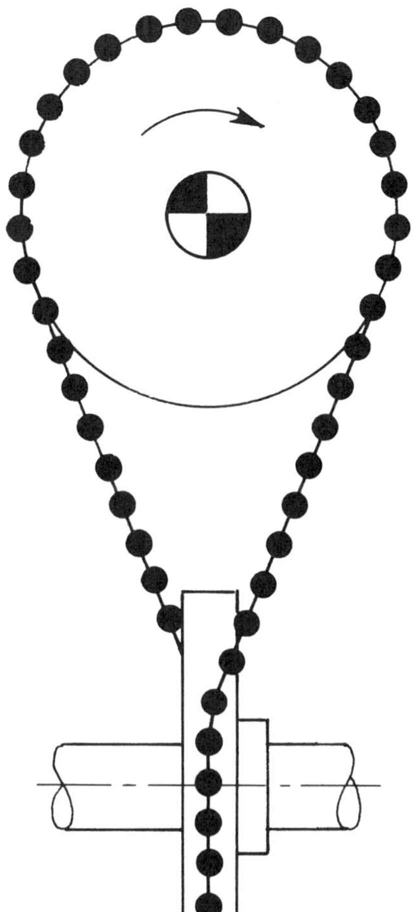

Figure 7.42 Transmission of motion between nonparallel shafts.

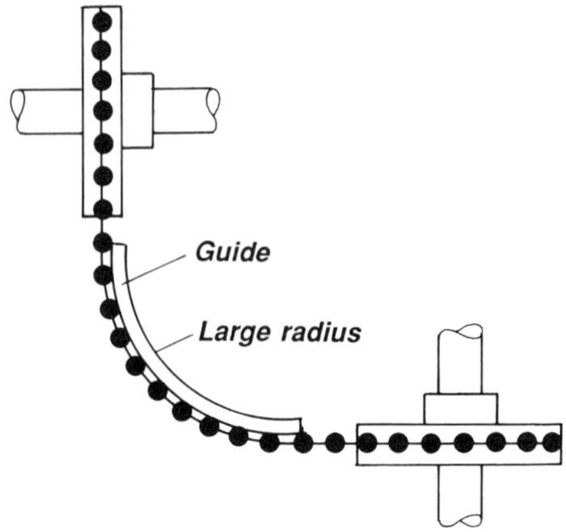

Figure 7.43 Right-angle drive between perpendicular shafts is achieved (but only for low torque) by using a guide with large radius.

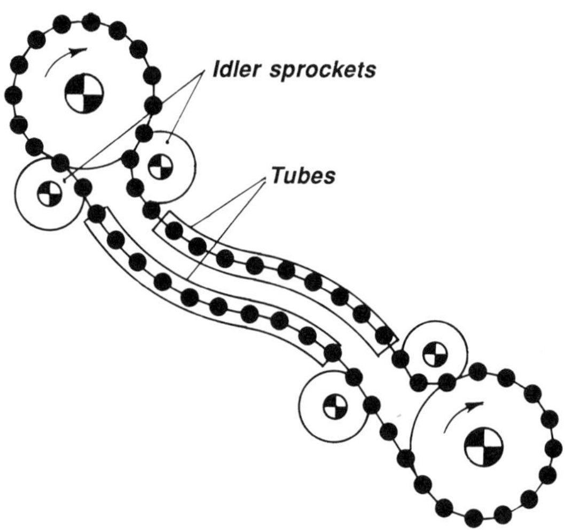

Figure 7.44 Remote motion around corners. Chain is guided through tubes to avoid interference with other parts.

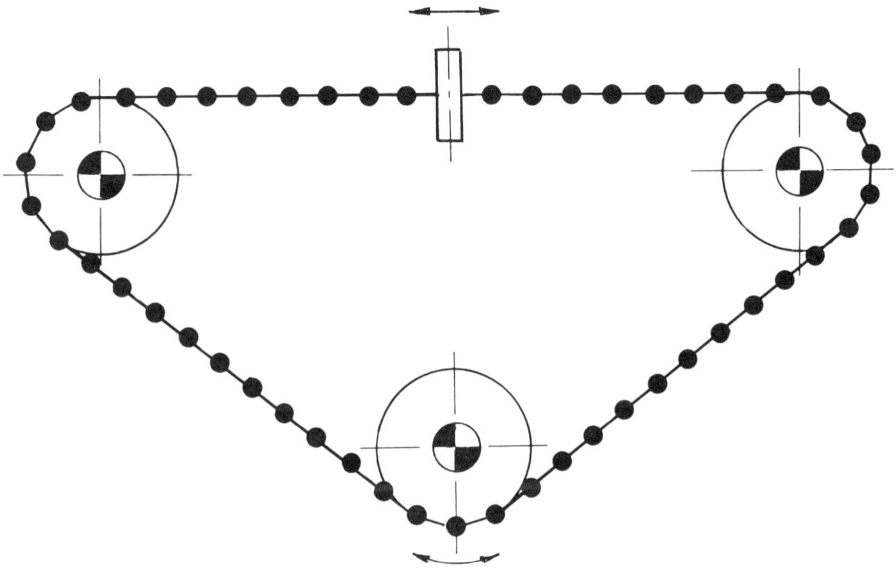

Figure 7.45 Oscillating input is converted to linear oscillating output or vice versa.

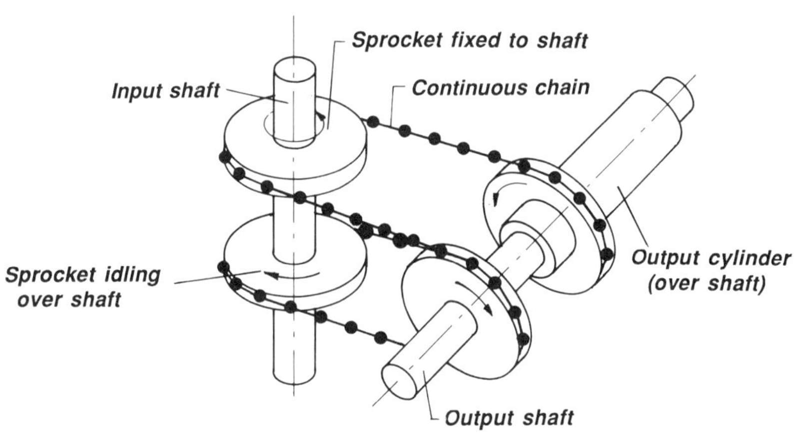

Figure 7.46 Counterrotating shafts. Input shaft uses a chain to drive counterrotating output shaft and output cylinder.

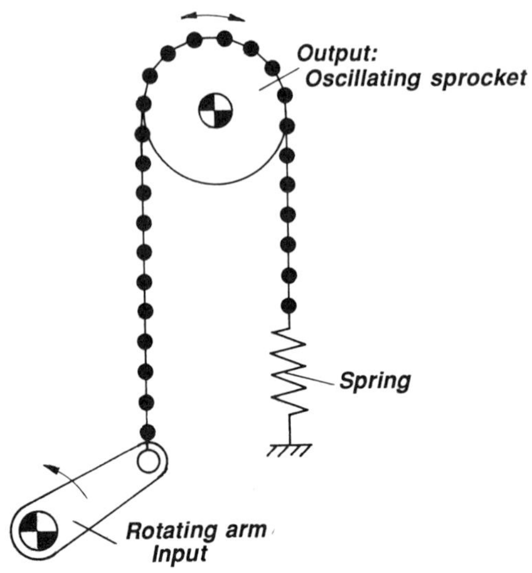

Figure 7.47 The rotary motion of the arm is converted to the oscillating motion of sprocket.

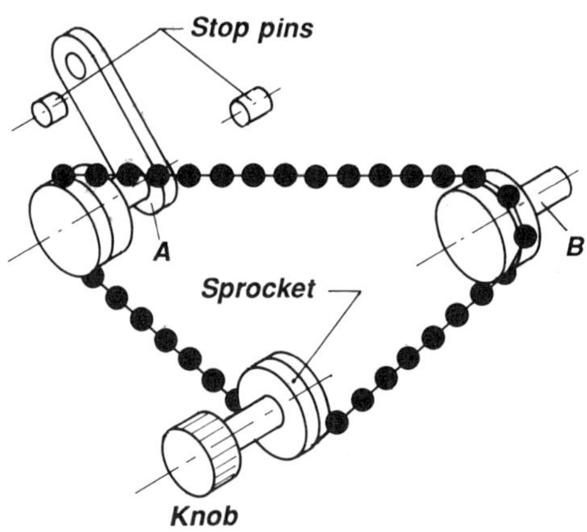

Figure 7.48 Restricted angular motion. Stopping pins limit the swinging motion of the arm, which is driven by the chain. Oscillation of the knob causes arm to oscillate, but only through a predetermined angle.

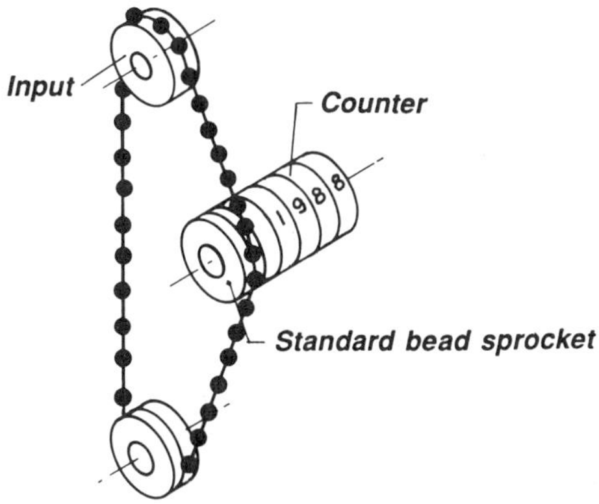

Figure 7.49 Remote control of counter. Rotary input of (distant) sprocket is transmitted to counter.

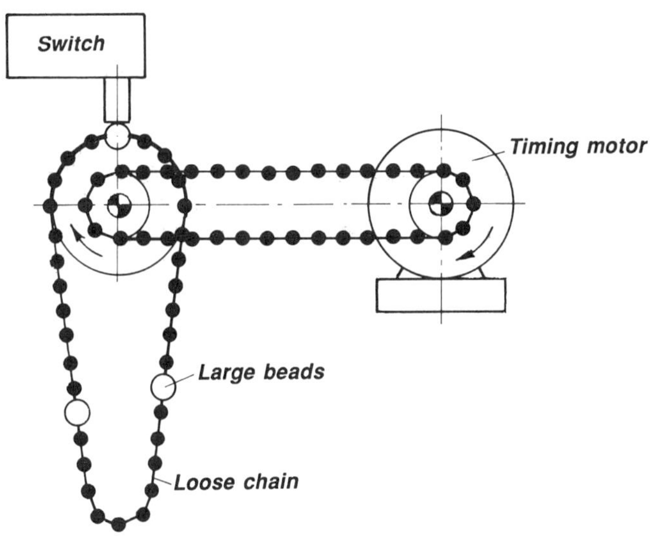

Figure 7.50 Timing chain. Timing motor drives sprocket, which in turn moves chain with large beads. Large beads activate the switch.

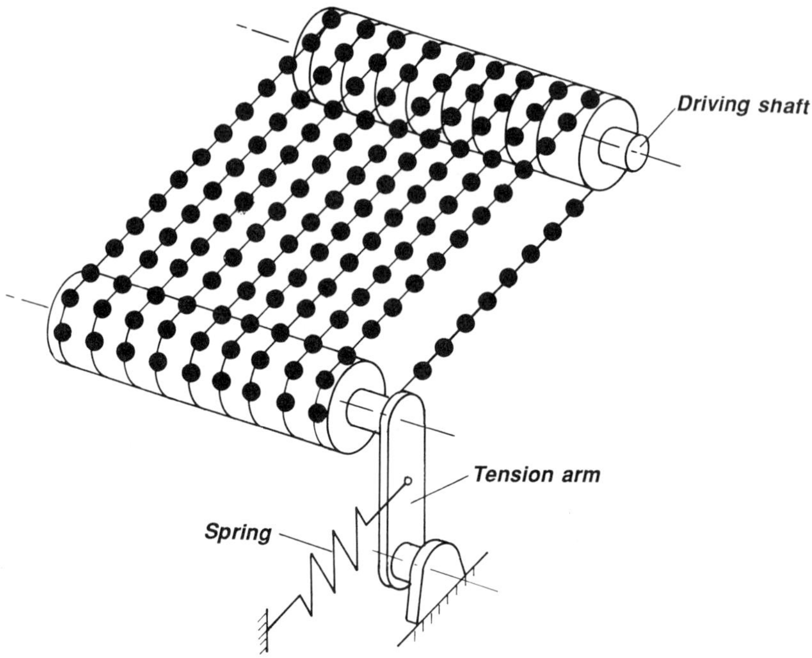

Figure 7.51 Bead chains combined to form conveyor belt. Tension arm maintains tension in chains.

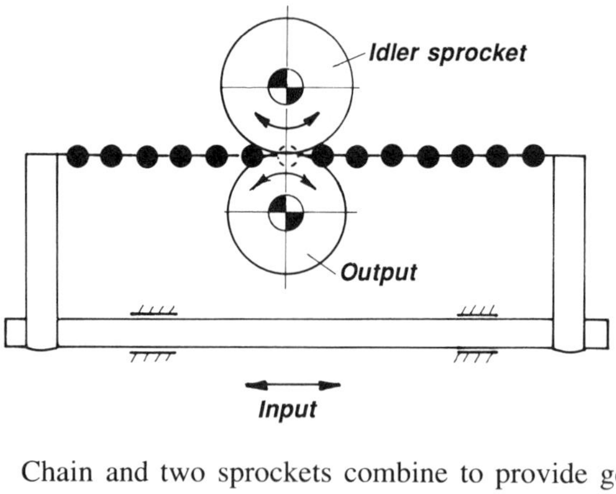

Figure 7.52 Chain and two sprockets combine to provide gear-rack motion. Idler sprocket maintains necessary counterpressure so that the chain does not escape.

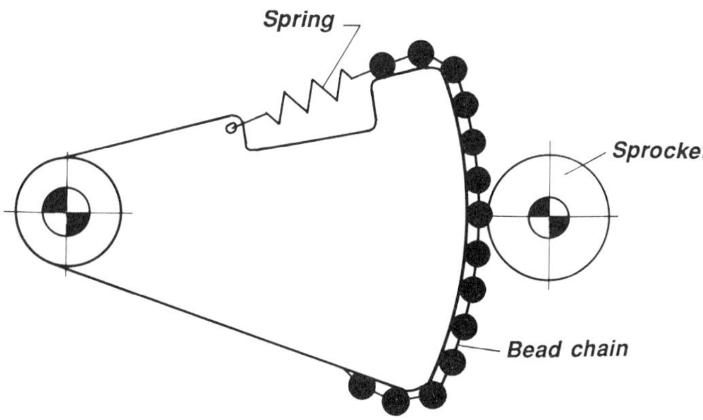

Figure 7.53 Gear and gear segment made inexpensively using sprocket and a bead chain wrapped around a sheet metal sector.

DETERMINATION OF VELOCITIES AND ACCELERATIONS

For a chain link moved around by a sprocket rotating with a constant angular velocity (Fig. 7.54), the velocity and acceleration of the link are determined by

$$V_A = r\omega \qquad (7.4)$$

$$A_A = r\omega^2 \qquad (7.5)$$

$$\omega = \frac{N(2\pi)}{60}$$

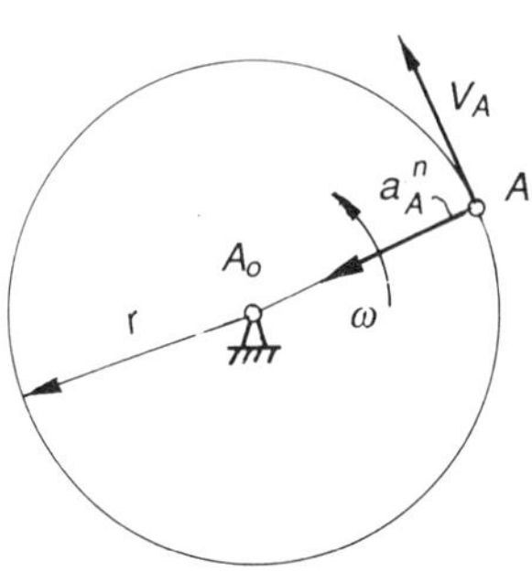

Figure 7.54 Velocity and acceleration of particle moving on a circular path with constant velocity.

where N is the speed in revolutions per minute. The velocity V_A of point A is always tangential to the circular path. The acceleration A_A of point A is determined by eq. 7.5 and is always directed towards the center of the sprocket if the sprocket rotates with constant angular velocity.

When the output member is moving along a straight line with a velocity V, and the pin is driving a Scotch yoke (Fig. 7.55), then the velocity of the yoke is found by projecting the velocity of the roller or chain link on the direction of motion of the yoke. The velocity of the chain is V and the velocity of the yoke is

$$V_{SY} = V\cos\alpha \tag{7.6}$$

where α is the angle between the direction of motion of the roller and the direction of motion of the yoke. Therefore, the velocity of the yoke is zero when the motion of the roller is perpendicular to the motion of the yoke, and the velocity is a maximum when the motion of the roller is perpendicular to the center line of the yoke, or when α = 0.

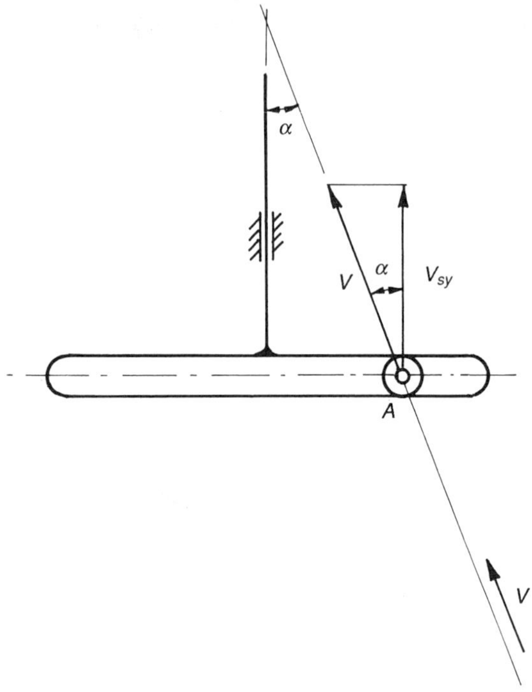

Figure 7.55 Velocities and accelerations by the scotch yoke when driven from a crank.

The acceleration of the yoke when the driving pin is guided by a sprocket (Fig. 7.56) is found by projecting a_A^n on the direction of motion of the yoke. The resultant acceleration of the yoke is

$$a_{sy} = a_A^n \sin\alpha \tag{7.7}$$

The acceleration of the yoke, in the position shown, is directed opposite its velocity, so that the velocity of the yoke is decreasing.

In Fig. 7.57 a sprocket drives a roller with the tangential velocity V. In the position shown, the velocity component V_c perpendicular to the slotted member is

$$V_c = V\cos\alpha \tag{7.8}$$

and the angular velocity of the slotted member is

$$\omega_s = \frac{V_c}{\ell} \tag{7.9}$$

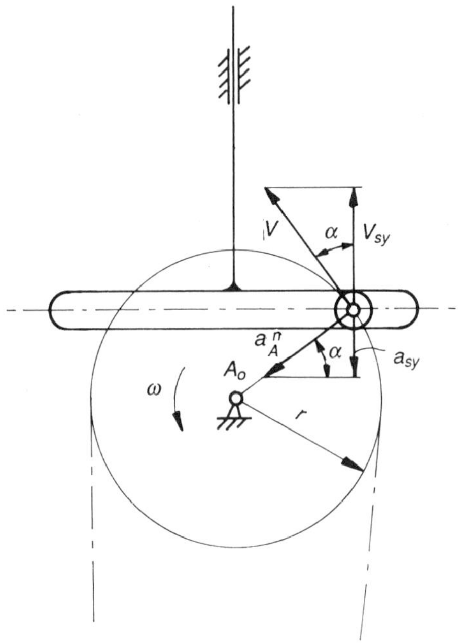

Figure 7.56 Determination of velocity of scotch yoke when driving pin is moving along a straight line with uniform velocity.

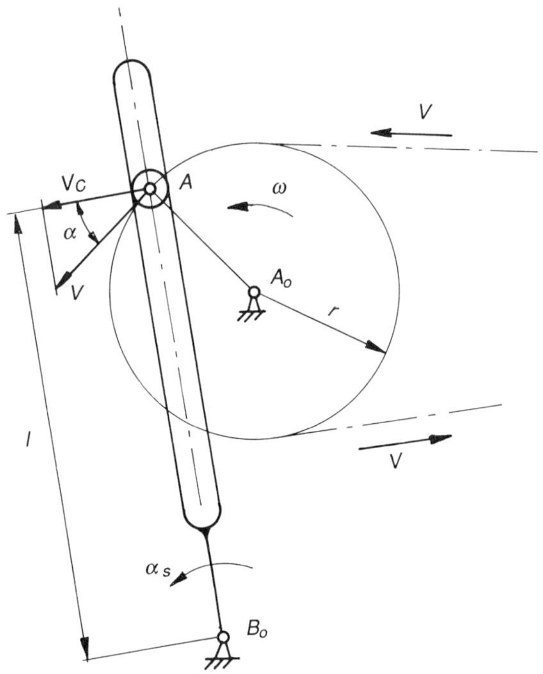

Figure 7.57 Determination of angular velocity and angular acceleration of slotted member when driving pin is moving on a circular path.

where ℓ is equal the length B_0A. The angular velocity of the sprocket is

$$\omega = \frac{V}{r} \tag{7.10}$$

To find the maximum angular acceleration of the slotted member (Fig. 7.58), draw the crank A_0A in the position where it is perpendicular to the center line of the slotted member. The maximum angular acceleration $(\alpha_s)_{max}$ is

$$(\alpha_s)_{max} = \frac{r\omega^2}{\ell} \tag{7.11}$$

The force between the roller and the slotted member in this position is

$$F = \frac{I_0(\alpha_s)_{max}}{\ell} \tag{7.12}$$

where I_0 is the moment of inertia of mass around B_0. To find the angular

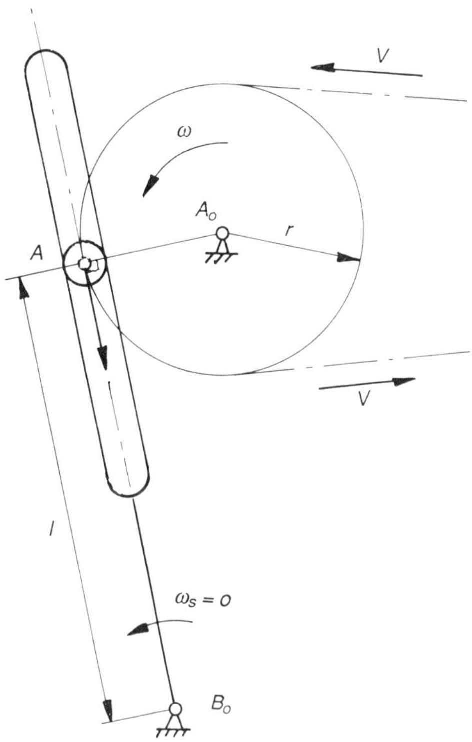

Figure 7.58 Maximum angular acceleration of slotted member occurs when crank radius is perpendicular to center line of slotted member.

acceleration of the slotted member in other positions, see Chapter 4 on Geneva mechanisms.

In Fig. 7.59 a slider is driven from A, which is on the sprocket. To find the angular velocity of the sprocket, divide the velocity V of the chain by r, the pitch radius of the sprocket. The kinematic equivalent mechanism is a slider crank. For velocities and accelerations of B, see Chapter 3 on the slider crank.

In Fig. 7.60 Pin A is moving on the sprocket, which rotates with the angular velocity ω and drives the slider B. To find the velocity and acceleration of the slider B, see Chapter 3 on the slider crank.

Figure 7.61 shows that when the output motion is swinging and pin A is on the sprocket, then A_0ABB_0 is exactly equivalent to a four-bar linkage. For angular velocities and angular accelerations of the output link B_0B, see Chapter 2 on motion and force transmission in linkages.

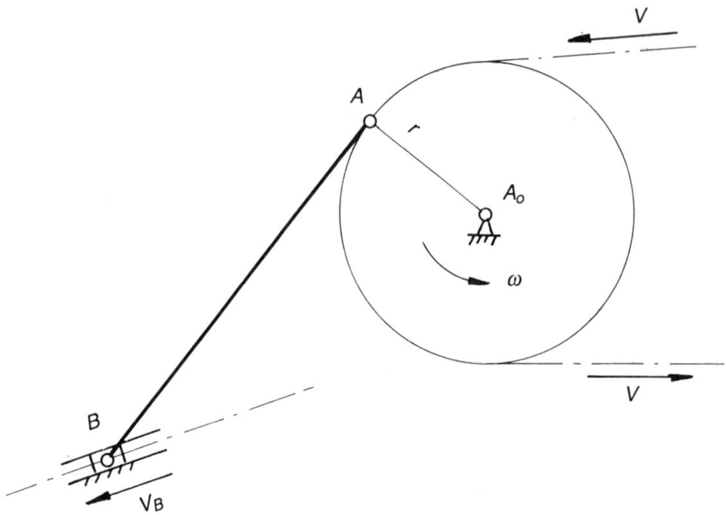

Figure 7.59 Pin A moves on a circular path and drives member B along a straight-line path.

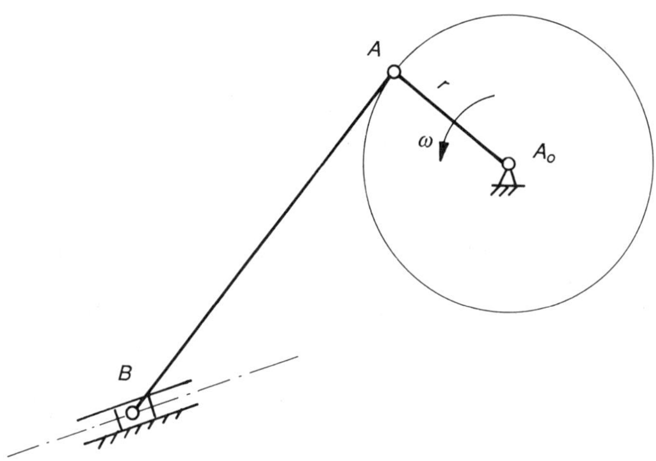

Figure 7.60 The kinematic equivalent mechanism to Fig. 7.32 is an offset slider crank.

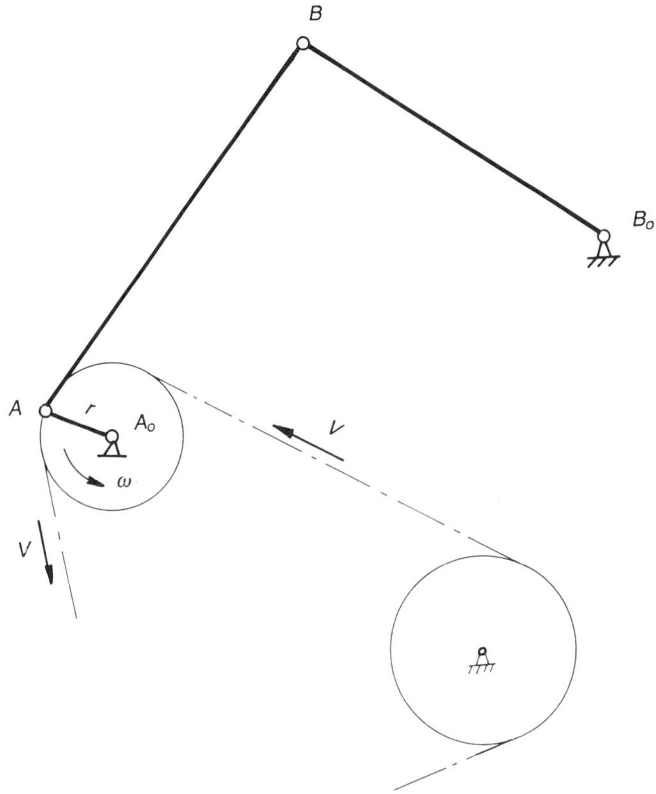

Figure 7.61 The kinematic equivalent mechanism to a pin A moving on a circular path and driving the oscillating member B_0B is a four-bar linkage A_0ABB_0.

Figure 7.62 is an example of linear-to-linear motion. If a rod is moved by two chains traveling in different directions, and the velocity V_A of pin A is known, then the velocity V_B of pin B is found by first locating the instant center O as the intersection of the perpendicular to the direction of motion of pins A and B. The magnitude of the velocity V_B is found from eq. (7.13).

$$V_B = \frac{\overline{OB}}{\overline{OA}} V_A \tag{7.13}$$

Figure 7.63 depicts velocity and acceleration diagrams for points A and B. To find the acceleration of pin B, find the velocity vector V_B of B:

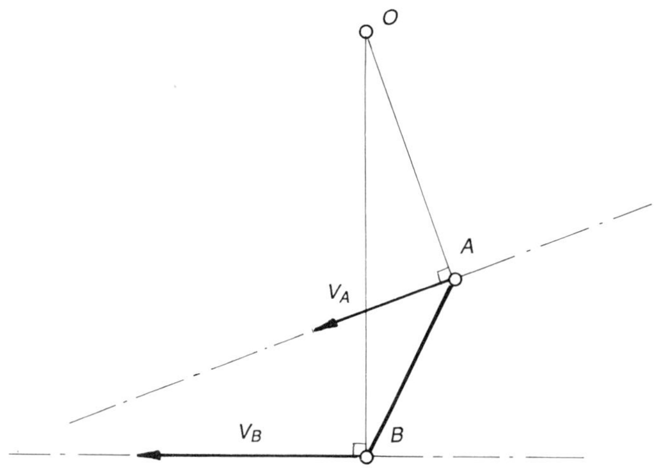

Figure 7.62 A pin A drives the output member B along a straight-line path. The instant center O is used to find the velocity V_B of B when the velocity V_A of A is known.

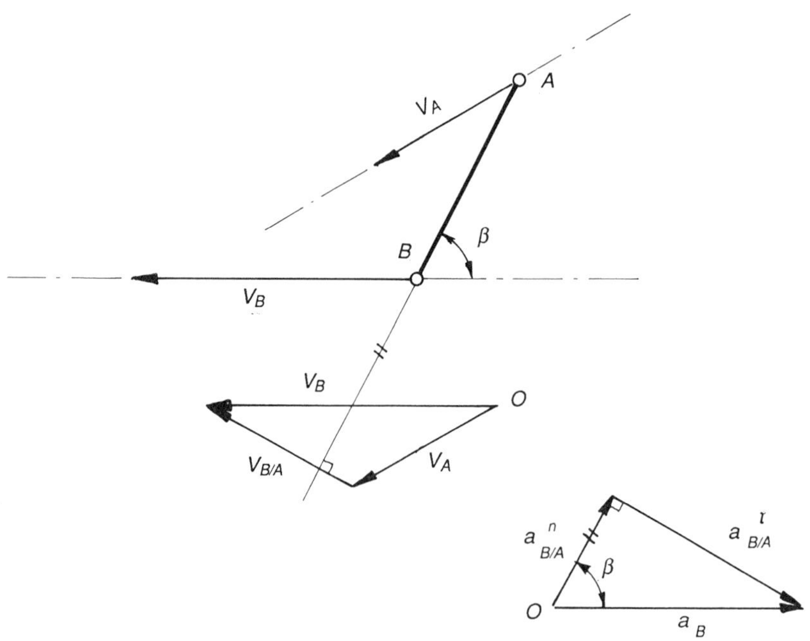

Figure 7.63 Velocity and acceleration diagrams for points A and B when both move along a straight line.

$$V_B = V_A \leftrightarrow V_{B/A} \tag{7.14}$$

The velocity of $V_{B/A}$ (the velocity of B relative to A) is perpendicular to AB, and the velocities V_B and $V_{B/A}$ can be found by scaling of the drawing.

The acceleration of B is directed along its linear path of motion, and

$$a_B = a_A \leftrightarrow a_{B/A}$$

$$\underset{\measuredangle}{a_B} = \underset{=}{a_A} \leftrightarrow \underset{\equiv}{a^n_{B/A}} \leftrightarrow \underset{\measuredangle}{a^t_{B/A}}$$

$$a^n_{B/A} = \frac{V^2_{B/A}}{AB} \qquad \text{directed from B towards A}$$

$$a_A = 0 \qquad V_A \text{ is constant and A is moving on a linear path}$$

The acceleration diagram is shown in Fig. 7.63 and because the angle β between $a^n_{B/A}$ and a_B in the same as the angle β between AB and the direction of motion of B, it follows that

$$a_B = a^n_{B/A} \tan\beta$$

or

$$a_B = \frac{V^2_{B/A}}{AB} \tan\beta \tag{7.15}$$

8
Screw Mechanisms

APPLICATIONS

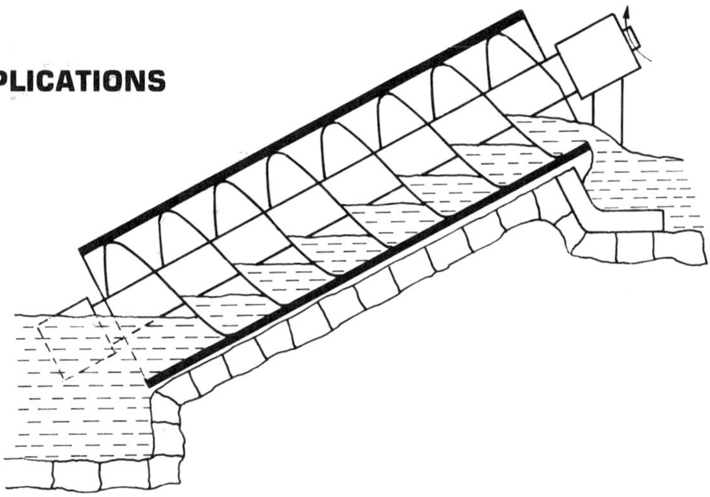

Figure 8.1 One of the oldest and, at the same time, most ingenious applications of a screw is the Archimedes' screw for lifting water. As the screw is rotated, each thread of the screw transports a certain amount of water. Theoretically there is no limit to the height the water can be lifted other than manufacturing capabilities of long length and the required torque on the input shaft.

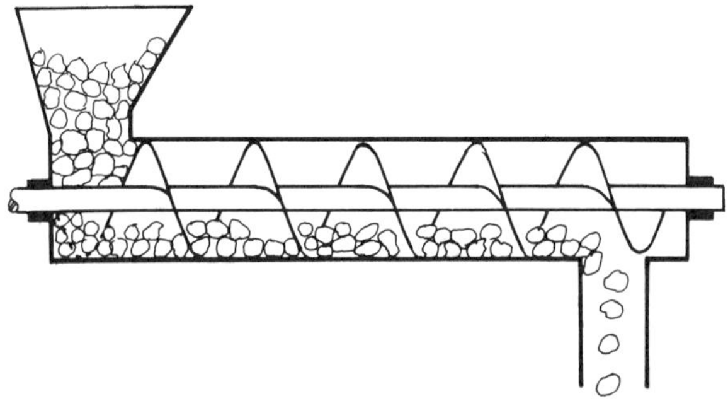

Figure 8.2 A similar device but of a much newer origin is a screw used to transport material.

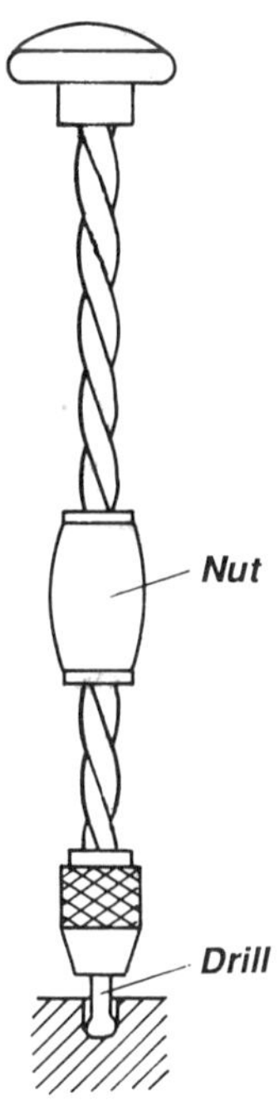

Figure 8.3 A hand-operated screw drive. Moving the nut up and down causes the drill to rotate.

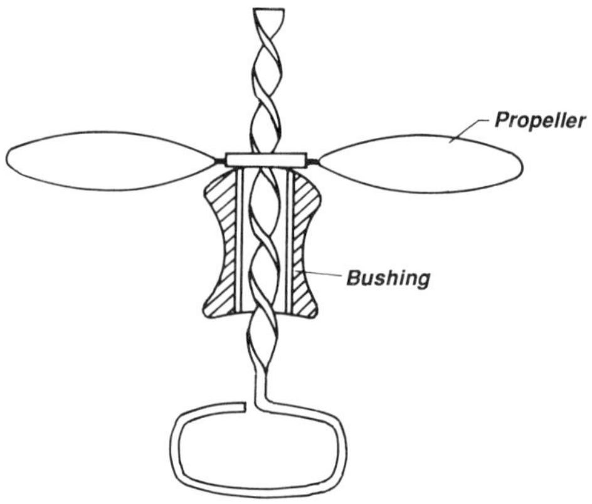

Figure 8.4 A flying propeller toy operated by pushing the bushing upward. The screw is made of flat material that has been twisted; the nut is just a rectangular opening in the propeller. From a structural point of view, the four mechanisms shown so far are open kinematic chains (see Chapter 1).

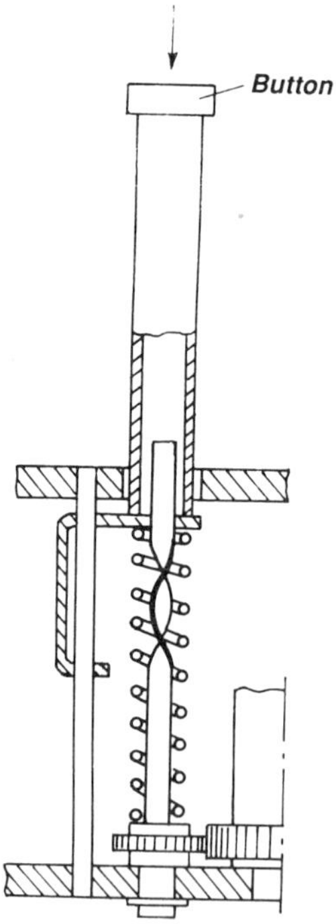

Figure 8.5 A metal strip or square rod may be twisted to make a long-lead thread, ideal for transforming linear into rotary motion. Here a push-button mechanism winds a camera. Note that the number of turns or dwell of the output gear is easily altered by changing the twist of the strip.

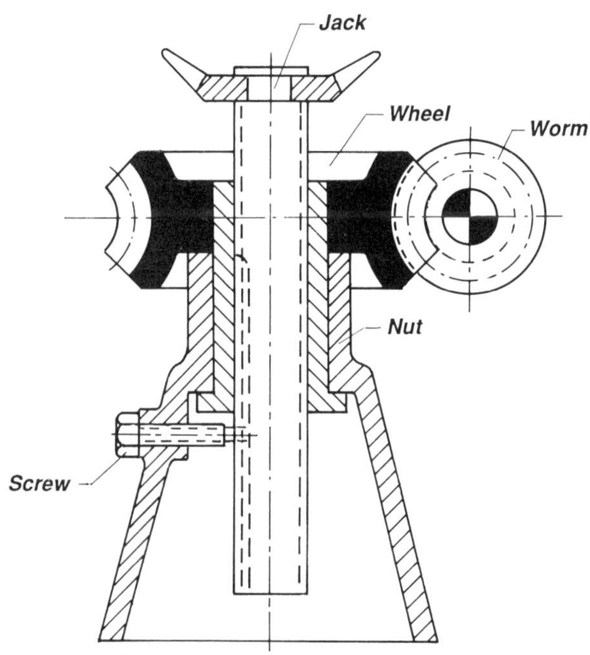

Figure 8.6 Shows a screw jack. The worm drives the wheel, which also acts as a nut. The rotation of the nut lowers or lifts the jack, which is prevented from rotating by a screw that fits into a keyway in the jack. The worm and worm wheel drive decreases the input torque.

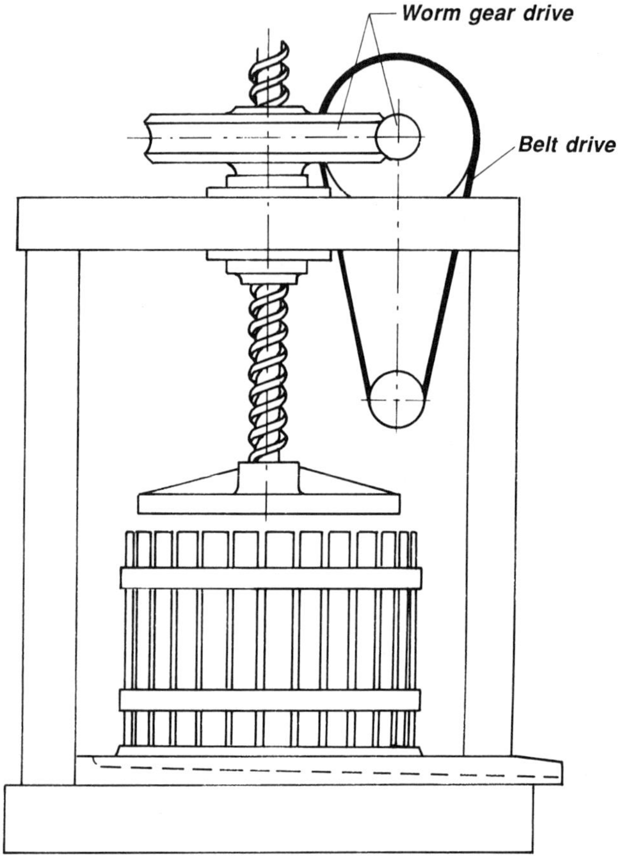

Figure 8.7 A wine press. The worm and worm wheel is driven by a belt drive. The circular plate is driven downwards and squeezes the grapes.

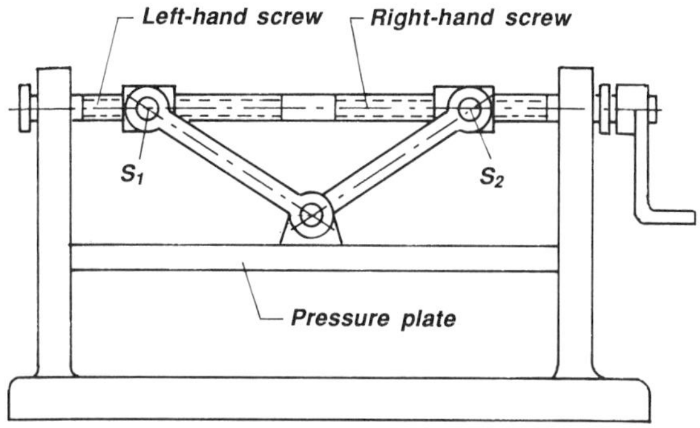

Figure 8.8 With a screw press one of the objectives is to apply a varying pressure, the maximum pressure being exerted when the pressure plate is in its lowest position. The screw has a right- and left-handed thread. When turned it moves the two nuts S_1 and S_2 closer to each other, thereby lowering the plate. As the two nuts get closer and closer, the plate moves slower and slower, thereby increasing the mechanical advantage.

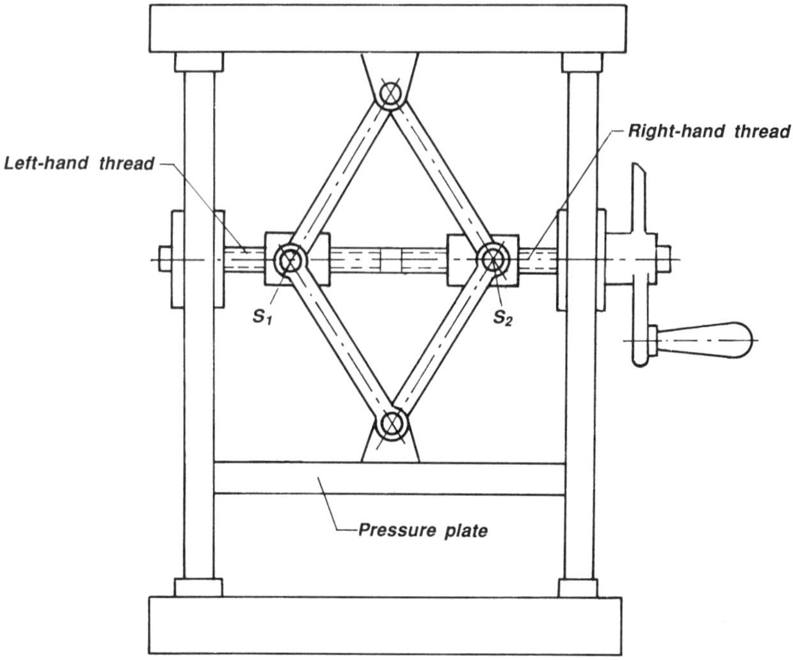

Figure 8.9 Using two more links as compared with Fig. 8.8 causes the hand-driven device to move the pressure plate down with double the speed, but the screw axis must be guided vertically so that it can move down, too, but only with half the speed of the pressure plate.

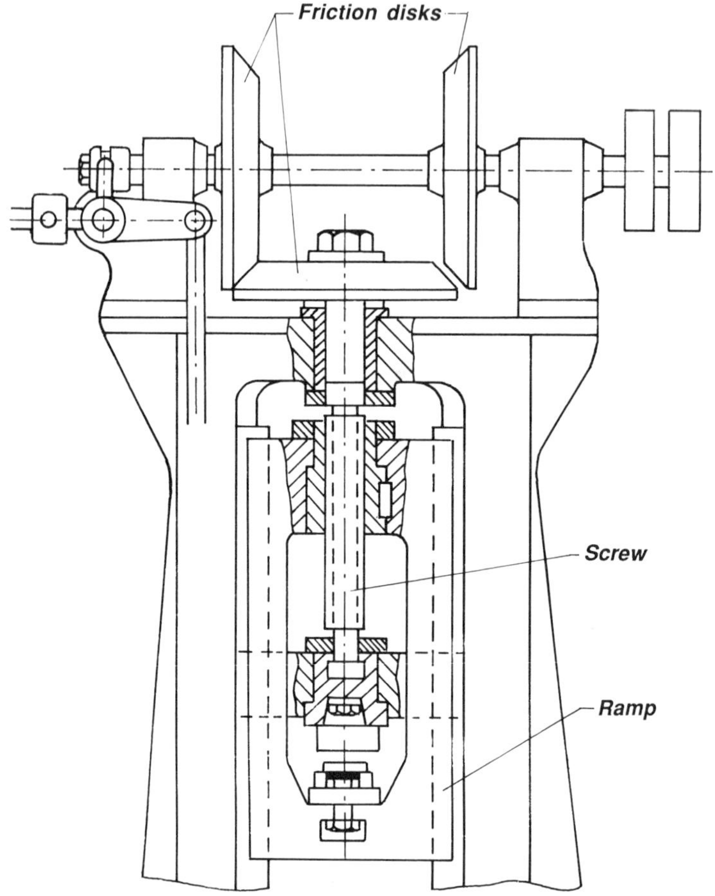

Figure 8.10 Quick reversal of motion is obtained with this screw mechanism. If the left conical friction disc is engaged with the center disc, the screw is driven in one direction, and if the right conical disc is engaged by moving the horizontal shaft to the left, the screw is driven in the opposite direction. The result is that the ramp can be moved up and down very fast, dependent on how fast the horizontal shaft is moved.

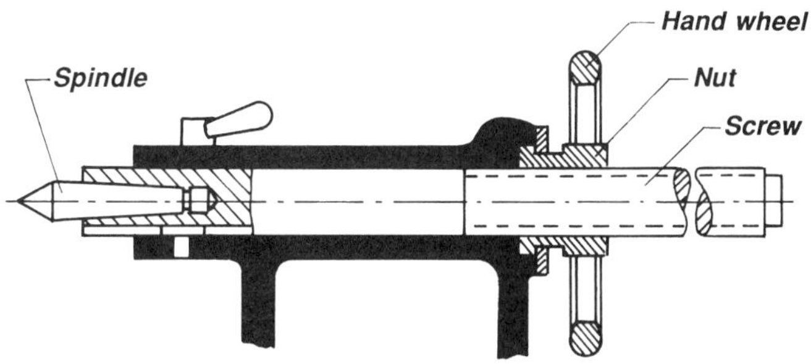

Figure 8.11 A lathe tailstock with a screw mechanism. Turning the hand wheel CW or CCW moves the spindle back and forth. The release of the spindle is not automatic (see Fig. 812).

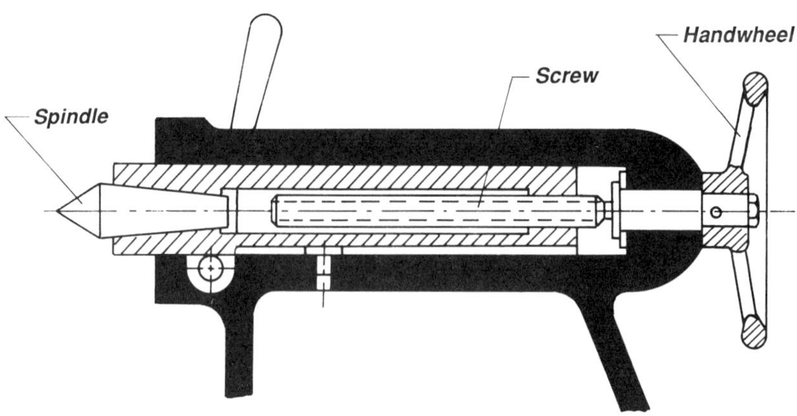

Figure 8.12 An improved arrangement of a tailstock as compared with Fig. 8.11. The retraction of the spindle to the extreme right position will cause the screw to press the spindle out and thereby release it. Comparing the two designs in Figs. 8.11 and 8.12 is a good example of how attention to seemingly small details can drastically improve a design.

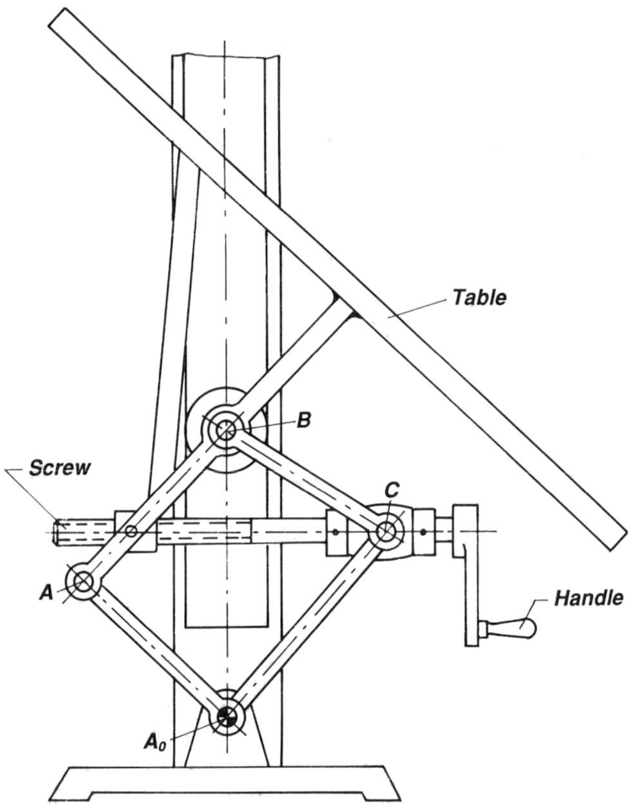

Figure 8.13 The interconnected links A_0A, AB, BC, and A_0C, when moved closer by the screw, tilt and lift the table at the same time. The input motion is to the handle that activates the screw.

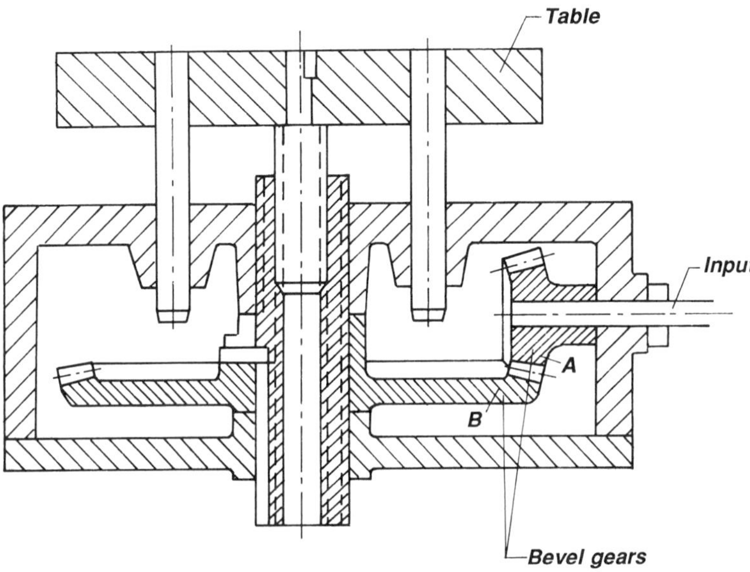

Figure 8.14 The vertical position of the table can be finely adjusted by means of the double screw that is rotated by the two bevel gears A and B, where B is driven from the input shaft.

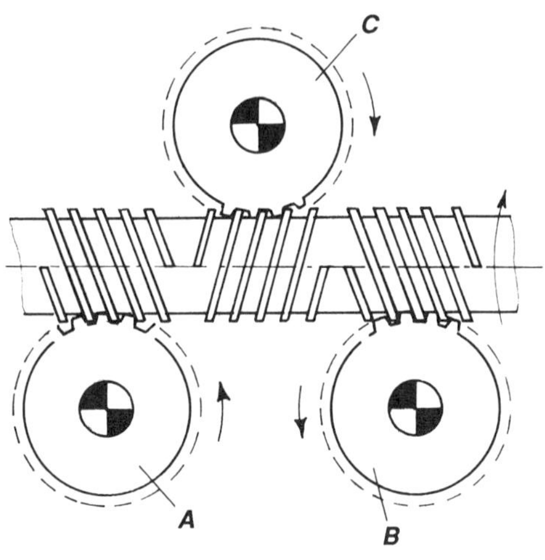

Figure 8.15 Three integral screws, the center screw left-handed and the other two right-handed, turn the three gears A, B, and C. The screws are made of square wire, resulting in a low-cost design.

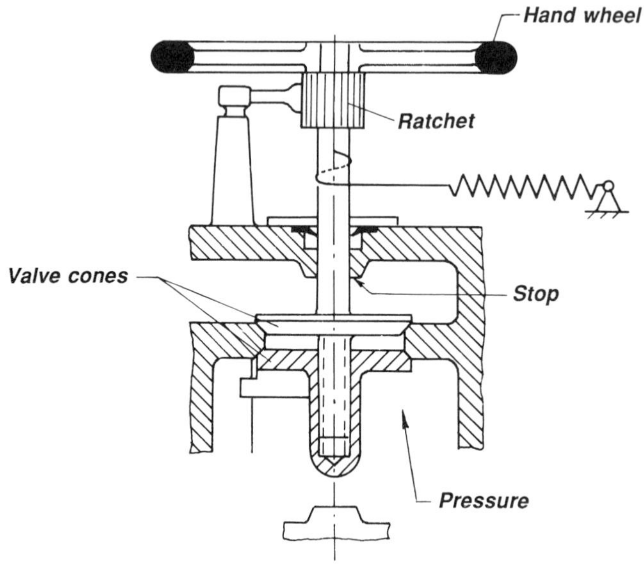

Figure 8.16 Valve stem has two opposite-moving valve cones. When opening, the upper cone moves up first, until it contacts a stop. Further turning of the wheel forces the lower cone out of its seat. The spring is wound up at the same time. When the rachet is released, spring pulls both cones into their seats.

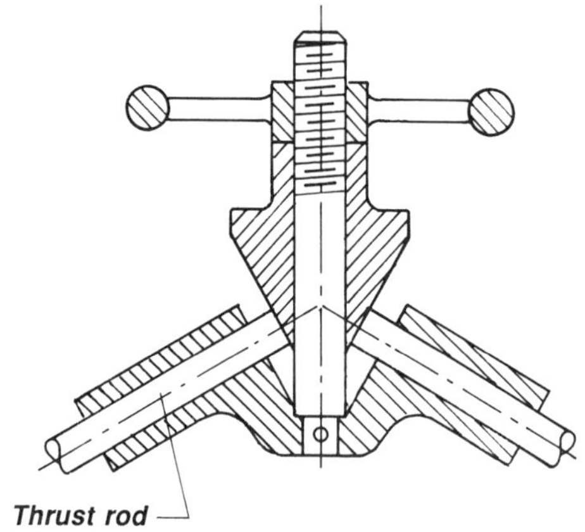

Figure 8.17 The conical-shaped member is operated by a screw mechanism and moves the two thrust rods in unison.

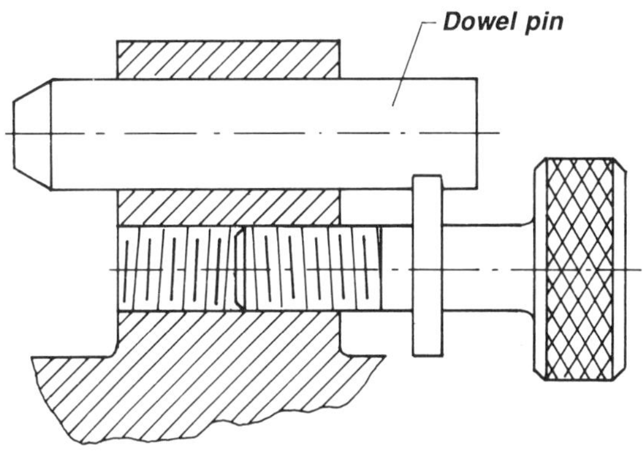

Figure 8.18 The screw motion moves the dowel pin back and forth for locking applications.

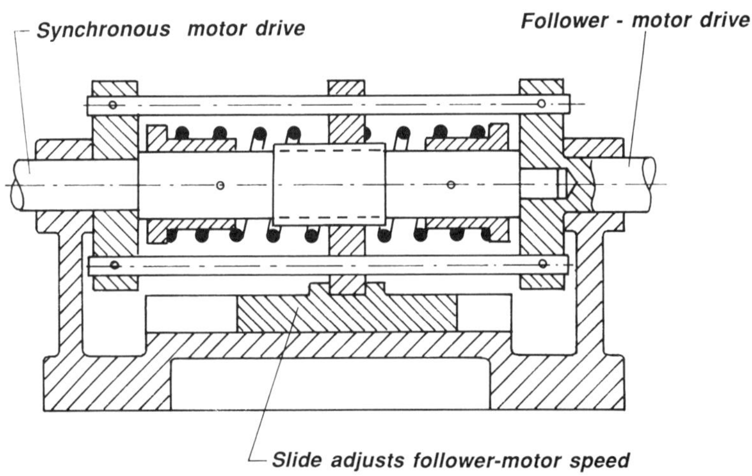

Figure 8.19 Any variable-speed motor can be made to follow a small synchronous motor by connecting them to the two shafts of this differential screw. Differences in number of revolutions between the two motors appear as motion of the traveling nut and slide so that an electrical speed compensation is made.

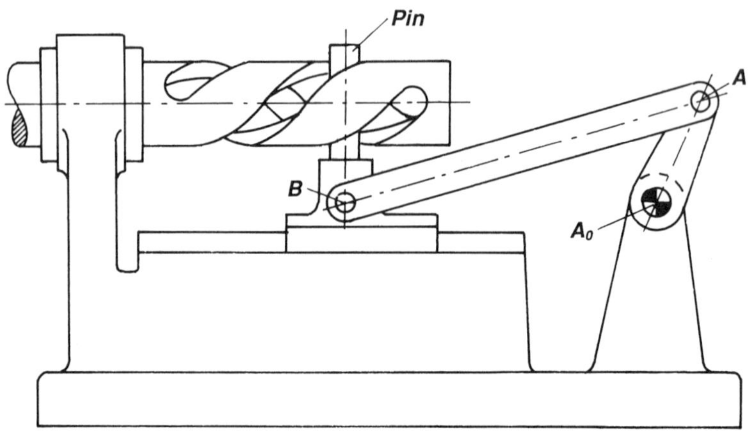

Figure 8.20 A screw thread does not always look like a screw thread, but the slotted member here performs exactly the same function as a screw. The rotation of crank A_0A of the slider crank A_0AB moves the slider with a linear motion. The pin on the slider fits into the slotted member and forces the screw-like slotted member to oscillate with excellent acceleration characteristics (continuous and close to simple harmonic motion).

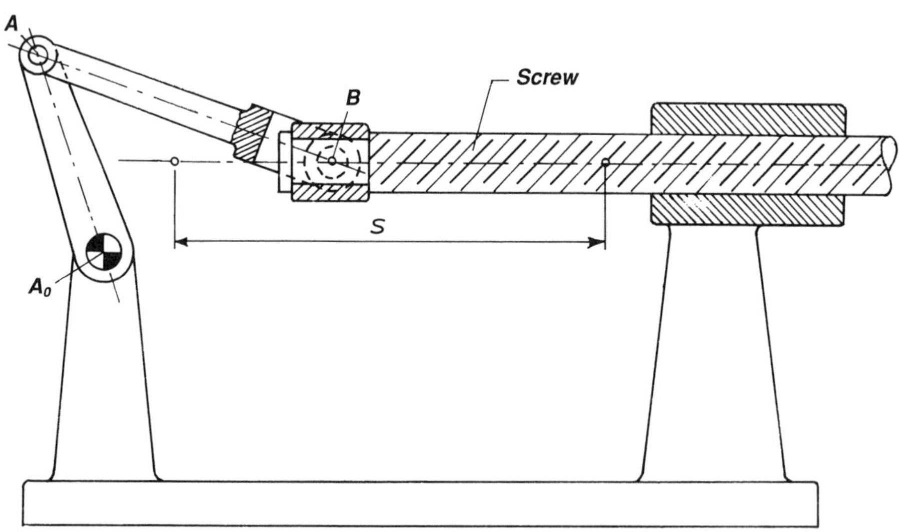

Figure 8.21 The slider crank A_0AB moves the slider B back and forth. The slider is a screw, and the translating motion of B is converted to the translating and rotating of the screw.

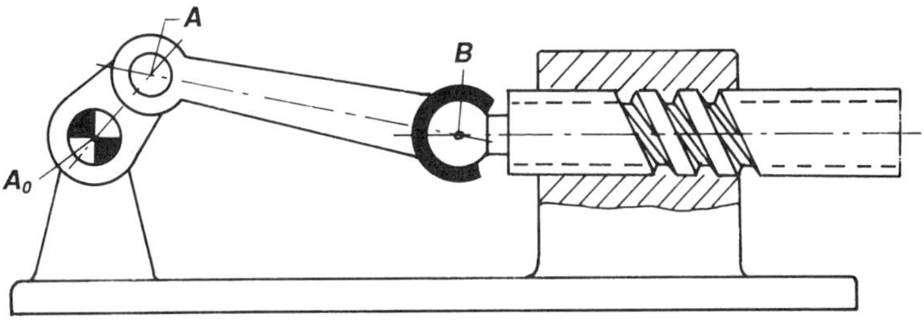

Figure 8.22 The slider crank A_0AB moves the slider back and forth. The slider is a screw, and the translating motion of B is converted to the translating and rotating of the screw. The joint at B is called a ball-and-socket joint and allows rotation in three directions: one around the axis AB, one in the plane of the paper, and one in the direction perpendicular to the paper.

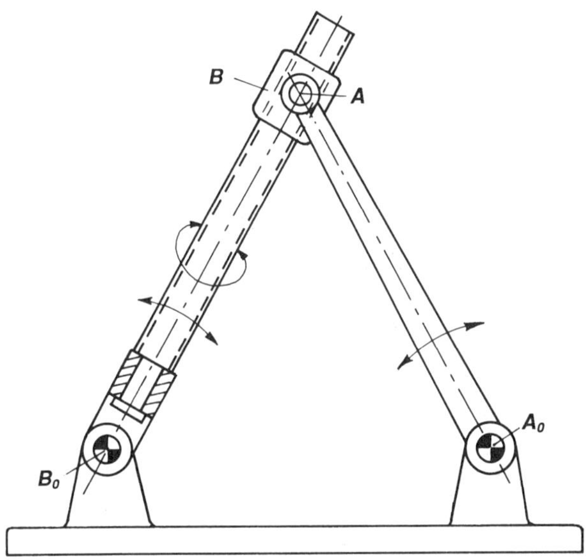

Figure 8.23 A shaft is made to oscillate around a fixed center and at the same time rotate around its own axis. The arm A_0A is made to oscillate. The slider B is connected to A_0A by a turning joint and a nut. As A_0A rotates, it imparts a rotary motion to B_0B. The screw rotates around its own axis and at the same time its axis oscillates around B_0.

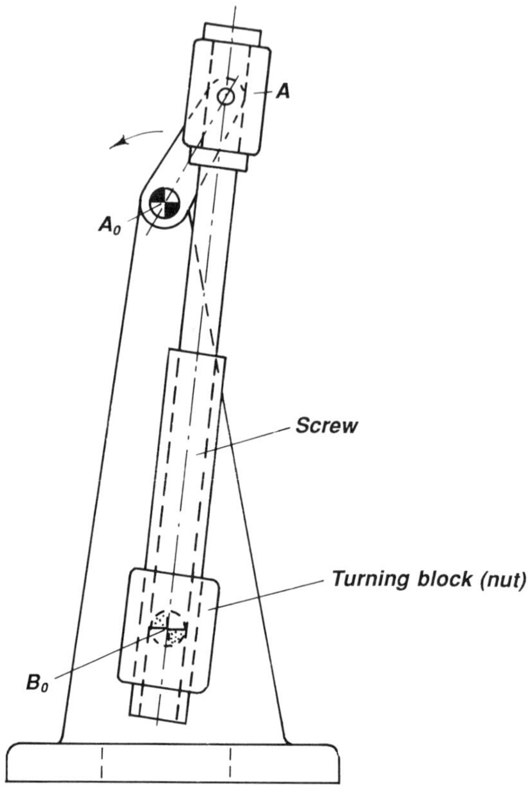

Figure 8.24 Here, the arm A_0A rotates and is connected to the screw by two turning joints. The screw is guided at B_0 by a turning block in the form of a nut. The nut has B_0 as a fixed center of rotation. The screw rotation and oscillation are as in Fig. 8.23.

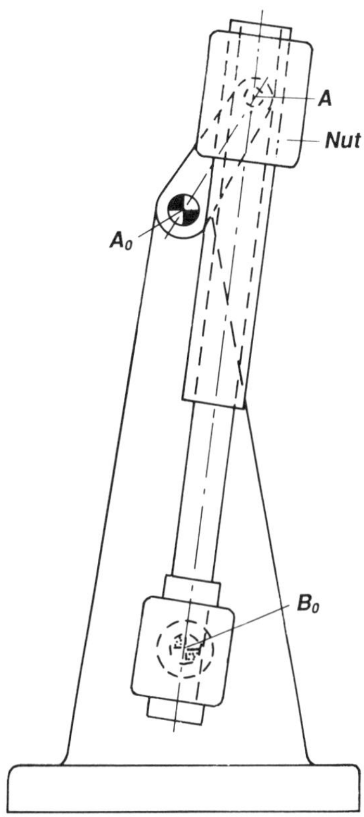

Figure 8.25 Here, the arm A_0A rotates and is connected to the nut by a turning joint. The screw is guided at B_0 by a turning block that rotates around B_0. The screw rotation and oscillation are as in Fig. 8.23.

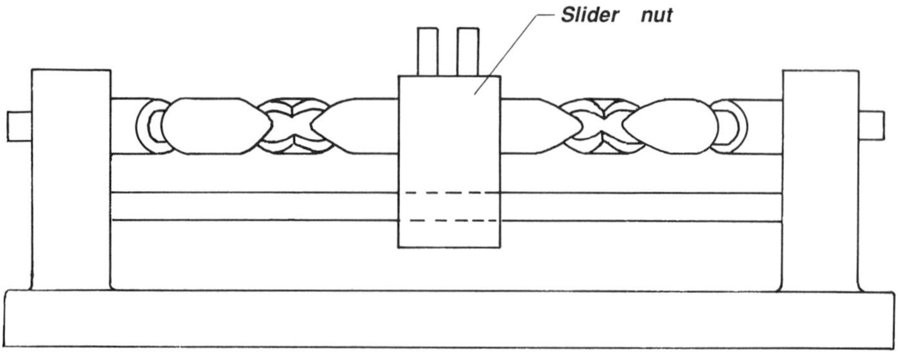

Figure 8.26 A double-helix screw provides motion back and forth from a rotating output. The slider nut follows the grooves so that it translates back and forth when the screw is rotated.

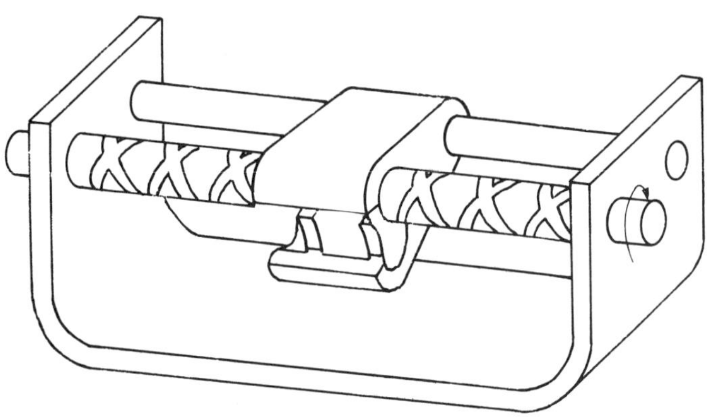

Figure 8.27 A nut with left- and right-handed thread provides linear oscillating motion. The nut fits either the left- or right-hand thread of the double thread on the shaft. (If the right-hand thread is engaged, the nut moves to the right if shaft rotates CW as seen from the right.) Shifting the nut from one side to the other causes a reversal of the linear travel of the nut.

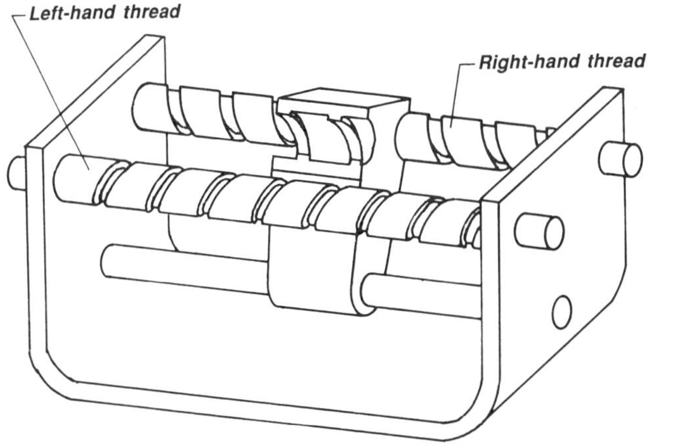

Figure 8.28 Again, the nut fits either the left- or right-hand thread of the two screws. Shifting the nut from one screw to the other causes a reversal of the linear travel of the nut.

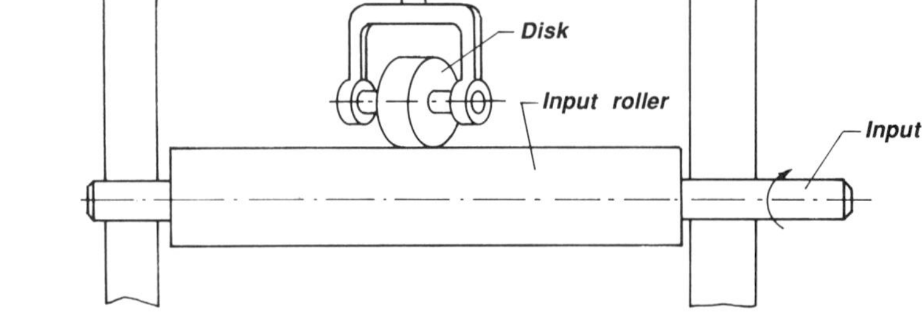

Figure 8.29 A reversal translating motion is obtained with this disk and roller driver. Here a hardened disk, riding at an angle to an input roller, transforms the rotary motion into linear motion *parallel to the axis of the input*. The roller is pressed against the input shaft with the help of flat spring, F. Feed rate is easily varied by changing the angle of the disk. Arrangement can produce an extremely slow feed with a built-in safety factor in case of possible jamming. Often the perimeter of the disc is made almost knife-edged, and consequently contact stresses are high.

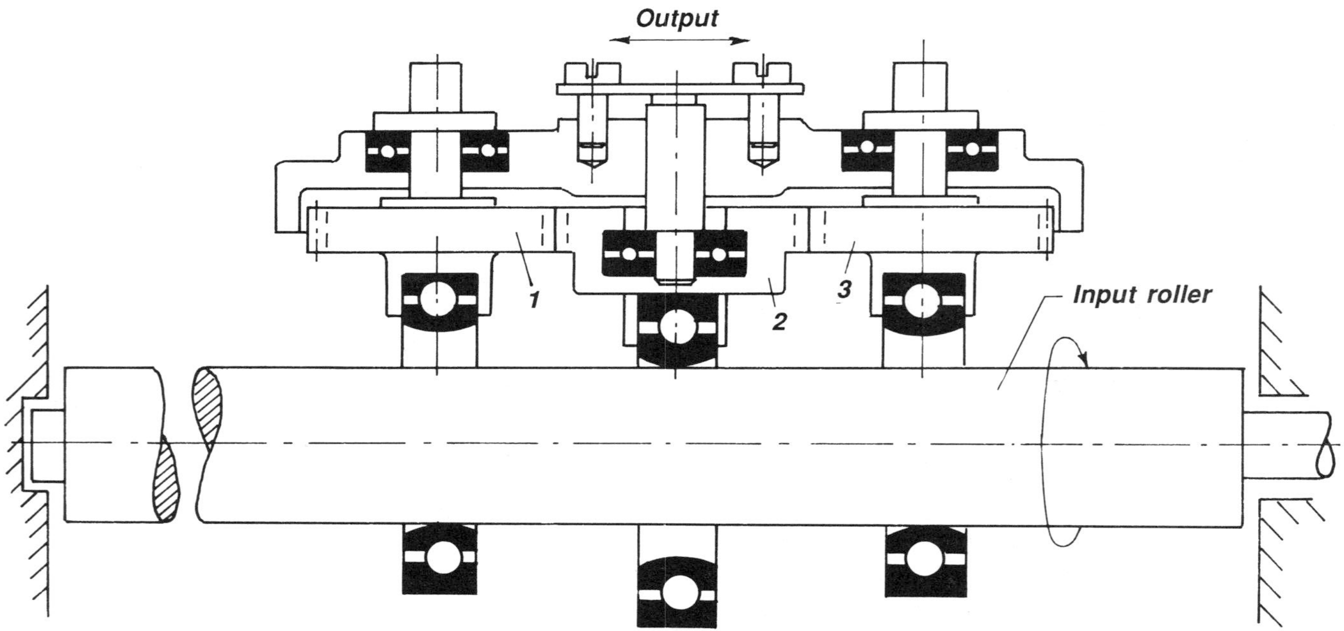

Figure 8.30 J. Uhing's roller drive. The high contact stresses and the bending moment on the shaft are considerably decreased as compared with Fig. 8.29 by having the inner, spherical-shaped rings of the ball bearings contact the input roller. When the center bearing, which contacts the upper side of the shaft, is rotated to obtain a screwed angle, the outer bearings, through gears 1, 2, and 3, are screwed in the opposite direction, but contact the input roller on the lower side of the shaft. The results is that all three bearings combine to translate the output member.

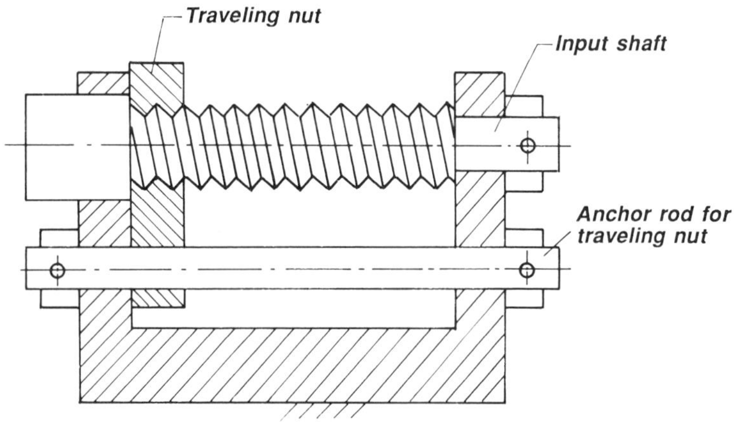

Figure 8.31 Traveling nut limits number of revolutions the input shaft can turn. Rotating the input shaft causes the nut to travel. The nut blocks the motion of the screw when in an extreme position, thereby limiting the number of revolutions of the input shaft. If the nut presses too hard towards the frame-stop, it might be difficult to reverse the motion. This can be remedied by increasing the screw pitch.

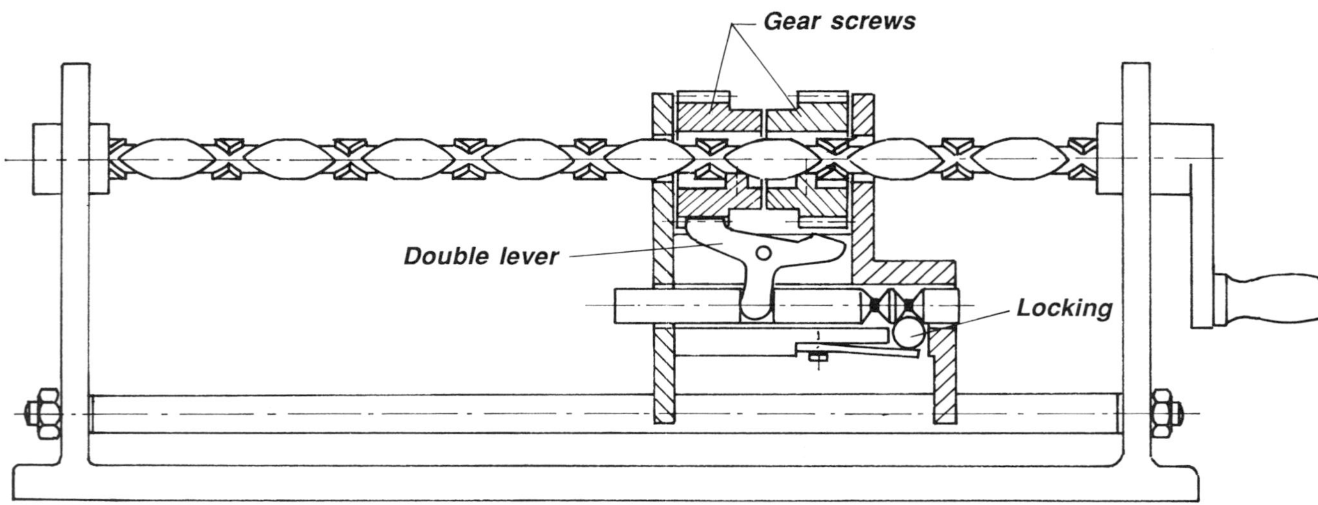

Figure 8.32 Reciprocating motion with partly constant velocity. The double lever shifts from, say, left gear-screw to right gear-screw when approaching the extreme left position, so that the motion of the carriage is reversed.

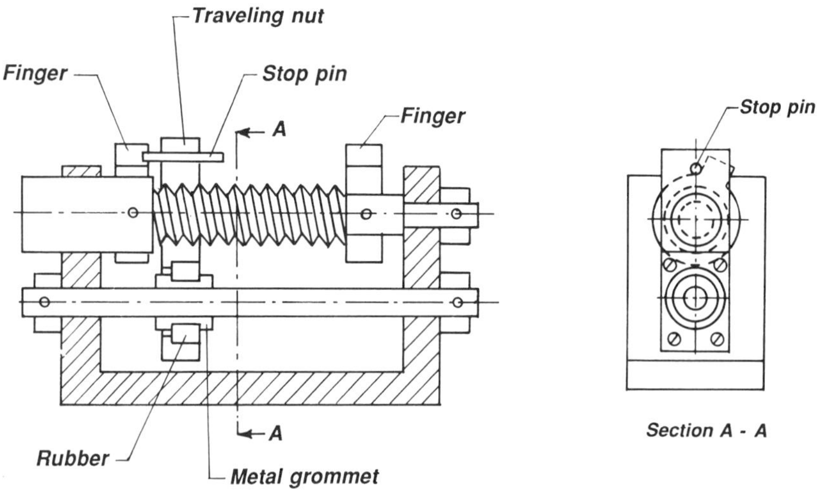

Figure 8.33 In this design a stop pin is provided in a traveling nut. Engagement between pin and rotating finger must be shorter than the thread pitch, so pin can clear finger on the first reverse turn. The rubber ring and grommet lessen impact, providing a sliding surface. The grommet can be oil-impregnated metal.

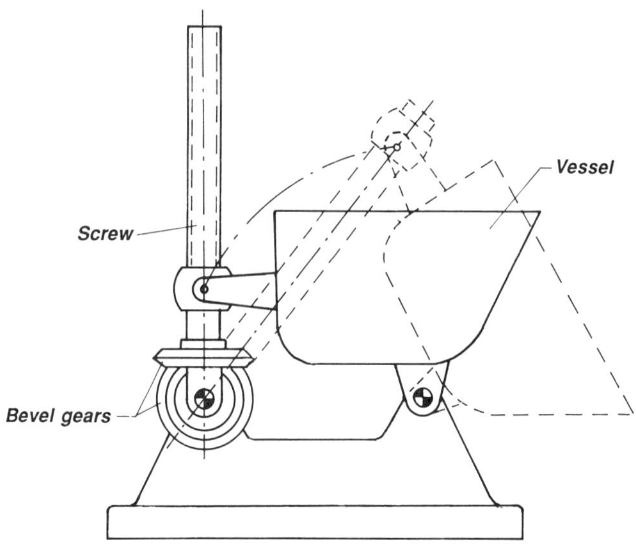

Figure 8.34 A lever screw is used to tip a container. The screw is driven by two bevel gears.

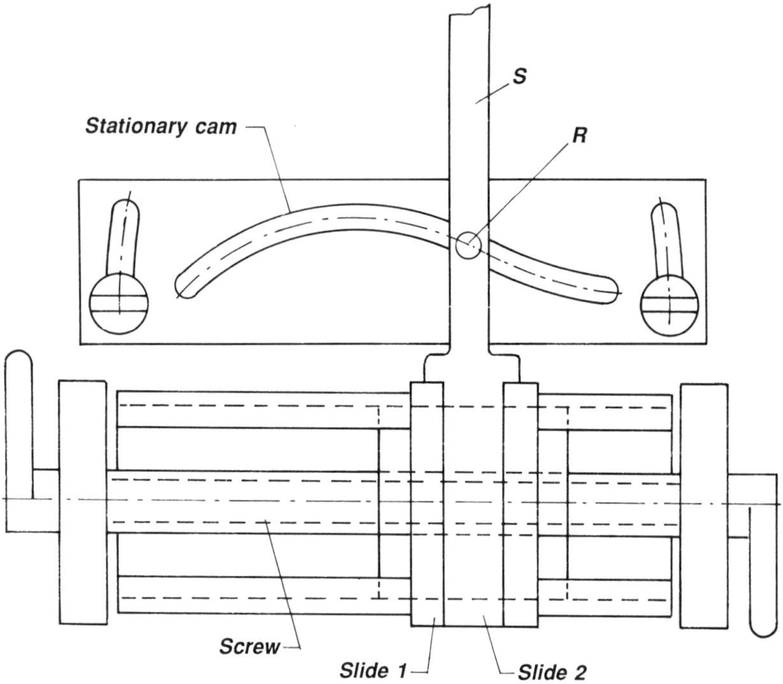

Figure 8.35 A motion in one direction is converted to a perpendicular motion. Slide 1 is moved by the screw in a horizontal direction. Slide 2 can slide in 1, but at the same time its motion is controlled by the stationary cam by means of the roller R. The position of the cam can be adjusted by the two screws.

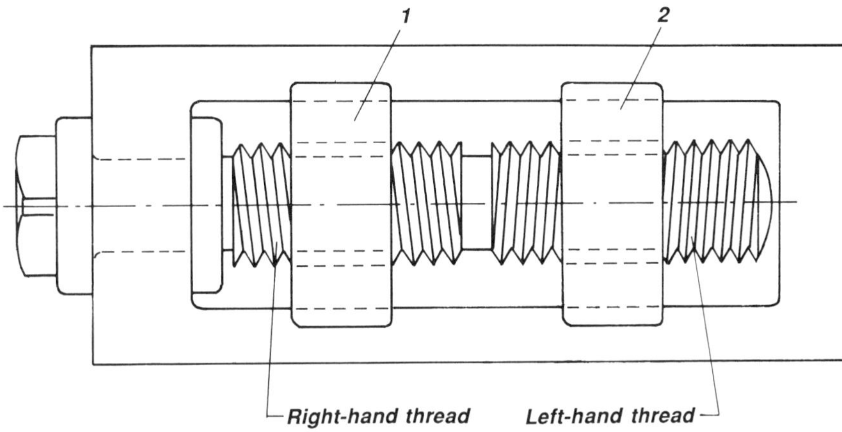

Figure 8.36 A double screw with a left- and right-hand thread forces the two members 1 and 2 towards or away from each other, dependent on the direction of motion of the double screw. This mechanism can be used as clamping device.

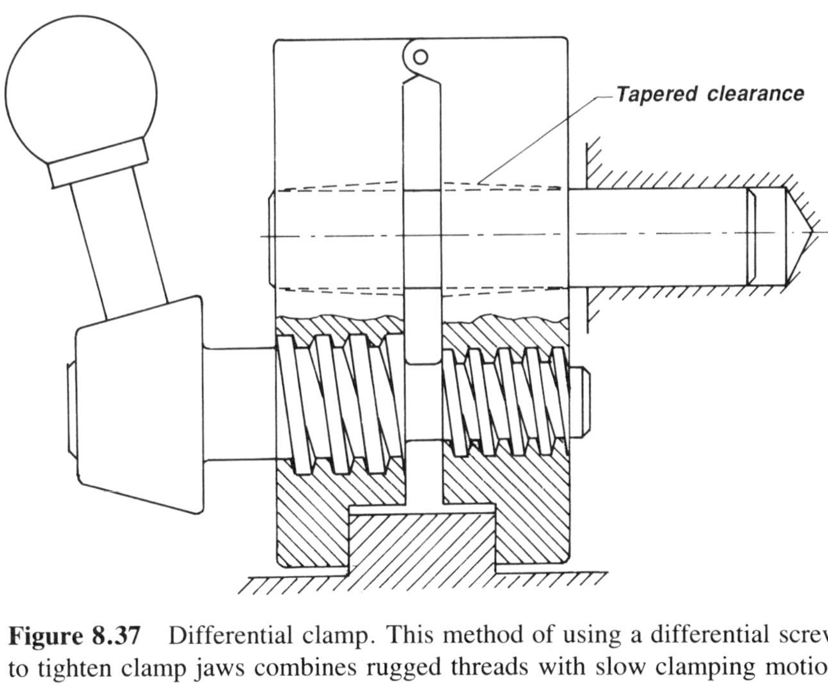

Figure 8.37 Differential clamp. This method of using a differential screw to tighten clamp jaws combines rugged threads with slow clamping motion and therefore clamping power.

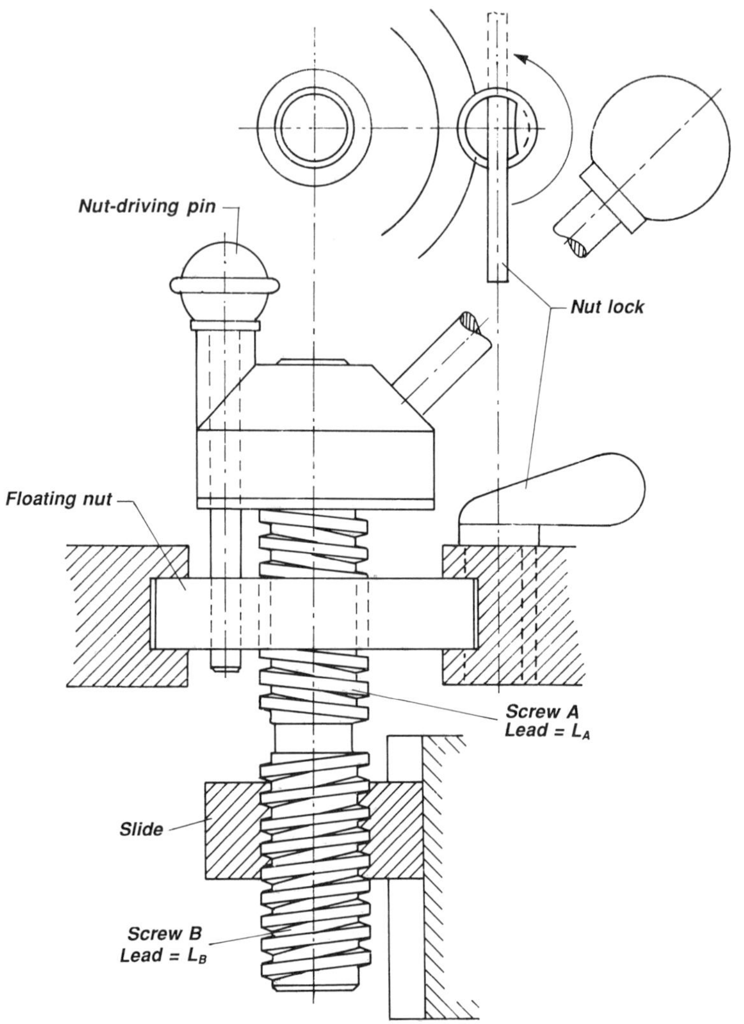

Figure 8.38 Rapid and slow feed mechanism. With left- and right-hand threads, slide motion with nut locked equals L_A plus L_B per turn; with nut floating, slide motion per turn equals L_B. This arrangement gets extremely fine feed with rapid return motion when threads are differential.

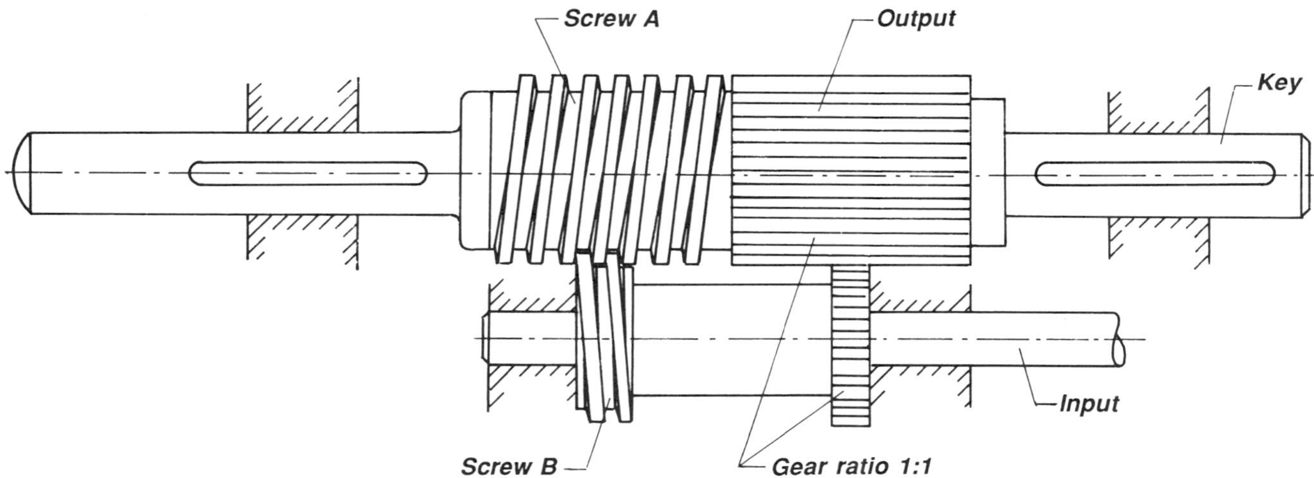

Figure 8.39 High reduction of rotary to fine linear motion is possible. Screws are left- and right-hand. PA = PB plus or minus a small increment. When PB = 1/10 in. and PA = 1/10.5 in., linear motion of screw A will be 0.05 in. per turn. This arrangement can be used only for small forces.

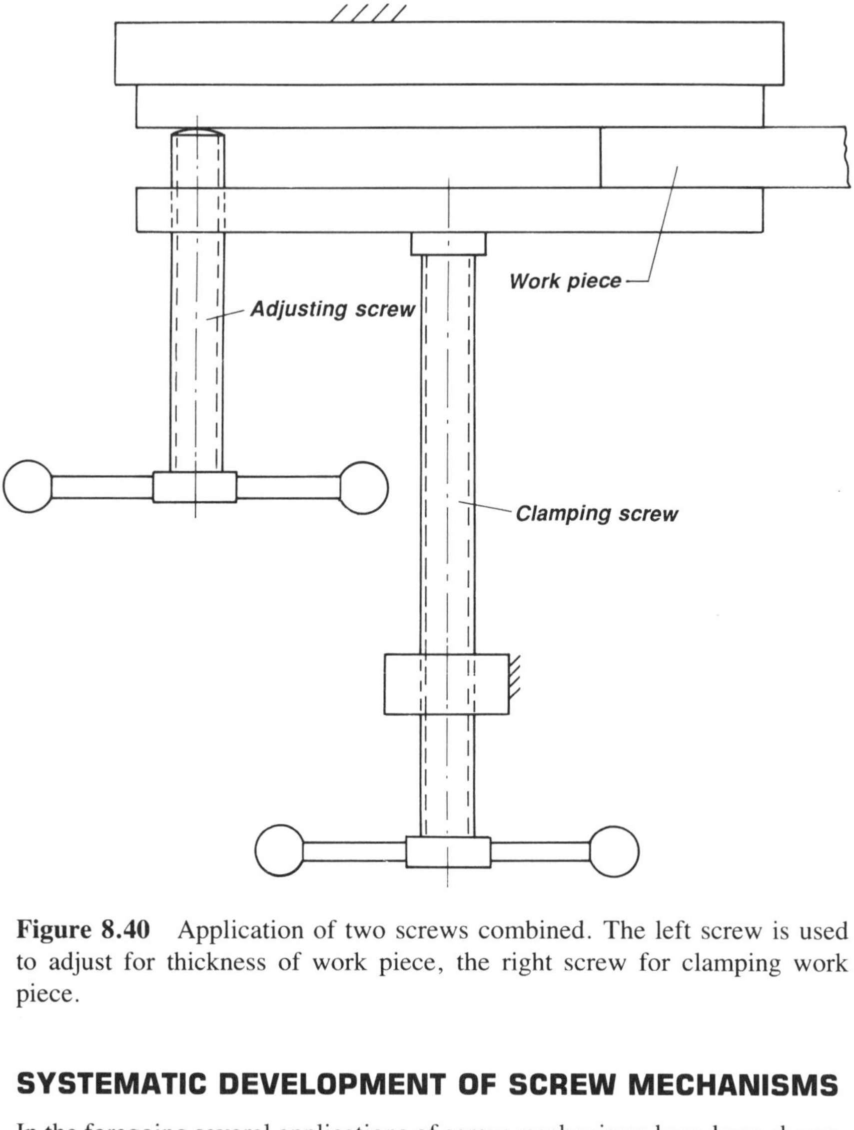

Figure 8.40 Application of two screws combined. The left screw is used to adjust for thickness of work piece, the right screw for clamping work piece.

SYSTEMATIC DEVELOPMENT OF SCREW MECHANISMS

In the foregoing several applications of screw mechanisms have been shown. Just as the systematic development of many mechanisms is carried out in Chapter 17, the same can be done with screw mechanisms. The basis for the following development of screw mechanisms will be the lower-pair connections and the associated screw mechanisms. Interspersed among the theoretical possibilities will be a number of applications along with references to some of the screw mechanisms already shown.

JOINTS

Joints, or kinematic pairs, connect the various, in general rigid, members of mechanism. As shown in Chapter 1, certain joints, however, should be considered as mechanisms in their own right. A kinematic pair has a number of degrees of freedom, which can be anywhere from one to five. A kinematic pair with surface contact is classified as a lower pair.

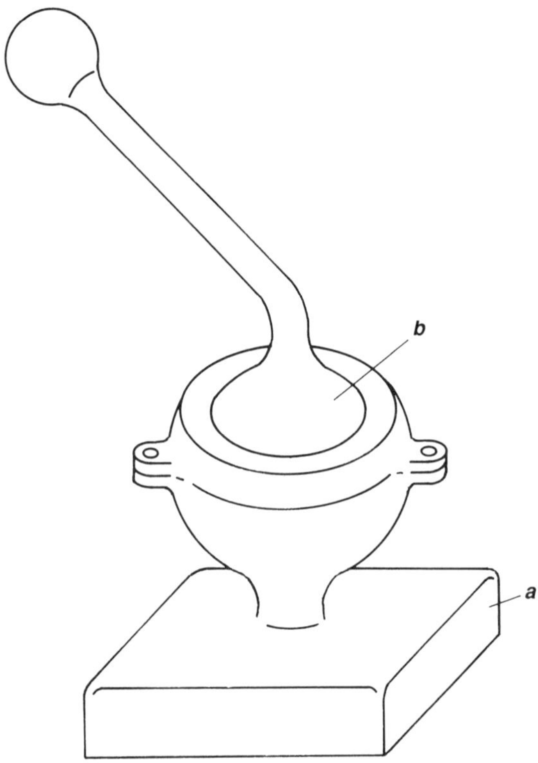

Figure 8.41 All ball-and-socket joint where a relative to b can make three rotations including the one around its own axis (three degrees of freedom). The kind of contact between two elements a and b is surface contact.

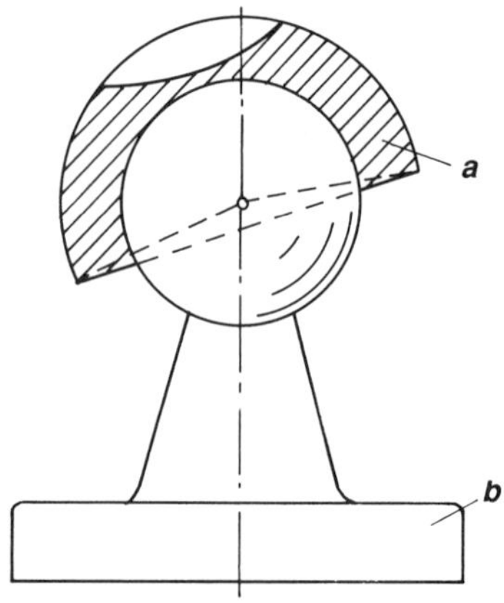

Figure 8.42 It is possible to interchange the solid sphere and the hollow sphere as shown without changing the relative motion (kinematic inversion).

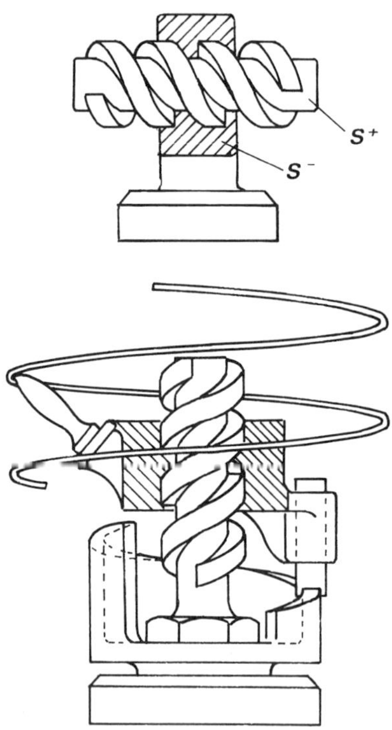

Figure 8.43 A screw pair. If the nut is turned, a point on the nut traces an helix. If the nut is kept stationary, the screw traces a helix, too.

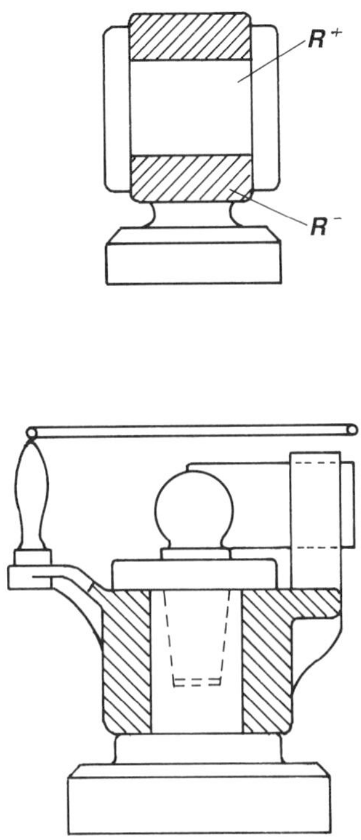

Figure 8.44 A turning, or revolute, pair. Whether the pin or the bearing is turned, the two elements trace circles relative to each other.

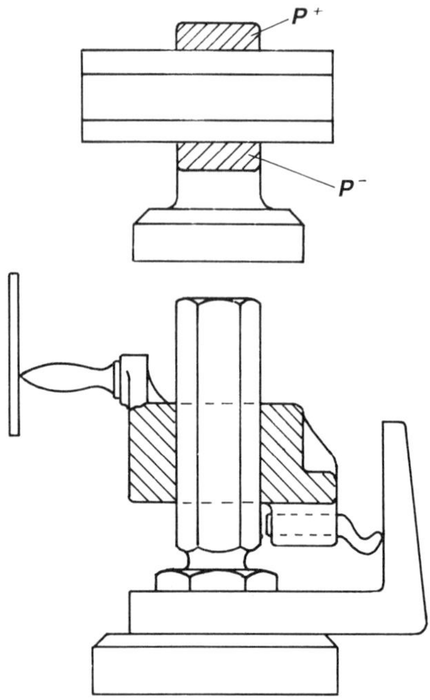

Figure 8.45 A sliding pair. If the prism is moved, it traces a straight line relative to its bearing. A kinematic inversion, keeping the prism stationary and moving the bearing, doesn't change the relative motion between the two elements.

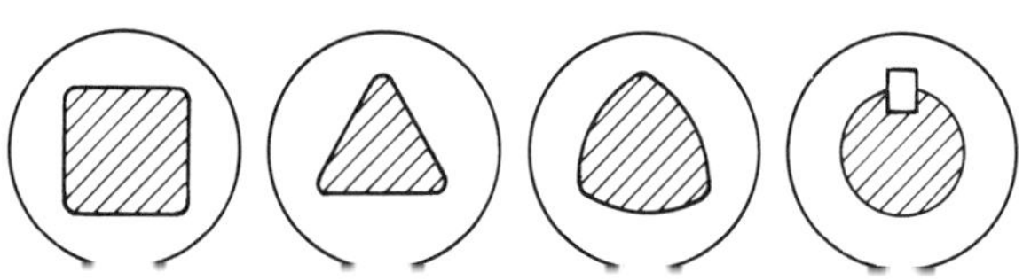

Figure 8.46 The cross-sections of a sliding pair can have any of the forms (and a host of others) shown.

Let S designate a screw pair, R a turning pair (revolute pair, turning joint), and P a sliding pair (prismatic pair). Each pair consists of two elements. Let R^- designate the hollow part and R^+ the solid part of the revolute pair. In the same way, the two other joints P and S can be identified by two parts, namely, P^+/P^- and S^+/S^-. These designations are shown in the top drawings of figs. 8.43–8.45.

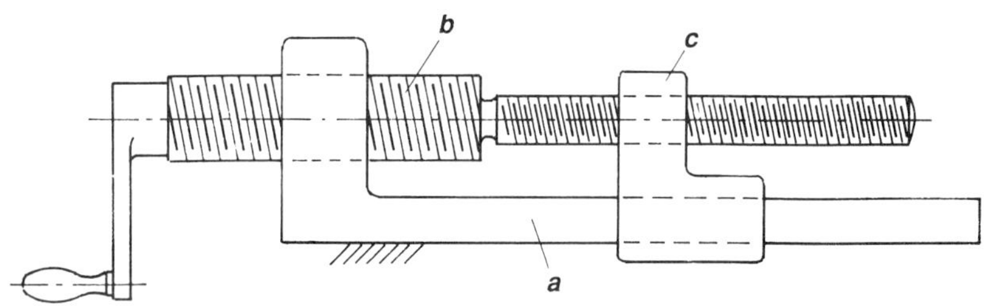

Figure 8.47 A differential three-joint mechanism with two screws and one sliding joint.

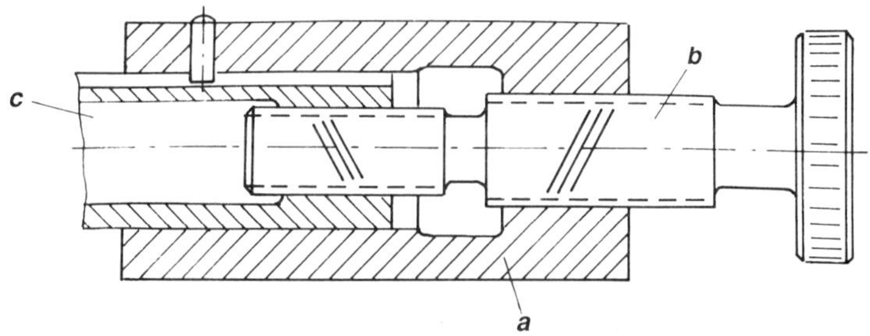

Figure 8.48 A differential screw mechanism (for fine feed) where the two screws are either both right- or left-handed, or one is left- and the other right-handed.

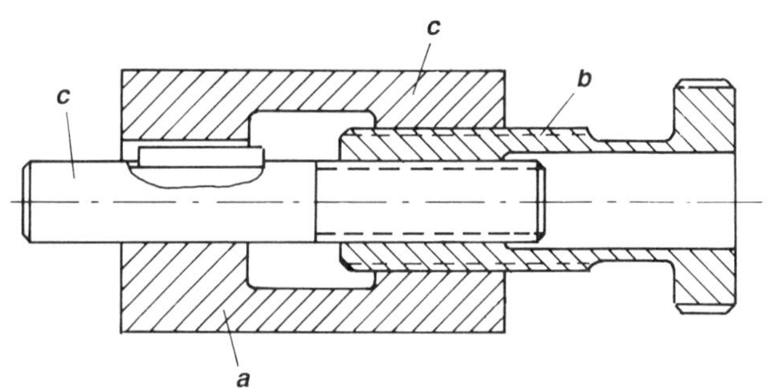

Figure 8.49 Another differential screw mechanism with same-hand or different-hand threads.

USING KINEMATIC NOTATION

Let us start the investigation of kinematic notation with a known mechanism with three parts. The mechanism (Fig. 8.50) consists of three screws with colinear axes. Designate the frame by a, and the two other members by c and d. The links a, b, and c are now written

$$\underline{a} \qquad b \qquad c$$

with the frame "a" underlined. If we look at the mechanism a little closer, we realize that the connection from a to b is S^- to S^+, because it is a screw joint that connects a and b with the hollow part on a. The connection from b to c is S^+ to S^-. This is expressed in symbolic form to the left in Fig. 8.50. A convenient way of writing the symbolic representation of the mechanism is to write

$$\underbrace{S^- \qquad S^-}_{\underline{a}} \quad \underbrace{S^+ \qquad S^-}_{b} \quad \underbrace{S^+ \qquad S^+}_{c}$$

if the mechanism is read CCW. A reading in a CW direction yields

$$\underbrace{S^- \qquad S^-}_{\underline{a}} \quad \underbrace{S^+ \qquad S^+}_{b} \quad \underbrace{S^- \qquad S^+}_{c}$$

In order to differentiate between the pitches of the different screws, we use subscripts 1, 2, and 3, so that the notation is written

$$\underbrace{S_3^- \qquad S_1^+}_{\underline{a}} \quad \underbrace{S_1^+ \qquad S_2^+}_{b} \quad \underbrace{S_2^- \qquad S_3^+}_{c}$$

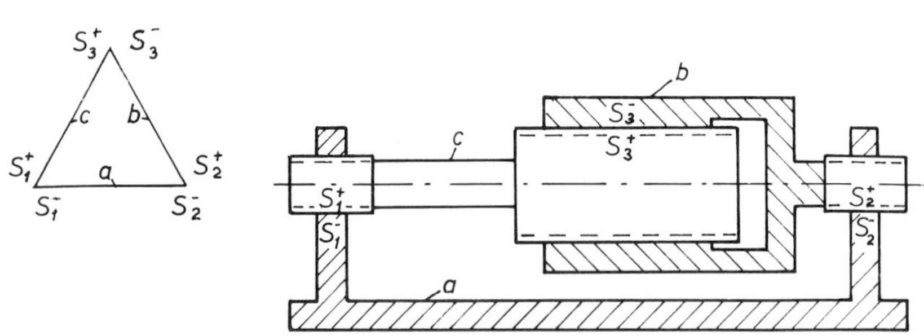

Figure 8.50 Variants of three-link, three-screw mechanism with symbolic notation to the left. (See also Figs. 8.51–8.53.)

The screw pitches are designated S_1, S_2, S_3 and are considered positive (+S1) if a right-hand thread and negative (−S1) if a left-hand thread. The various designations with the symbolic representation are shown to the left, and the actual mechanism to the right.

Based on the notation described above, three more versions can be developed by interchanging hollow and solid elements Figs. 8.51, 8.52, and 8.53.

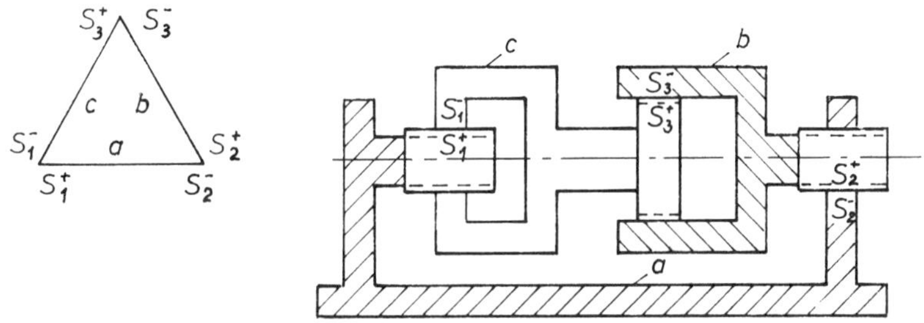

Figure 8.51

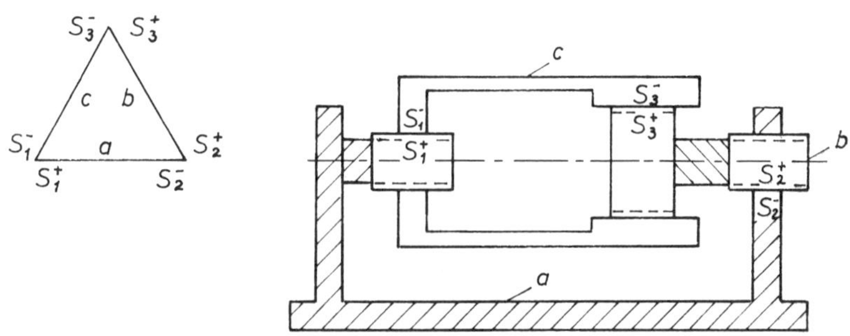

Figure 8.52

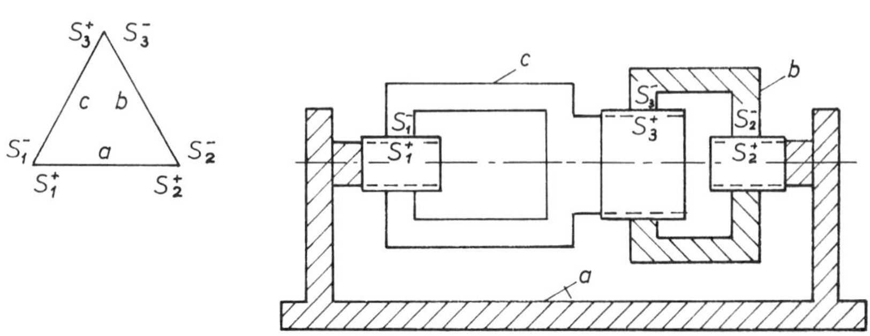

Figure 8.53

To develop formulas relating the rotations and displacements of the various members (Fig. 8.54), let

$$\left.\begin{matrix}S_1\\S_2\\S_3\end{matrix}\right\} = \text{pitches of the screw joints}$$

a = frame member

$$\left.\begin{matrix}b\\c\end{matrix}\right\} = \text{moveable members}$$

$$\left.\begin{matrix}n_b\\n_c\end{matrix}\right\} = \text{number of revolutions of members a and b}$$

and let us cut the frame so that a stationary left part a is obtained and a right part a′, free to rotate. First, rotate b, c, and a′ through n_c. The linear displacement of part a′ is

$$X_1 = n_c S_1 \tag{8.1}$$

Now keep c stationary and rotate b and a′ through $-n_c + n_b$ so that member b has rotated a total of $n_c + (-n_c + n_b) = n_b$. The corresponding linear displacement of a′ is

$$X_2 = (-n_c + n_b)S_3 \tag{8.2}$$

Finally, keeping b and c stationary, a′ is rotated through n_b and its linear displacement is

$$X_3 = n_b S_2 \tag{8.3}$$

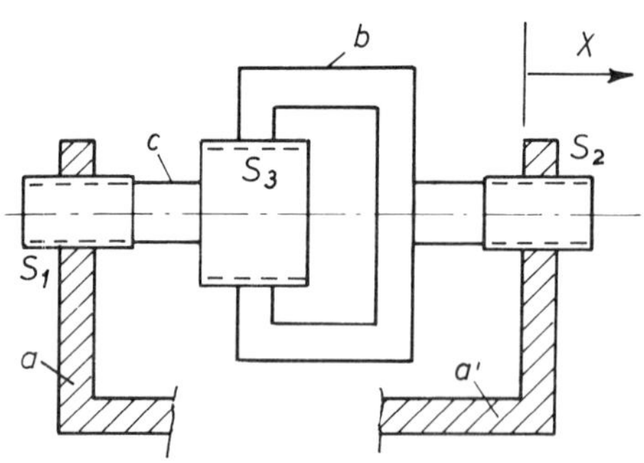

Figure 8.54 Arrangement to develop displacement formula for three-link, three-screw mechanism.

but

$$X_3 = X_1 + X_2 \tag{8.4}$$

$$n_b S_2 = -n_c S_3 + n_b S_3 + n_c S_1 \tag{8.5}$$

or

$$S_1 = RS_2 + (1 - R)S_3 \tag{8.6}$$

where

$$R = \frac{n_b}{n_c} \tag{8.7}$$

If S_1 is right-handed, use $+S_1$; if left-handed, $-S_1$. The same relation holds for the other screw pairs.

Figures 8.55 and 8.56 show a mechanism consisting of two screw pairs and one rotary joint.

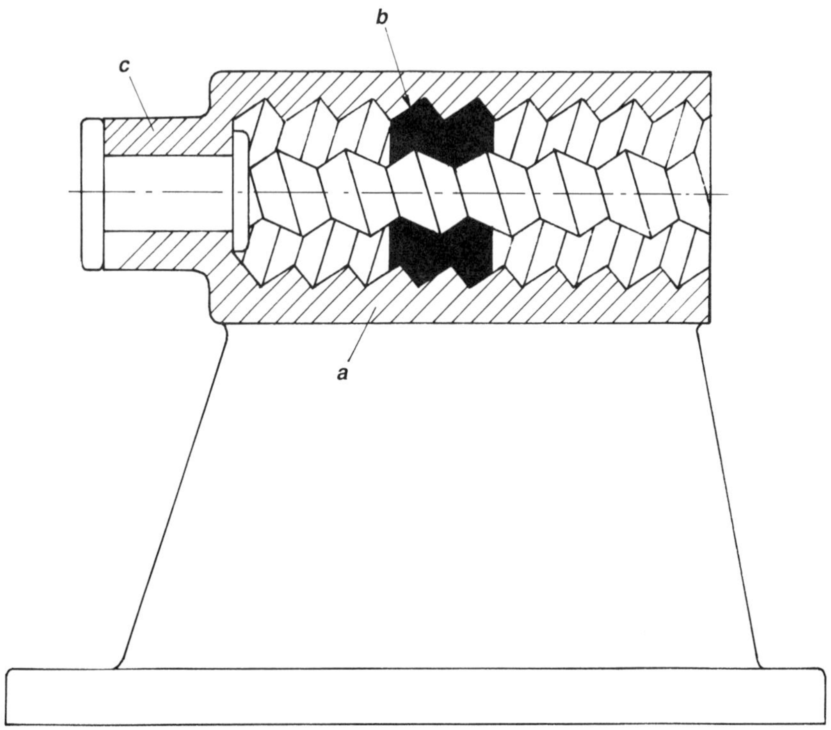

Figure 8.55 Differential screw mechanism. Same-hand or different-hand threads on two different shafts. (See also Fig. 8.56.)

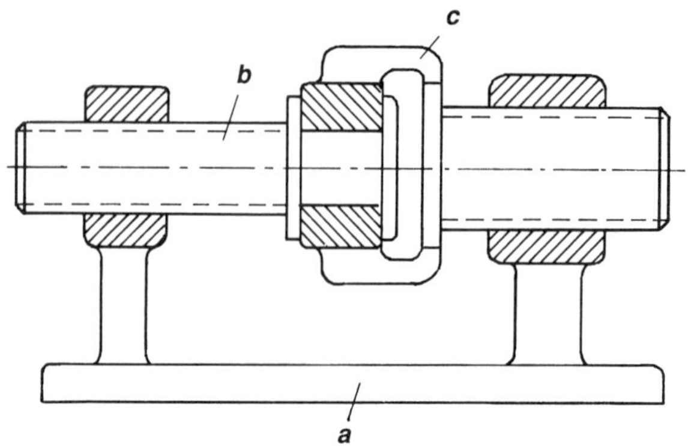

Figure 8.56

Figure 8.57 shows a mechanism consisting of two screw pairs and one sliding joint. A systematic development of this type of mechanism can be carried out in a manner as already described. Let the investigation be started with one of the screw pairs being replaced by a sliding or prismatic joint. The following solutions are possible.

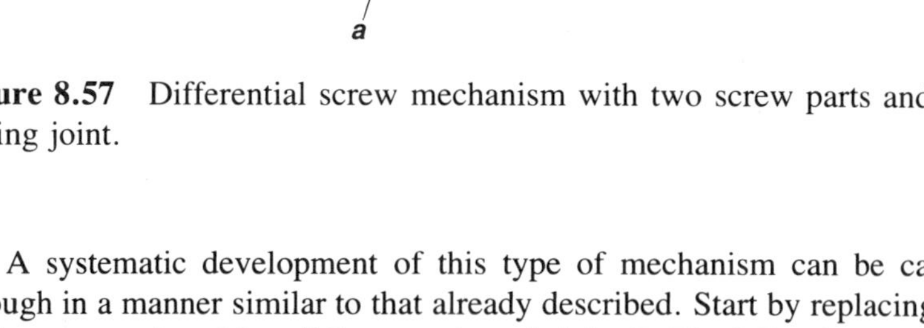

Figure 8.57 Differential screw mechanism with two screw parts and one sliding joint.

A systematic development of this type of mechanism can be carried through in a manner similar to that already described. Start by replacing one of the screw pairs with a sliding or prismatic joint P. The following solutions are possible:

$$\underbrace{P^{+} \quad S^{+}}_{a} \quad \underbrace{S^{-} \quad S^{+}}_{b} \quad \underbrace{S^{-} \quad P^{-}}_{c}$$

$$PSS \left[\begin{array}{lll}
\underbrace{P^+ \qquad S^+}_{\underline{a}} & \underbrace{S^- \qquad S^-}_{b} & \underbrace{S^+ \qquad P^-}_{c} \\
\underbrace{P^+ \qquad S^-}_{\underline{a}} & \underbrace{S^+ \qquad S^+}_{b} & \underbrace{S^- \qquad P^-}_{c} \\
\underbrace{P^+ \qquad S^-}_{\underline{a}} & \underbrace{S^+ \qquad S^-}_{b} & \underbrace{S^+ \qquad P^-}_{c} \\
\underbrace{P^- \qquad S^+}_{\underline{a}} & \underbrace{S^- \qquad S^+}_{b} & \underbrace{S^- \qquad P^+}_{c} \\
\underbrace{P^- \qquad S^+}_{\underline{a}} & \underbrace{S^- \qquad S^-}_{b} & \underbrace{S^+ \qquad P^+}_{c} \\
\underbrace{P^- \qquad S^-}_{\underline{a}} & \underbrace{S^+ \qquad S^+}_{b} & \underbrace{S^- \qquad P^+}_{c} \\
\underbrace{P^- \qquad S^-}_{\underline{a}} & \underbrace{S^+ \qquad S^-}_{b} & \underbrace{S^+ \qquad P^+}_{c}
\end{array} \right.$$

$$SSP \left[\begin{array}{lll}
\underbrace{S^+ \qquad S^+}_{\underline{a}} & \underbrace{S^- \qquad P^+}_{b} & \underbrace{P^- \qquad S^-}_{c} \\
\underbrace{S^+ \qquad S^-}_{\underline{a}} & \underbrace{S^+ \qquad P^+}_{b} & \underbrace{P^- \qquad S^-}_{c} \\
\underbrace{S^- \qquad S^+}_{\underline{a}} & \underbrace{S^- \qquad P^+}_{b} & \underbrace{P^- \qquad S^+}_{c} \\
\underbrace{S^- \qquad S^-}_{\underline{a}} & \underbrace{S^+ \qquad P^+}_{b} & \underbrace{P^- \qquad S^+}_{c}
\end{array} \right.$$

Using exactly the same notations, we can obtain the following by replacing a sliding joint P by a turning joint R:

$$\left[\begin{array}{lll}
\underbrace{R^+ \qquad S^+}_{\underline{a}} & \underbrace{S^- \qquad S^+}_{b} & \underbrace{S^- \qquad R^-}_{c}
\end{array} \right.$$

RSS

$$\underbrace{R^+ \quad S^+}_{\underline{a}} \quad \underbrace{S^- \quad S^-}_{b} \quad \underbrace{S^+ \quad R^-}_{c}$$

$$\underbrace{R^+ \quad S^-}_{\underline{a}} \quad \underbrace{S^+ \quad S^+}_{b} \quad \underbrace{S^- \quad R^-}_{c}$$

$$\underbrace{R^+ \quad S^-}_{\underline{a}} \quad \underbrace{S^+ \quad S^-}_{b} \quad \underbrace{S^+ \quad R^-}_{c}$$

$$\underbrace{R^- \quad S^+}_{\underline{a}} \quad \underbrace{S^- \quad S^-}_{b} \quad \underbrace{S^+ \quad R^+}_{c}$$

$$\underbrace{R^- \quad S^-}_{\underline{a}} \quad \underbrace{S^+ \quad S^+}_{b} \quad \underbrace{S^- \quad R^+}_{c}$$

$$\underbrace{R^- \quad S^-}_{\underline{a}} \quad \underbrace{S^+ \quad S^-}_{b} \quad \underbrace{S^- \quad R^+}_{c}$$

$$\underbrace{S^+ \quad S^+}_{\underline{a}} \quad \underbrace{S^- \quad R^+}_{b} \quad \underbrace{R^- \quad S^-}_{c}$$

$$\underbrace{S^+ \quad S^-}_{\underline{a}} \quad \underbrace{S^+ \quad R^+}_{b} \quad \underbrace{R^- \quad S^-}_{c}$$

$$\underbrace{S^- \quad S^+}_{\underline{a}} \quad \underbrace{S^- \quad R^+}_{b} \quad \underbrace{R^- \quad S^+}_{c}$$

$$\underbrace{S^- \quad S^-}_{\underline{a}} \quad \underbrace{S^+ \quad R^+}_{b} \quad \underbrace{R^- \quad S^+}_{c}$$

Finally, two screw pairs can be replaced by an R pair and a P pair, resulting in the following solutions:

$$\underbrace{P^+ \quad S^+}_{\underline{a}} \quad \underbrace{S^- \quad R^+}_{b} \quad \underbrace{R^- \quad P^-}_{c} \qquad \text{(Fig. 8.58)}$$

$$\underbrace{P^+ \quad S^+}_{\underline{a}} \quad \underbrace{S^- \quad R^-}_{b} \quad \underbrace{R^+ \quad P^-}_{c} \qquad \text{(Fig. 8.59)}$$

$$\underbrace{P^+ \quad S^-}_{\underline{a}} \quad \underset{b}{S^+ \quad R^+} \quad \underbrace{R^- \quad P^-}_{c} \qquad \text{(Fig. 8.60)}$$

PSR

$$\underbrace{P^{+} \quad S^{-}}_{\underline{a}} \quad \underbrace{S^{+} \quad R^{-}}_{b} \quad \underbrace{R^{+} \quad P^{-}}_{c} \qquad \text{(Fig. 8.61)}$$

$$\underbrace{P^{-} \quad S^{+}}_{\underline{a}} \quad \underbrace{S^{-} \quad R^{+}}_{b} \quad \underbrace{R^{-} \quad P^{+}}_{c} \qquad \text{(Fig. 8.62)}$$

$$\underbrace{P^{-} \quad S^{+}}_{\underline{a}} \quad \underbrace{S^{-} \quad R^{-}}_{b} \quad \underbrace{R^{+} \quad P^{+}}_{c} \qquad \text{(Fig. 8.63)}$$

$$\underbrace{P^{-} \quad S^{-}}_{\underline{a}} \quad \underbrace{S^{+} \quad R^{+}}_{b} \quad \underbrace{R^{-} \quad P^{+}}_{c} \qquad \text{(Fig. 8.64)}$$

$$\underbrace{P^{-} \quad S^{-}}_{\underline{a}} \quad \underbrace{S^{+} \quad R^{-}}_{b} \quad \underbrace{R^{+} \quad P^{+}}_{c} \qquad \text{(Fig. 8.65)}$$

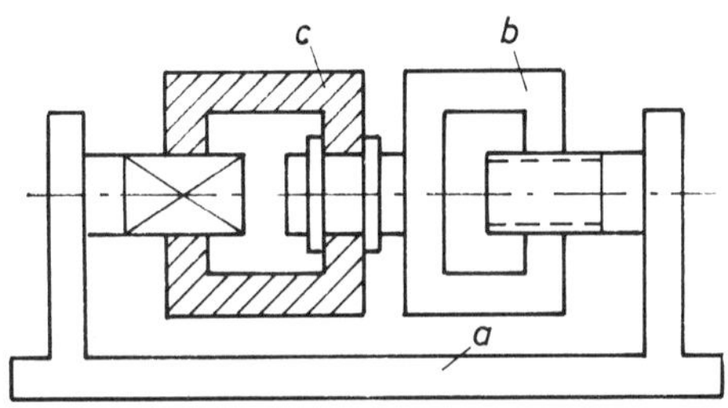

Figure 8.58–.81 Three-link, one screw joint S, one turning joint R, and one sliding joint P. The possible solutions when interchanging male and female members, say, S^{+} and S^{-}, respectively.

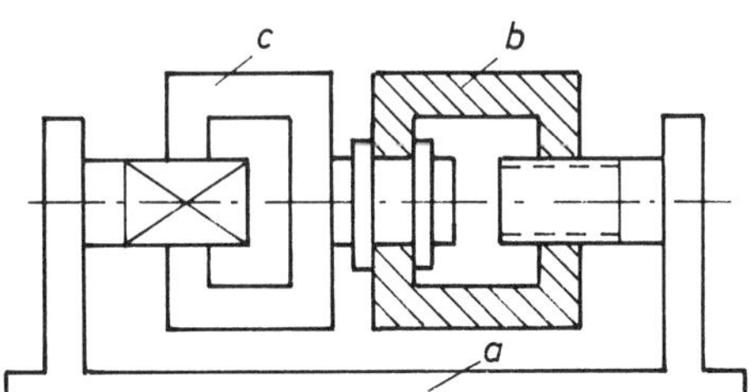

Figure 8.59

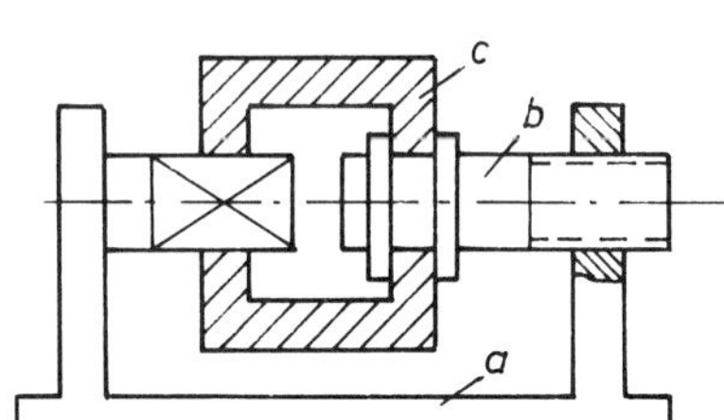

Figure 8.60

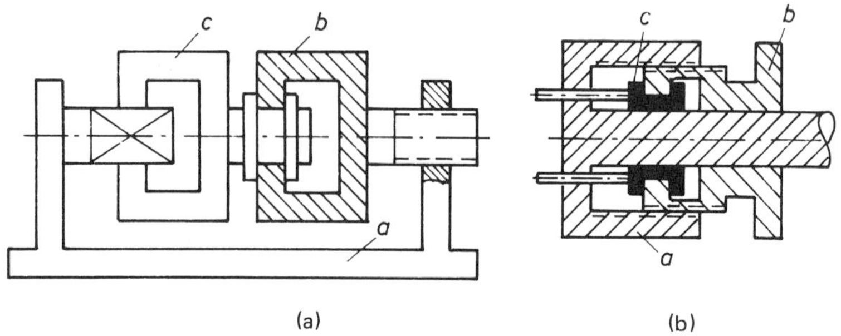

Figure 8.61

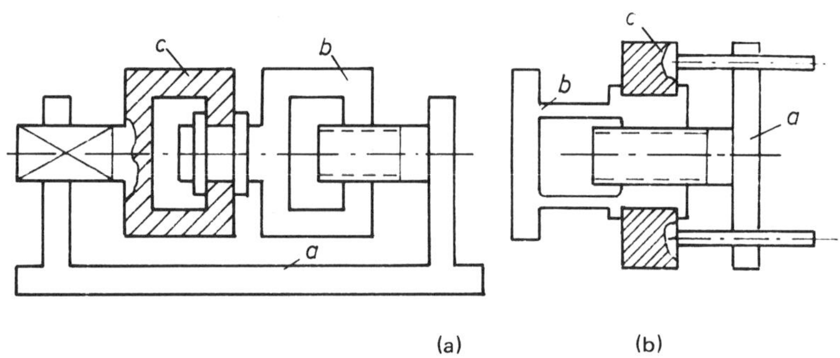

Figure 8.62

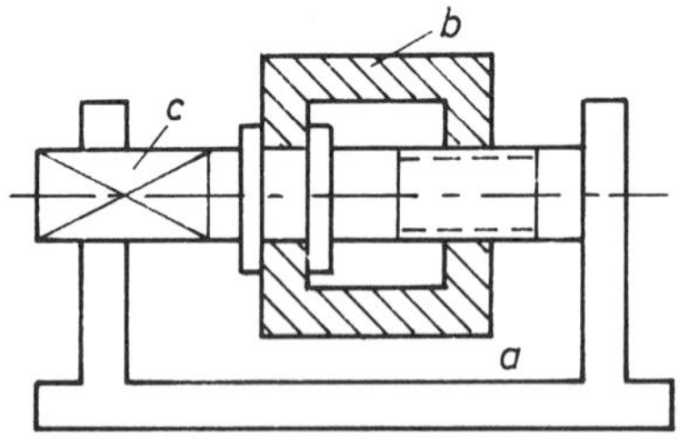

Figure 8.63

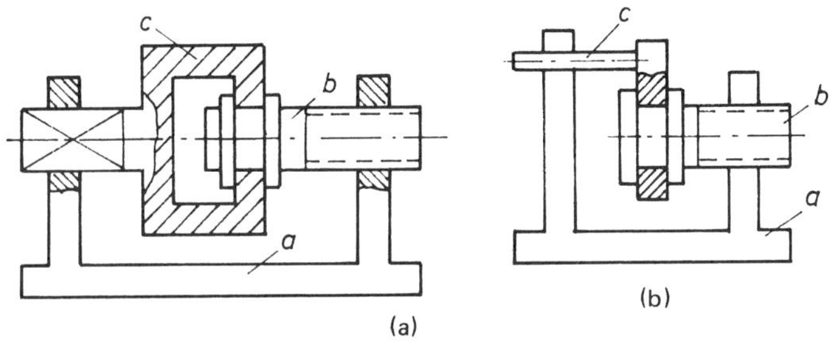

Figure 8.64

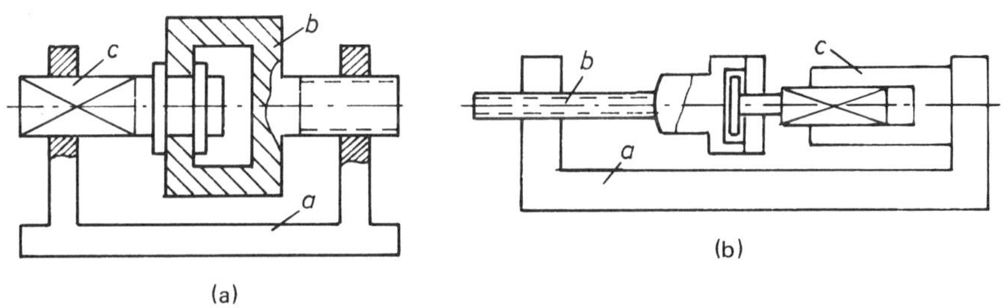

Figure 8.65

$$
PRS \left[
\begin{array}{llll}
\underbrace{P^{+} \quad R^{+}}_{\underline{a}} & \underbrace{R^{-} \quad S^{+}}_{b} & \underbrace{S^{-} \quad P^{-}}_{c} & \text{(Fig. 8.66)} \\
\underbrace{P^{+} \quad R^{+}}_{\underline{a}} & \underbrace{R^{-} \quad S^{-}}_{b} & \underbrace{S^{+} \quad P^{-}}_{c} & \text{(Fig.8.67)} \\
\underbrace{P^{+} \quad R^{-}}_{\underline{a}} & \underbrace{R^{+} \quad S^{+}}_{b} & \underbrace{S^{-} \quad P^{-}}_{c} & \text{(Fig. 8.68)} \\
\underbrace{P^{+} \quad R^{-}}_{\underline{a}} & \underbrace{R^{+} \quad S^{+}}_{b} & \underbrace{S^{+} \quad P^{-}}_{c} & \text{(Fig. 8.69)} \\
\underbrace{P^{-} \quad R^{+}}_{\underline{a}} & \underbrace{R^{-} \quad S^{+}}_{b} & \underbrace{S^{-} \quad P^{+}}_{c} & \text{(Fig. 8.70)} \\
\underbrace{P^{-} \quad R^{+}}_{\underline{a}} & \underbrace{R^{-} \quad S^{-}}_{b} & \underbrace{S^{+} \quad P^{+}}_{c} & \text{(Fig. 8.71)}
\end{array}
\right.
$$

$$\left|\begin{array}{l} \underbrace{P^- \quad R^-}_{\underline{a}} \quad \underbrace{R^+ \quad S^+}_{b} \quad \underbrace{S^- \quad P^+}_{c} \qquad \text{(Fig. 8.72)} \\ \underbrace{P^- \quad R^-}_{\underline{a}} \quad \underbrace{R^+ \quad S^-}_{b} \quad \underbrace{S^+ \quad P^+}_{c} \qquad \text{(Fig. 8.73)} \end{array}\right.$$

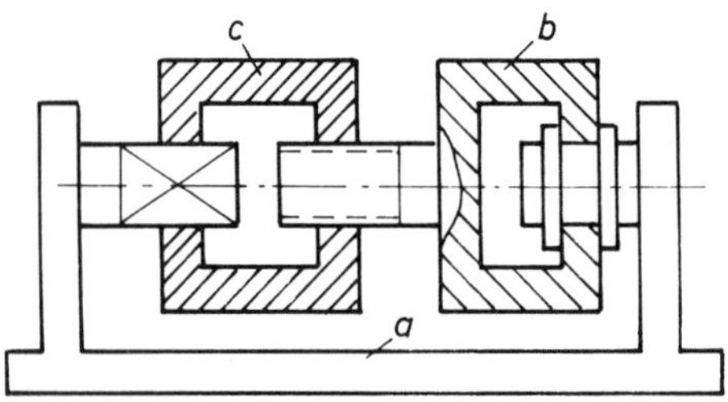

Figure 8.66

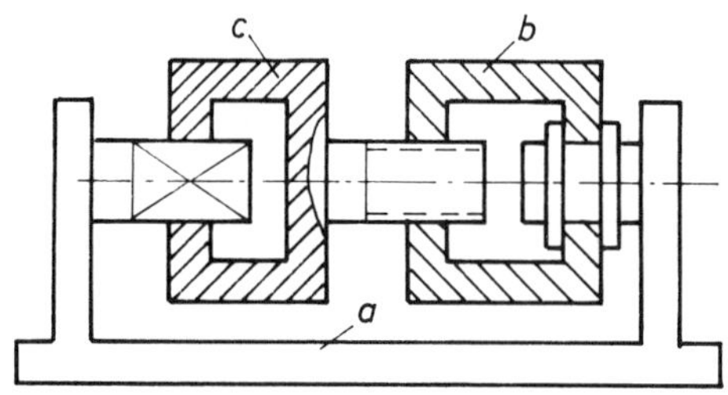

Figure 8.67

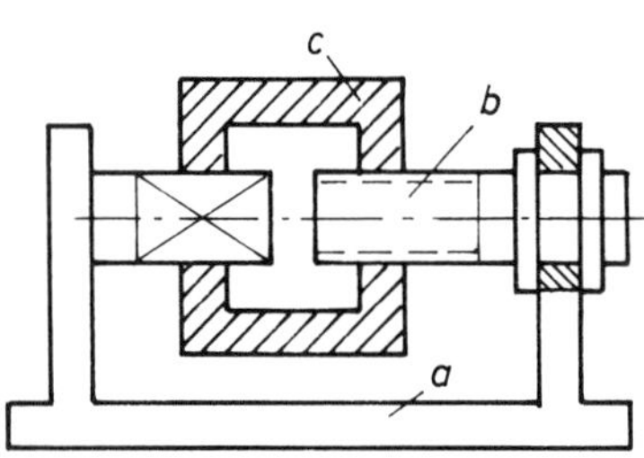

Figure 8.68

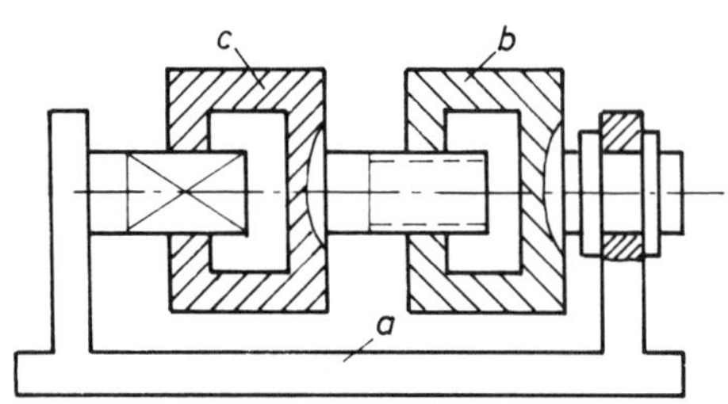

Figure 8.69

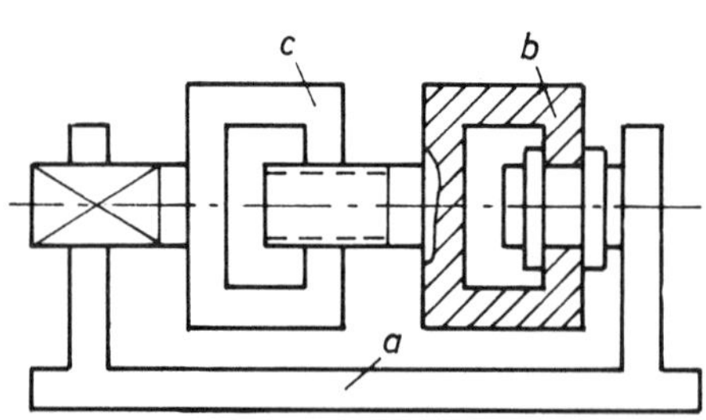

Figure 8.70

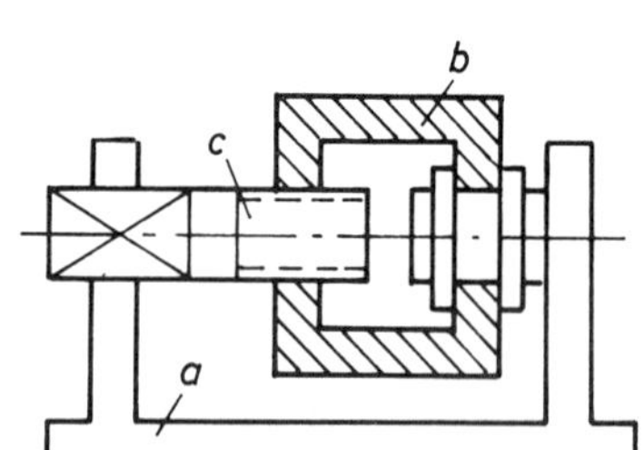

Figure 8.71

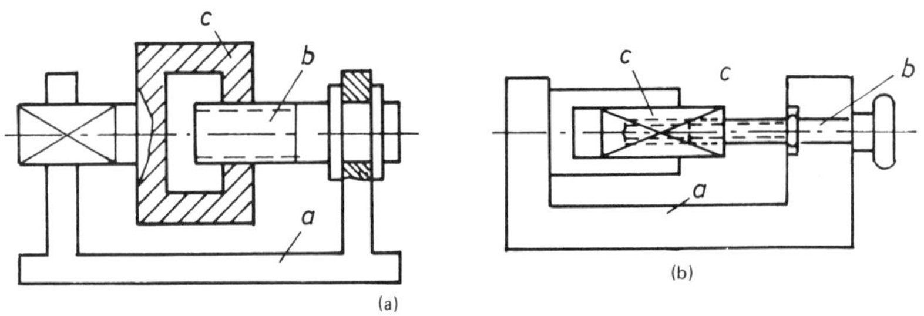

Figure 8.72

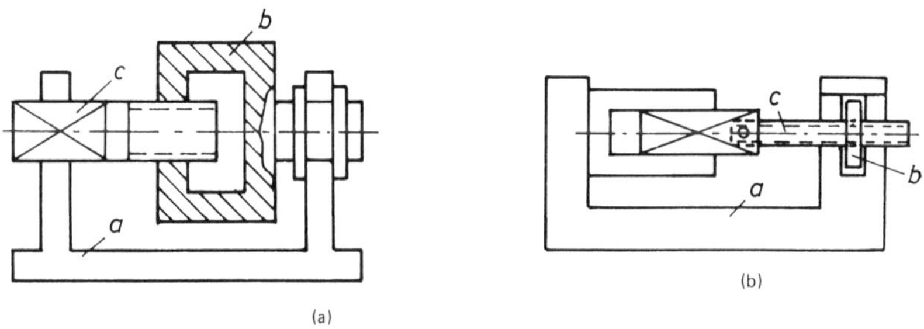

Figure 8.73

$$
SRP\left[\begin{array}{llll}
\underbrace{S^{+}\quad R^{+}}_{\underline{a}} & \underbrace{R^{-}\quad P^{+}}_{b} & \underbrace{P^{-}\quad S^{-}}_{c} & \text{(Fig. 8.74)}\\
\underbrace{S^{+}\quad R^{+}}_{\underline{a}} & \underbrace{R^{-}\quad P^{-}}_{b} & \underbrace{P^{+}\quad S^{-}}_{c} & \text{(Fig. 8.75)}\\
\underbrace{S^{+}\quad R^{-}}_{\underline{a}} & \underbrace{R^{+}\quad P^{+}}_{b} & \underbrace{P^{-}\quad S^{-}}_{c} & \text{(Fig. 8.76)}\\
\underbrace{S^{+}\quad R^{-}}_{\underline{a}} & \underbrace{R^{+}\quad P^{-}}_{b} & \underbrace{P^{+}\quad S^{-}}_{c} & \text{(Fig. 8.77)}\\
\underbrace{S^{-}\quad R^{+}}_{\underline{a}} & \underbrace{R^{-}\quad P^{+}}_{b} & \underbrace{P^{-}\quad S^{+}}_{c} & \text{(Fig. 8.78)}\\
\underbrace{S^{-}\quad R^{+}}_{\underline{a}} & \underbrace{R^{-}\quad P^{-}}_{b} & \underbrace{P^{+}\quad S^{+}}_{c} & \text{(Fig. 8.79)}
\end{array}\right.
$$

$$\underbrace{S^- \qquad R^-}_{\underline{a}} \quad \underbrace{R^+ \qquad P^+}_{b} \quad \underbrace{P^- \qquad S^+}_{c} \qquad \text{(Fig.8.80)}$$

$$\underbrace{S^- \qquad R^-}_{\underline{a}} \quad \underbrace{R^+ \qquad P^-}_{b} \quad \underbrace{P^+ \qquad S^+}_{c} \qquad \text{(Fig. 8.81)}$$

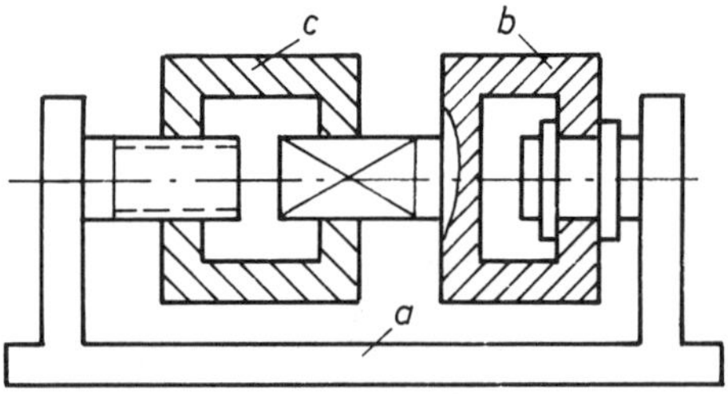

Figure 8.74

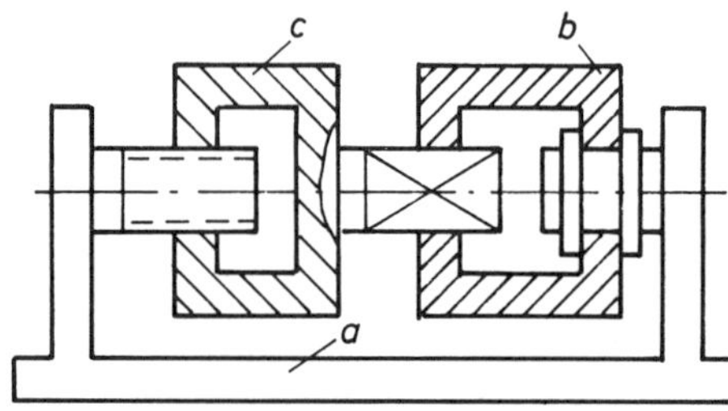

Figure 8.75

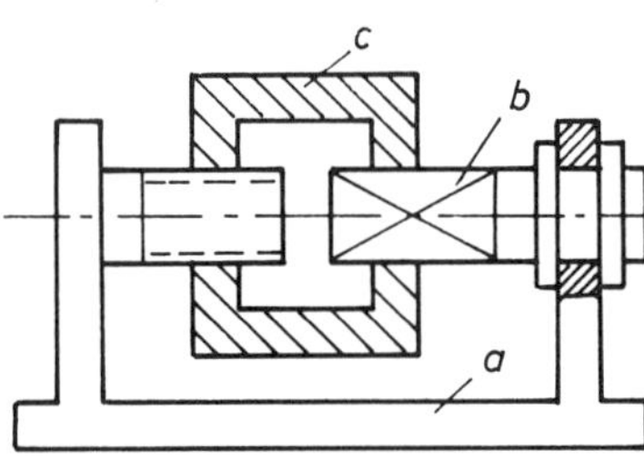

Figure 8.76

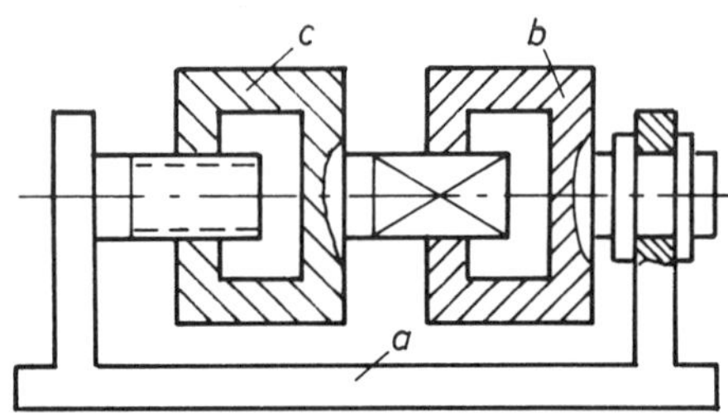

Figure 8.77

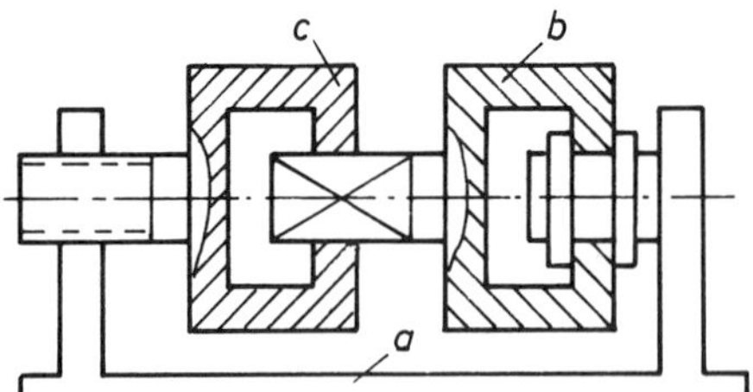

Figure 8.78

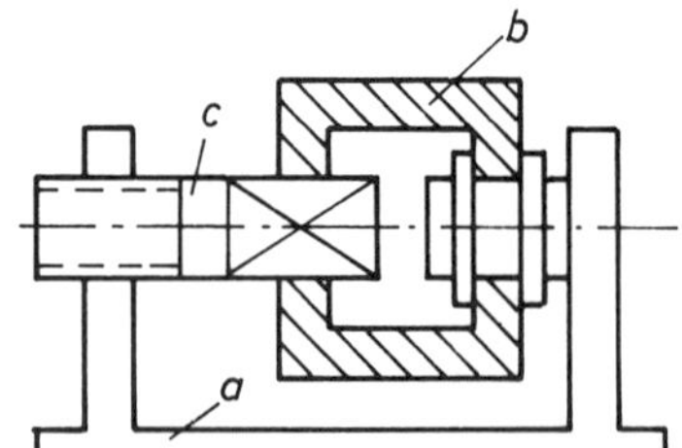

Figure 8.79

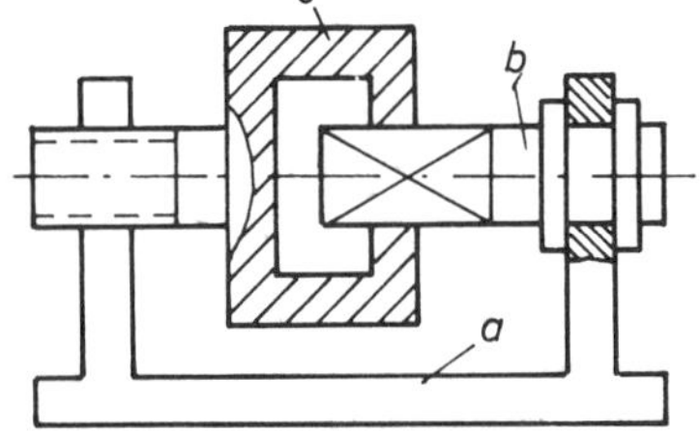

Figure 8.80

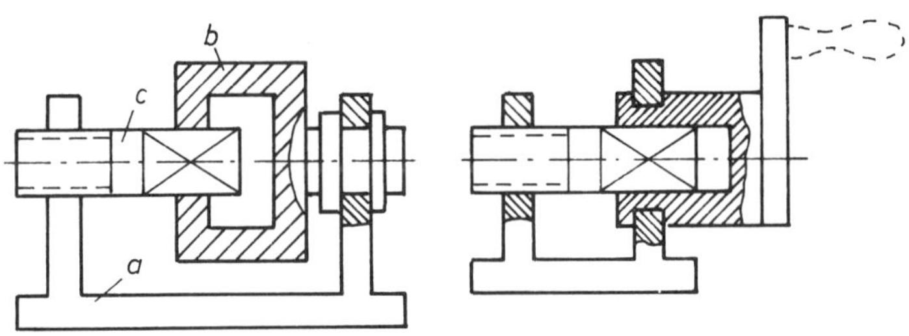

Figure 8.81

The number of solutions is rather large, but the sketches, and in some cases the alternative solutions shown to the right, should be helpful in picking a solution. Notice that input can be to either of the two movable members.

POWER SCREWS

A screw mechanism can be used for motion and/or force transmission (Fig. 8.82). The equations for motion transmission were developed for a three-screw mechanism and the other cases can be developed by inspection.

When screws are used for force transmission, it is desirable to find the torque to resist or overcome a certain load. Let

α = helix angle of thread, deg.
e = efficiency of power screw
p = pitch of screw, in.
r_c = mean radius of collar = 1/2 (inner radius + outside radius)
r_t = mean radius of thread = 1/2 (root radius + outside radius)
T_1 = torque required to motion opposed to W
T_2 = torque required to motion assisted by W
T_3 = torque required to resist the overhauling screw

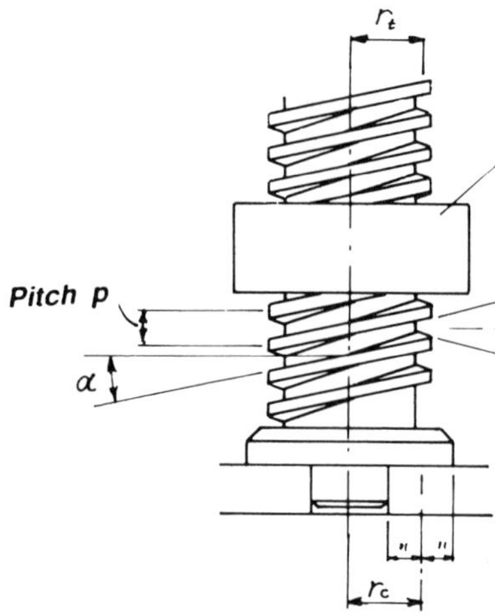

Figure 8.82 Power screw.

θ = half angle between the thread faces, deg.
W = axial load, lb
μ_1 = coefficient of thread friction
μ_2 = coefficient of friction of base or collar

The torque required motion opposed to the local W is

$$T_1 = Wr_t \left(\frac{\cos\theta_n \tan\alpha + \mu_1}{\cos\theta_n - \mu_1 \tan\alpha} + \frac{r_c}{r_t} \mu_2 \right) \tag{8.8}$$

where

$$\theta_n = \tan^{-1}(\tan\theta \cos\alpha). \tag{8.9}$$

If the collar is an antifriction bearing, then $\mu_2 \approx 0$ and can be neglected.

The torque required to motion assisted by W is

$$T_2 = Wr_t \left(- \frac{\cos\theta_n \tan\alpha - \mu_1}{\cos\theta_n - \mu_1 \tan\alpha} + \frac{r_c}{r_t} \mu_2 \right) \tag{8.10}$$

If the helix angle is sufficiently great, the screw will overhaul, or the weight will revolve the screw.

The torque required to resist the overhauling screw is

$$T_3 = Wr_t \left(\frac{\cos\theta_n \tan\alpha - \mu_1}{\cos\theta_n + \mu_1 \tan\alpha} - \frac{r_c}{r_t} \mu_2 \right)$$

In case all friction can be eliminated ($\mu_1 = \mu_2 = 0$), then the torque required to raise the load would be

$$T_1' = Wr_t \tan\alpha$$

The efficiency of the power screw, therefore, is

$$e = \frac{T_1'}{T_1}$$

If collar friction is negligible,

$$e = \frac{\cos\theta_n - \mu_1 \tan\alpha}{\cos\theta_n + \mu_1 \cot\alpha}$$

The efficiency is a maximum when the helix angle is 45° but deviates relatively little for

$$25° < \alpha < 55°$$

but is very low when close to 0° or 90°.

9
Clamping Mechanisms

Clamps are very useful when it is desirable to attach quickly and easily two machine parts to each other so that they can be separated without damage. The two parts may be held together with what is designated *force closure* or *form closure*. These methods of closure will be described in the following, but bear in mind that a distinct separation between the two is not always possible.

Clamping devices are employed extensively. For example, a lid on a coffee can is a form closure clamping device, where the lid has a close fit; when it is pressed onto the can, friction holds the two parts together. Cover catches constitute one of the most common clamping devices.

FRICTION CLAMPING DEVICES

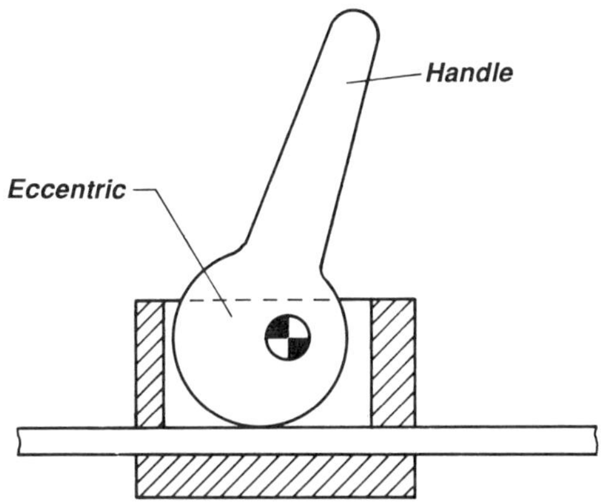

Figure 9.1 Eccentric clamp. An eccentric with a handle presses down on a round- or square-shaped member. The eccentricity is chosen so small that friction suffices to lock the members together.

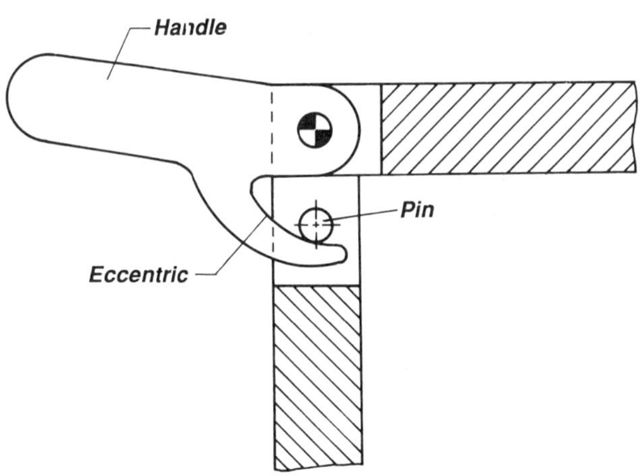

Figure 9.2 Eccentric clamp. An eccentric is used for a cover catch. The eccentric presses on the pin when turned in a CCW direction.

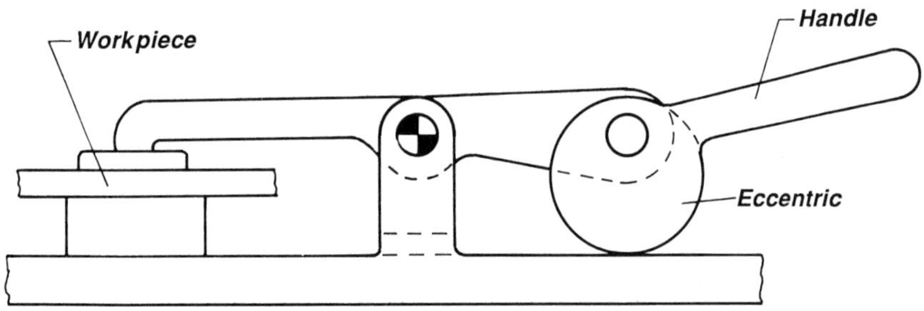

Figure 9.3 Eccentric clamp. The clamping handle is connected to an eccentric, and through an arm the locking action keeps the workpiece in place.

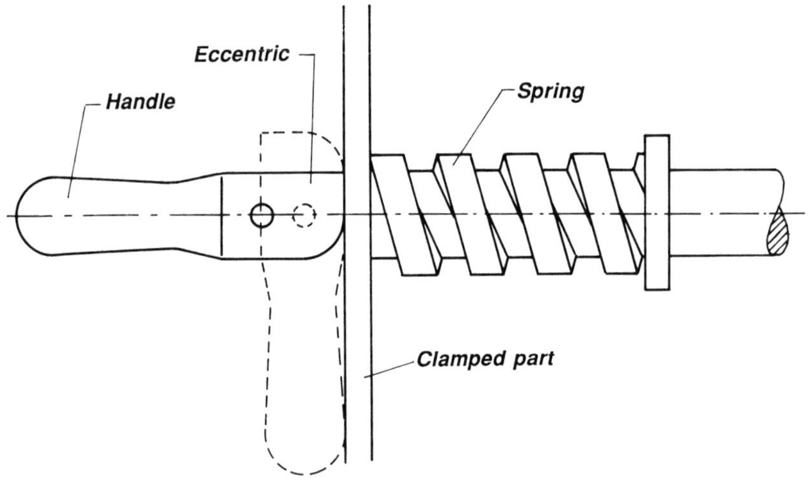

Figure 9.4 Eccentric clamp. A spring ensures contact between the clamp and the clamped part.

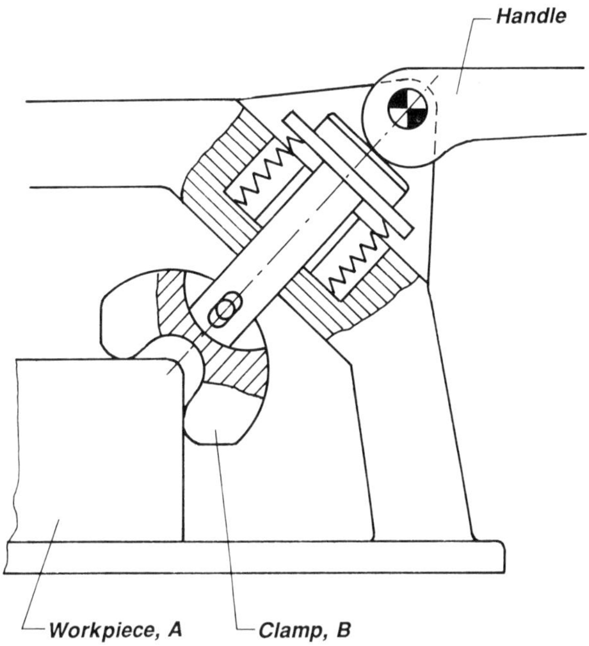

Figure 9.5 Eccentric clamp. A time-efficient clamping device used to clamp the workpiece A in position. The shape of member B ensures a good holding grip and equal force transmitted to the workpiece at the two points of contact.

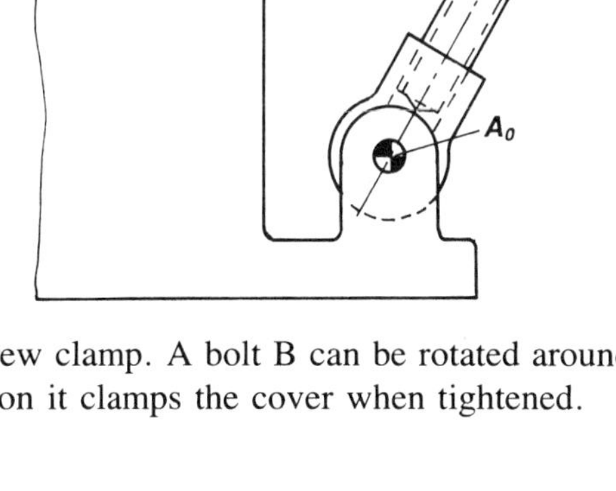

Figure 9.6 Screw clamp. A bolt B can be rotated around A_0 and when in an upright position it clamps the cover when tightened.

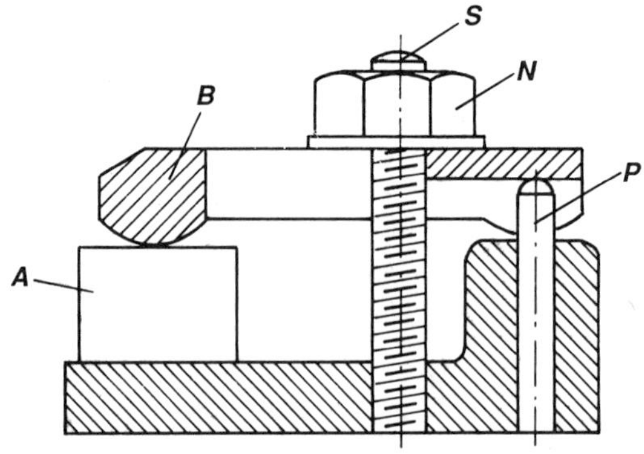

Figure 9.7 Screw clamp. The workpiece A is clamped by a member B, which in turn is clamped by the screw S and a nut N. Pin P ensures a horizontal position of member B.

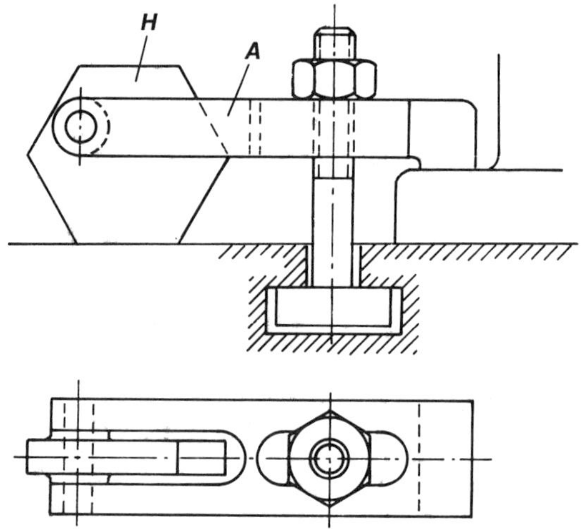

Figure 9.8 Adjustable screw clamp. The hexagonal-shaped member H is connected to the arm A, which can be clamped by the screw mechanism. The hexagonal member is connected to A by a rotary joint, but the center of this joint is placed asymmetrically on the hexagon so that it is possible to vary the clamping height.

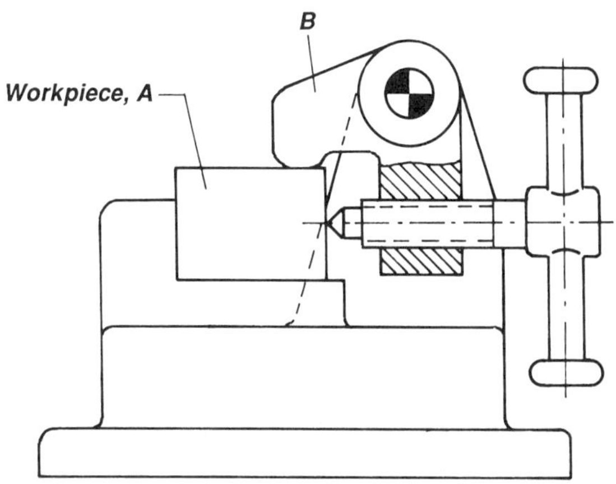

Figure 9.9 The clamping member B in the form of a bell crank clamps the workpiece A securely against two surfaces when the screw mechanism is activated.

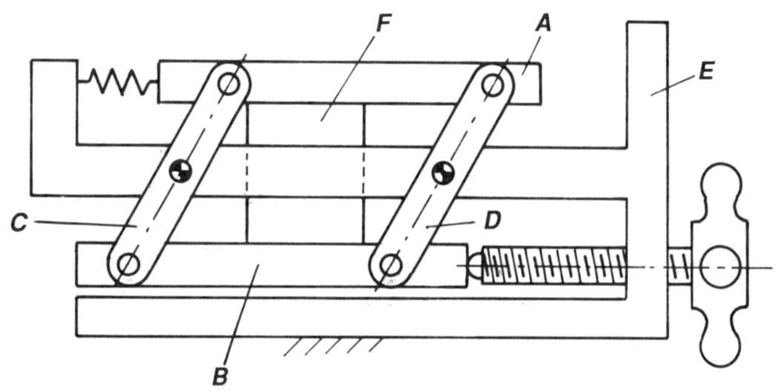

Figure 9.10 The two parallel plates A and B are moved in parallel motion by the two links C and D that rotate in joints in member E. When the screw mechanism is activated, the two parallel members move towards each other, thereby clamping the work or workpieces F.

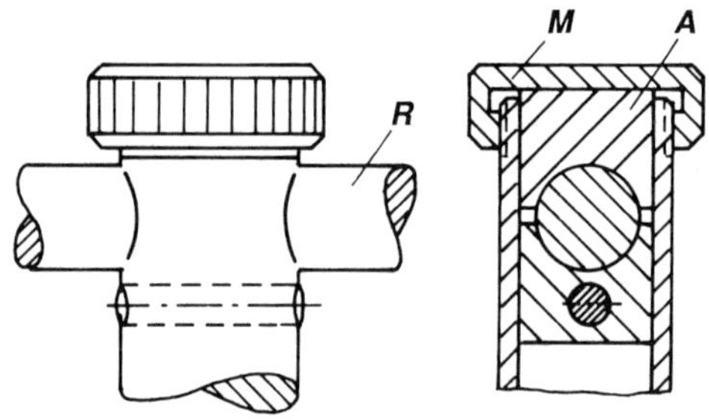

Figure 9.11 The nut M presses A downwards, which in turn presses against the rod R, thereby clamping it.

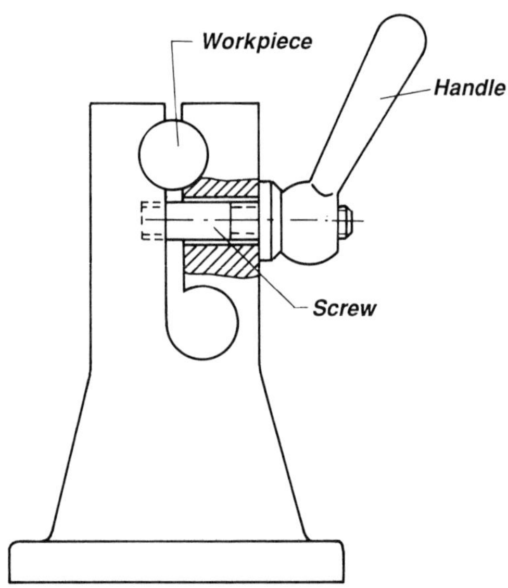

Figure 9.12 To avoid having a loose clamp, here the clamp is made flexible, and when the screw mechanism is activated the rod is clamped by friction.

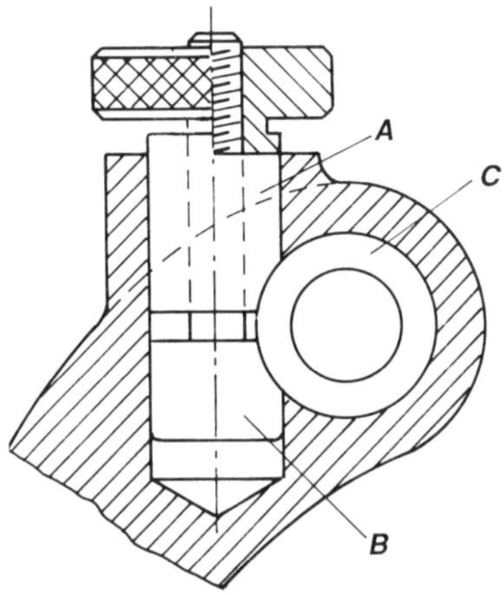

Figure 9.13 Two members A and B are pulled together by a screw mechanism. Because of the wedging action, the clamping force on member C is increased.

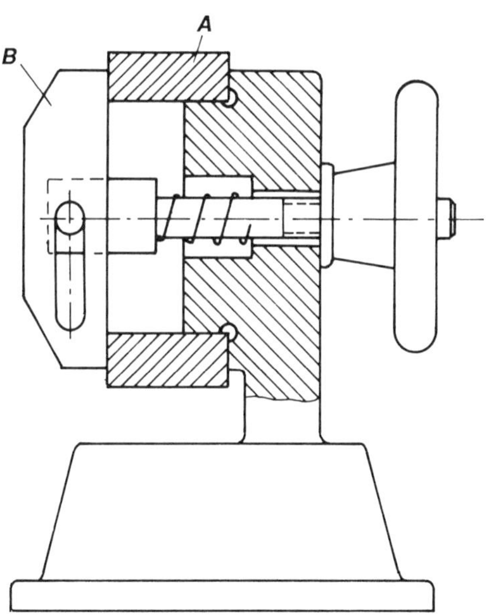

Figure 9.14 A ring-shaped workpiece A is clamped by a yoke B. The clamping force is distributed over two points of contact.

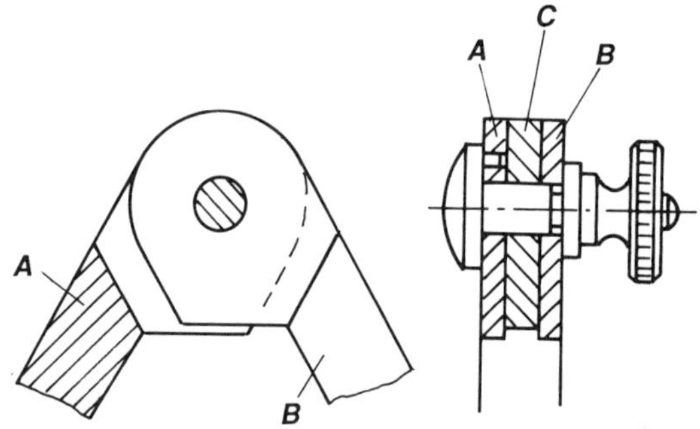

Figure 9.15 Two members, A and B, are clamped when the screw mechanism is activated. Member C is an intermediate ring that increases friction between the two members.

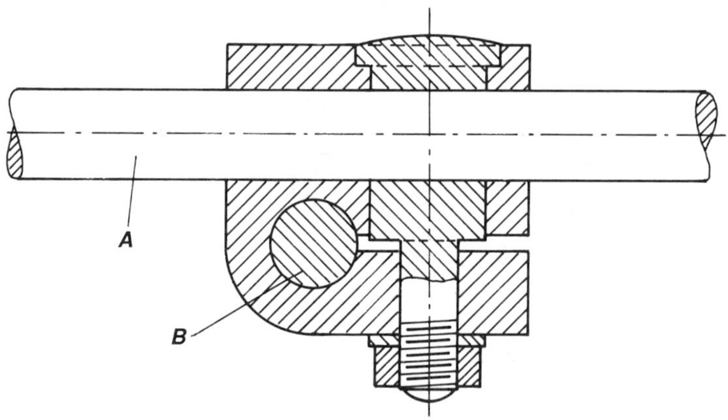

Figure 9.16 Two crossing rods, A and B, can be clamped with this device.

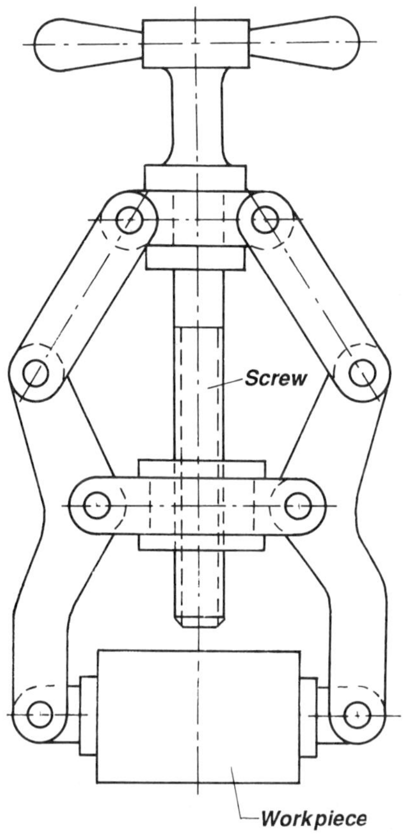

Figure 9.17 A large clamping force can be obtained with this clamping device, which combines screw and links and makes use of a toggle-like motion. This principle is used to remove bearings from shafts.

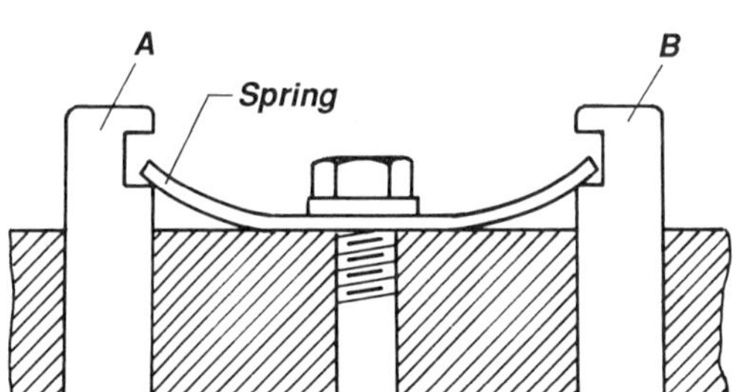

Figure 9.18 A spring used as a clamp. Two members, pin A and B, can be fastened simultaneously by one screw.

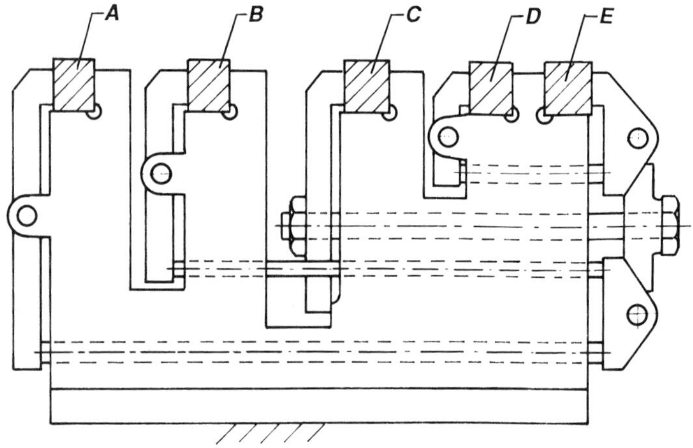

Figure 9.19 Five workpieces, A, B, C, D, and E, can be clamped at the same time with this screw clamp device. This arrangement ensures uniform pressure on all the workpieces.

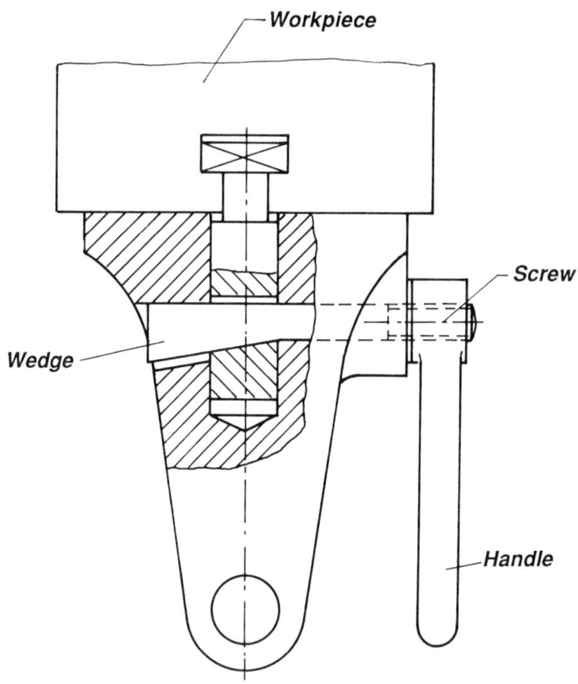

Figure 9.20 Wedge and screw mechanism in series lock workpiece securely.

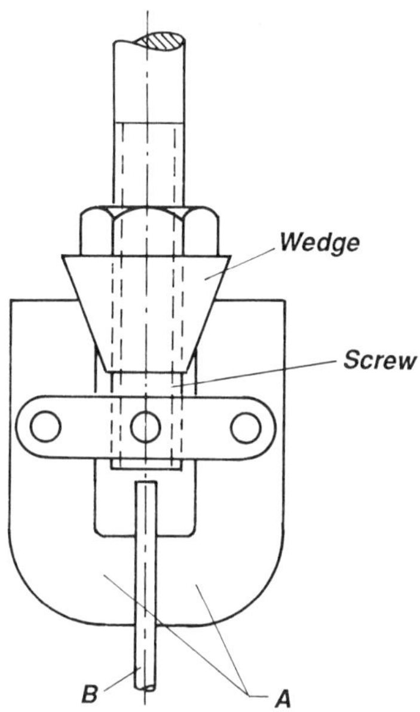

Figure 9.21 Wedge and screw mechanism combine to press jaws A together to lock rod or wire B.

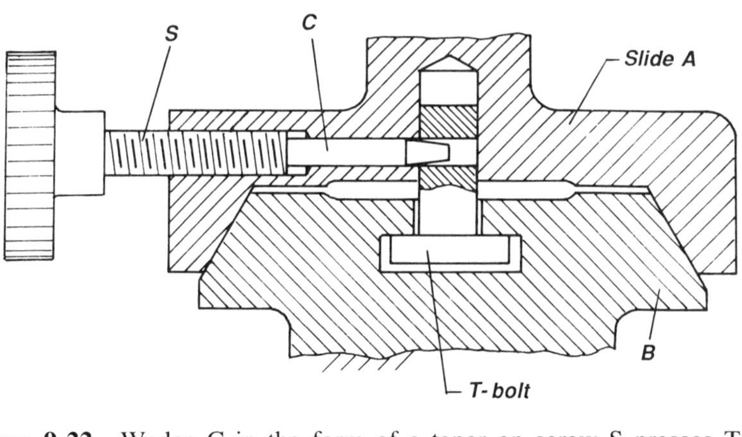

Figure 9.22 Wedge C in the form of a taper on screw S presses T-bolt upwards and presses slide A against bed B.

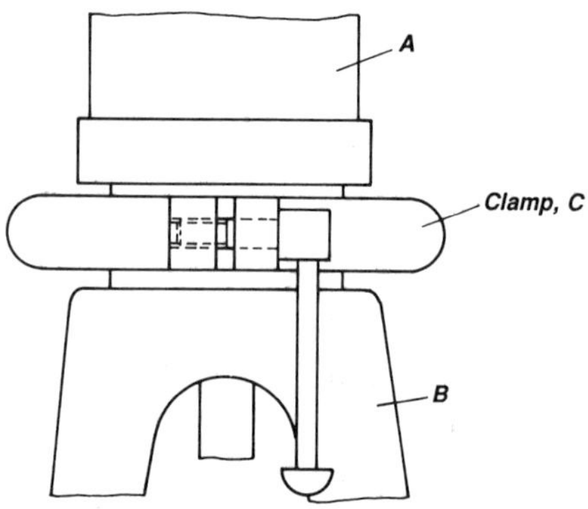

Figure 9.23 The ring-shaped clamp C fits around the two members A and B to be clamped. When tightened by the screw, it locks to the structure. Before clamping, the ring-shaped member C and member A can be rotated to the desired position.

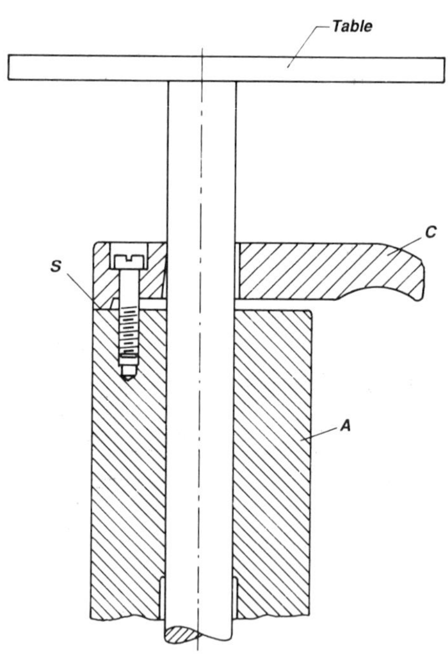

Figure 9.24 Clamp C can rotate a little, and when the table is loaded the clamp will rotate around its support S and prevent the table from moving down.

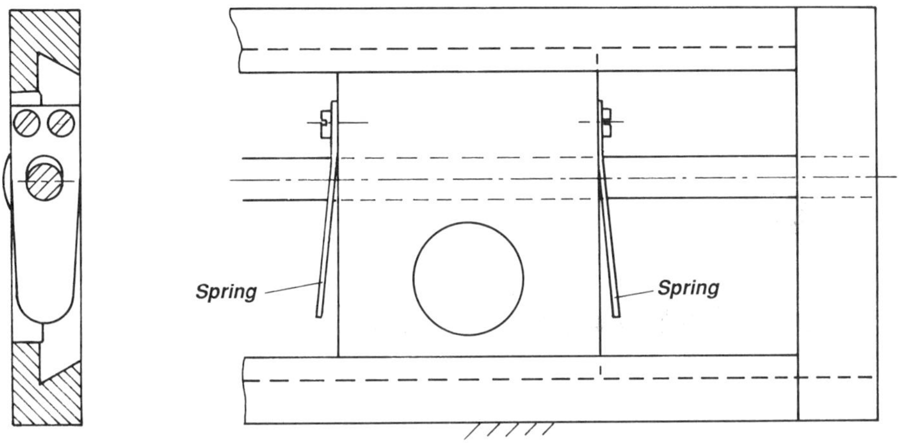

Figure 9.25 The same action as in Fig. 9.24 is achieved here with two springs. One spring will lock the shaft in one direction, and the other spring in the other direction. Pressing the two springs together will release the device.

10

Antibacklash Devices

Backlash is the play resulting from loose connections between gears or other mechanical elements. There are various ways to compensate for backlash. For instance the action of springs may be used to automatically compensate for backlash, or screws working together may be used to adjust the amount of permissible backlash.

DEVICES EMPLOYING SPRINGS

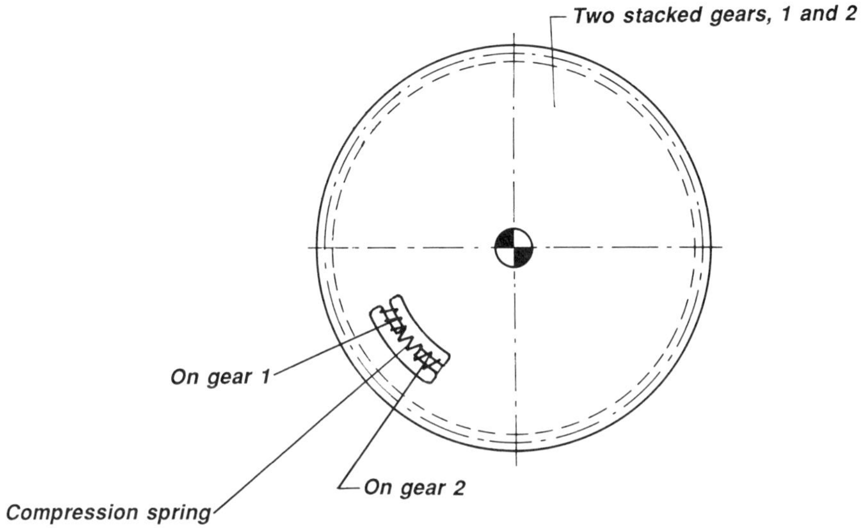

Figure 10.1 In precision instruments where the motion between two shafts has to be very accurate, two gears are used to compensate for backlash. The two gears are pressed against each other by a compression spring. Because the spring rotates together with the gears, unlimited rotation is possible.

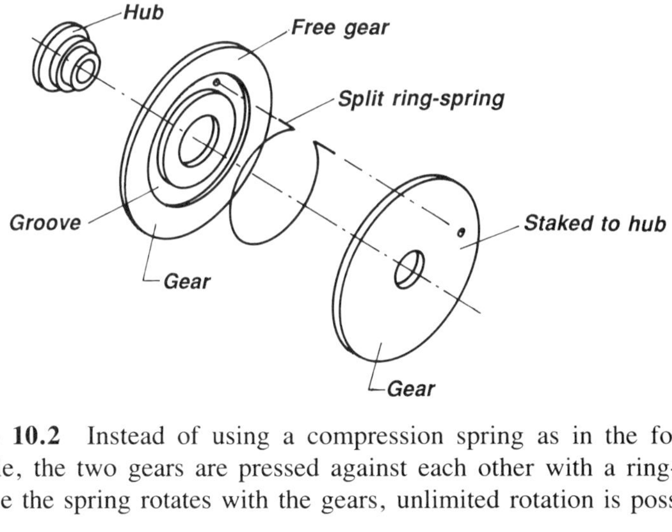

Figure 10.2 Instead of using a compression spring as in the foregoing example, the two gears are pressed against each other with a ring-spring. Because the spring rotates with the gears, unlimited rotation is possible.

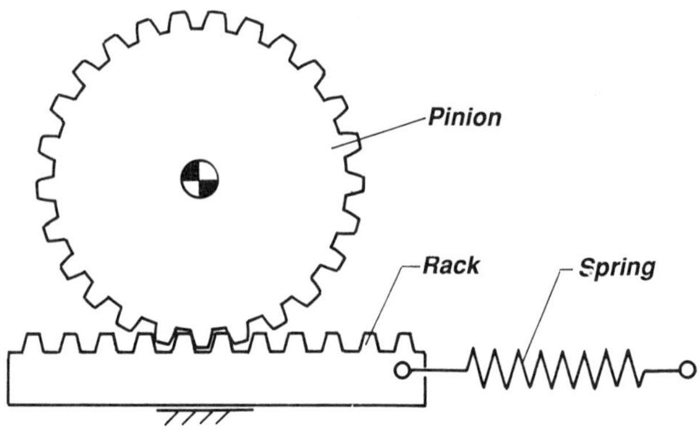

Figure 10.3 If a gear and rack are in mesh and the members make a limited motion, then a spring attached to the rack may be used to compensate for backlash.

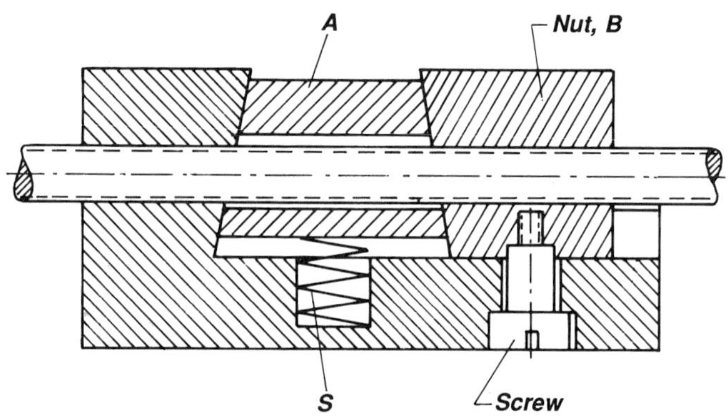

Figure 10.4 A self-compensating antibacklash device is obtained with the help of spring S, which presses block A against the nut B of a screw mechanism.

DEVICES EMPLOYING TWO NUTS

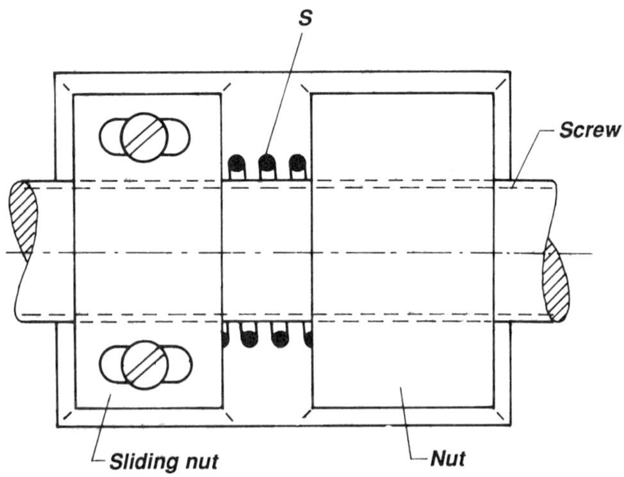

Figure 10.5 The antibacklash in this screw mechanism is obtained using a sliding nut that is spring loaded by a compression spring; it is permitted to make only a sliding motion, being held in position by the two setscrews and being permitted to move in an axial direction by the slots.

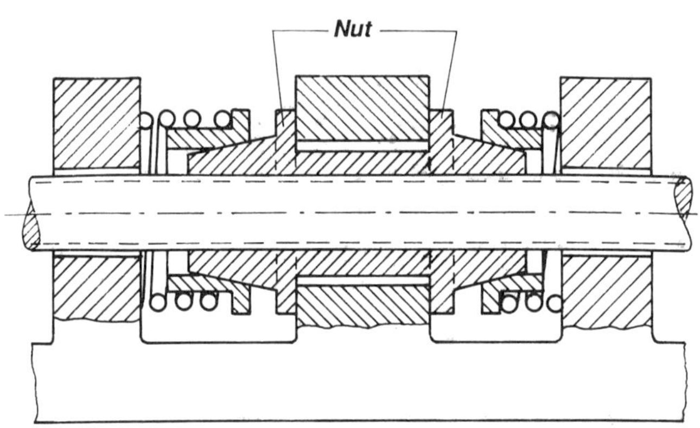

Figure 10.6 Automatic adjustment for backlash. The nut is flanged on each end and has a square outer section between the flanges and slots cut in the tapered sections. Spring forces have components that push slotted sections radially inwards and take up backlash.

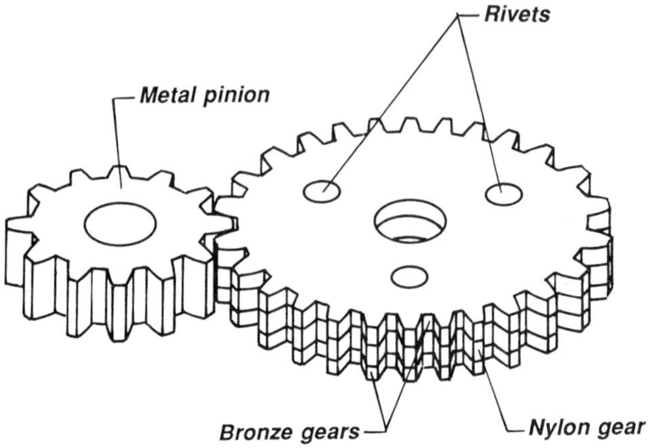

Figure 10.7 The three gears shown to the right are machined to the same dimensions, but the nylon lamination swells slightly, thereby creating a tighter mesh between the three gears and the metal pinion.

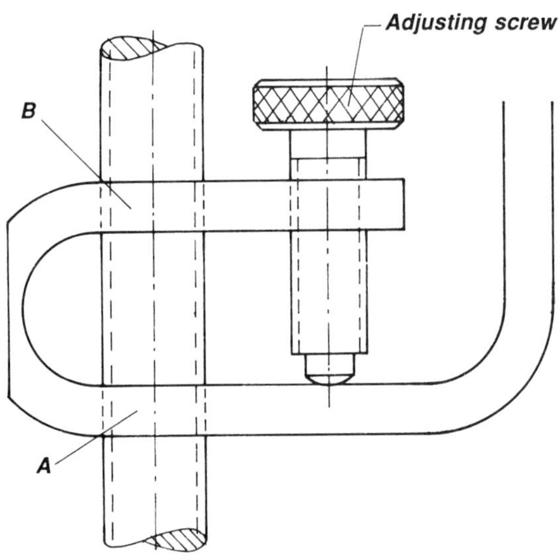

Figure 10.8 The adjusting screw presses the bent member against the long screw. The bent member is threaded at both A and B, thereby acting as a double nut, the parts of which we press against each other by the adjusting screw.

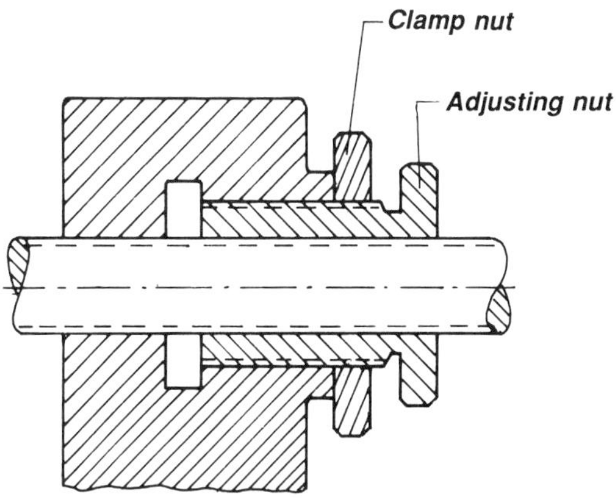

Figure 10.9 Again, two nuts are used to eliminate backlash. The adjusting nut is turned (clamped) to neutralize backlash between double nut and screw.

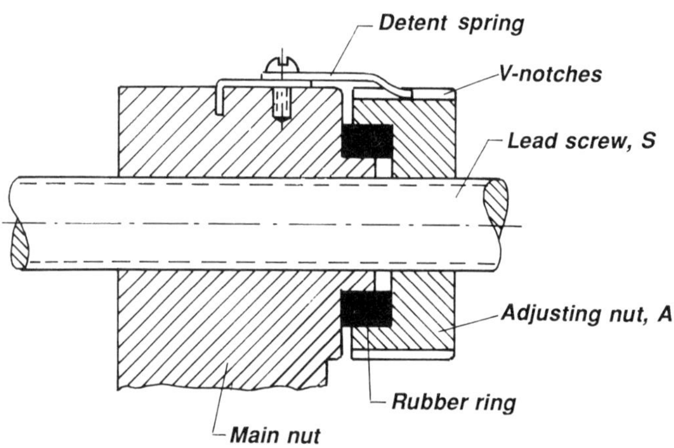

Figure 10.10 The adjusting nut A is kept in position by a detent spring that fits into notches around the periphery of A. A rubber ring permits exact adjustment between the nut A and the screw S.

DEVICE EMPLOYING ADJUSTING SCREW AND NUT

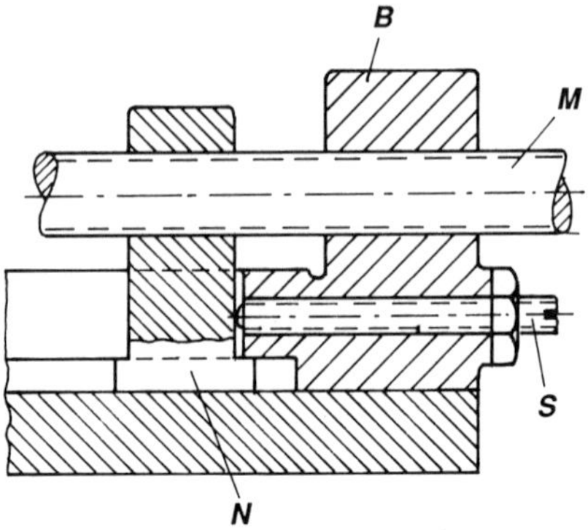

Figure 10.11 The compensation for backlash is obtained here by screw S, which presses against an auxiliary nut N, thereby pressing the screw M against the working nut B.

11

Infinite-Variable-Speed Drives

Infinite-variable-speed drives have been the subject of investigation for many years, and many ingenious devices have emerged. They all rely on some form of force closure if they are to be infinite variable, that is, have any accurate speed ratio. If transmission of forces is done by form closure, then the output speed is not exactly constant, although in some cases it is very close.

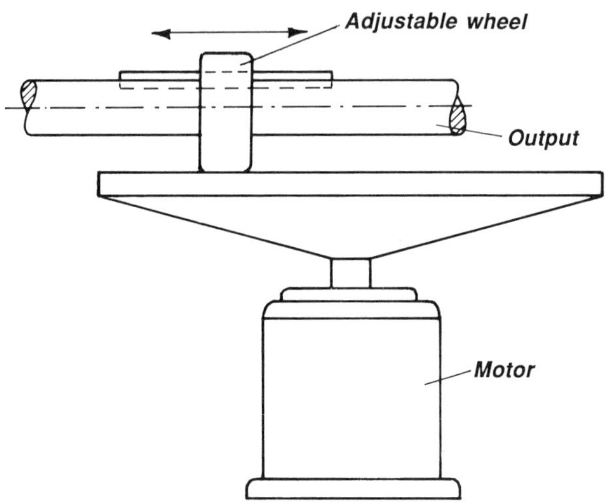

Figure 11.1 In this infinite-variable-speed drive the circular disc is in contact with a steel friction roller which can slide on the output shaft but which rotates with it by means of a key and keyway. If the roller is at the center of the circular disc, then the output motion is zero, but as the roller is moved farther away from the center, the speed of the output shaft increases proportionately. Because pure rolling cannot take place between the roller and the disc—due to varying speeds along the contact line of the disc versus the constant circumferential speed of the roller—the roller is often made spherical and often is covered with a material that increases friction and wear resistance.

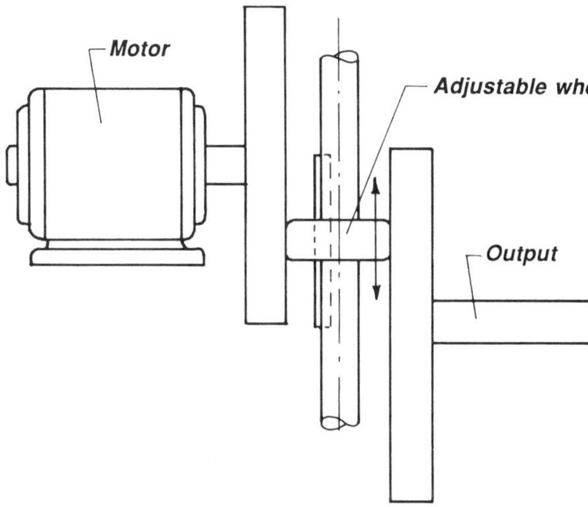

Figure 11.2 A friction roller interposed between two circular discs converts rotational speeds between parallel shafts but otherwise has the same characteristic as that the drive in Fig. 11.1.

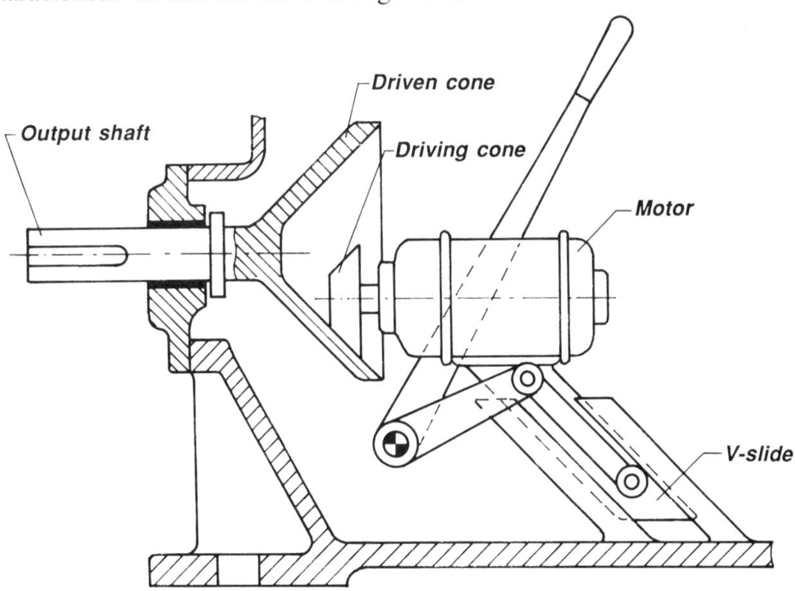

Figure 11.3 Cone-in-cone drive. This unit is driven from the conical roller, which is in contact with the internal cone. There is line contact, but sliding due to different velocities of the cone and roller along the line of contact causes wear.

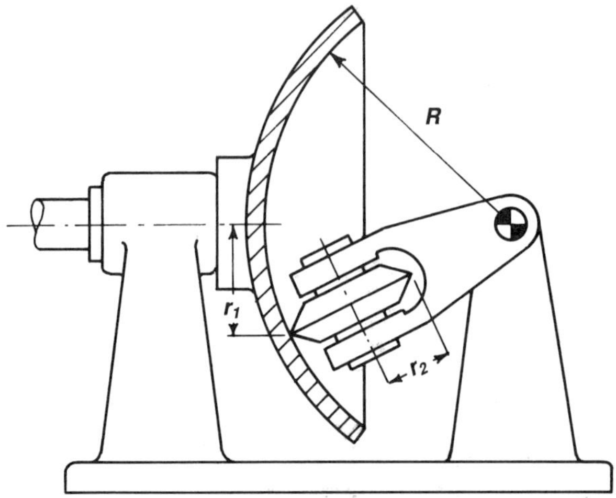

Figure 11.4 Hollow sphere and sharp-edged roller combine to give an infinite-variable reduction. The roller has been made sharp-edged in order to reduce wear due to the different circumferential speeds of the roller and sphere. However, as wear takes place, the contact pressure between the two members is reduced unless compensated for by spring forces (not shown). The output speed can be reversed.

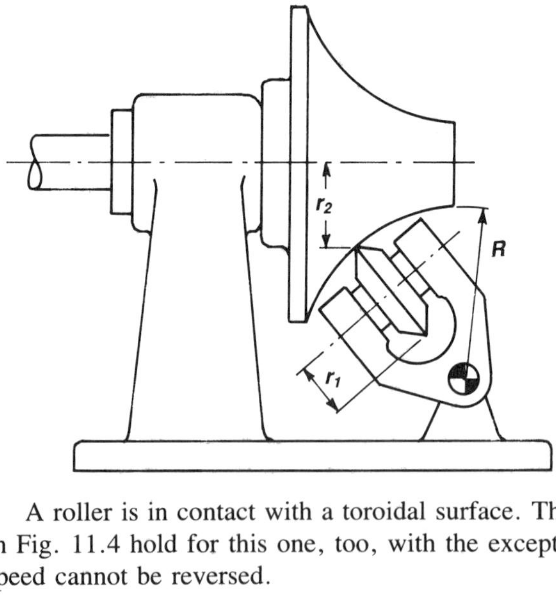

Figure 11.5 A roller is in contact with a toroidal surface. The features of the device in Fig. 11.4 hold for this one, too, with the exception that here the output speed cannot be reversed.

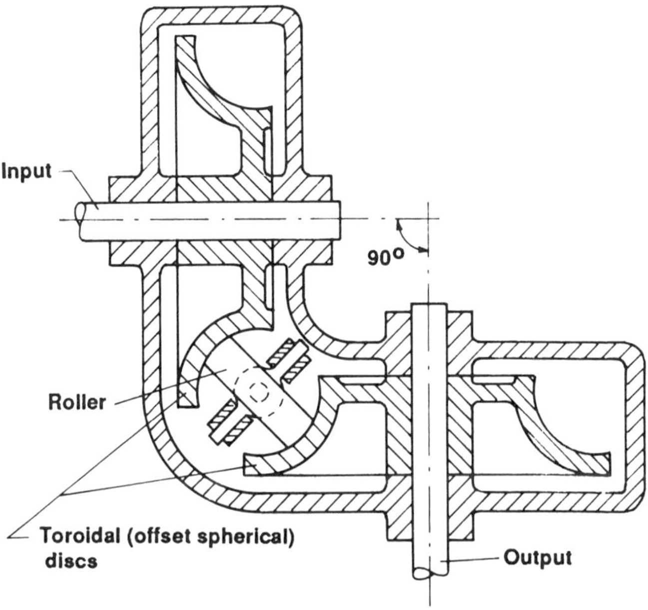

Figure 11.6 The contact of two toroidal surfaces with one friction roller results in a wider range of speed reduction. The input and output shafts can be perpendicular to each other as shown, but the angle between the shafts can be other than 90°.

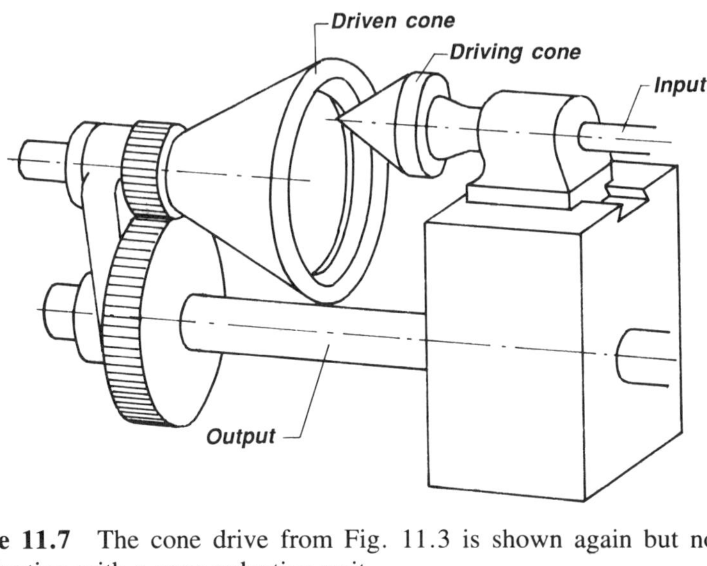

Figure 11.7 The cone drive from Fig. 11.3 is shown again but now in combination with a gear reduction unit.

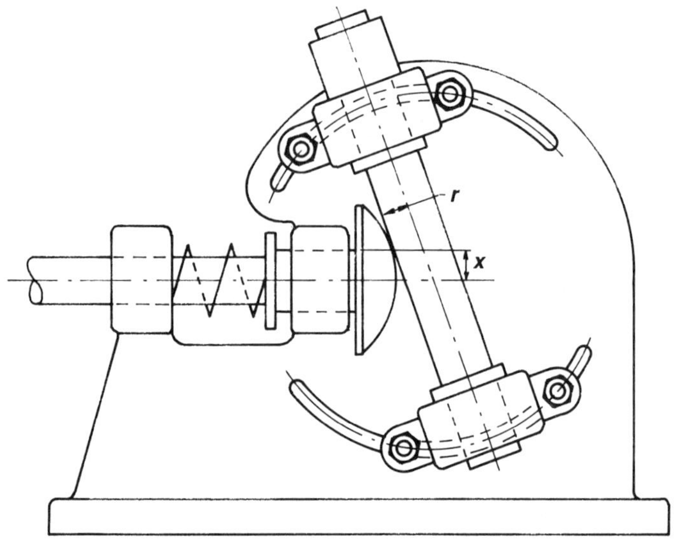

Figure 11.8 A sphere in contact with a shaft is another possibility. A spring is employed to ensure contact between the two members. If speed change is required often, then an arrangement other than the four bolts in two slots can be used.

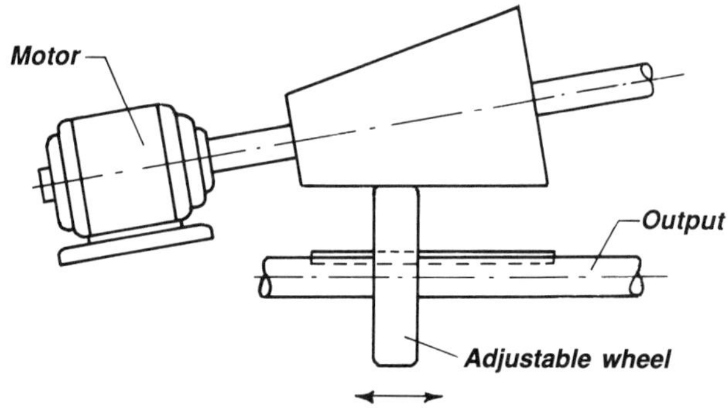

Figure 11.9 A roller in contact with a cone. Input and output shafts are nonparallel but lie in the same plane.

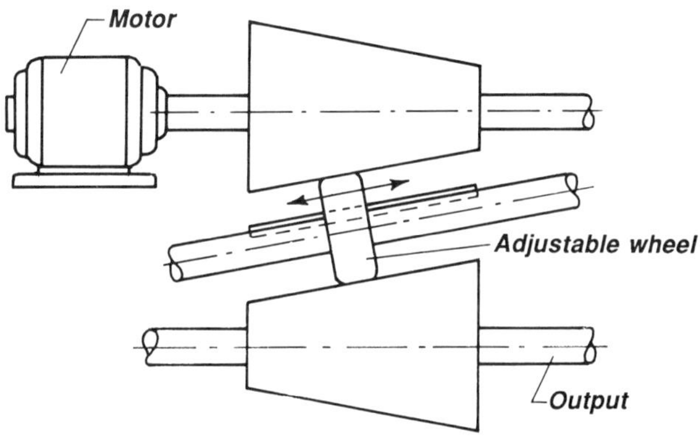

Figure 11.10 One roller in contact with two cones increases the adjustable speed range. Input and output shafts are parallel.

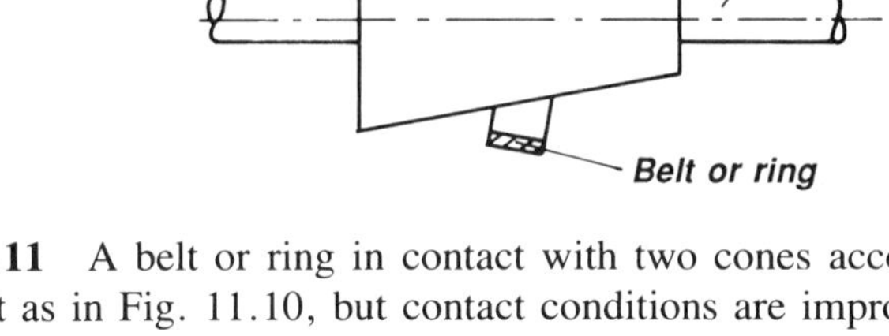

Figure 11.11 A belt or ring in contact with two cones accomplishes the same result as in Fig. 11.10, but contact conditions are improved.

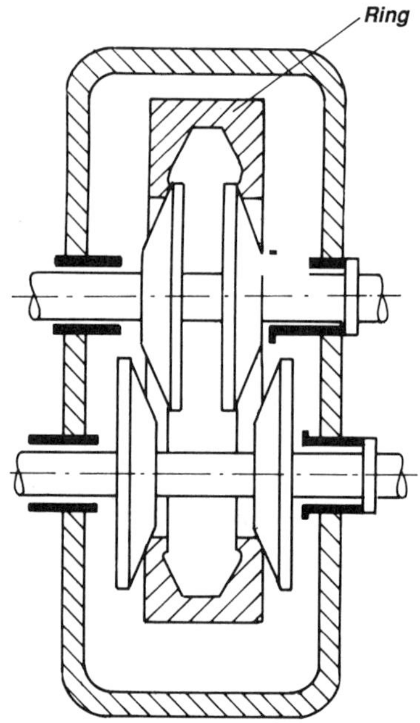

Figure 11.12 A variation of the ring combination of Fig. 11.11. Specially shaped rings transmit motion between the cones. The twin arrangement doubles the power that can be transmitted.

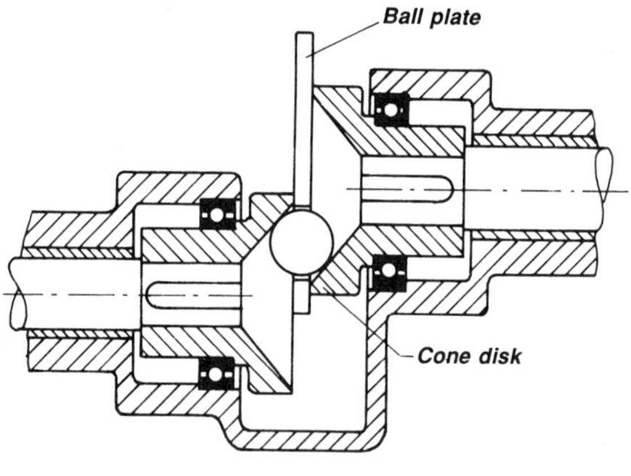

Figure 11.13 A ball-and-cone drive transmits motion between two parallel shafts. Moving the ball up and down increases and decreases the speed reduction ratio, respectively. The two shafts rotate in opposite directions.

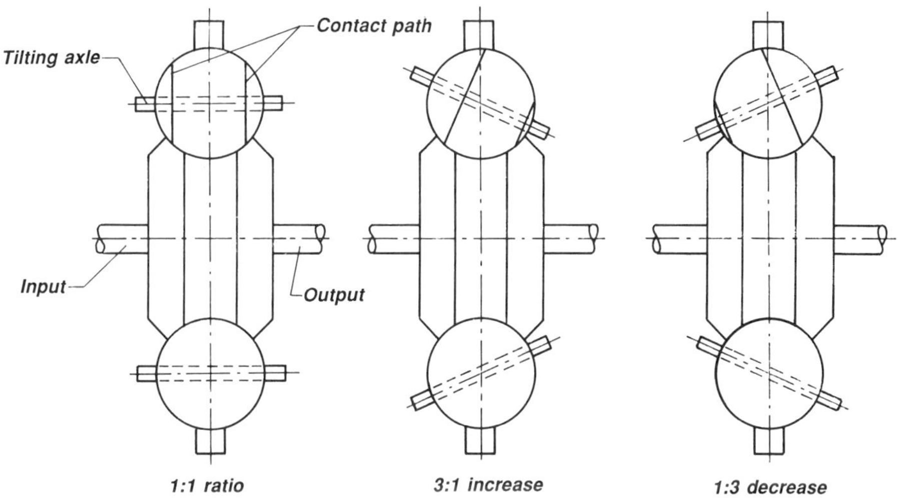

Figure 11.14 This drive can be considered a variation of the drive in Fig. 11.5. Dependent on the position of the ball shafts, the transmission ratio can be varied around +1, meaning that the shafts rotate in the same direction.

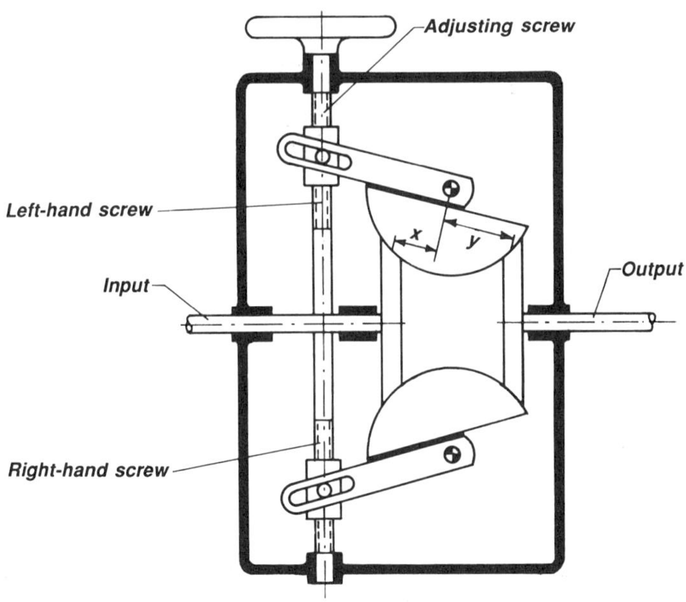

Figure 11.15 A variation of the principle in Fig. 11.14. The values of x and y determine the reduction ratio.

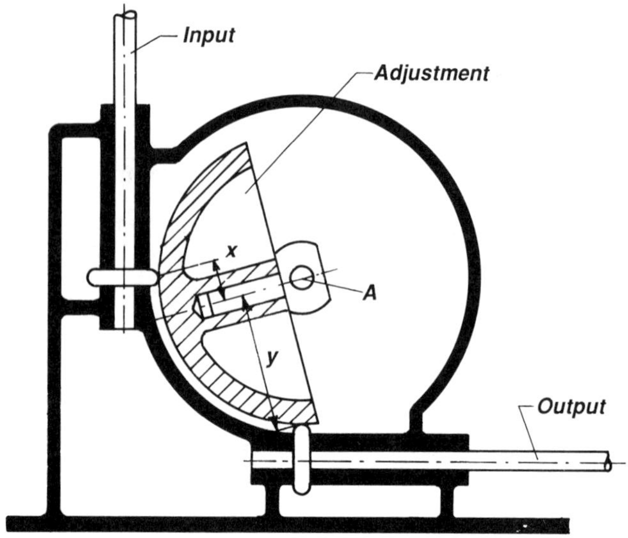

Figure 11.16 The sphere can be rotated around A but can also rotate around its own axis, thereby transmitting a variable-speed ratio between the input and output shafts.

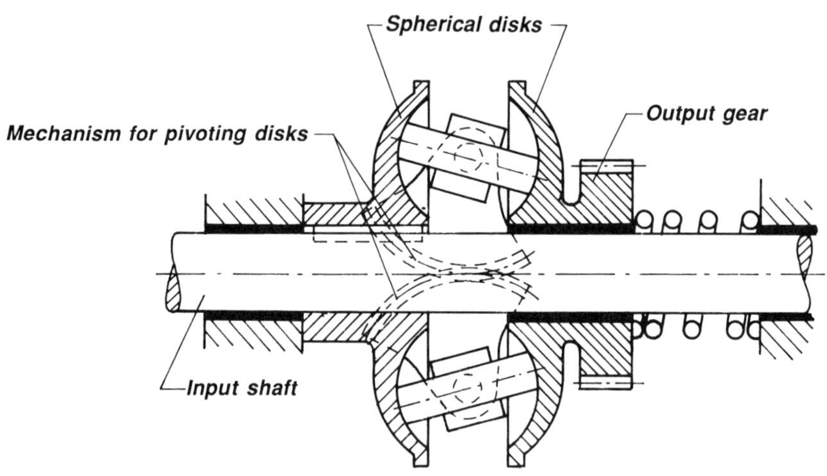

Figure 11.17 The twin arrangement of two rollers in contact with offset spheres allows doubling of power transmitted. Notice the spring that ensures contact between the rotating elements.

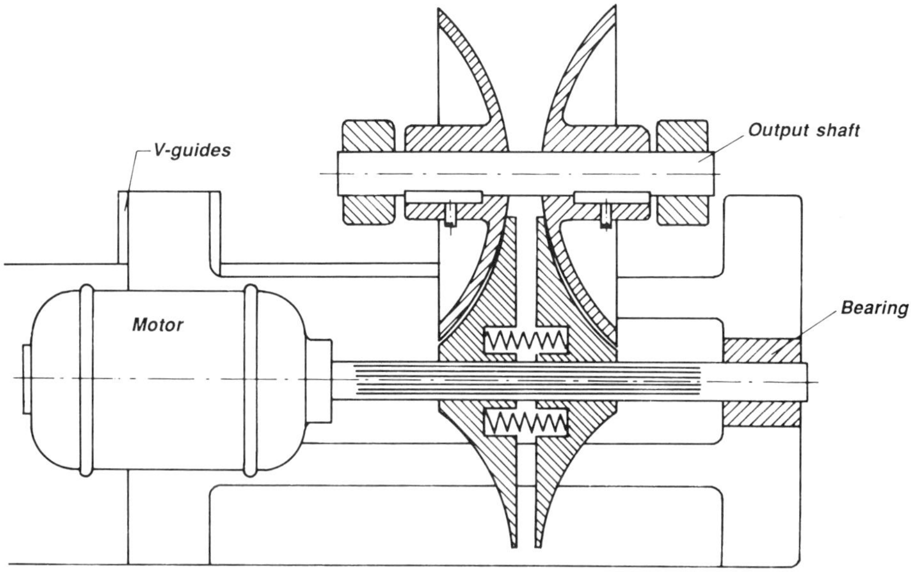

Figure 11.18 Two half spheres are in contact with toroidal elements that are pressed up against the spheres by springs. A motion up and down of the spheres changes the transmission ratio.

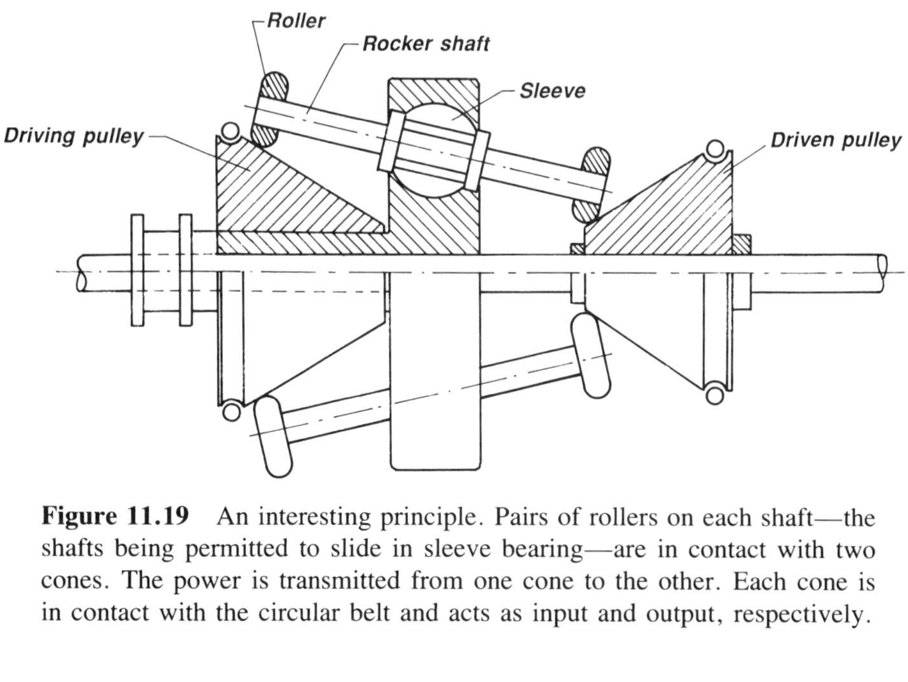

Figure 11.19 An interesting principle. Pairs of rollers on each shaft—the shafts being permitted to slide in sleeve bearing—are in contact with two cones. The power is transmitted from one cone to the other. Each cone is in contact with the circular belt and acts as input and output, respectively.

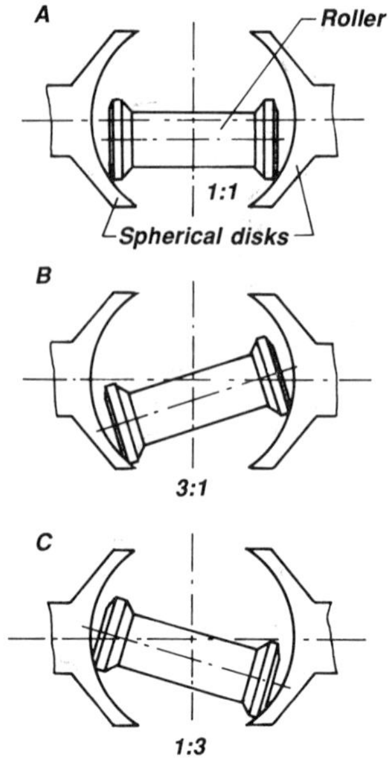

Figure 11.20 Almost a twin arrangement of Fig. 11.4. The roller is actually a sphere, and changing its angular position changes the speed ratio between the input and output shafts.

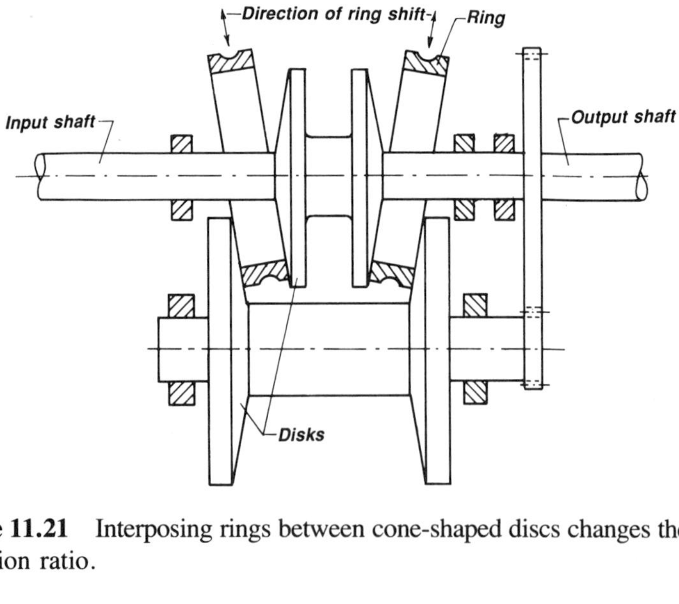

Figure 11.21 Interposing rings between cone-shaped discs changes the speed-reduction ratio.

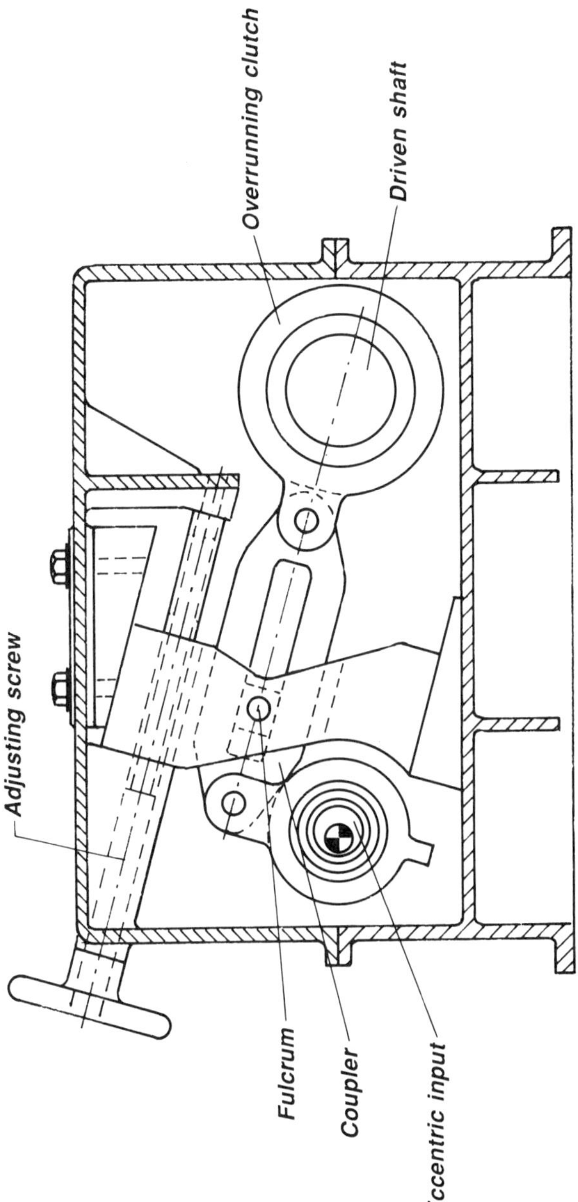

Figure 11.22 The position of the fulcrum of this inverted slider crank can be adjusted so that the angle of oscillation of the output member can be varied. The one-way, or overrunning, clutch ensures rotation of the output shaft in the same direction. Employing several parallel units results in a very smooth output rotation.

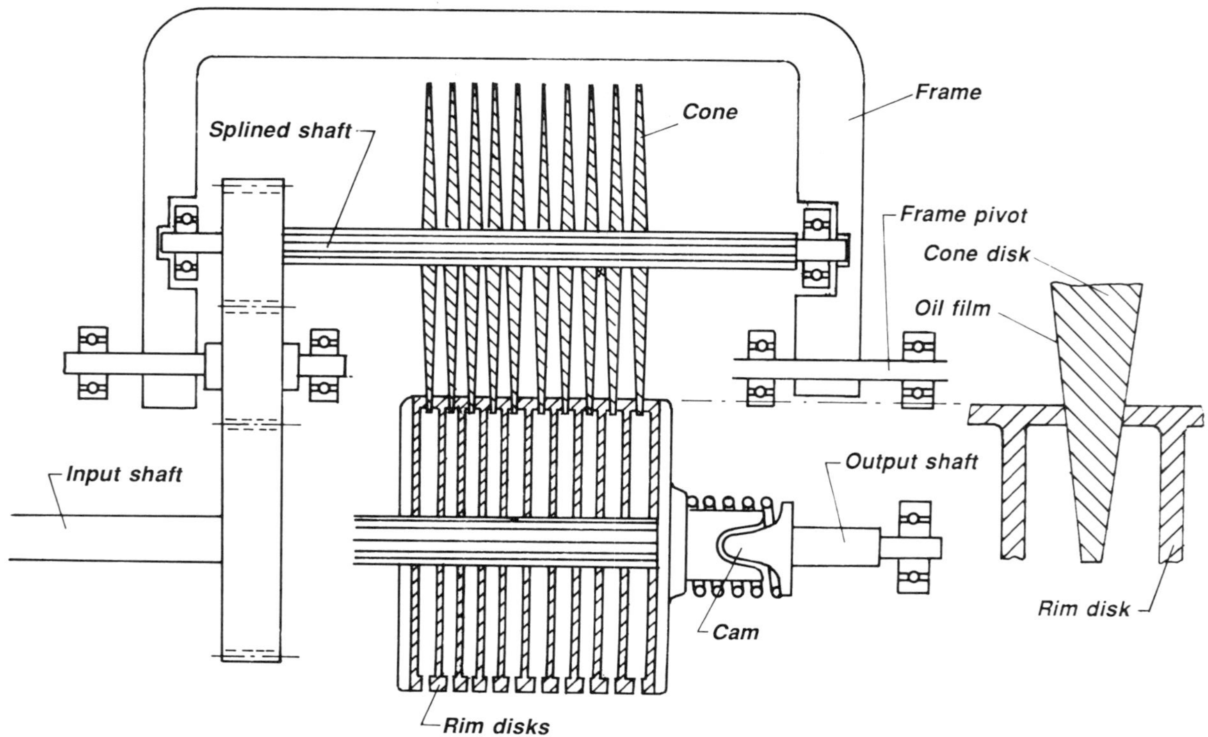

Fig. 11.23 The transmission of power to the output shaft is done by slightly conical discs that press against each other. The drive is available up to 60 hp with negligible slip (from 1% to 3%).

12

Snap-Action Switching Mechanisms

Snap action results when a force applied to a mechanism exceeds a certain predetermined level, causing a sudden motion to occur. Often the snap action is caused by a spring that suddenly forces the mechanism to another position. Electrical switches are one area where these mechanisms are employed. Some of these mechanisms are a delight for a mechanism designer.

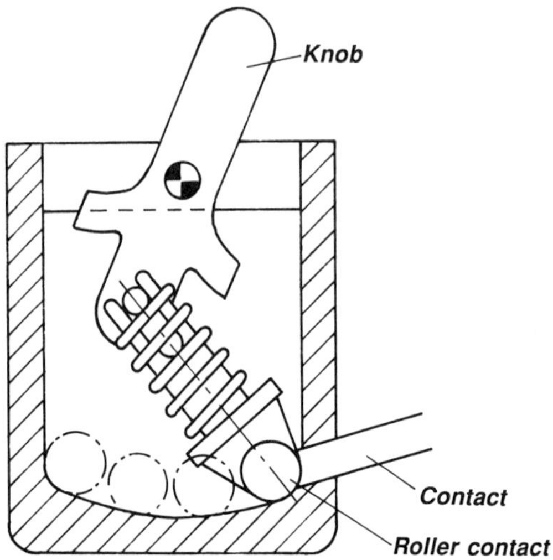

Figure 12.1 Moving the knob in a CCW direction will cause the spring-loaded member to shift to the left when the knob has passed its center position. The action takes place suddenly.

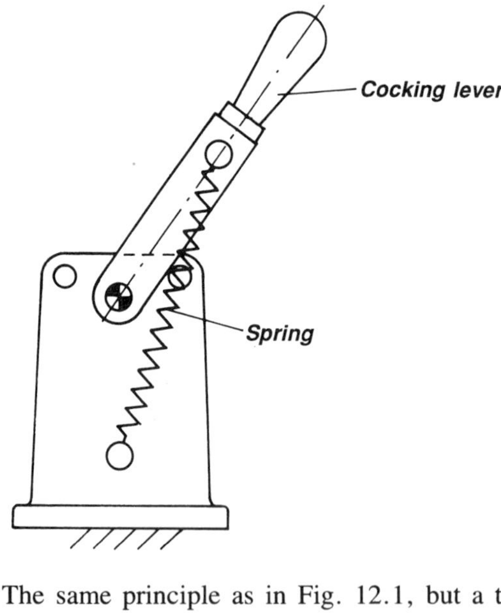

Figure 12.2 The same principle as in Fig. 12.1, but a tension spring is used.

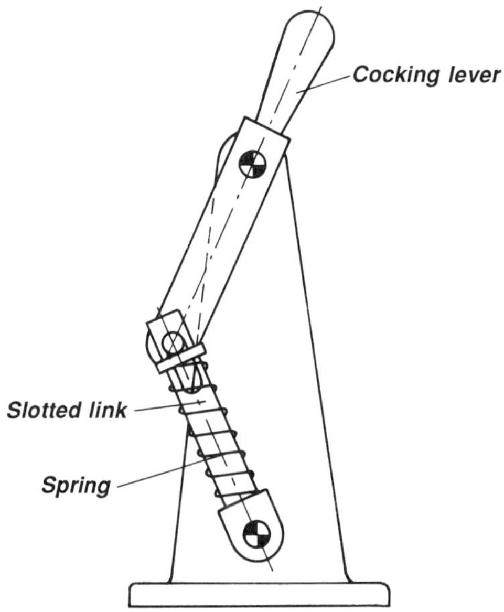

Figure 12.3 When lever is moved CCW, an over-center position snap action takes place.

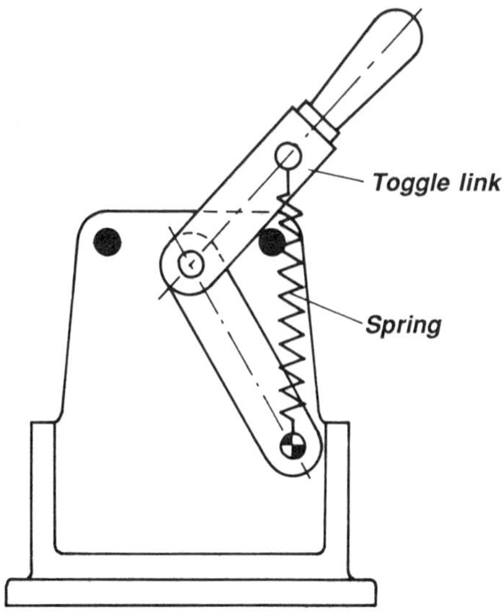

Figure 12.4 A double link, spring loaded against each other, provides snap action.

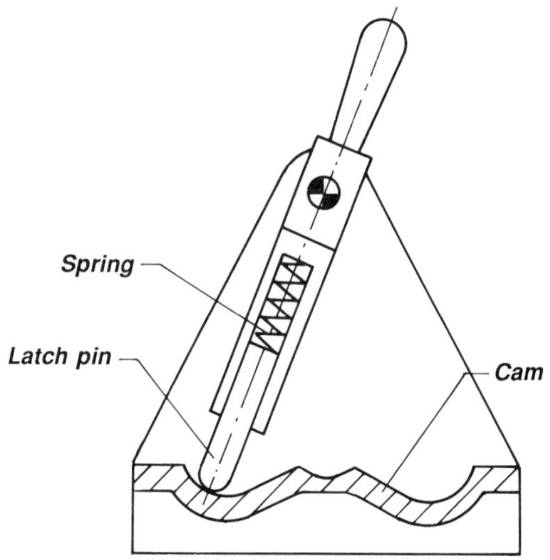

Figure 12.5 Using a cam-guided link results in three distinct positions of the latch.

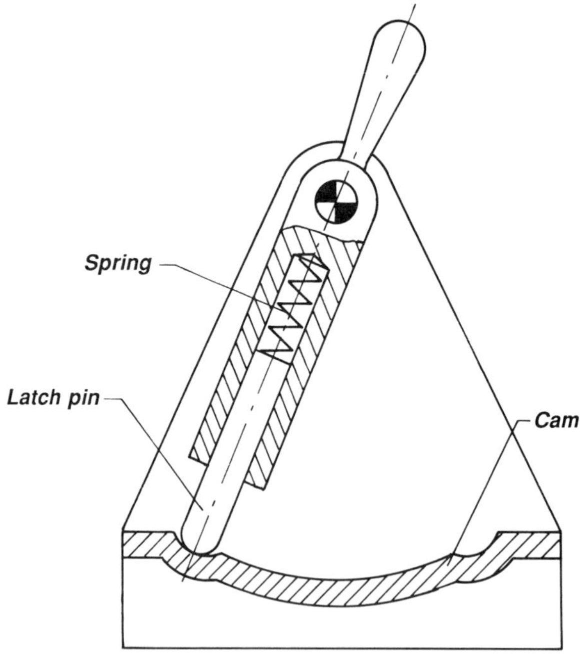

Figure 12.6 Using a cam-guided link results in two distinct positions of the latch. (See also Figs. 12.7 and 12.8.)

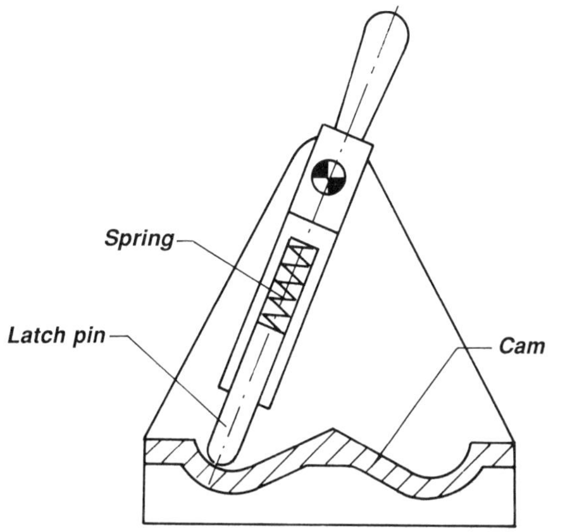

Figure 12.7

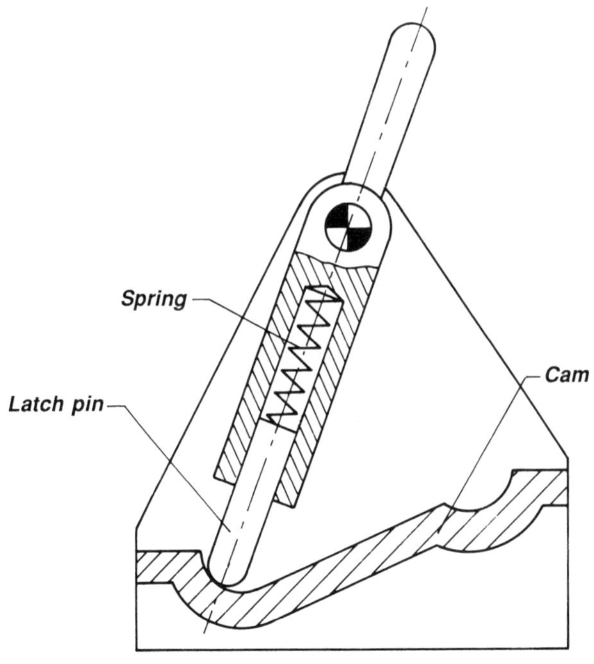

Figure 12.8

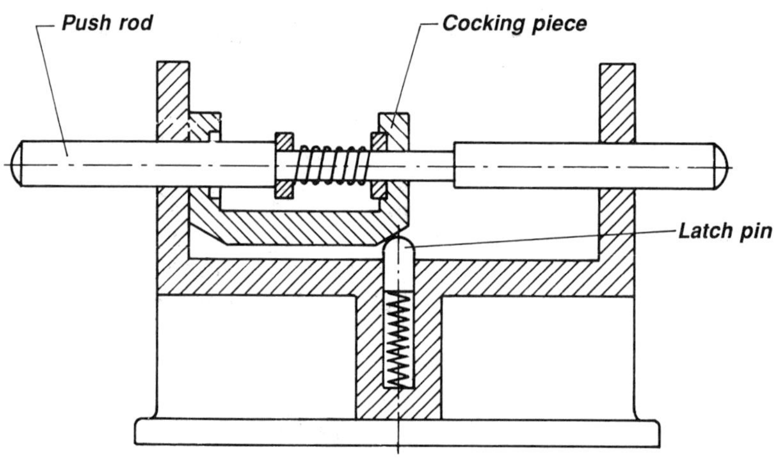

Figure 12.9 Moving the push rod from left to right causes the cocking piece with chamfers to slide over the latch pin. When left chamfer reaches latch pin, cocking piece moves to the extreme right position. Now the same is repeated for motion from right to left.

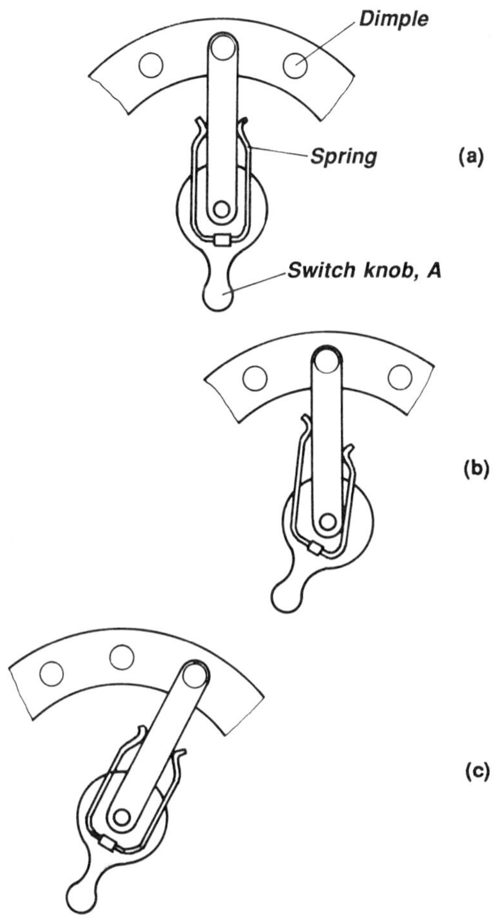

Figure 12.10 Moving the switch knob A CW results in the build-up of pressure, through spring force, on the lever to be switched, but its motion is prevented by the dimple, which is a combination of form and force closure. When spring force is sufficiently great, lever switches.

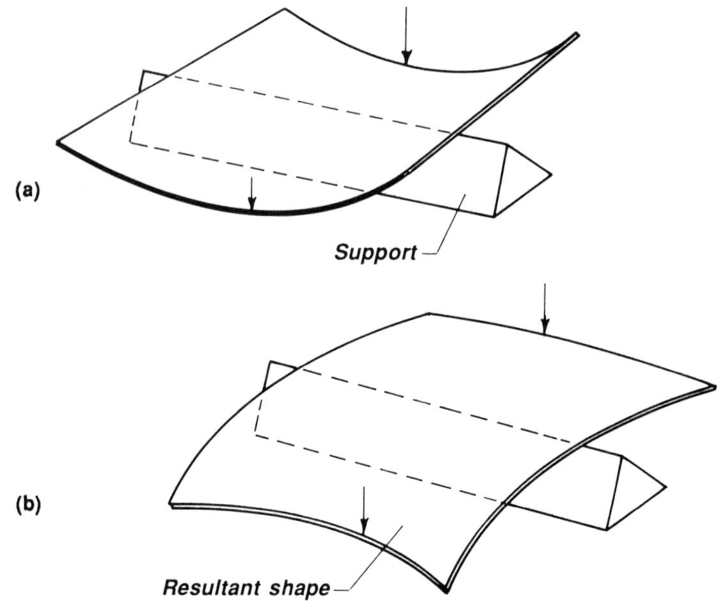

Figure 12.11 A curved, flat spring provides snap action because of its shape. At *a* is shown the unloaded curved spring. When loaded it suddenly snaps into the shape shown in *b*.

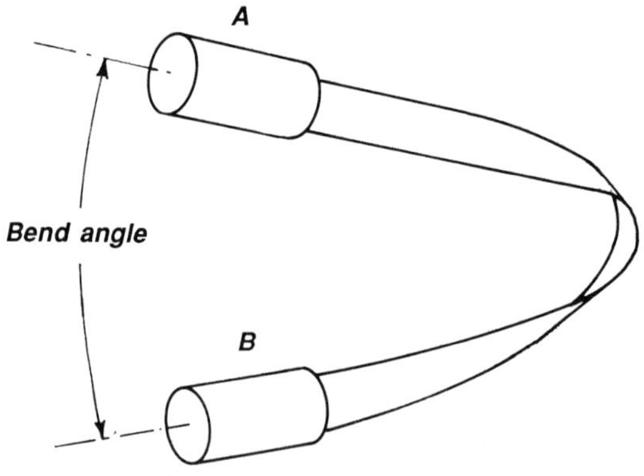

Figure 12.12 Flat strip spring provides snap action. When knob A is turned, the spring twists and suddenly it snaps over and rotates the other knob B half a revolution.

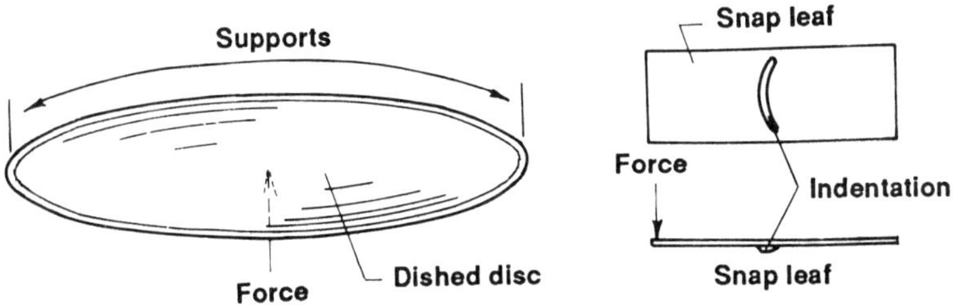

Figure 12.13 Indented flat springs provide snap action.

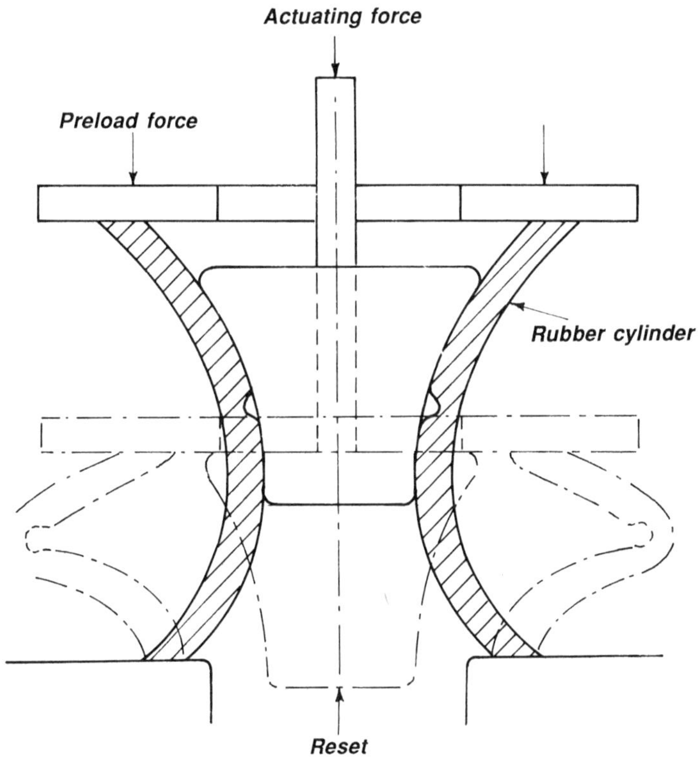

Figure 12.14 In much the same way as in Fig. 12.13, the rubber cylinder collapses when subjected to the actuating force.

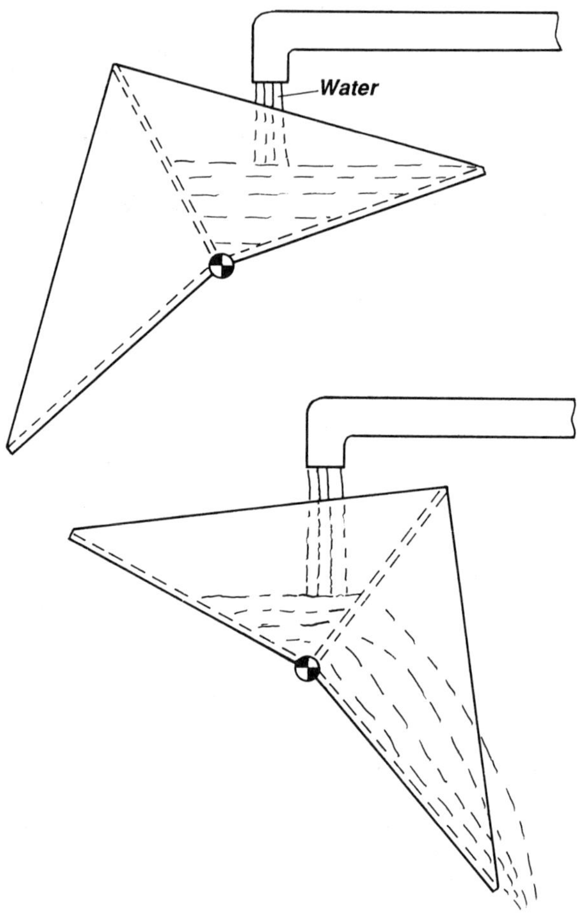

Figure 12.15 Gravity is used here to provide snap action. When one side of the trough is filled, it tips over and empties while the other side starts to get filled.

13
Parallelogram Mechanisms

Parallelogram mechanisms are a special form of four-bar linkage, the four links of which form a parallelogram. Despite its simplicity, parallelogram mechanisms are often employed in real applications. For the mechanism designer it is important to know the peculiarities and the design changes that are possible without changing the motion itself. One has to watch out for dead-center positions and find out how to overcome this disadvantage. A parallelogram mechanism transmits a uniform angular velocity ratio of +1 : 1 between input and output shafts.

PARALLELOGRAM LINKAGES

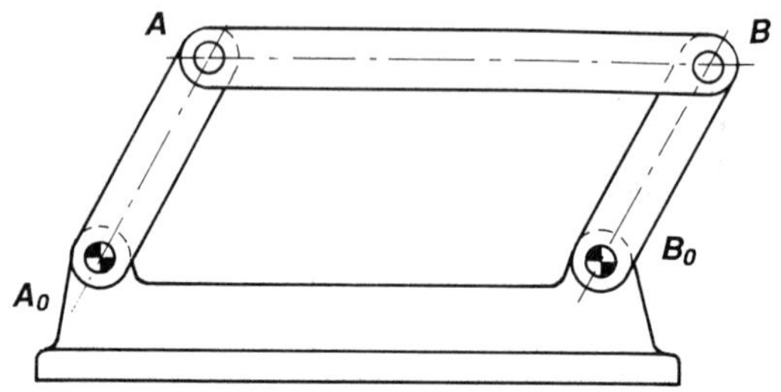

Figure 13.1 This is the basic parallelogram mechanism. The length of opposing links is equal. When link A_0A rotates, link B_0B remains parallel to link A_0A.

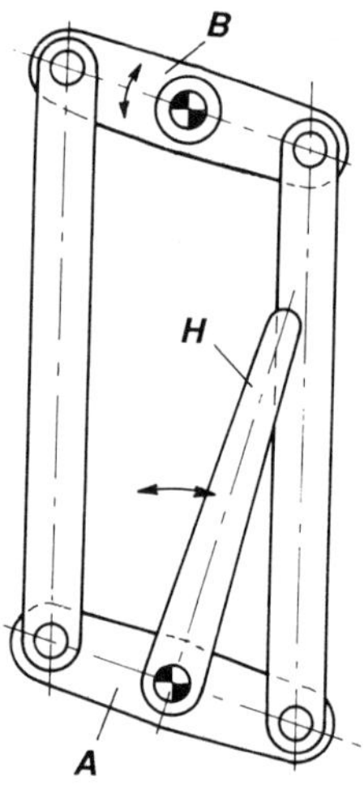

Figure 13.2 The swinging motion of handle H is transferred to exactly the same swinging motion of link B by using a parallelogram mechanism. Link A and handle H are rigidly connected.

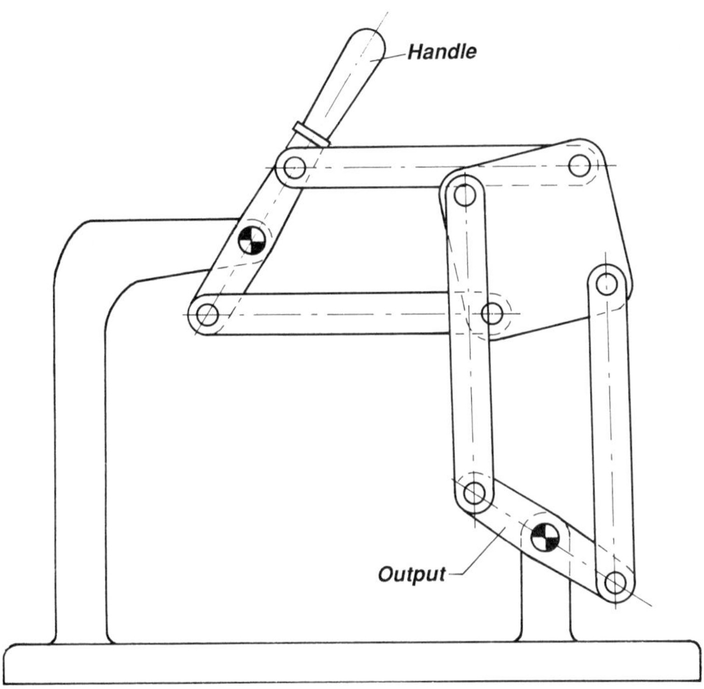

Figure 13.3 Motion around corners. Using two parallelogram mechanisms in series can transmit motion around corners.

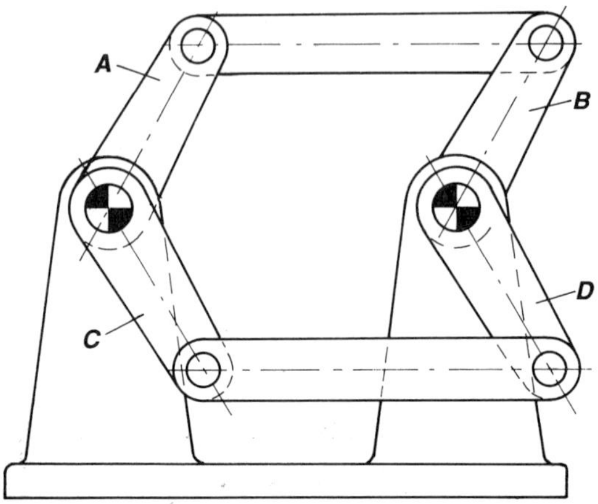

Figure 13.4 To overcome dead-center positions, instead of having one crank only, bell cranks AC and BD are used. Dead-center positions occur when a floating link and its two cranks are in line. In that position the output link is able to move either CW or CCW, but because one crank of the bell crank is not in dead-center position the dead-center position of the other crank is neutralized. With this design it is not possible to have through-going shafts because the floating links would cut the shafts.

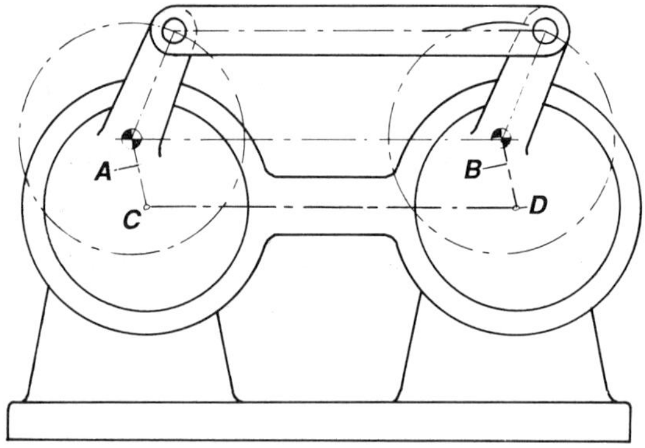

Figure 13.5 By using pin enlargement, i.e., enlarging the pins at C and D (Fig. 13.4) so that they can be replaced by eccentrics on the drive shafts, results in a parallelogram linkage where through-going shafts can be used.

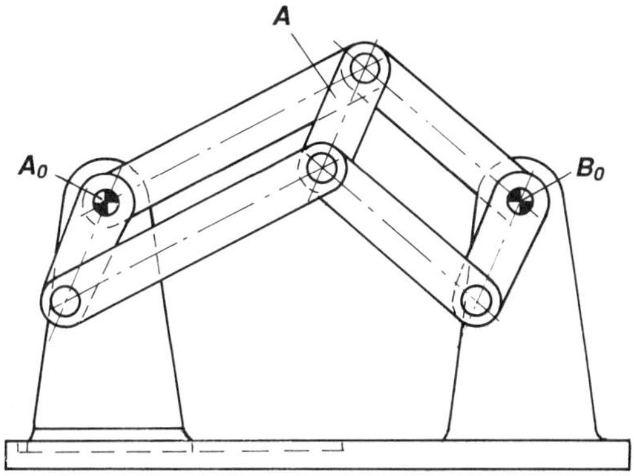

Figure 13.6 By employing two parallelogram mechanisms with a floating link A, dead-center positions are avoided. The distance between the two shafts A_0 and B_0 can be varied. Motion can be transmitted around corners.

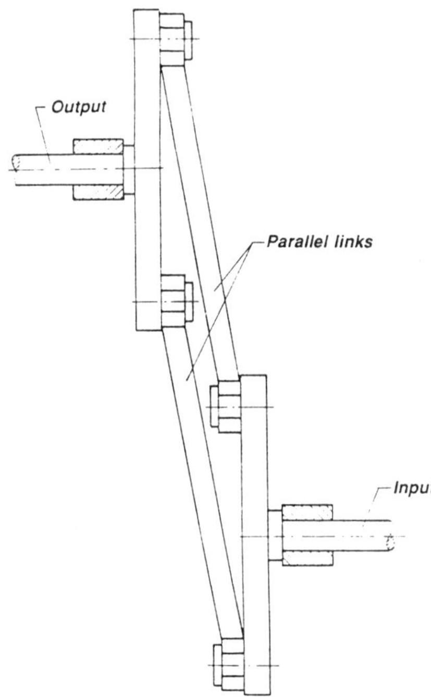

Figure 13.7 Two or more connecting rods (equal in length) are used to overcome dead-center positions. The stepped design of the rods enables the rods to pass each other.

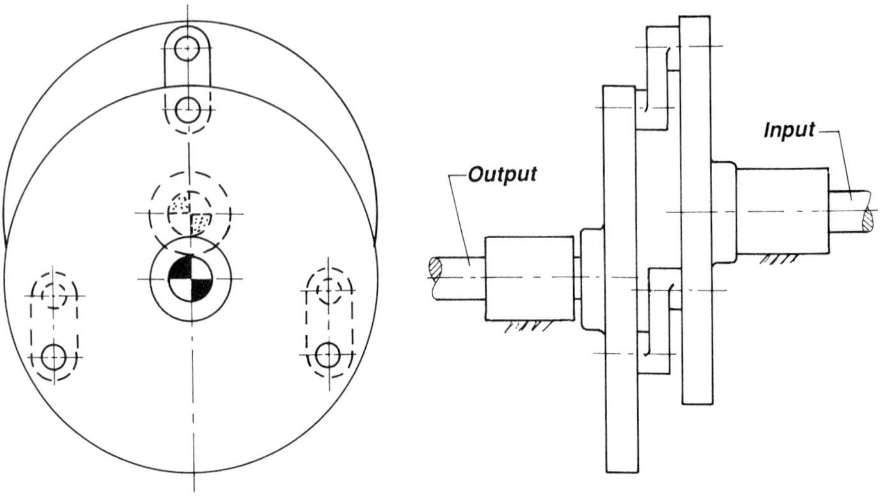

Figure 13.8 Here the connecting rods (equal in length) are made so small that they can pass each other.

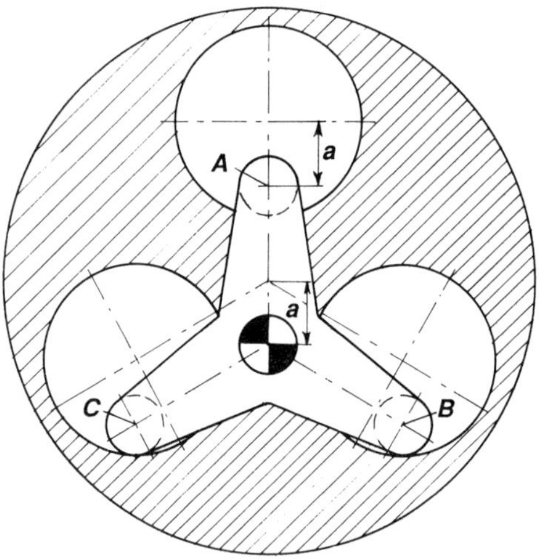

Figure 13.9 Connecting rods can be omitted all together in this design where *a* corresponds to the length of a connecting rod. The rods are replaced by holes, the radii of which are equal to the length of the omitted links. Rollers are at A, B, and C. Unit builds very compact.

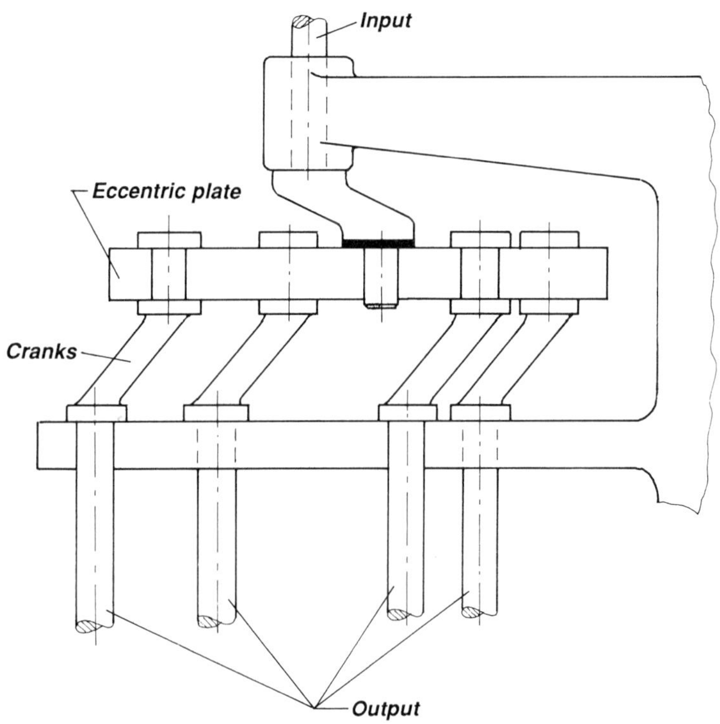

Figure 13.10 Combined parallelogram mechanism is used here to drive a number of shafts with the same angular velocity. The input crank drives the eccentric plate. Output shafts can be placed very close together.

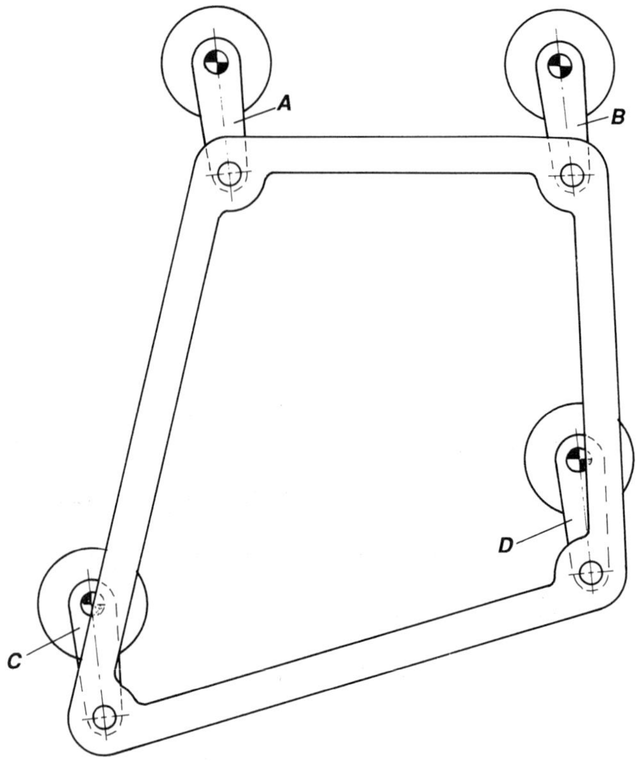

Figure 13.11 Motion is transmitted from any of the shafts A, B, C, or D to the other shafts with the same angular velocity. All cranks must be equal in length and parallel.

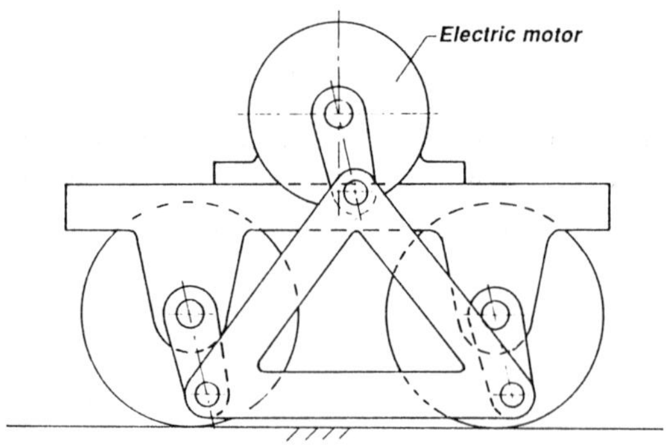

Figure 13.12 Parallel motion is transferred from the motor to the wheels of this electric locomotive.

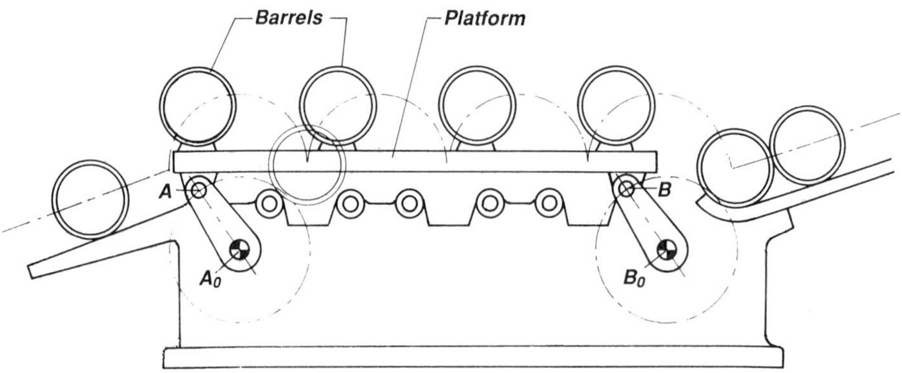

Figure 13.13 A parallelogram mechanism A_0ABB_0 is used here to transport several barrels at the same time from one station to another. Dead-center positions should be avoided here.

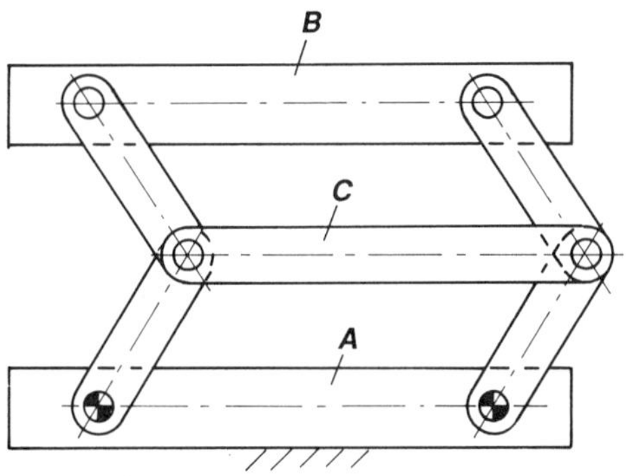

Figure 13.14 The moving member B remains parallel with the fixed member A, or, both moving members remain parallel at all times.

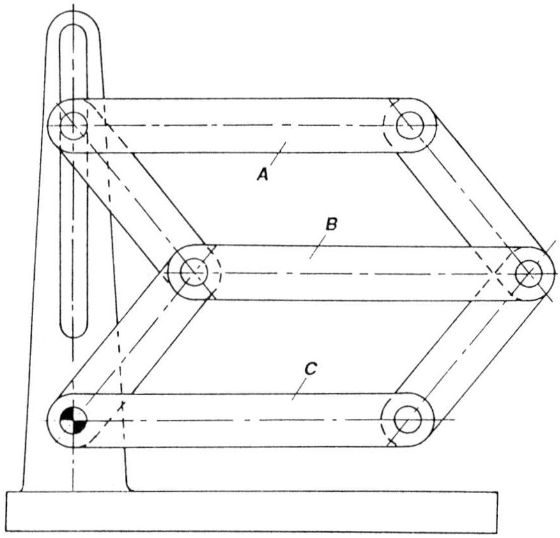

Figure 13.15 Members A, B, and C remain parallel at all times as member A is guided along a straight line by the slotted member.

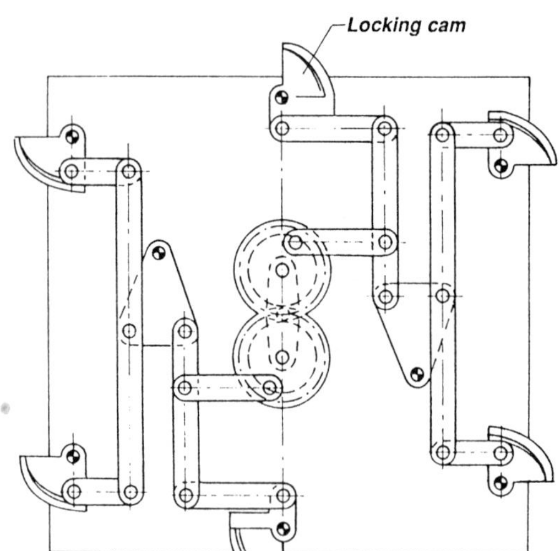

Figure 13.16 This safe is closed by parallelogram mechanisms operated in unison, all locking cam members being moved in and out of locking position at the same time.

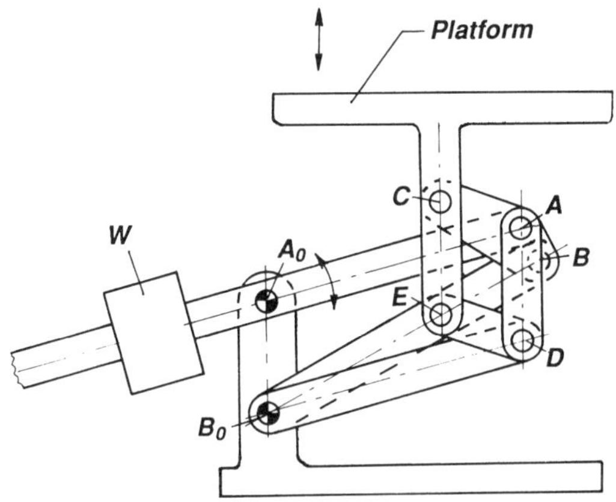

Figure 13.17 Link AD in this interesting arrangement will always be parallel to EC, and link ED parallel to AC. Hence CE will always be parallel to link AD. The links of the four-bar linkage A_0ABB_0 are so proportioned that point C moves in approximately a straight line. Final result is that the output plate will be kept horizontal while moving almost straight up or down. The counterweight W permits this device to function as a disappearing platform in a stage.

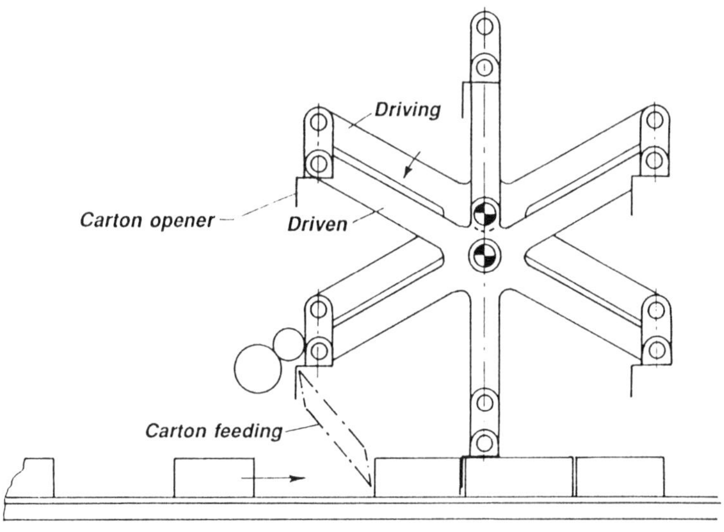

Figure 13.18 All six short links are kept in a vertical position while rotating. The center link is the driver. This particular application feeds and opens cartons, but the device is capable of many diverse applications.

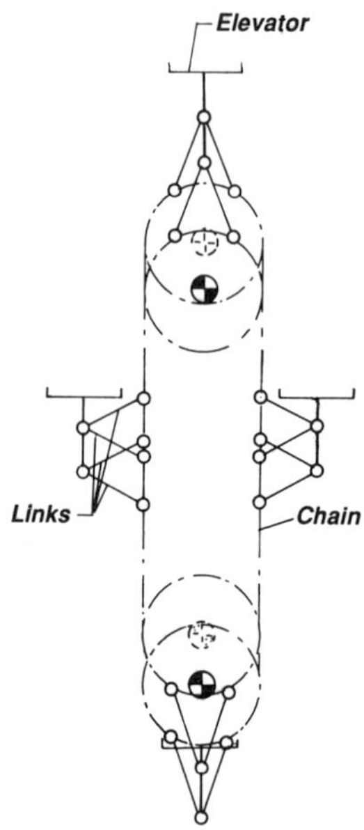

Figure 13.19 Chain-driven parallelogram linkages guide elevators in parallel, vertical position through the entire motion. Principle used in elevators that constantly circulate, so-called paternosters.

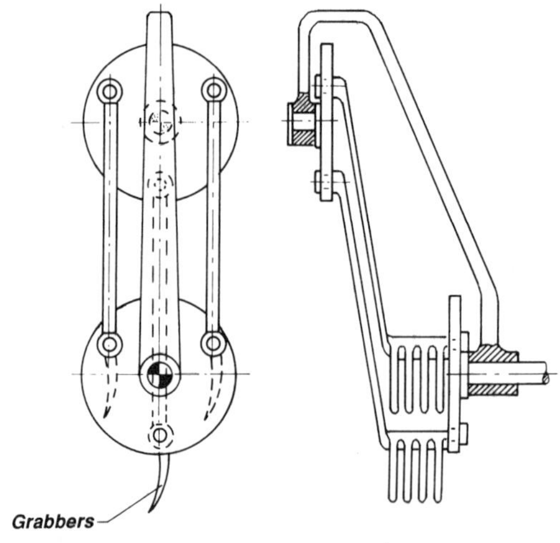

Figure 13.20 Parallelogram linkages used to move the grabbers in circular parallel motion. Used for harvesting potatoes.

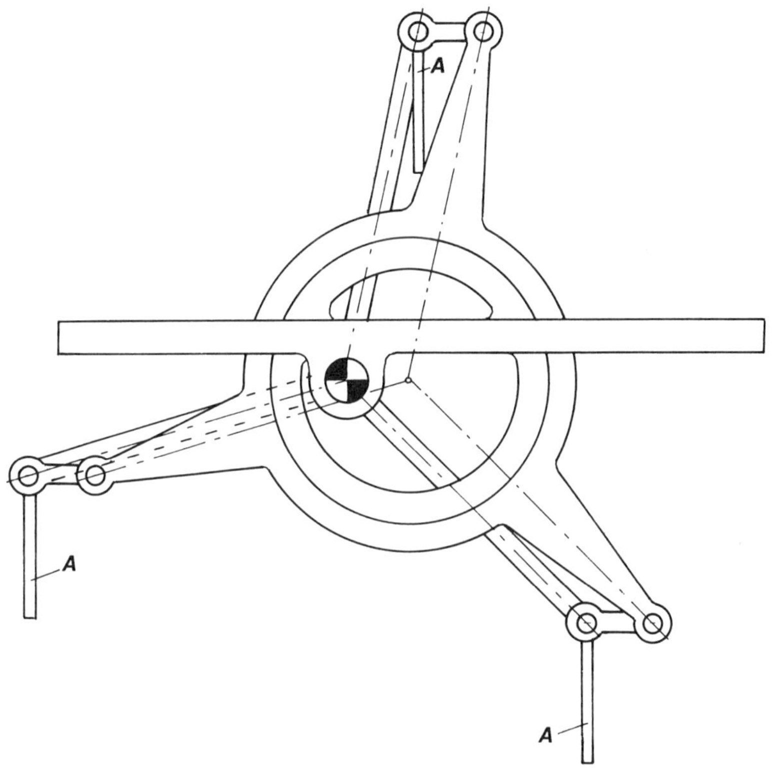

Figure 13.21 Parallelogram linkages A are moved in vertical position at all times along a circular path. Used in hay tedder.

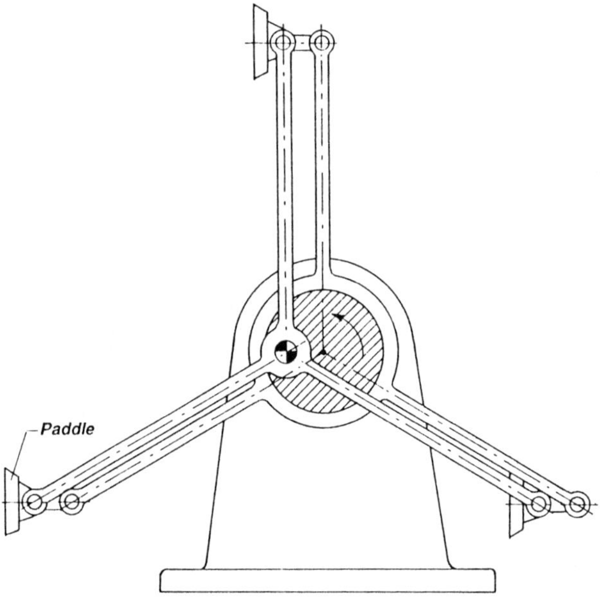

Figure 13.22 The paddles are kept parallel by means of parallelogram linkages.

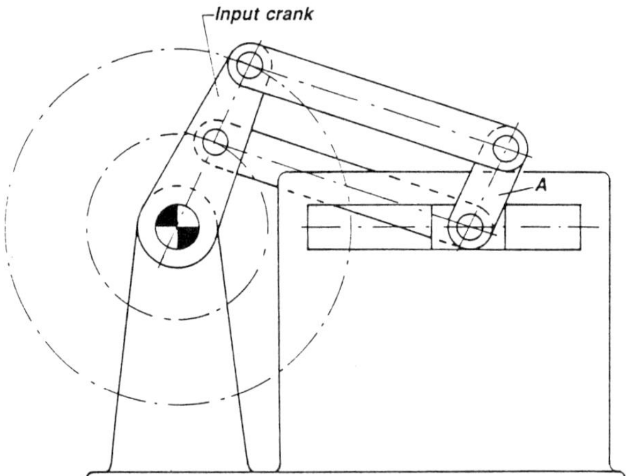

Figure 13.23 Link A moves on the guide but at the same time it rotates in unison with the input crank, which imparts a rather complex motion, but A remains parallel to the input crank at all times.

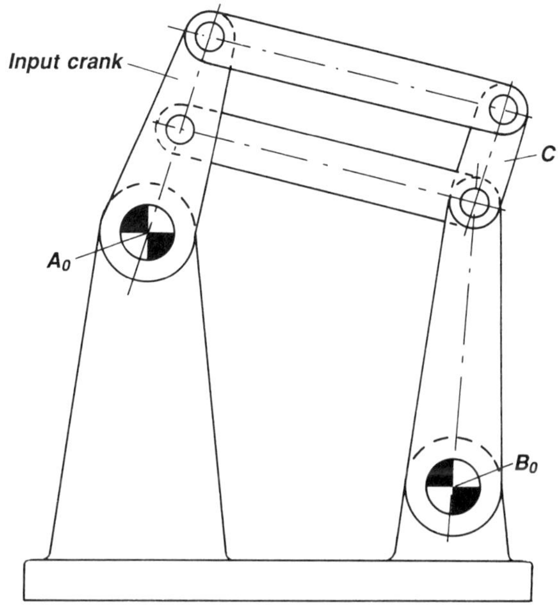

Figure 13.24 Link C is here guided on a circular arc with center at B_0 but rotates in unison with the input crank. Watch out for dead-center positions. Similar to Fig. 13.23.

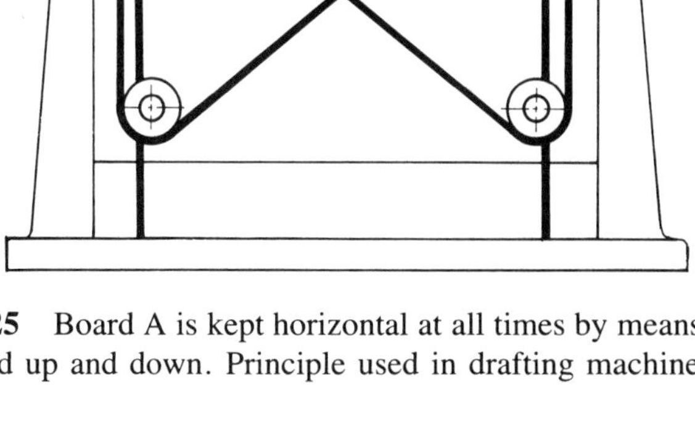

Figure 13.25 Board A is kept horizontal at all times by means of the cords when moved up and down. Principle used in drafting machines.

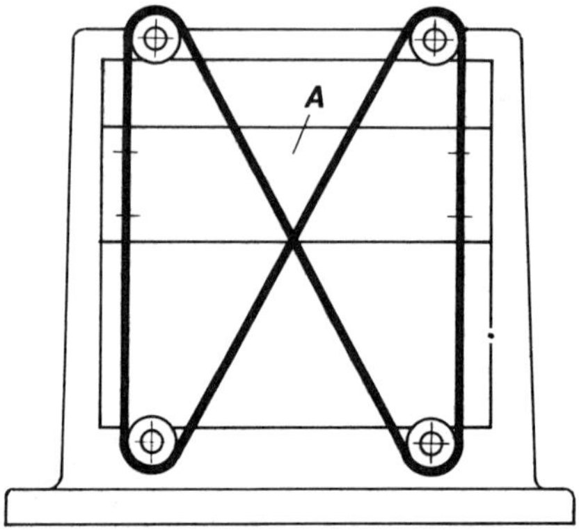

Figure 13.26 A slightly different arrangement of that in Fig. 13.25, but the purpose is the same.

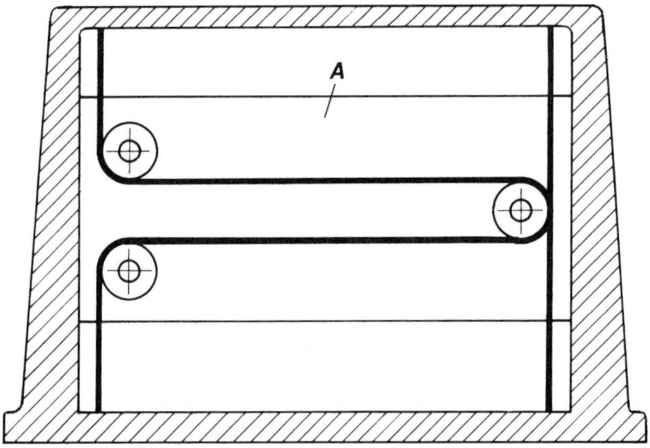

Figure 13.27 Yet still another arrangement, but otherwise the same purpose as that in Fig. 13.25.

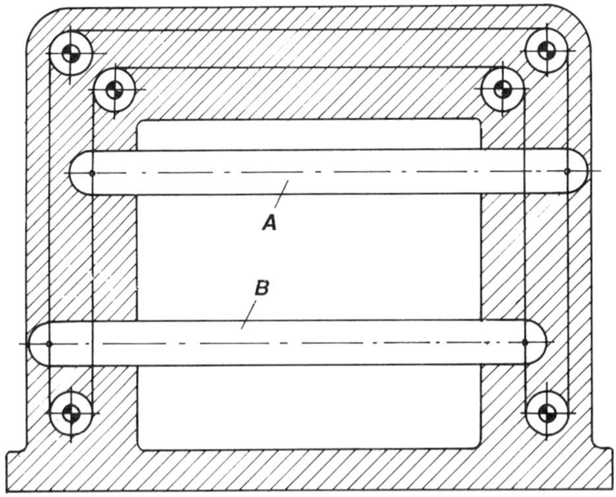

Figure 13.28 What goes up must come down. Theatrical decorations A and B are moved up and down with this device. The motion is parallel and when one decoration A goes up, the other decoration B comes down.

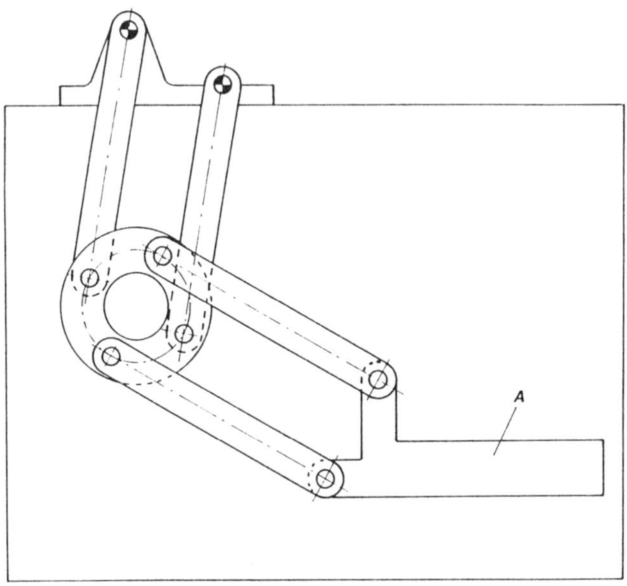

Figure 13.29 Here, two parallelogram mechanisms in series ensure parallel motion of the straight edge of a drafting machine.

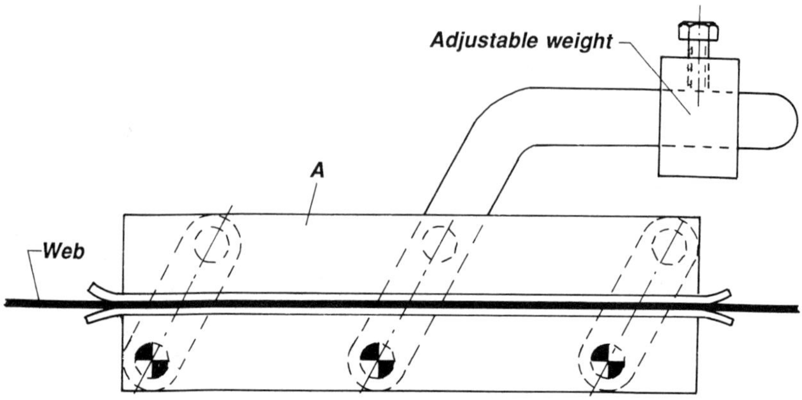

Figure 13.30 A simple tensioning mechanism is obtained by use of a parallelogram mechanism. Member A remains horizontal. Adjusting the weight varies the drag.

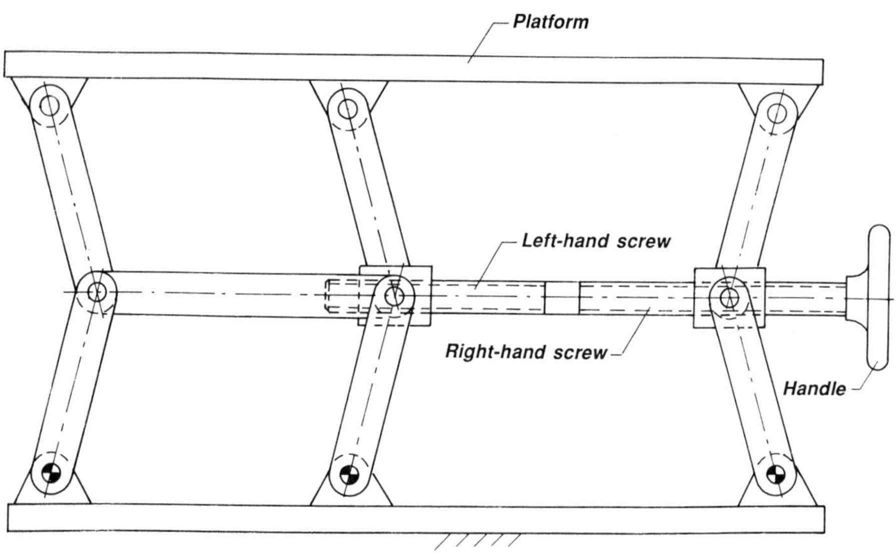

Figure 13.31 Turning the double-handed screw spreads or contracts the linkage to raise or lower the table.

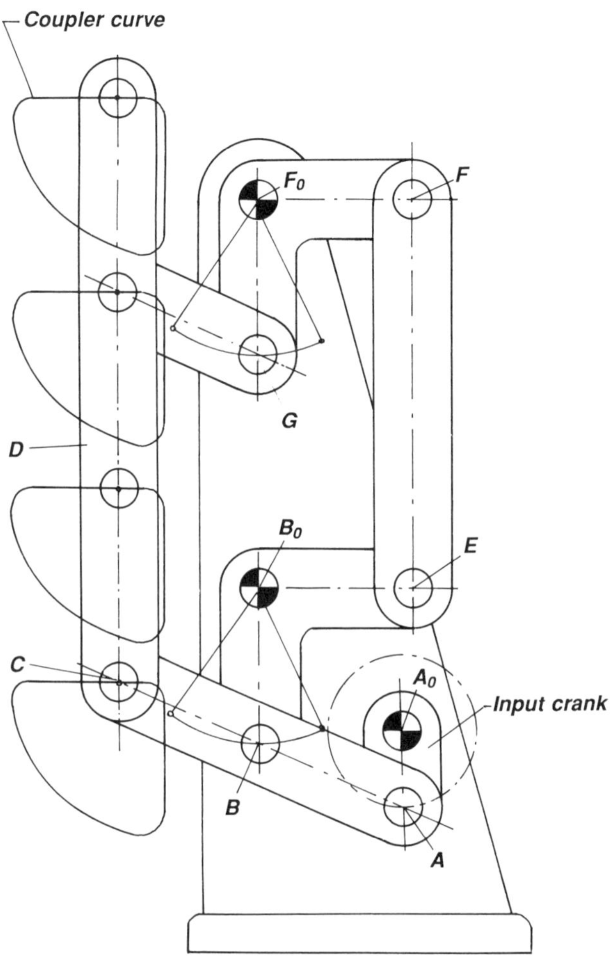

Figure 13.32 The four-bar linkage A_0ABB_0 with the coupler point C guides the link D along the coupler course of C, its parallel motion being maintained by the auxiliary linkage EFF_0. Used in transport mechanisms.

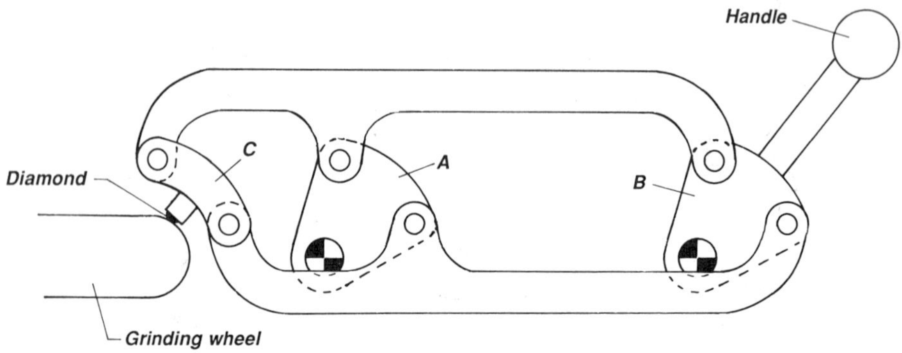

Figure 13.33 Two plates A and B move around fixed centers of rotation. Because of the parallelogram linkages, the diamond on link C describes a circular arc (with center inside the grinding wheel). Built for rounding out grinding wheels.

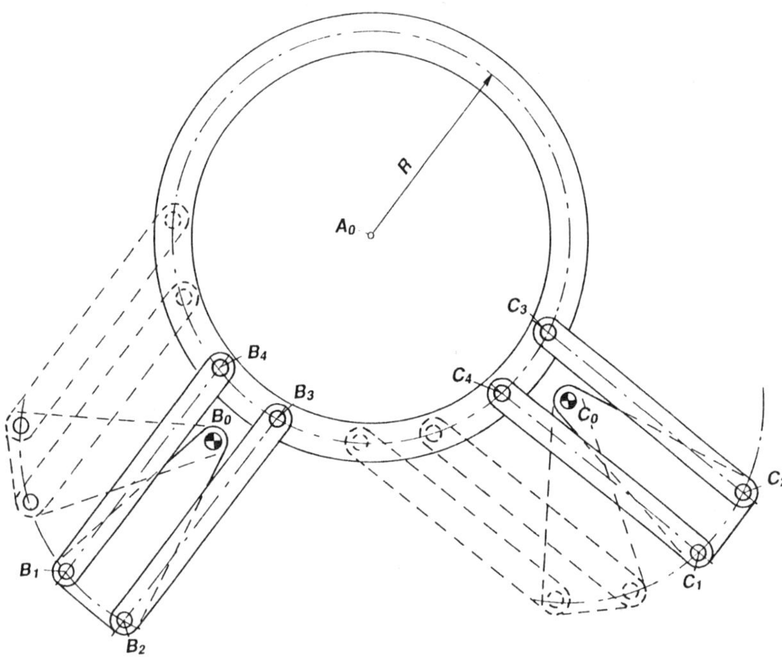

Figure 13.34 The two parallelogram linkages are proportional so that they guide the ring around the center at A_0. The purpose is to clear the inside area of the large ring of any obstructions.

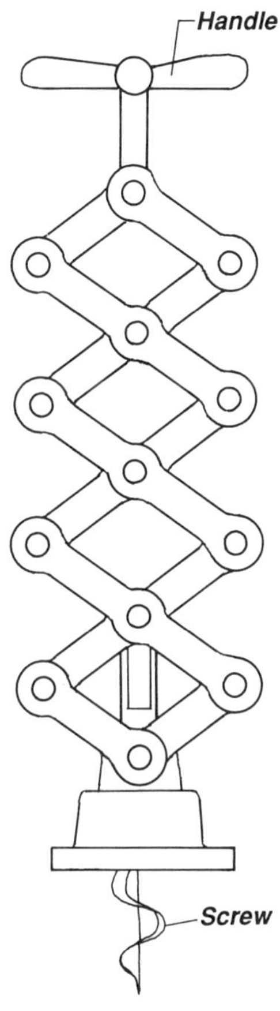

Figure 13.35 Motion reducer or amplifier. When pulling the handle the screw which has been inserted into the cork of a bottle is pulled upwards. Because of the many parallelograms the motion of the screw is very slow, but on the other hand the pulling force on the cork is amplified. Can also be used for motion amplification.

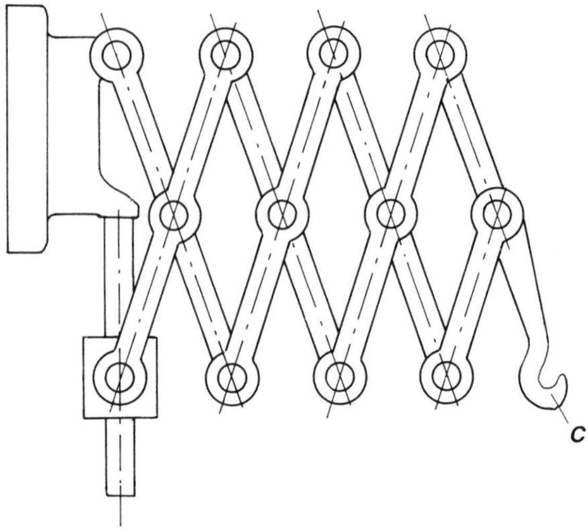

Figure 13.36 A series of parallelogram linkages goes under the name of "Nuremberg scissors." Here they are used to guide a hook for lamps or telephones along a straight path.

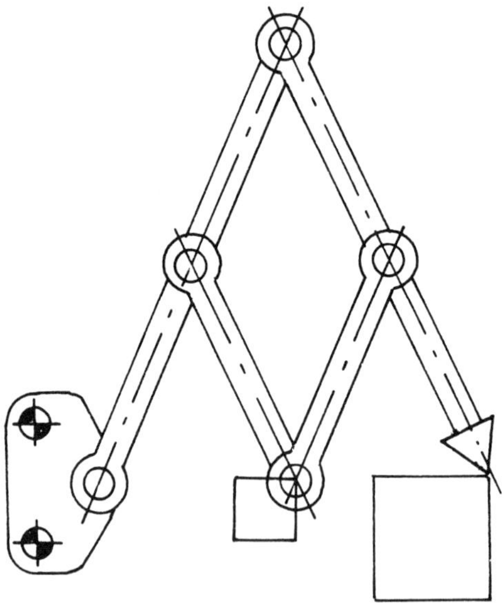

Figure 13.37 Here is shown a so-called pantograph used to enlarge or reduce drawings.

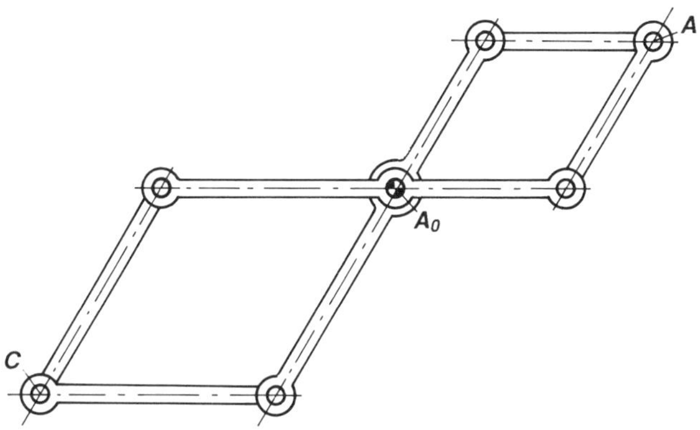

Figure 13.38 A little different arrangement from that of Fig. 13.37, but it does exactly the same, namely reduce or enlarge motion.

ANTIPARALLELOGRAM LINKAGES

If the links of a parallelogram linkages are in a crossed position the result is a so-called antiparallelogram linkage. These have characteristics different from those of parallelogram linkages, as seen in the following.

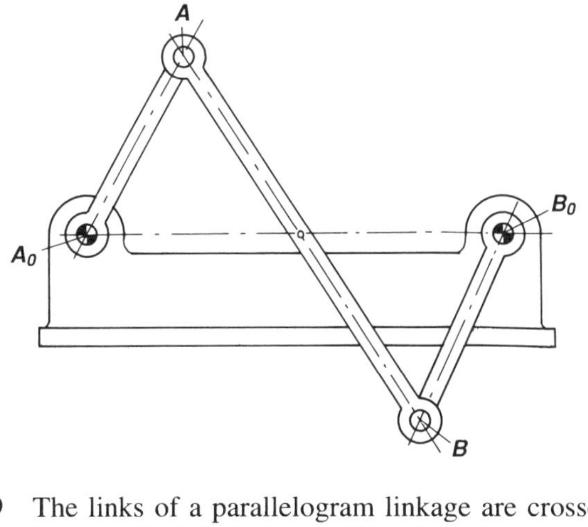

Figure 13.39 The links of a parallelogram linkage are crossed. Although a complete revolution of both links is possible, they do not move in unison; they rotate in opposite directions, and in the dead-center position they might cross back to a parallelogram.

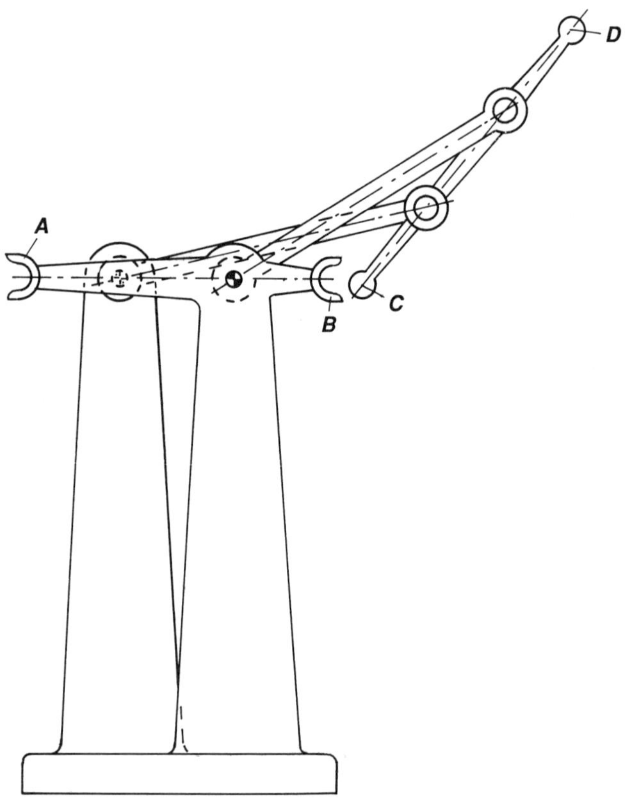

Figure 13.40 The "forks" A and B help the links of this anti-parallelogram linkage to move through the dead-center positions. The pins C and D together with the forks act as pin gearing.

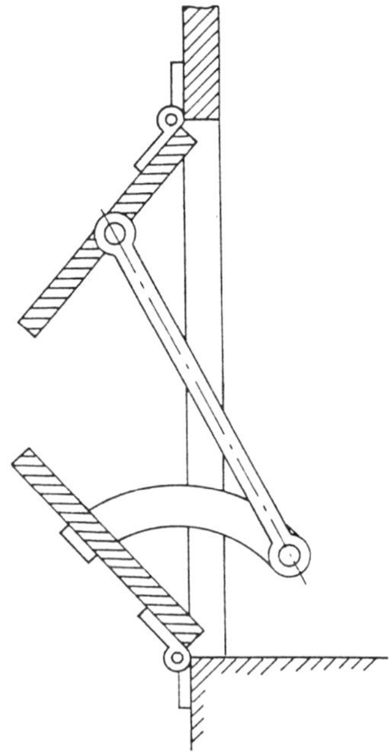

Figure 13.41 An anti-parallelogram linkage is here used to open the two halves of a door at the same time.

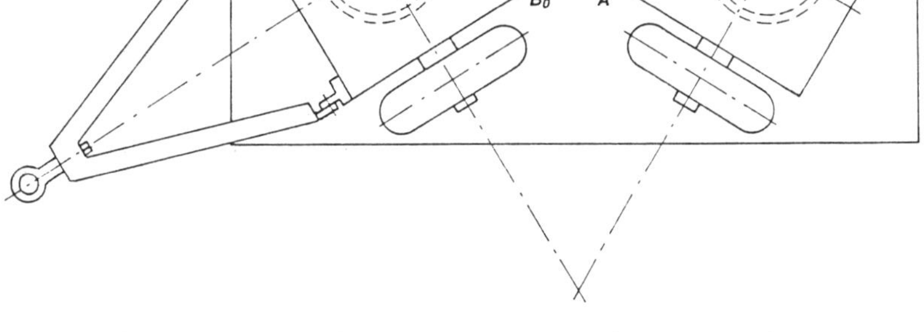

Figure 13.42 An anti-parallelogram linkage A_0ABB_0 is used here to rotate the wheels of a cart so that the radius of curvature described by the wheels is small.

14

Gears, Gearing, and Noncircular Gears

Gears are used widely and constitute one of the most important machine elements. Here, gears are considered motion transformers as opposed to power transmitting elements. Noncircular gears will be considered together with other forms of gears. Special tooth shapes will also be considered. Gears can transmit a constant or varying transmission ratio. When the transmission ratio is varying, noncircular gears are employed. The reader will find it helpful to refer to Chapter 5, "Planetary Gears Systems."

GEARING WITH CONSTANT TRANSMISSION RATIO

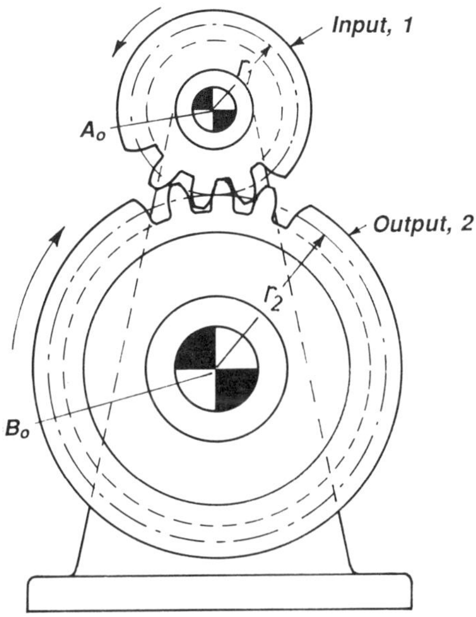

Figure 14.1 Two external gears in mesh. Pinion 1 and gear 2 have fixed centers of rotation A_0 and B_0. The radii of the pitch circles are r_1 and r_2. The pitch circles can be thought of as cylinders rolling on each other without sliding. If the number of teeth is designated N_1 and N_2, the following holds:

$$R = \frac{\omega_{out}}{\omega_{in}} = -\frac{r_1}{r_2} = -\frac{N_1}{N_2}$$

R is called the transmission ratio, or train value. The transmission ratio is negative because rotation is reversed between gears 1 and 2. Motion is transmitted between two parallel axes.

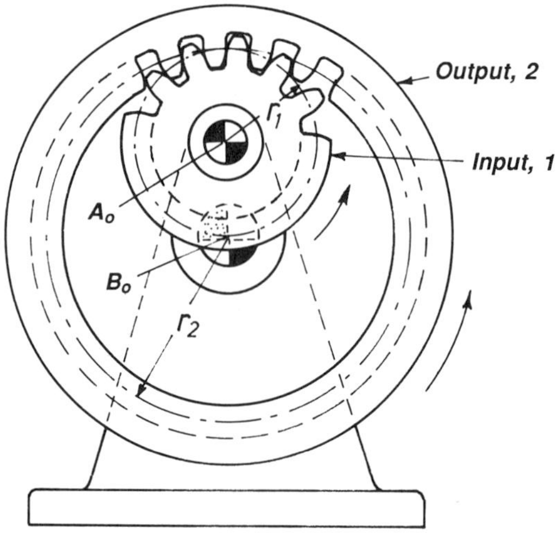

Figure 14.2 Here is shown an external gear 1 in mesh with internal gear 2. The transmission ratio is

$$R = \frac{\omega_{out}}{\omega_{in}} = \frac{r_1}{r_2} = \frac{N_1}{N_2}$$

The train value is positive because gears 1 and 2 turn in the same direction.

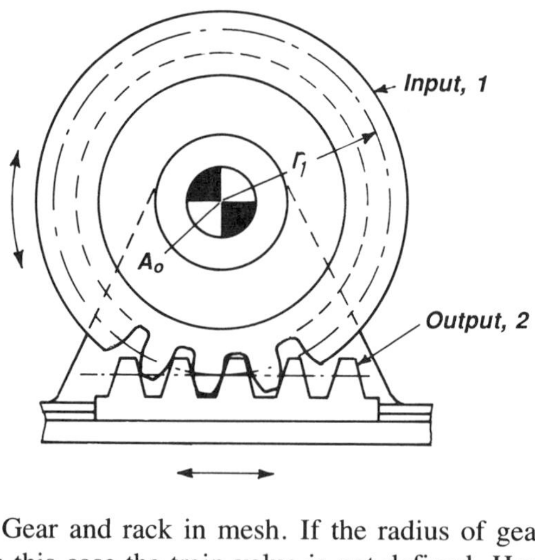

Figure 14.3 Gear and rack in mesh. If the radius of gear 2 is infinite, a rack results. In this case the train value is not defined. Here, rotary motion can be converted to linear, or linear motion can be converted to rotary.

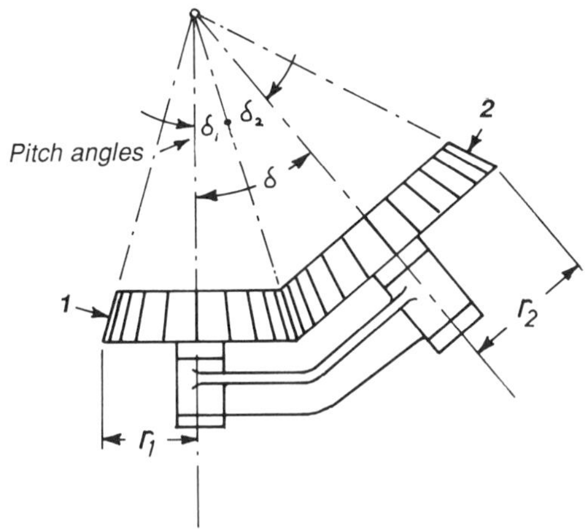

Figure 14.4 Two external straight bevel (conical) gears (shown as cones) are in mesh. Motion can be transmitted between two axes that are not parallel. The transmission ratio is defined by

$$R = \frac{\omega_{out}}{\omega_{in}} = -\frac{r_1}{r_2} = -\frac{\tan\delta_2}{\tan\delta_1} = -\frac{N_1}{N_2}$$

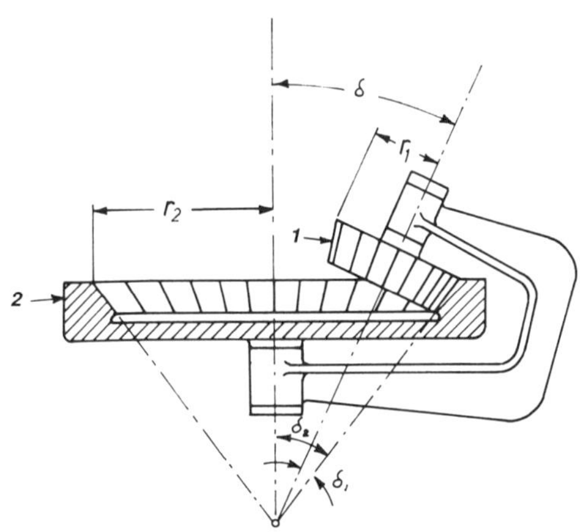

Figure 14.5 An external and an internal straight bevel gear are in mesh. The transmission ratio is positive and is defined by

$$R = \frac{\omega_{out}}{\omega_{in}} = \frac{r_1}{r_2} = \frac{\tan\delta_2}{\tan\delta_1} = \frac{N_1}{N_2}$$

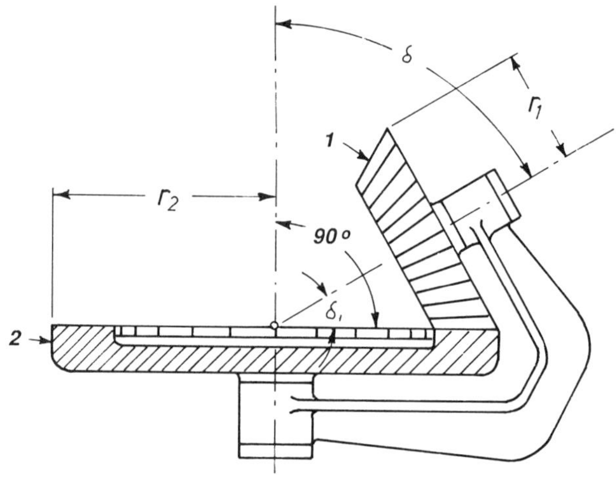

Figure 14.6 It is possible to make the pitch angle of one of the bevel gears ninety degrees as shown.

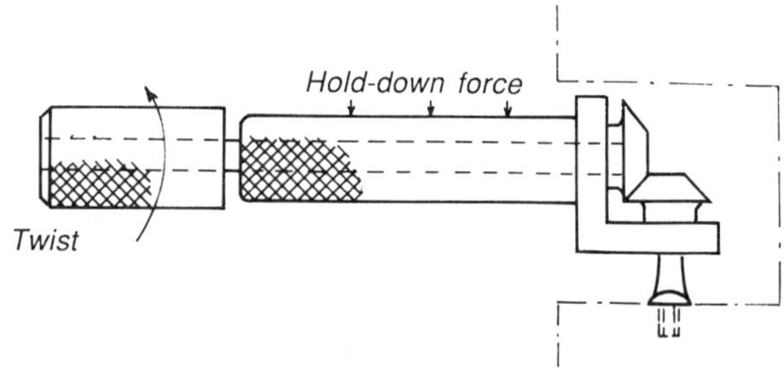

Figure 14.7 A practical application of straight bevel gears is a right-angle screwdriver.

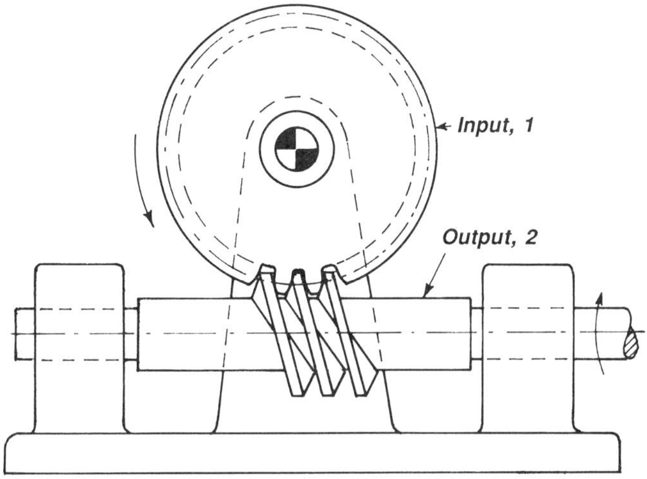

Figure 14.8 Worm gearing. Motion is transmitted between nonintersecting shafts, subtending an angle of 90°.

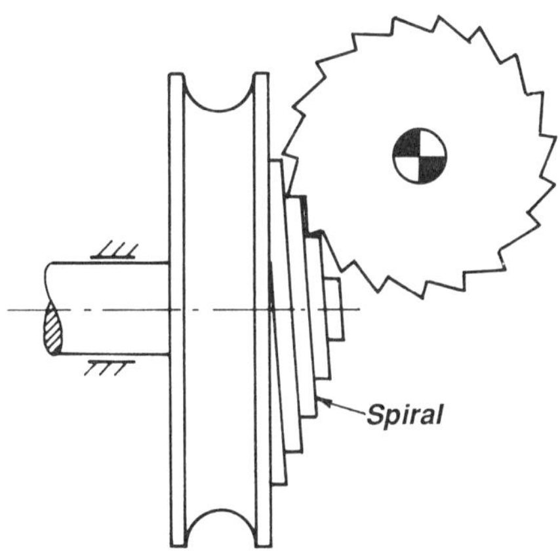

Figure 14.9 The worm gear is shaped like a spiral.

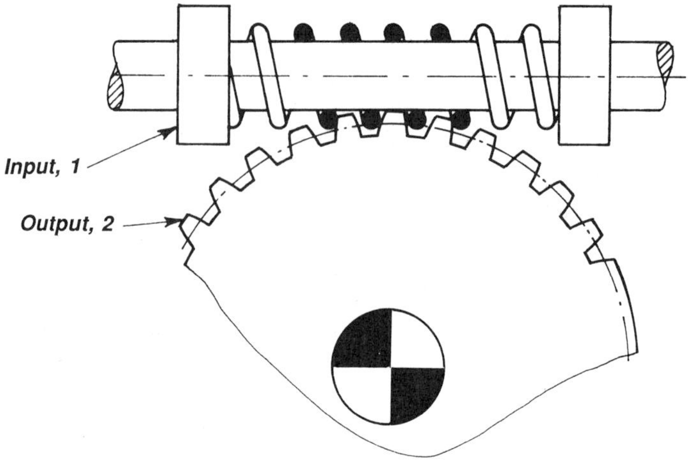

Figure 14.10 The worm gear can be made of a steel wire wrapped around a shaft. The spring has an additional function, namely, acting as a shock absorber.

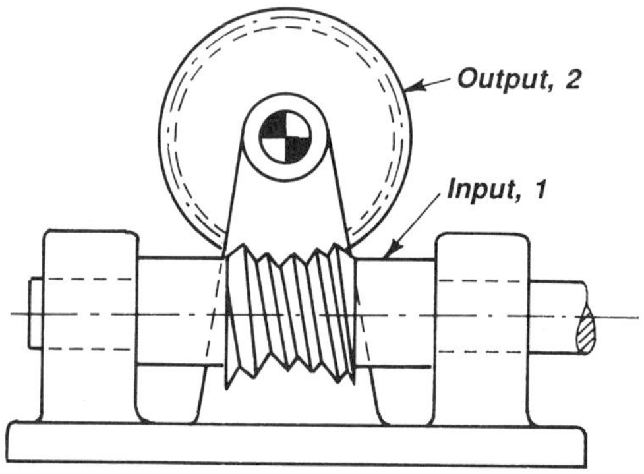

Figure 14.11 Double-enveloping worm gear drive. The "wraparound" of the worm ensures improved force transmission.

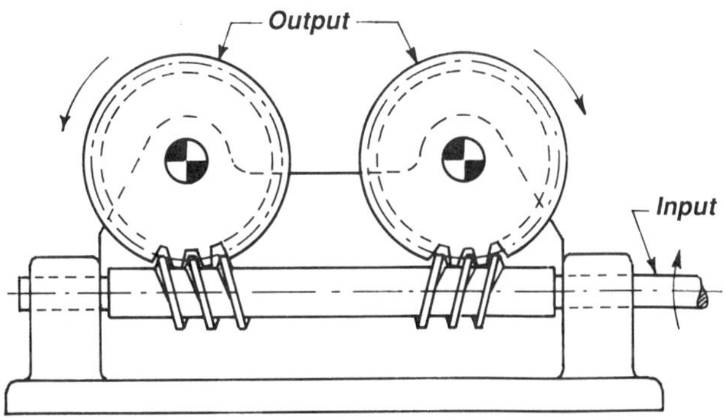

Figure 14.12 Two worm gears, left-handed and right-handed, transmit motion to two parallel shafts rotating in opposite directions.

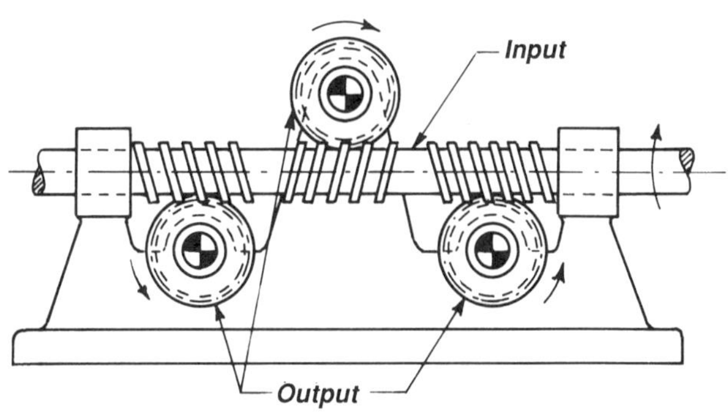

Figure 14.13 Triple worm gear. Motion is transmitted to three parallel shafts. The direction of motion of each individual shaft is dependent on whether the worm gear is left-handed or right-handed.

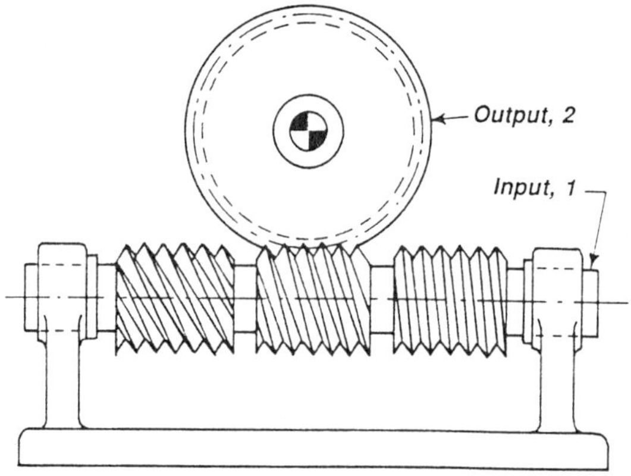

Figure 14.14 Three worm gears with different leads transmit a constant but different transmission ratio from shaft 1 to shaft 2, dependent on which worm gear is in mesh.

SOME HISTORICAL GEARING SYSTEMS

Although some of the following gearing systems are familiar, their presentation here, it is hoped, can inspire the reader to realize creative, yet practical, applications.

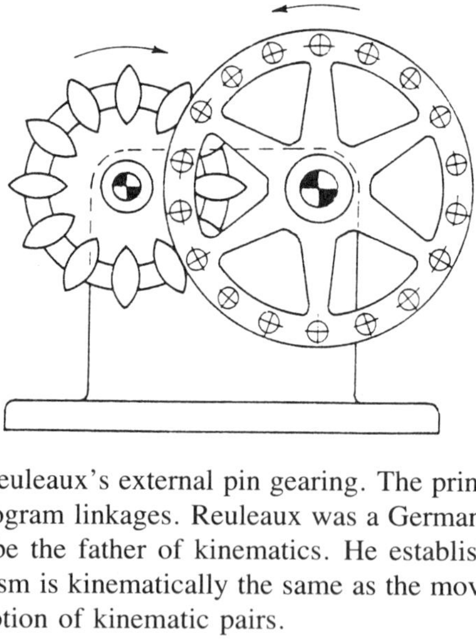

Figure 14.15 Reuleaux's external pin gearing. The principle of gearing is based on parallelogram linkages. Reuleaux was a German kinematician and is considered to be the father of kinematics. He established that the fixed link of a mechanism is kinematically the same as the moving links. He also established the notion of kinematic pairs.

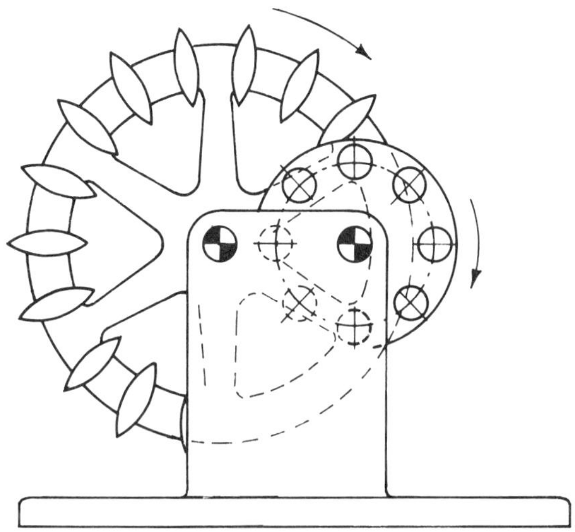

Figure 14.16 Reuleaux's internal pin gearing. (See Fig. 14.15 and legend.)

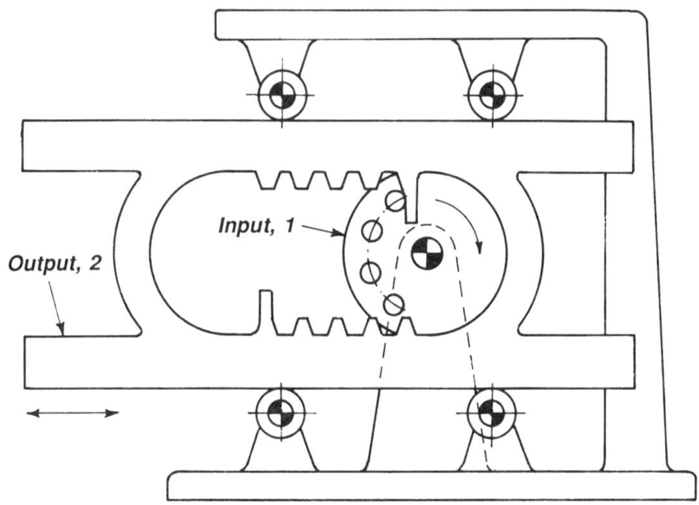

Figure 14.17 Rotary input to intermittent, oscillating pump. Pin gearing transmits motion to a rack that moves back and forth. The pin wheel 1, which is input, moves the rack to the right. At a point, this motion stops but then the pin wheel engages the lower teeth and starts moving the rack to the left.

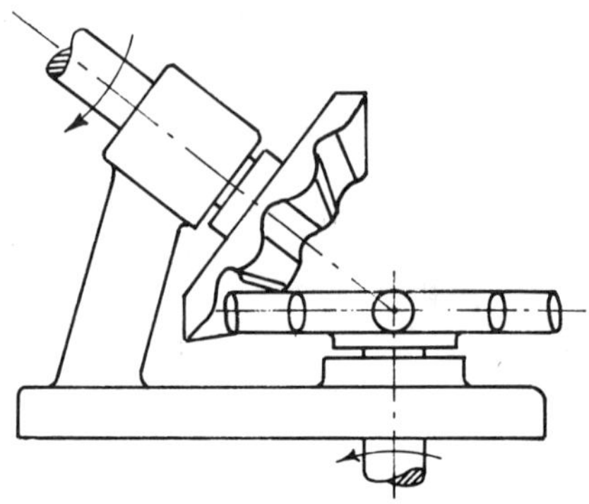

Figure 14.18 Pin gearing with bevel gears. The pins are located on a flat cone.

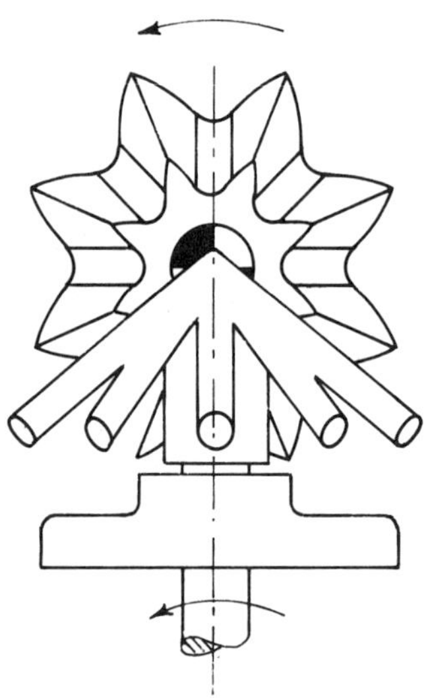

Figure 14.19 In this pin gearing with bevel gears, the pins form a cone.

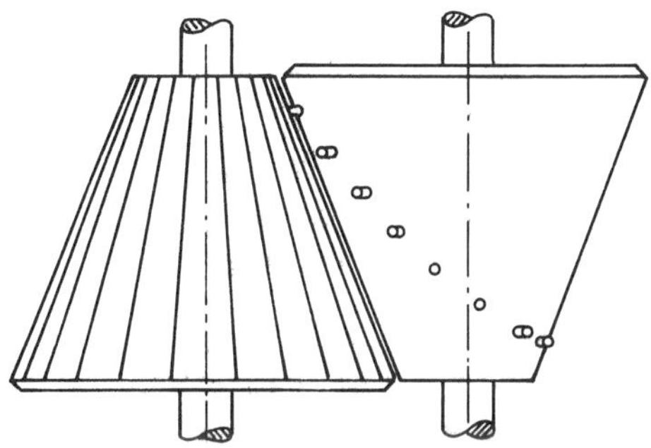

Figure 14.20 Roman tapered pin gearing. There is a varying transmission ratio (which is negative) between the two parallel shafts.

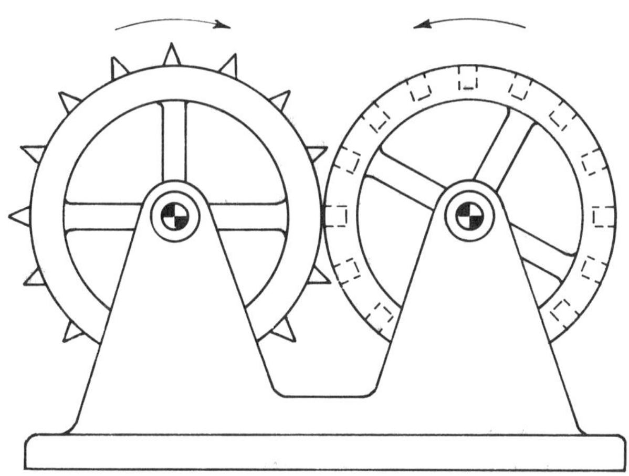

Figure 14.21 Pin gearing with pointed pins. Transmission ratio is constant overall, but due to the shape of the pins the transmission ratio varies slightly.

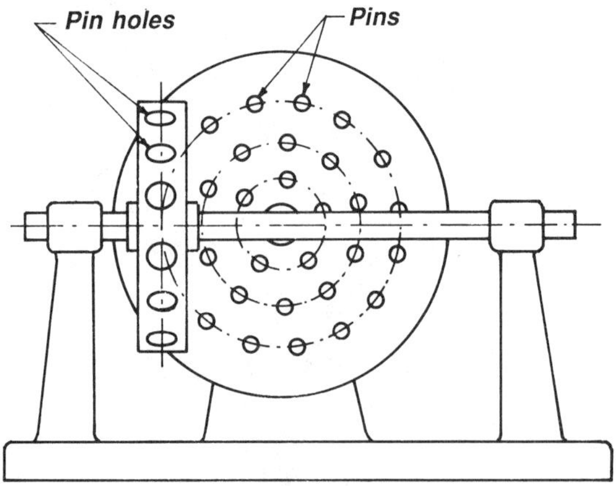

Figure 14.22 Pin gearing between intersecting shafts. Dependent on the position of the gear with the holes, a different, but constant, transmission ratio is obtained.

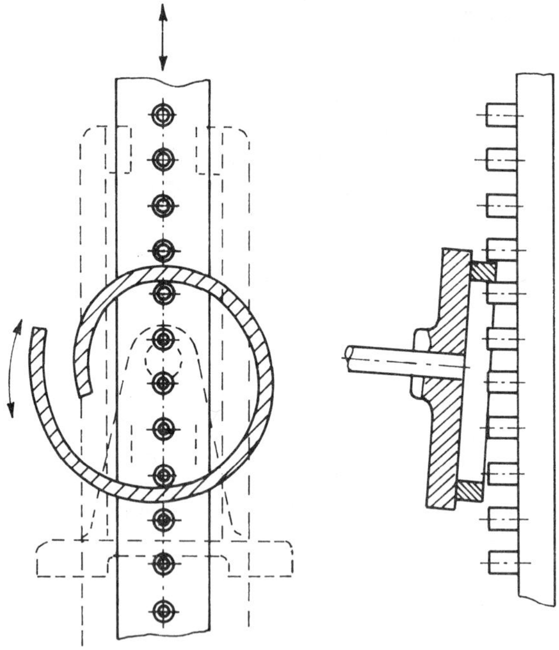

Figure 14.23 Slanted spiral gear transmits a constant velocity ratio between the gear and the rack.

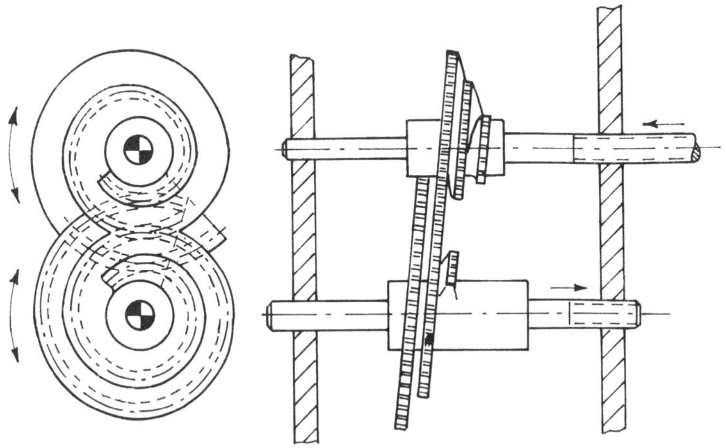

Figure 14.24 Slanted spiral gears transmit a varying velocity ratio between the two shafts.

NONCIRCULAR GEAR MECHANISMS

Noncircular gears transmit a varying transmission ratio between the input gear and the output gear. They can be manufactured on NC machines so that they are not as costly as they have been and they can be made accurately.

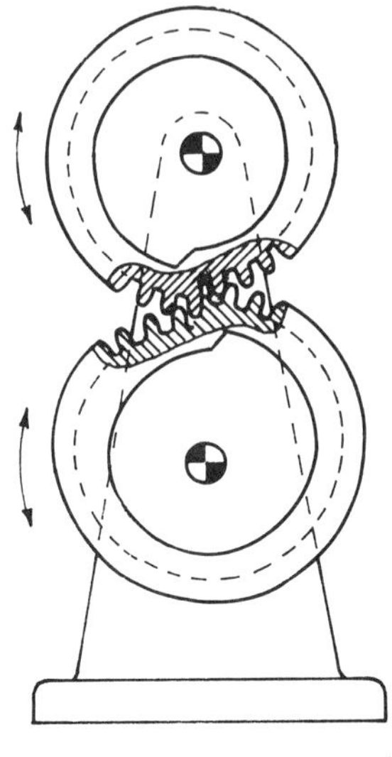

Figure 14.25 Two spiral gears in mesh transmit a varying transmission ratio. The gears cannot make a full revolution.

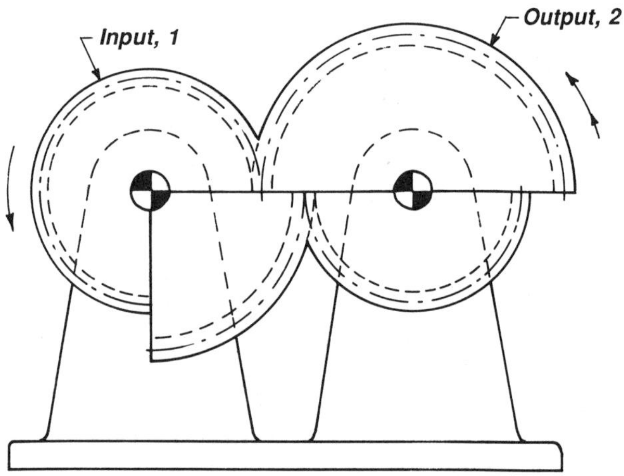

Figure 14.26 Two segmented circular gears transmit a constant transmission ratio over a limited angle of rotation, then another constant transmission ratio for the remainder of the rotation. They can be used only for low speeds because of the sudden change in transmission ratio.

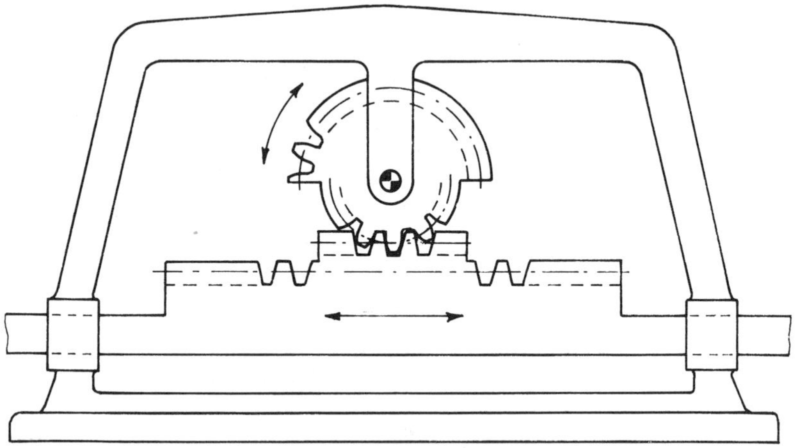

Figure 14.27 Different gear segments on rack-and-gear member provide for a varying transmission ratio.

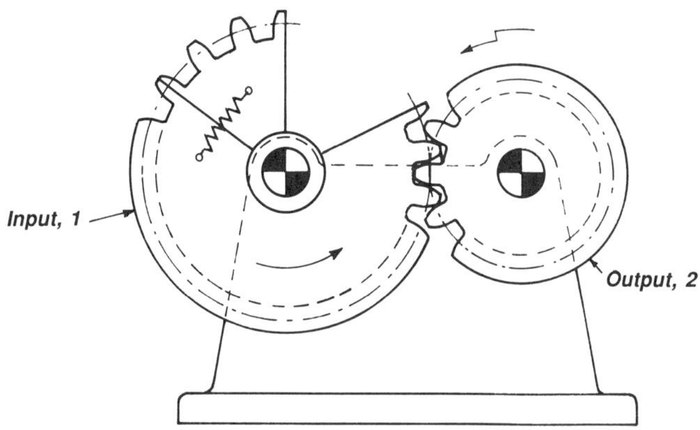

Figure 14.28 Because of the spring-loaded gear segment, the constant angular velocity of input gear 1 is converted to a constant angular velocity of gear 2, interrupted by a prolonged dwell. Suitable for low speeds only. The resisting torque on gear 2 must also be sufficient to overcome the spring force.

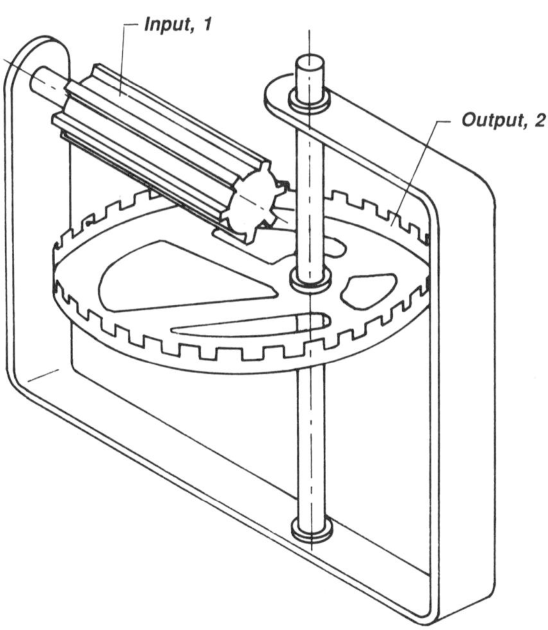

Figure 14.29 Input gear 1 meshes with an eccentric circular gear 2. Because of the eccentricity a varying transmission ratio is transmitted between the two shafts. Used in mechanical toys.

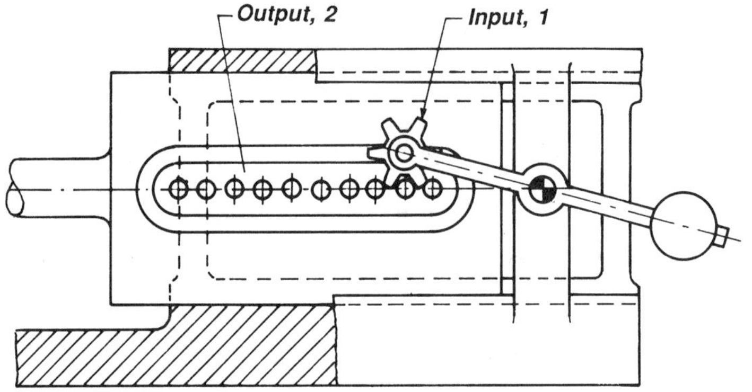

Figure 14.30 This mechanism converts the rotary motion of gear 1 to the translating back-and-forth motion of rack 2.

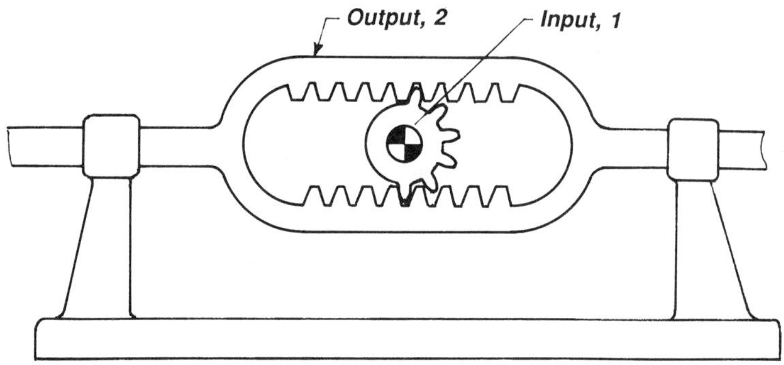

Figure 14.31 Pinion 1 drives rack 2 back and forth. The teeth of pinion 1 alternately engage the two opposing racks, imparting an oscillating sliding motion to the racks. The rack has a short dwell at each of its extreme positions.

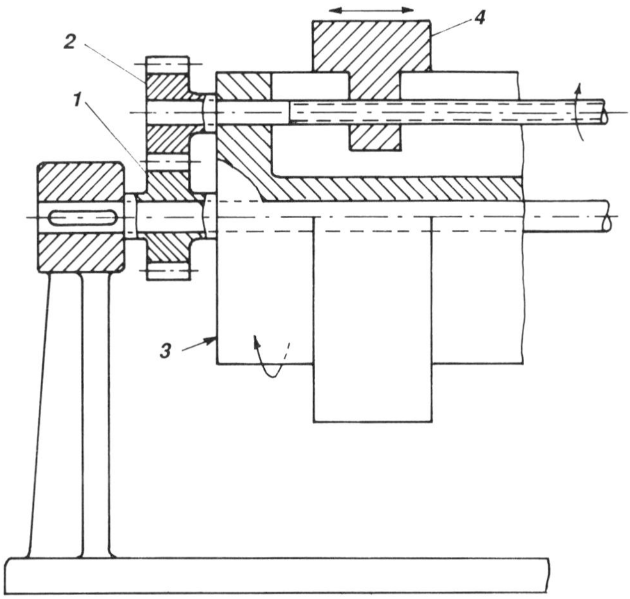

Figure 14.32 Drum 3, which carries planet gear 2, revolves around the shaft of sun gear 1, thereby moving the toolhead 4 in axial direction while it rotates around the drum shaft.

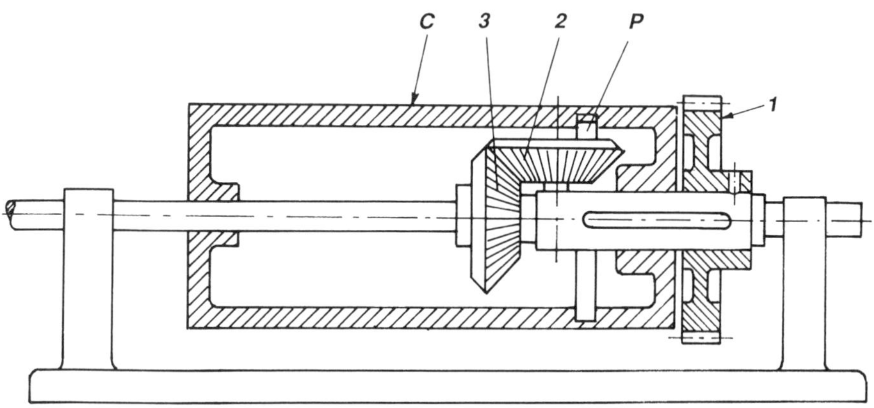

Figure 14.33 Cylinder C, which is rotated from gear 1, is given an axial motion due to the pin P on bevel gear 2, which is rotated from bevel gear 3. Bevel gear 2 is stationary but can rotate around its own axis. Used in printing machines.

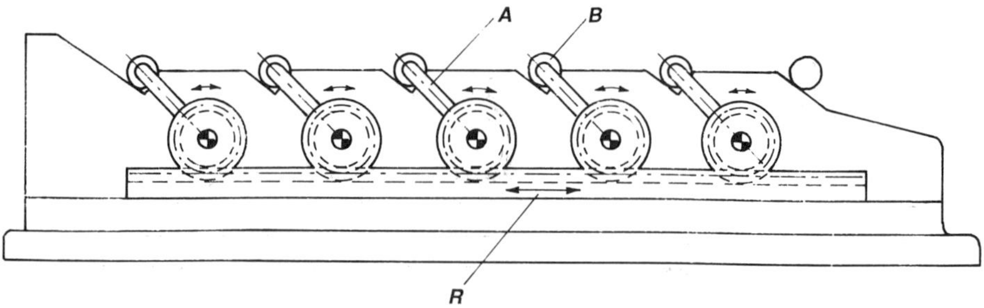

Figure 14.34 The rack R drives the five spur gears. When the rack moves back and forth, the gears oscillate. Each gear is provided with an arm A, which grips the workpieces B and transports them from one station to the next.

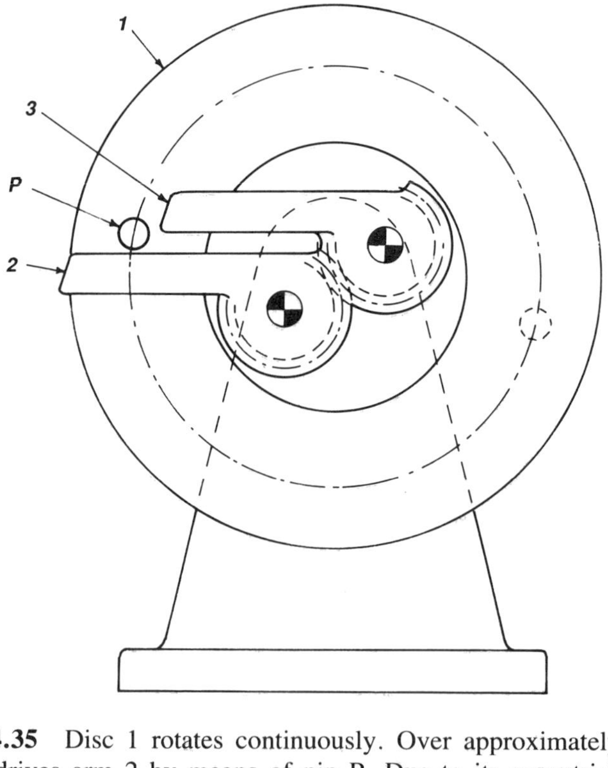

Figure 14.35 Disc 1 rotates continuously. Over approximately 180° rotation, it drives arm 2 by means of pin P. Due to its eccentricity, arm 2 gradually gets out of contact with pin P. In the meantime arm 2 is in a position where it can be driven by pin P. The resultant motion of arms 2 and 3 is an oscillation with dwells in between.

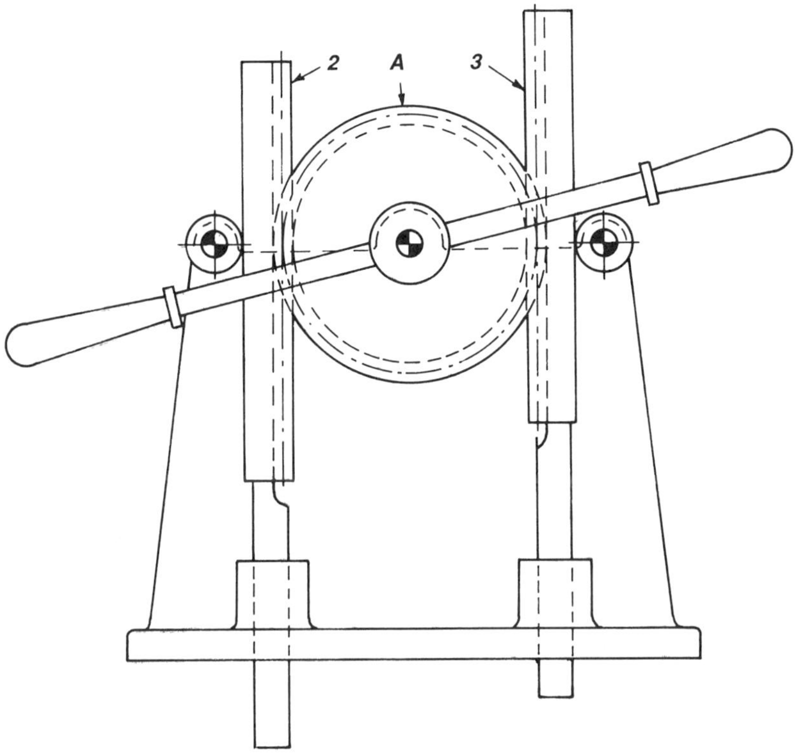

Figure 14.36 When gear A is oscillated, racks 2 and 3 are moved up and down 180° out of phase with each other.

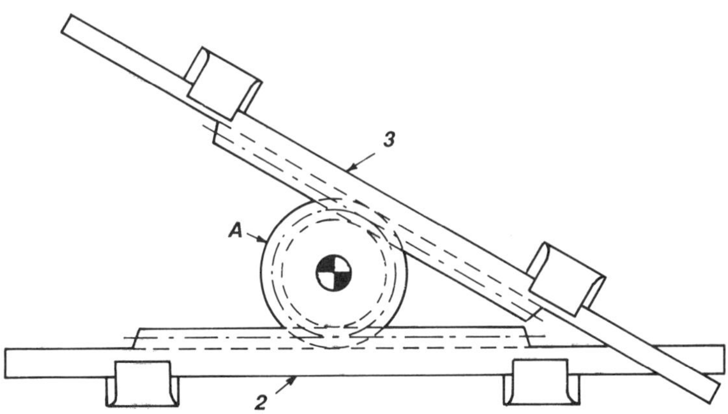

Figure 14.37 Same arrangement as in Fig. 14.36, but the motions of the two racks are not parallel.

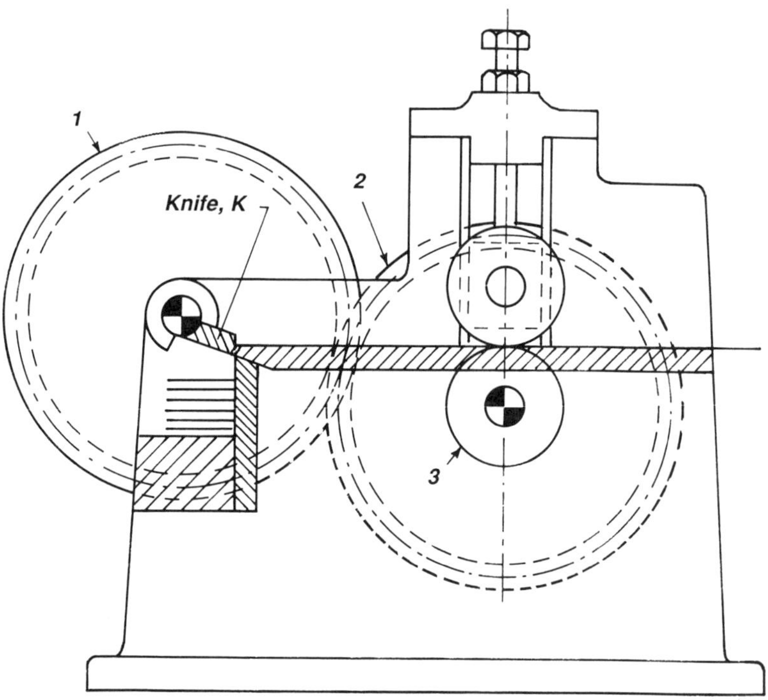

Figure 14.38 A rotating knife K driven by gear 1 cuts sheet metal continuously. Gear 1 is in mesh with gear 2, which drives the roller 3 that transports the material.

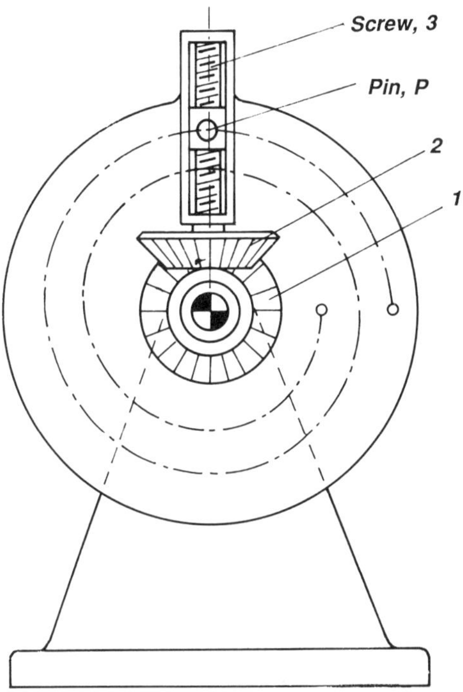

Figure 14.39 The two bevel gears 1 and 2 act as a planetary gear train. Gear 1 is stationary and gear 2 rolls on and rotates about gear 1. The rotary motion of gear 2 about its own axis causes the screw 3 to turn, thereby moving pin P on a spiral curve as shown.

NONCIRCULAR GEARS FOR CONTINUOUS TRANSMISSION OF MOTION

Circular gears transmit rotary motion to rotary motion with a constant transmission ratio. The transmission ratio varies for noncircular gears, which are not really popular with designers. It may be that a designer does not think in terms of noncircular gears because he or she has read little about them and because they are not easy to calculate. However, once they are understood—and it really does not take much effort to learn—they can get a designer out of a bind. It must however be remembered that the varying angular velocity gives rise to angular accelerations, and inertia forces may be difficult to control.

GEOMETRICAL CONSTRUCTION OF NONCIRCULAR GEARS

First, a geometrical treatment is given because it deepens the understanding of the motion itself.

Both Gears Are Noncircular

Figure 14.40(a) shows two circular gears in mesh. They are represented by their pitch curves, shown as circles. The two pitch circles rolls on each other without sliding. In Figure 14.40(b) a more elaborate presentation of two gears in mesh is shown, where an external gear is in mesh with an internal. When designing noncircular gears it is the pitch curves that are of primary importance, but once they have been found, questions about the cutting of gear teeth arise.

Figure 14.41 shows the angular displacements of two shafts that are to rotate in a certain manner relative to each other, and Fig. 14.42 shows the corresponding noncircular gears with center at O_1 and O_2. In Fig. 14.41 the rotation of the left gear with center at O_1 is the abscissa and that of the right gear with center at O_2 is the ordinate. The curve gives the relationship between the angular positions of the two gears. The curves may be given in the form of an equation or may be established based on practical requirements of the mechanism.

The curve 0–7 is divided into a number of sections. Here we have chosen 7 sections, but once the method is understood many more points can be chosen and all calculations made on a computer. The sectioning determines the angles θ_1 to θ_7 and ϕ_1 to ϕ_7. The curve is straight from 3 to 4 and no subdivisions in between these two points are required.

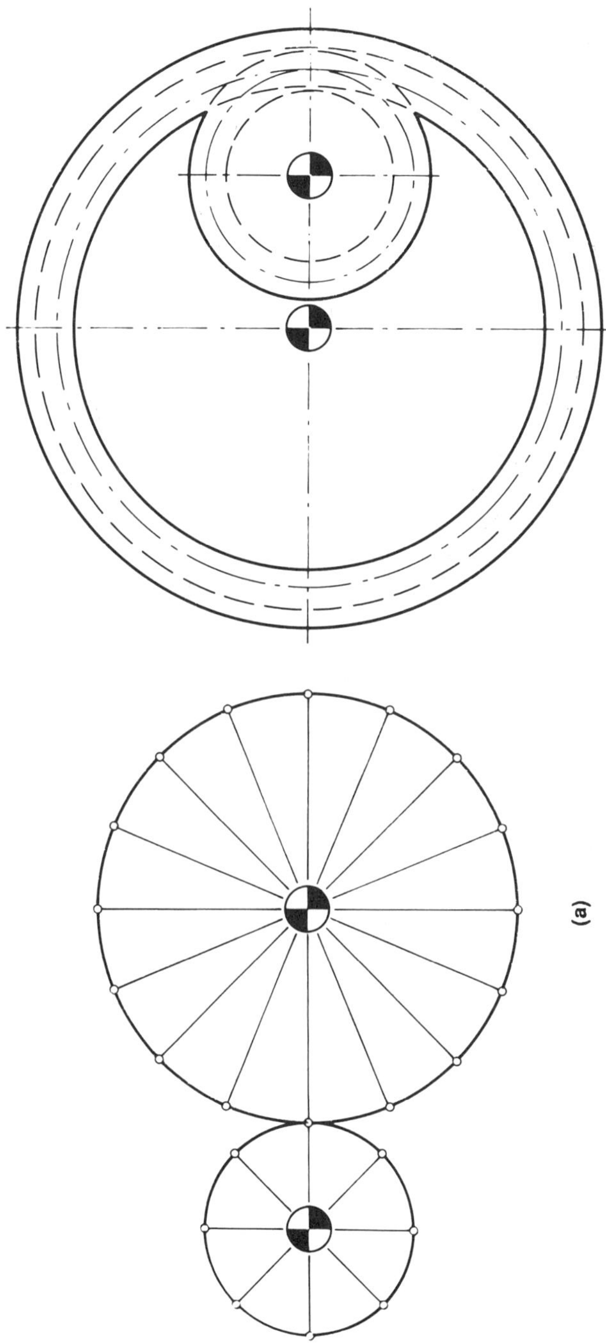

Figure 14.40 (a) Two circular pitch curves roll on each other without sliding. (b) An internal and external gear in mesh.

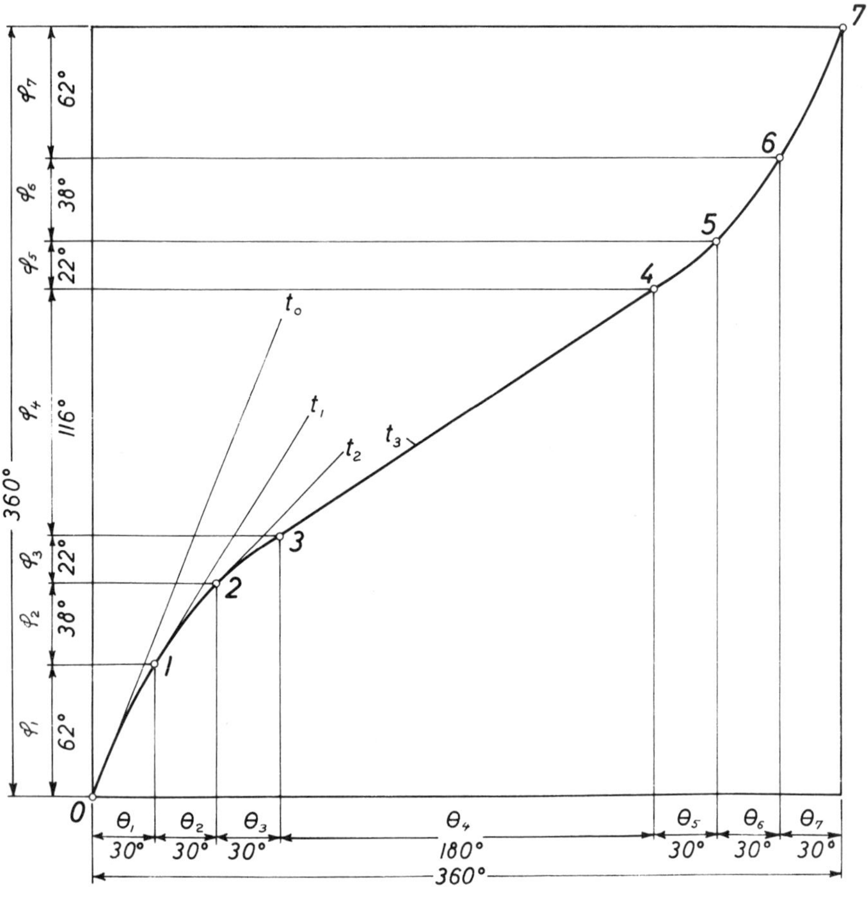

Figure 14.41 Time-displacement curve for two shafts.

In Fig. 14.42 the velocity of the pitch point P is determined from

$$V_P = R\omega_{in} = r\omega_{out}$$

or

$$\frac{d\phi}{d\theta} = \frac{\omega_{out}}{\omega_{in}} = \frac{R}{r} \tag{14.1}$$

where R is the distance from O_1 to the pitch point P, and r the distance from O_2 to P. Furthermore,

$$O_1O_2 = R + r \tag{14.2}$$

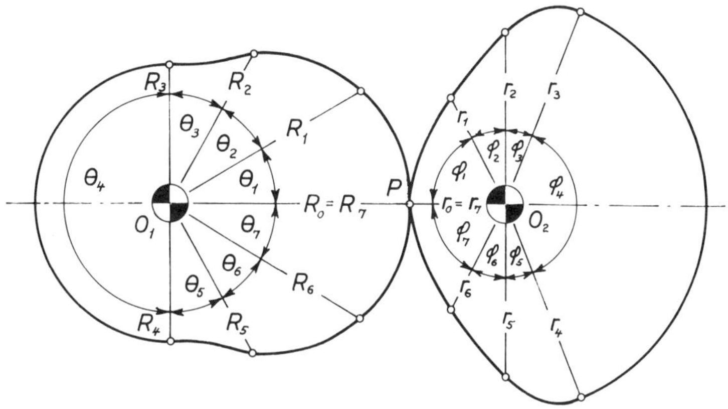

Figure 14.42 Pitch curves for two noncircular gears. Transmission ratio is −1 : 1.

The instantaneous angular velocity ratio $d\phi/d\theta$ between the two gears can be found from the diagram in Fig. 14.41. The tangents t_0, t_1, etc., are a measure for the instantaneous velocity ratio between the two gears. The values obtained are listed in Table 14.1. The resultant pitch curves are shown in Fig. 14.42. If it is desired to let the driving gear rotate twice for each revolution of the driven gear, but for the same displacement diagram as in Fig. 14.41, then all ϕ-angles and all measured slopes should be divided by two. The resultant values are listed in Table 14.2 and the resultant pitch curves are shown in Fig. 14.43.

Table 14.1 Instantaneous Velocity Ratio Between the Two Gears in Fig. 14.41

Point	$d\phi/d\theta = R/r$	R + r	R	r
0	2.5		71.4	28.6
1	1.9	—	65.5	34.5
2	1.0		50.0	50.0
3	0.66		39.8	60.2
4	0.66	100	39.8	60.2
5	1.0		50.0	50.0
6	1.9		65.5	34.5
7	2.5		71.4	28.6

Table 14.2 Instantaneous Velocity Ratio (see Table 14.1) for Two Driving Gear Rotations per Rotation of Driven Gear

Point			$d\phi/d\theta = R/r$	R + r	R	r
1	$\theta_1 = 30°$	$\phi_1 = 31°$	1.25		55.6	44.4
2	$\theta_2 = 30°$	$\phi_2 = 19°$	0.95		48.7	51.3
3	$\theta_3 = 30°$	$\phi_3 = 11°$	0.50		33.3	66.7
4	$\theta_4 = 180°$	$\phi_4 = 58°$	0.33	100	25.0	75.0
5	$\theta_5 = 30°$	$\phi_5 = 11°$	0.50		33.3	66.7
6	$\theta_6 = 30°$	$\phi_6 = 19°$	0.95		48.7	51.3
7	$\theta_7 = 30°$	$\phi_7 = 31°$	1.25		55.6	44.4

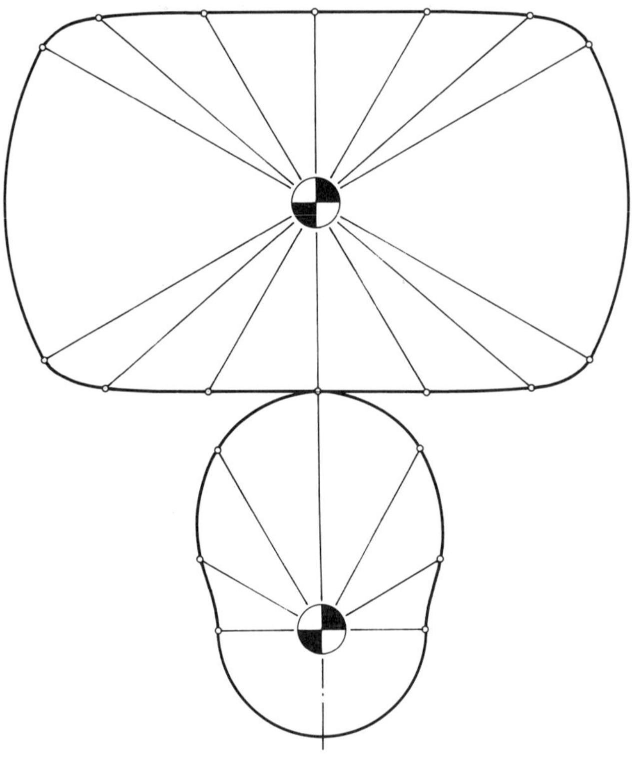

Figure 14.43 Pitch curves for two noncircular gears. Transmission ratio is $-1:2$.

One Gear Is Circular but Eccentrically Supported and the Other Gear Is Noncircular

Figure 14.44 shows a circular gear with off-center rotation. It is desired to find the pitch curve of the mating driven gear. The center distance between the two gears cannot be chosen freely but is governed by the equation

$$\frac{c}{r} \approx (v + 1) - \frac{(v + 1)(v - 2)(e/r)^2}{4v} \tag{14.3}$$

where

r = radius of circular gear
e = eccentricity of circular gear
c = distance O_1O_2
v = rotation ratio. If gear 1 rotates one revolution for each revolution of gear 2, then v = 1. If gear 1 rotates two revolutions for each revolution of gear 2, then v = 2, etc.

Table 14.3 lists some of these values.

CONSTRUCTION OF PITCH CURVE FOR DRIVEN GEAR

In Figure 14.44 r, e, and v are chosen and c = O_1O_2 is found from Eq. 14.3. The eccentric gear is drawn so that its line of symmetry lies on O_1O_2

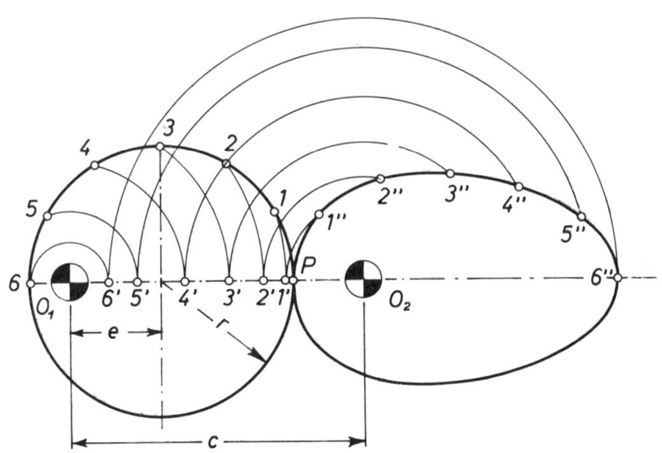

Figure 14.44 Circular eccentric in mesh with noncircular gear shown by their pitch curves only. Transmission ratio −1 : 1.

Table 14.3 Pitch Curve Values for the Driven Gear in Fig. 14.41

v	e/r						
	0.1	0.2	0.3	0.4	0.5	0.6	0.7
1	2.0	2.02	4.05	2.08	2.13	2.18	2.25
2	3.0·	3.0	3.0	3.0	3.0	3.0	3.0
3	4.00	3.99	3.97	3.95	3.92	3.88	3.84
4	4.99	4.98	4.94	4.90	4.84	4.78	4.69
5	5.99	5.96	5.92	5.86	5.78	5.67	5.56
6	6.99	6.95	6.90	6.81	6.71	6.58	6.43

as shown. Half the circumference of the gear is divided into six equal parts with designations P, 1, 2, . . . 6. With O_1 as center, circles are drawn through points 1–6, the points of intersection with O_1O_2 being 1′, 2′, . . . 6′, respectively. Through these points, circles are drawn with O_2 as center. The distances P1′, 1′2′, 2′3′, etc., are made equal to P1, 12, 23, etc., where points 1′, 2′, 3′, etc., lie on the circles with center at O_2. Although the construction is not mathematically accurate because the cord lengths P1, 12, etc., have been used instead of the arc lengths, the method is quite accurate and can be used when writing a program. The following values were used (Fig. 14.44):

$$r = 31.5 \text{ mm}$$
$$e = 22.0 \text{ mm}$$
$$v = 1$$

and from eq. (14.3)

$$\frac{c}{31.5} = 2 - \frac{2(-1)(22/31.5)}{4} = 2.244$$
$$c = 70.5 \text{ mm}$$

From Fig. 14.45 the following values were used

$$r = 31.5 \text{ mm}$$
$$e = 22.0 \text{ mm}$$
$$v = 2$$

and from eq. (14.3)

$$c = 94.5 \text{ mm}$$

From Fig. 14.46 the following values were used

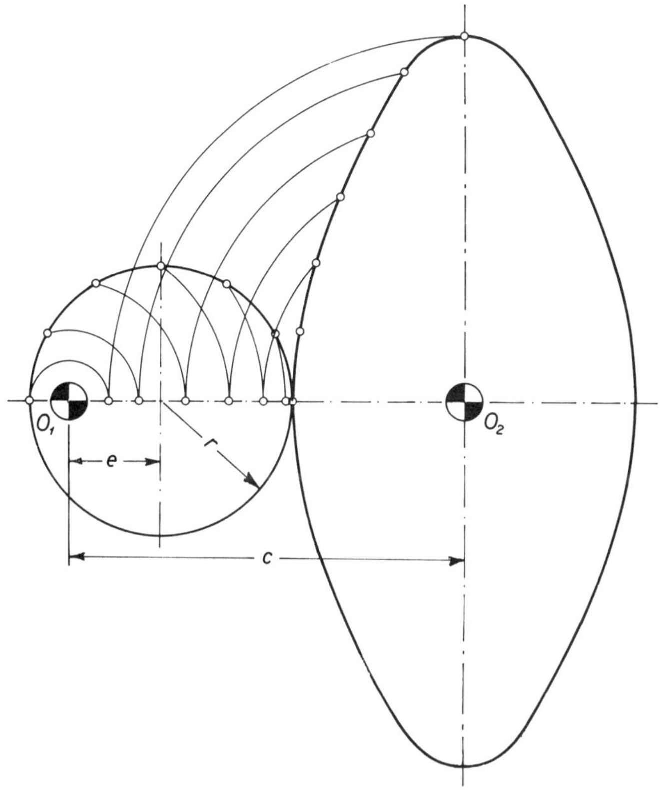

Figure 14.45 Circular eccentric arrangement as in Fig. 14.44. Transmission ratio −1 : 2.

$$r = 31.5 \text{ mm}$$
$$e = 22.0 \text{ mm}$$
$$v = 4$$

and from eq. (14.3)

$$c = 4.66 \text{ mm}$$

ELLIPTICAL GEARS

Elliptical gears have been treated extensively in the literature. They have been preferred because of the relative ease with which they can be calculated. Mechanical devices were also available to cut this type of gear, but

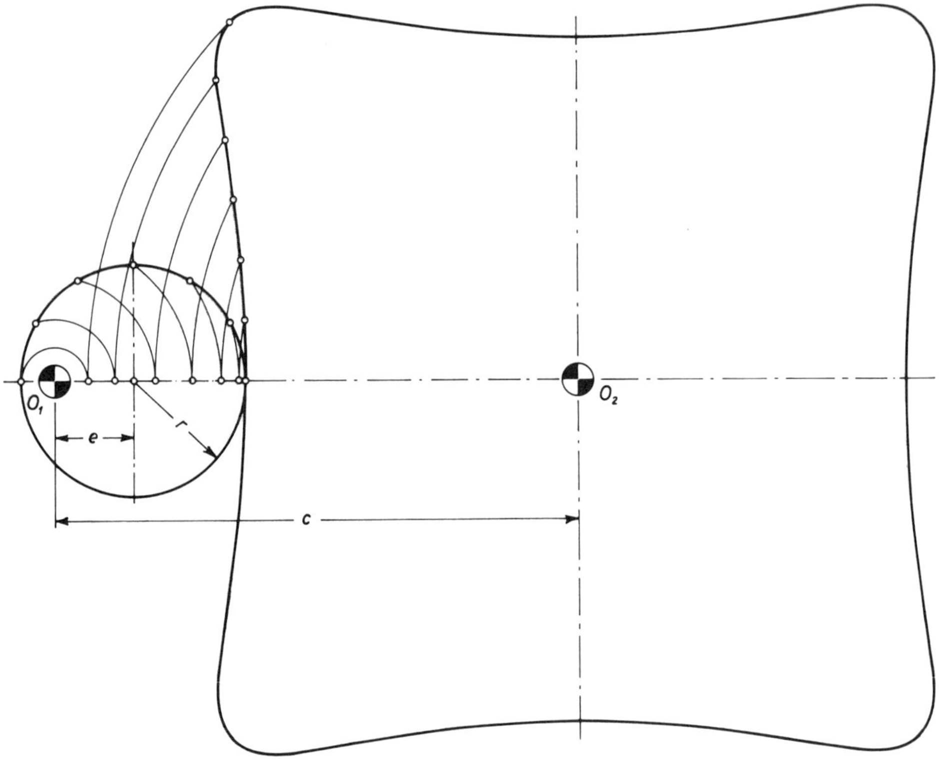

Figure 14.46 Circular eccentric arrangement; same principle as in Fig. 14.44. Transmission ratio −1 : 4.

these gear-cutting devices did not really function properly. Today no special shape of noncircular gears is preferred because they can now be manufactured on NC machines. But the subject of noncircular gears remains fascinating and elliptical gears are part of the picture.

Two identical ellipses are shown in Fig. 14.47. The requirements that two gears mesh correctly at any one instant are that the pitch point P is on the center line $F_1'F_2'$, that the sum of the lengths from F_1' and F_2' to the point P of contact between the two pitch curves is constant, and that equal arcs are traversed on the pitch curves. The length $F_1'F_2' = 2a$, the major diameter. Choosing a point P_1 on the left ellipse and P_2 on the right, both are symmetrical with respect to a tangent at P. Therefore, the arc $PP_1 = PP_2$, and $F_1'P_1 + F_2'P_2 = R_1 + R_2 = 2a$, and the above requirements are fulfilled.

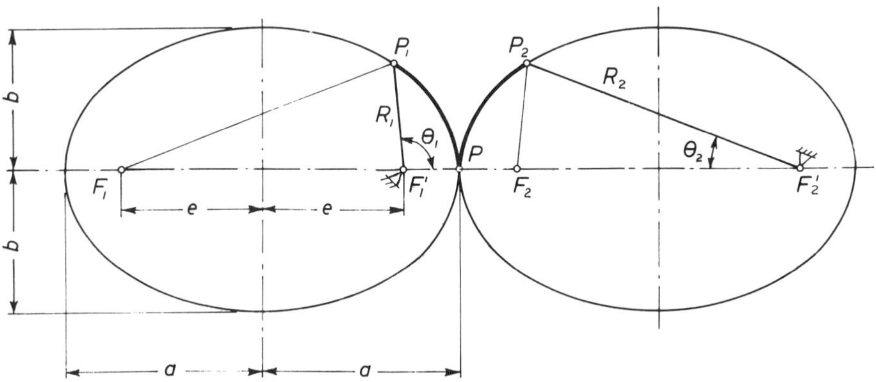

Figure 14.47 Two ellipses roll on each other.

Motion Relationship

The ratio of the instantaneous angular velocities of the two gears is inversely proportional to the distances from the center of rotation to the pitch point P. Therefore, if the two gears in Fig. 14.47 make contact at P_1 and P_2, then

$$\frac{\omega_1}{\omega_2} = -\frac{R_1}{R_2} \tag{14.4}$$

The pitch curve of the left ellipse expressed in polar coordinates is

$$R_1 = \frac{a(1 - \epsilon^2)}{1 + \epsilon\cos\theta_1} \tag{14.5}$$

where

$$\epsilon = \frac{a}{b}$$

but

$$R_1 + R_2 = 2a$$

$$R_2 = a(1 + 2\epsilon\cos\theta_1) = \frac{\epsilon^2}{1 + \epsilon\cos\theta_1} \tag{14.6}$$

From eqs. (14.4, 14.5, and 14.6):

$$\frac{d\theta_2}{ds} = \omega_2 = \omega_1 \frac{1 - \epsilon^2}{1 + \epsilon^2 + 2\epsilon\cos\theta_1} \tag{14.7}$$

The instantaneous transmission ratio is now defined as

$$i = \frac{\omega_2}{\omega_1}$$

and

$$i_{max} = (a + e)(1 - e)$$
$$i_{min} = (a - e)(a + e)$$
$$\frac{i_{max}}{i_{min}} = \left(\frac{a - e}{a - e}\right)^2 \qquad (14.8)$$

Number of Teeth

Figure 14.48 shows two elliptical gears in mesh. The number of teeth is uneven, and the two gears can therefore be made identical. One of the gears is just turned upside down. In other words, it is not necessary, as it is in Fig. 14.48, to ensure that a tooth and a tooth space is halved by the line of symmetry. If the number of teeth is uneven and it is desired that the two

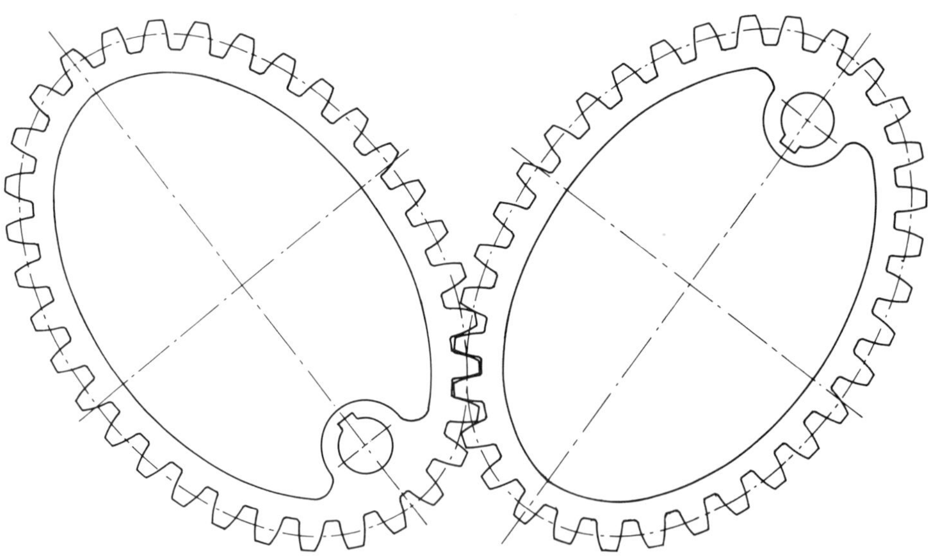

Figure 14.48 Two identical elliptical gears in mesh.

gears be identical, then it is necessary to move the teeth of one of the gears half a tooth width measured along the pitch curve.

It is important to remember that the two pitch curves of mating gears, for a transmission ratio of $-1:1$, must be equal in length and that the length must allow the required number of teeth and tooth spaces along the circumference.

A kinematic equivalent mechanism to elliptical gears is shown in Figure 14.49. If the opposing links are equal in length, the result is an anti-parallelogram linkage, which is described in Chapter 13 "Parallelogram Mechanisms." The motion of links $F_2'F_2$ and $F_1'F_1$ is exactly the same as the motion of the right and left elliptical gear.

Figure 14.50 shows that coordination of angles is possible to a certain extent. The center distance between the elliptical gears is 2a (= major axis). When the left gear turns through the angle θ out from the position of symmetry, the right gear is to move through the angle ϕ. These two angles determine point O, and making $OF_1' = OF_2$ and $OF_2' = OF_1$ as shown, determines points F_1 and F_2. Distance $F_1' = F_2' = 2e$. Knowing both a and e lets one determine the ellipses.

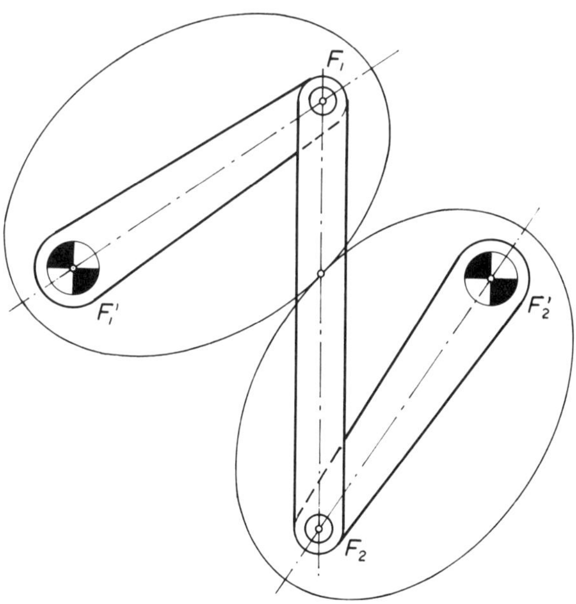

Figure 14.49 The motion of two elliptical gears in mesh can be replaced by an antiparallelogram linkage.

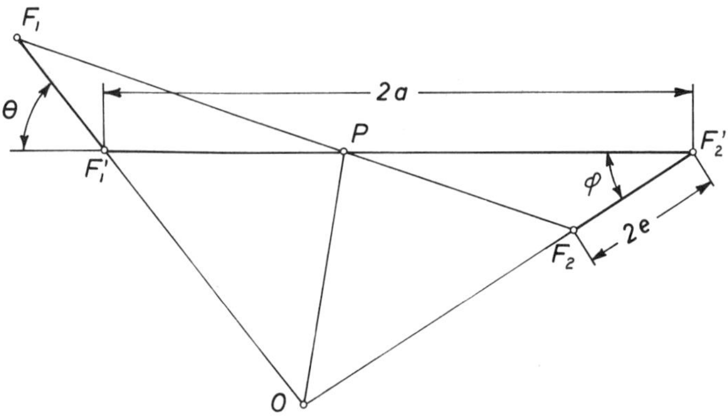

Figure 14.50 Coordination of angular motion of elliptical gears.

Both Gears Are Eccentric

In Fig. 14.51 if two circular but eccentric gears are used, the mechanism is lower in cost but at the same time subject to limitations. From a theoretical point of view the two gears are not meshing correctly, but the error can be neglected if the ratio of eccentricity e over the gear radius r is less than

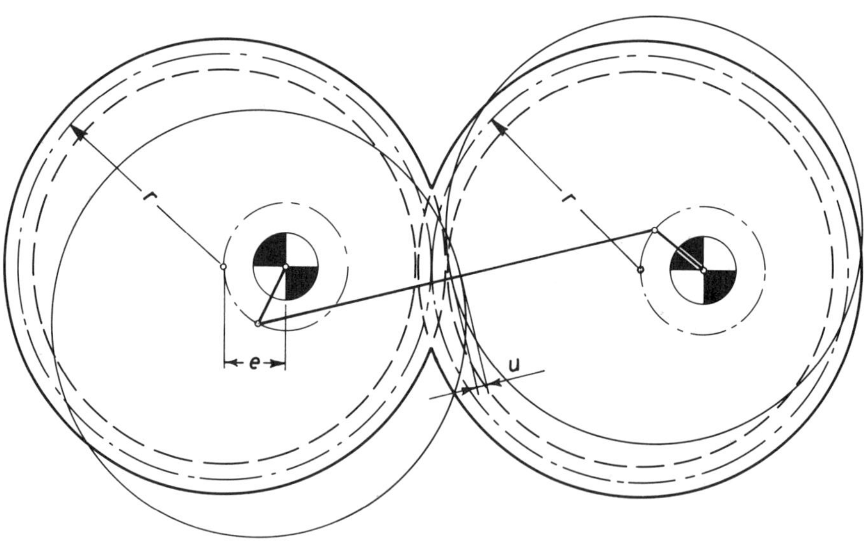

Figure 14.51 Two eccentric circular gears in mesh.

0.15. It can be shown that if the two gears are in correct mesh when the cranks lie on the line connecting the two gear centers, then there is an interference u in the position shown, and the maximum interference can be found from

$$u_{max} = 2r - 2r\left[1 - \left(\frac{e}{r}\right)^2\right]^{-1/2} \tag{14.9}$$

If a value of $e/r > 0.15$ is desired (Fig. 14.52), then the distance between the points of rotations A_0 and B_0 of the two gears can no longer be constant but must be made variable. This can be achieved by using an Oldham coupling to transmit motion to the output shaft.

EXACT CALCULATION OF NONCIRCULAR GEARS

So far, geometrical methods have been used to determine the pitch curves of noncircular gears in order to provide an understanding of the geometry of motion. Once the geometry is understood, calculations based on the geometrical method can be carried out.

The relationship between the rotation of two shafts was given by the curve 0–7 (Fig. 14.41). The pitch curves were constructed using values determined geometrically, (Fig. 14.42). This method cannot, however, be used to make the gears, however, because it is required that the lengths of

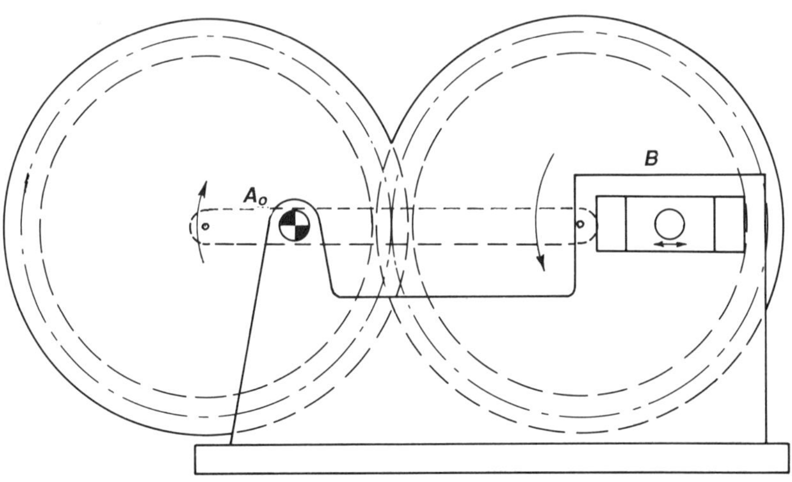

Figure 14.52 Two eccentric circular gears transmit a varying transmission ratio to the moving by means of an Oldham coupling (not shown).

the pitch curves are such that an integer number of teeth can be made according to the formula

$$S_v = P_C N \tag{14.10}$$

where

S_V = length of pitch curve
P_C = distance between the center lines of two adjacent teeth measured on the pitch curve
N = number of teeth (integer)

The length of a curve given in polar coordinates, is

$$S_v = \int_0^{2\pi} \sqrt{R^2 + (R')^2}\, d\,\theta \tag{14.11}$$

where R is radius vector, $R' = dR/d\theta$, and θ is the polar angle.

From Fig. 14.53, besides calculating the length of the pitch curve, it is also necessary to calculate ϕ, v, l, and h, but once the length of the pitch curve has been calculated, these values can be found relatively easy.

The relationship between the angular rotation of the shafts was given graphically (Fig. 14.41), but in order to calculate the pitch curves the relationship must be given mathematically. This is shown in the following.

Use of Plynomials

In Fig. 14.41 the relationship of angular rotation between shafts was given by a curve. In order to convert this to a mathematical expression, polyno-

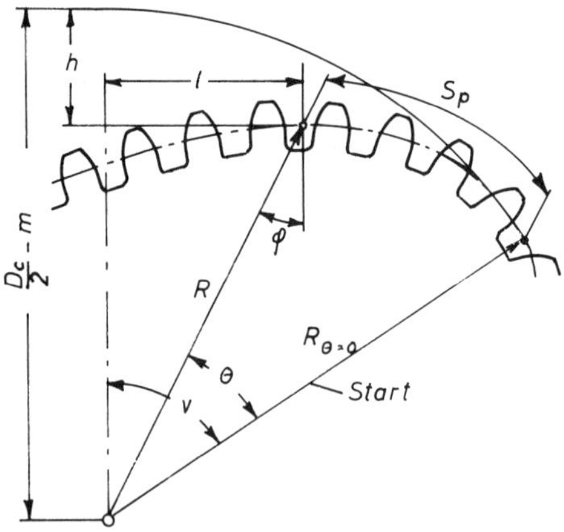

Figure 14.53 Designation by a noncircular gear.

mials can be used. The order of the polynomial is dependent on the number of conditions imposed on the curve. In Fig. 14.41 the curve is a straight line from 3 to 4 corresponding to the fact that the pitch curves are circular arcs. The curve 4–5–6–7 can be obtained by rotating 0–1–2–3–4 through 180°. Therefore, only the latter part is deciding. It may be required that the numbered shaft positions be given and, furthermore, that the slopes at these points, corresponding to the relative angular velocities, be given. Finally, it is required that the angular acceleration corresponding to point 3 be zero. The assumed requirements are listed in Table 14.2.

Table 14.4 lists 9 conditions, namely, four points, the slope at each of these points, and the angular acceleration at point 3. These conditions can be fulfilled by the polynomial

$$y = a_8x^8 + a_7x^7 + a_6x^6 + a_5^5 + a_4x^4 + a_3x^3 + a_2x^2 + a_1x + a_0$$

The constants in the above equation can be found by using Table 14.4. How many conditions are really necessary depends upon the problem, but this will be left to the individual reader, because it is the art of the engineer to make the right simplifications.

A center distance of $O_1O_2 = 100$ mm is assumed. S_v is found from eq. (14.11) by breaking the integral into three integrals, namely,

$$S_v = S_1 + S_2 + S_3$$

where

$$S_1 = \int_0^{\pi/4} \sqrt{R^2 + (R')^2}\, d\theta$$

$$S_2 = \int_{\pi/4}^{3\pi/4} \sqrt{R^2 + (R')^2}\, d\theta$$

$$S_3 = \int_{3\pi/4}^{\pi} \sqrt{R^2 + (R')^2}\, d\theta$$

Table 14.4

Point	x	y	$dy/dx = \frac{R}{r}$	d^2y/dx^2
0	0	0	2.5	
1	1/12	31/180	1.9	
2	1/6	5/18	1.0	
3	1/4	61/180	0.66	0

In our case,

$$S_1 = S_3, \text{ and } S_2 = \int_{\pi/4}^{3\pi/4} R \, d\theta$$

The integral S_1 can be calculated with the required degree of accuracy using Weddle's formula (see following section on numerical integration). If S_v is not of the required length, then O_1O_2 is multiplied with a factor of proportionality, namely, the desired S_v over the calculated S_v. The appropriate center distance having been found, new integrations must be performed using eq. (14.11), but this time with variable limit, to find the angle θ (Fig. 14.53). For manufacturing purposes, angle ϕ must be found; it is determined by the equation

$$\phi = \tan^{-1} \frac{R}{R'}$$

Numerical Integration

Sometimes an integral can be found only by numerical integration. A numerical integration (Fig. 14.54) can be performed to the desired degree of accuracy if the function is given mathematically. A function $y = f(x)$ is to be integrated. The corresponding area is divided into six equal parts of same width h. The function value of f(x) is calculated at the seven points. The function value at point i is designated f_i. Weddle's formula states that

$$\int_{x_{i-3}}^{x_{i+3}} f(x)\, dx = A = \frac{3h}{10} (f_{i-3} + 5f_{i-2} + f_{i-1} + 6f_i + f_{i+1} + 5f_{i+2} + f_{i+3})$$

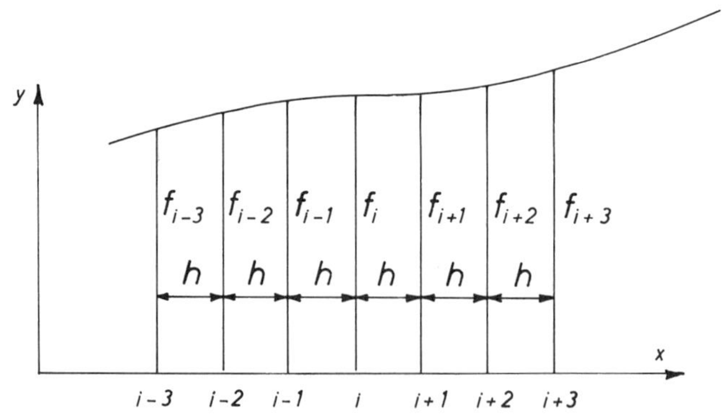

Figure 14.54 Designations for purpose of integration.

15
Detent, Indexing, and Ratchet Mechanisms

Detent mechanisms provide a means of locating single or multiple positions of sliding or rotating members. The detenting action is overruled by applying a force greater than that applied by the restraining member.

DETENT MECHANISMS

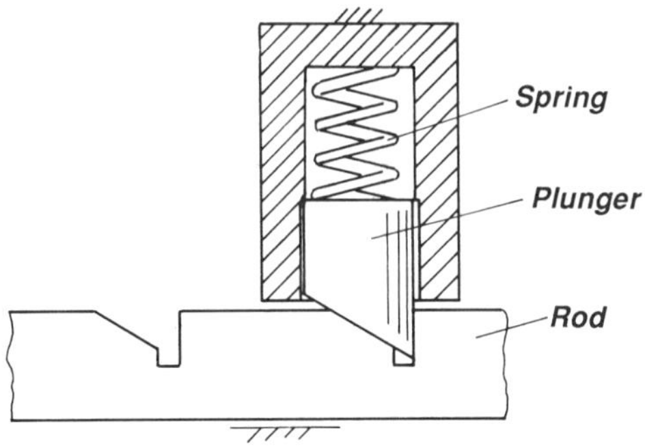

Figure 15.1 The spring-loaded plunger prevents motion of rod in one direction.

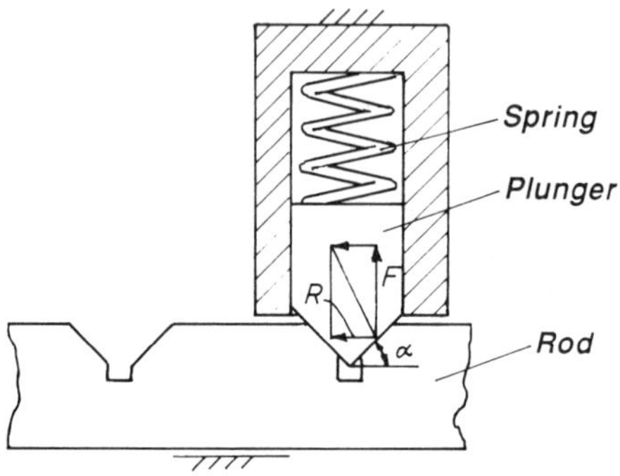

Figure 15.2 The motion to the left or right is possible only if a certain force is applied. The force R required to move the rod is

$$R = F(\tan\alpha + \mu)$$

where μ is the coefficient of friction, and F is the spring force.

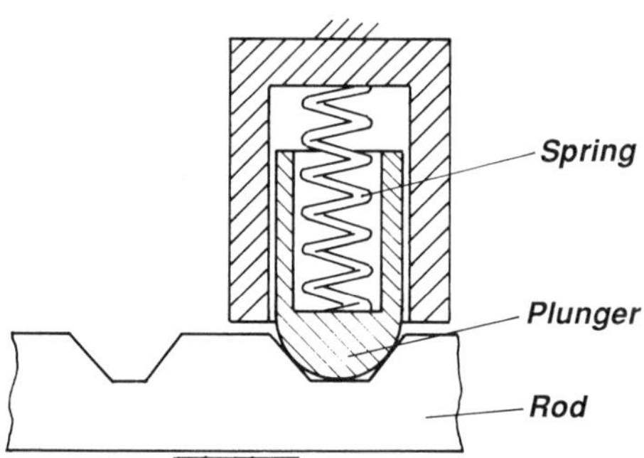

Figure 15.3 The plunger has a hole to accomodate a spring. Same equation applies as in Fig. 15.2

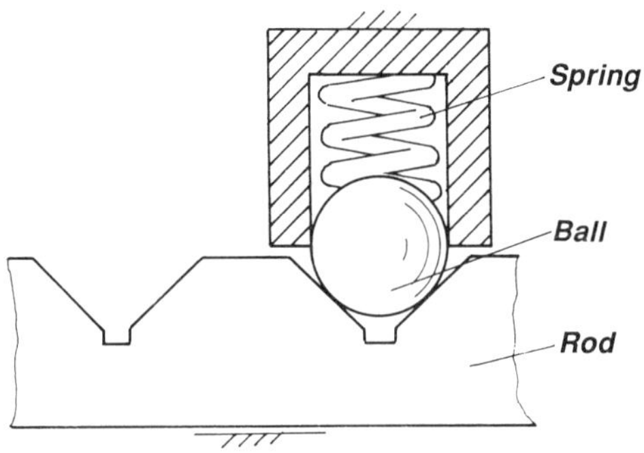

Figure 15.4 The detent is a ball that is spring loaded. A ball is cheaper than the plunger of the type shown in Fig. 15.2.

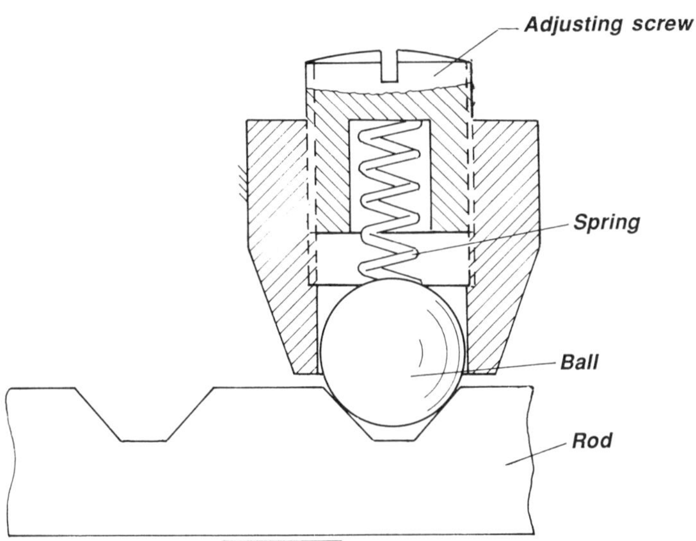

Figure 15.5 The force required to overcome the detenting action is made adjustable by means of an adjusting screw.

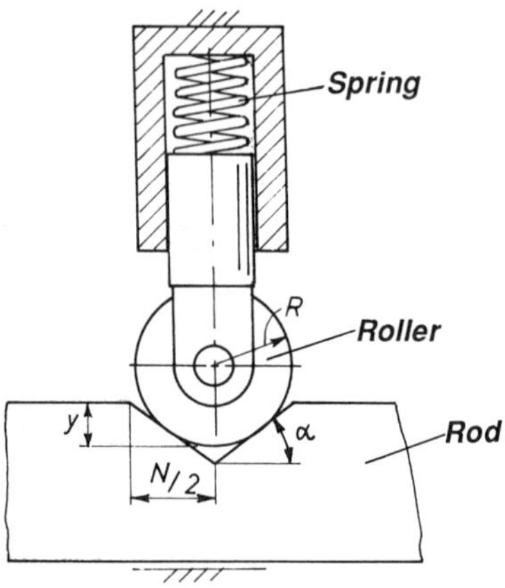

Figure 15.6 A roller is used as plunger. The rise y can be calculated from

$$y = N\tan\alpha/2 - R\,\frac{1 - \cos\alpha}{\cos\alpha}$$

The roller radius R can be calculated from

$$R = [(N/2)\tan\alpha - y]\left(\frac{\cos\alpha}{1 - \cos\alpha}\right)$$

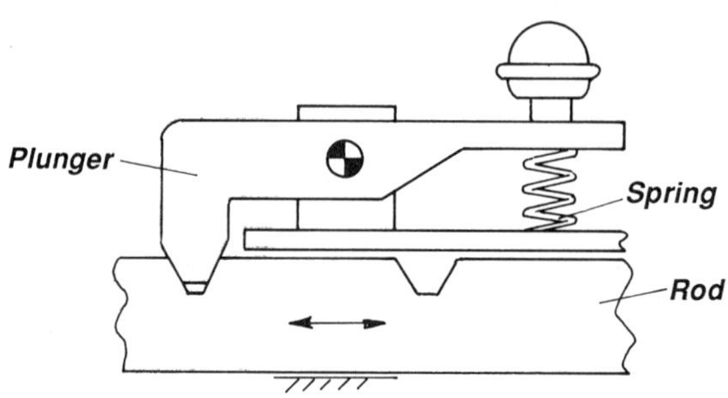

Figure 15.7 Rod motion is partially prevented by the spring-loaded detent. The locking action is overcome by manual release.

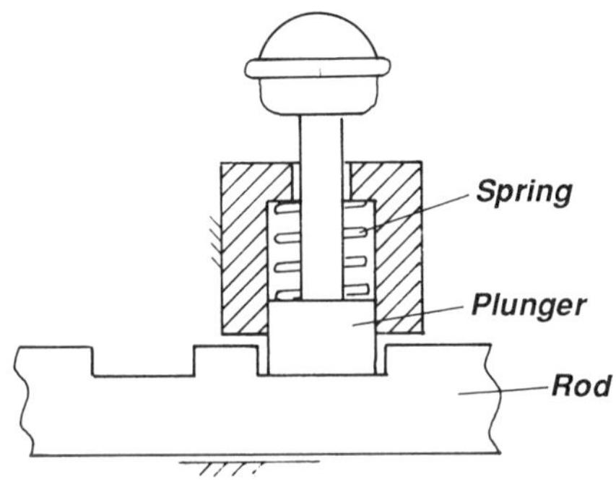

Figure 15.8 Movement in both directions is prevented unless manual release is activated.

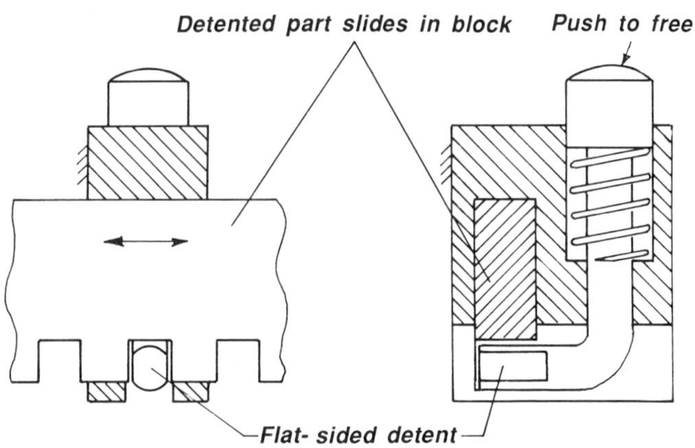

Figure 15.9 The detent has positive locking action, preventing motion in either direction unless manual release is activated.

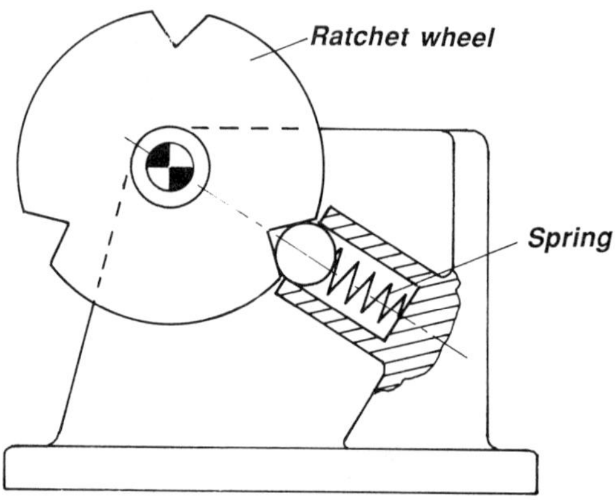

Figure 15.10 Here, rotary motion is prevented by a spring-loaded steel ball.

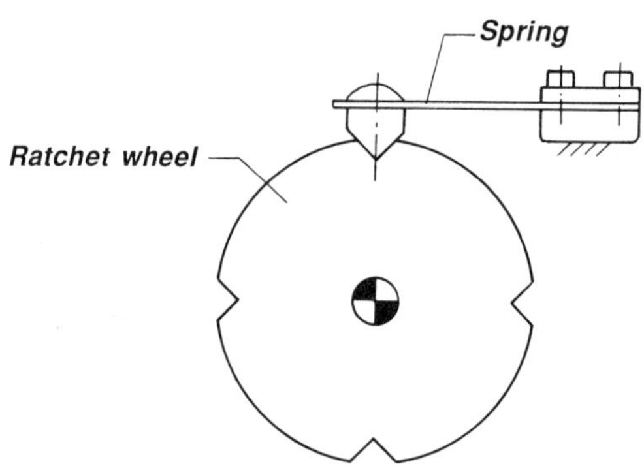

Figure 15.11 The detent and the spring are integrated in this design to prevent rotary motion.

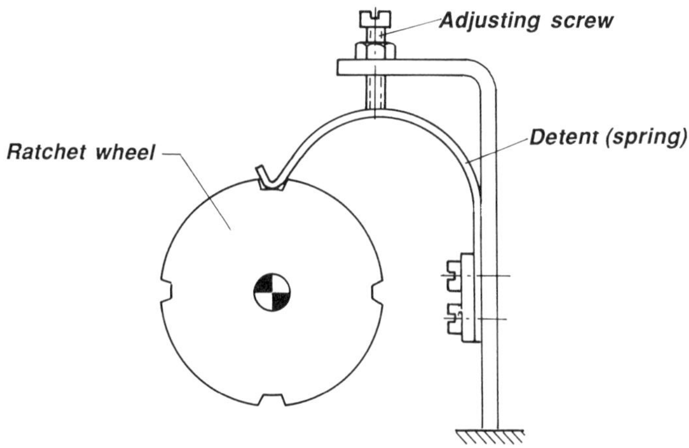

Figure 15.12 The force of the spring detent is made adjustable.

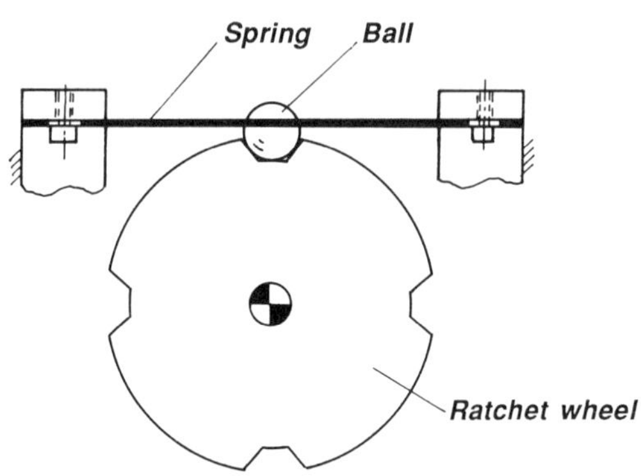

Figure 15.13 Compared with the previous design, the spring characteristic is changed to provide stronger locking action.

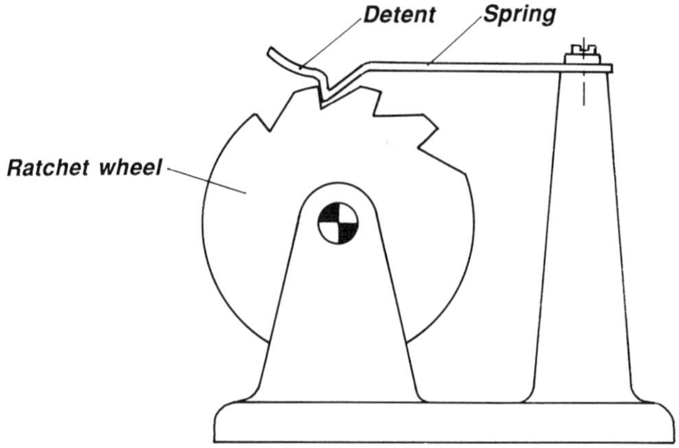

Figure 15.14 The pawl itself provides integrated detent and spring force.

Figure 15.15 The pawl provides integrated detent and spring force, but the spring characteristic is changed compared with Fig. 15.14.

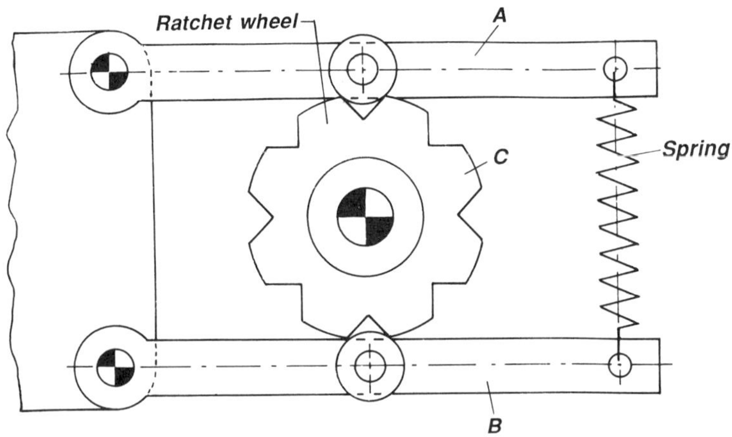

Figure 15.16 One spring is used to supply detenting action to two levers, A and B, to keep C in position.

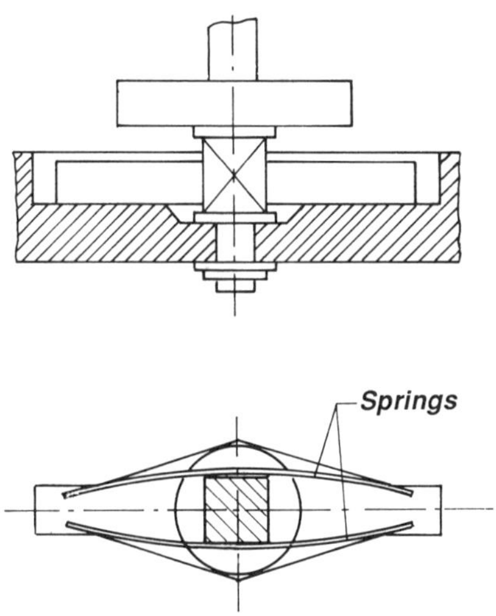

Figure 15.17 Two flat springs are used for detenting action. Provides for low cost.

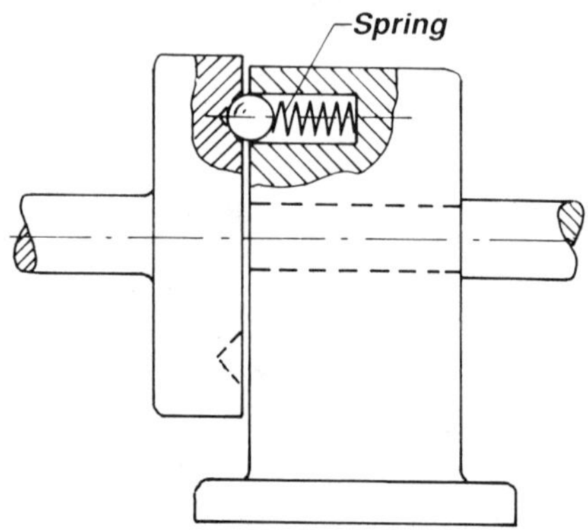

Figure 15.18 An axial-operating spring-loaded detent.

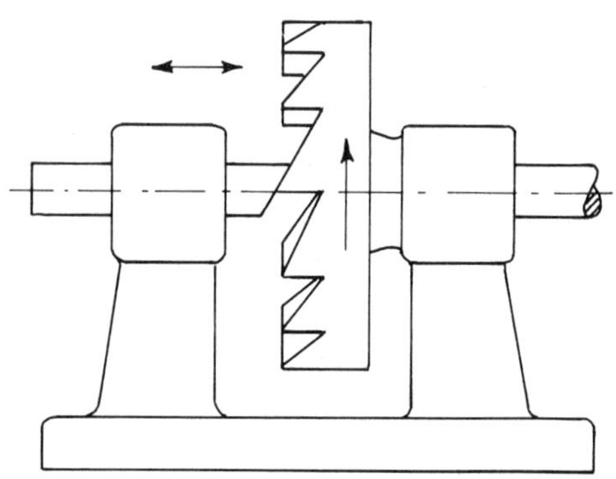

Figure 15.19 An axial-acting translating pawl, but note: sliding motion does not have as good friction characteristics as swinging pawls. Sliding motion also costs more in manufacturing.

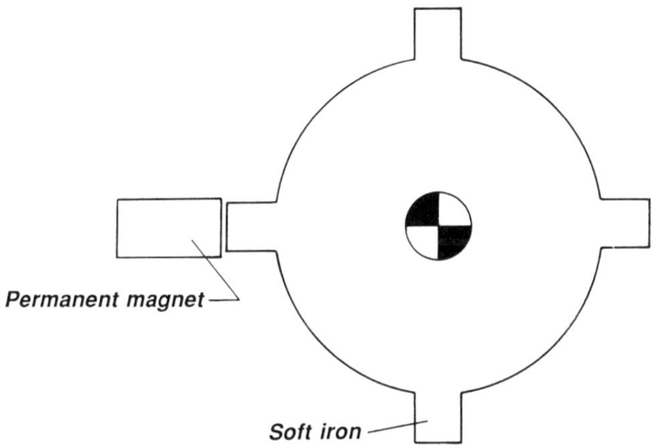

Figure 15.20 Angular four-position magnetic detent mechanism.

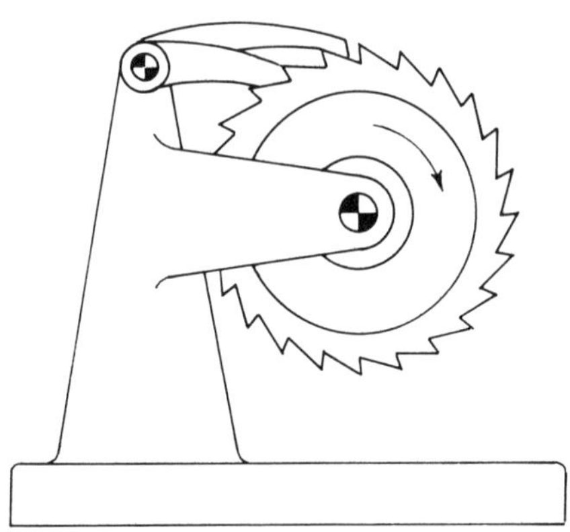

Figure 15.21 Often the teeth on the ratchet becomes so small that it is necessary to use two pawls. They are displaced half a circular pitch relative to each other.

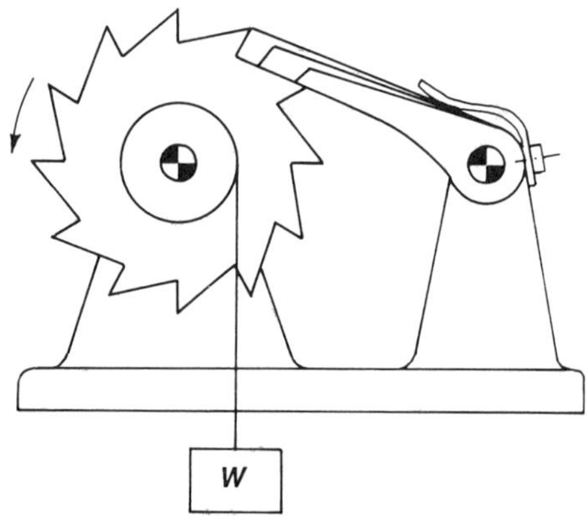

Figure 15.22 Three pawls on the same shaft permit the circular pitch on the ratchet to be three times as great.

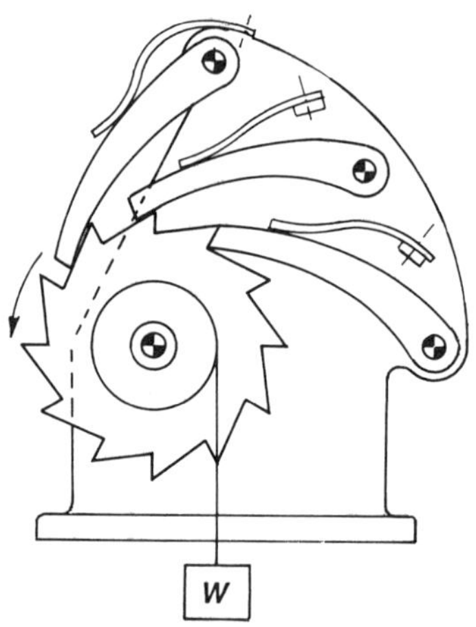

Figure 15.23 Three identical pawls displaced relative to each other permit three times as great a circular pitch on the ratchet.

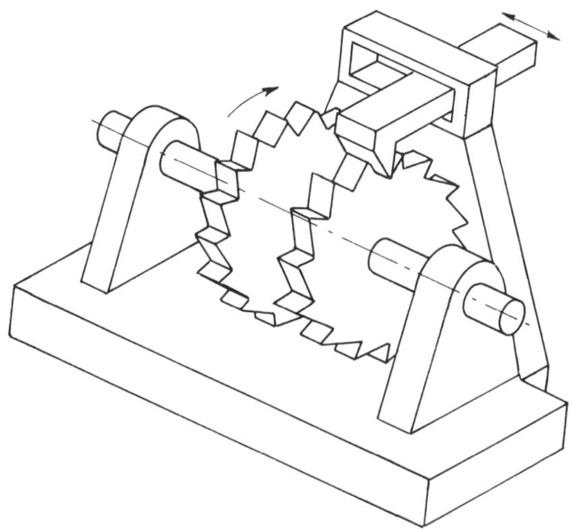

Figure 15.24 Instead of having two pawls and one ratchet, it is possible to use one pawl and two ratchets to obtain double the number of detents for each revolution, as shown here for a sliding pawl.

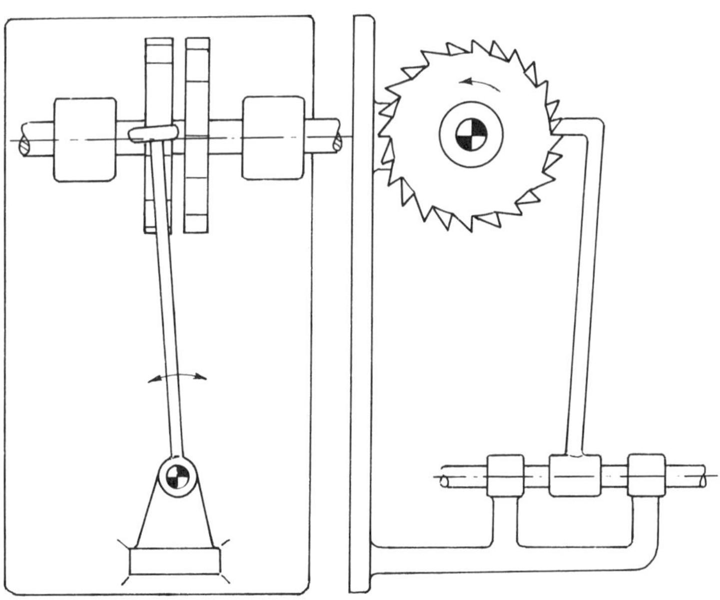

Figure 15.25 The principle of Fig. 15.24 is applied here to an oscillating pawl that engages two ratchets alternately.

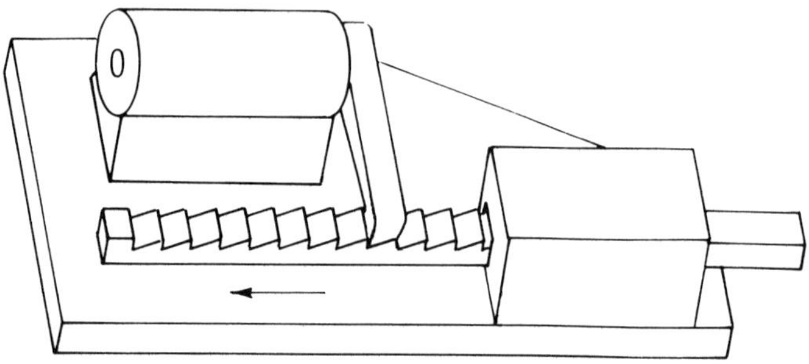

Figure 15.26 A swinging pawl design is cheaper to make than a sliding pawl.

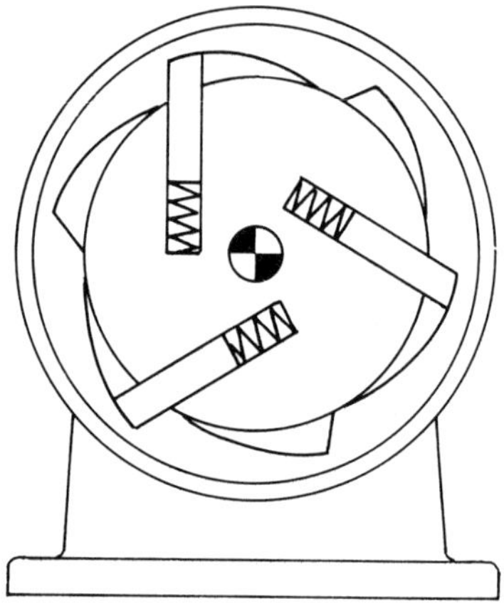

Figure 15.27 Ratchets can be made internal, but remember that although internal ratchets cost more than external ratchets, internal ratchets offer a more compact design and can be better protected against the environment.

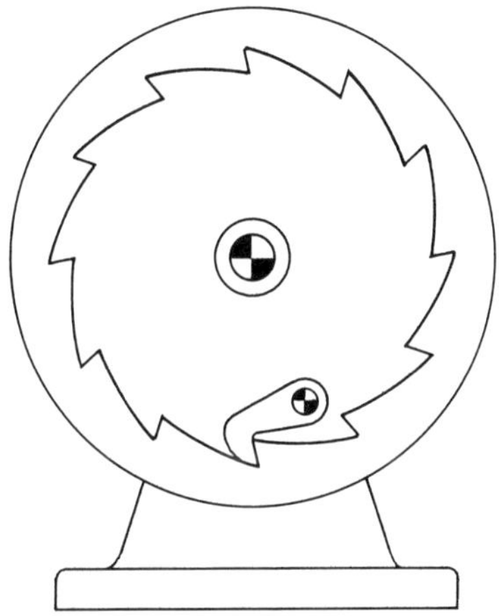

Figure 15.28 Another internal ratchet. Circular pitches are smaller with internal ratchets than with external ratchets for same indexing angle.

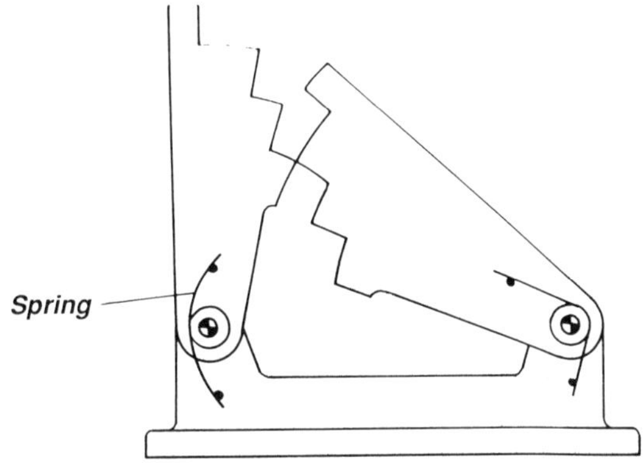

Figure 15.29 Interlocking pawls can be made to occupy a number of positions relative to each other.

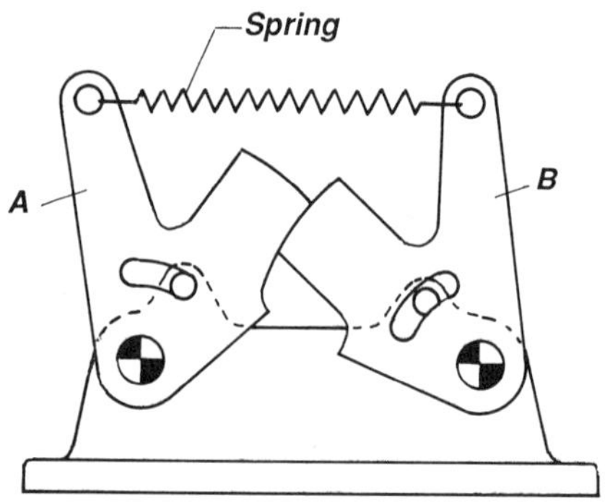

Figure 15.30 Interlocking device. Member A cannot be moved without being released by member B, and vice versa.

Reversible-Action Detents

Reversible-action detents allows change of direction in which a motion is being locked.

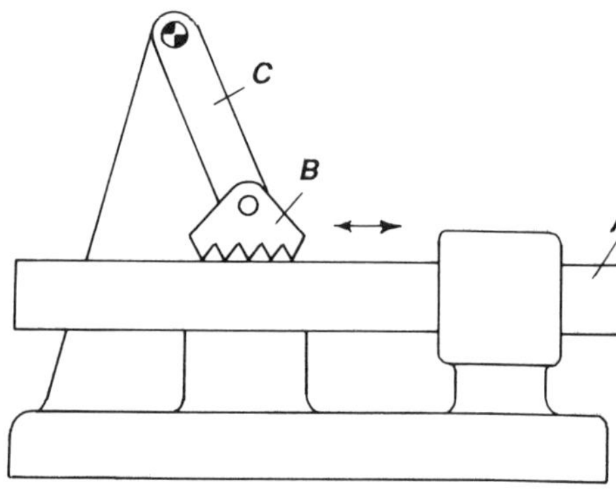

Figure 15.31 Member A is locked by jaw B, which is carried by arm C. By swinging arm C to the other side, locking action of A is in the opposite direction.

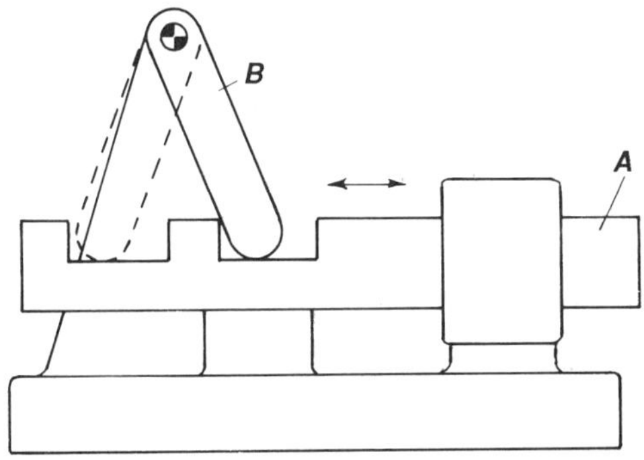

Figure 15.32 Rod A is locked by the action of pawl B. Swinging pawl B to the other side (shown in dashed lines) locks rod A in the opposite direction.

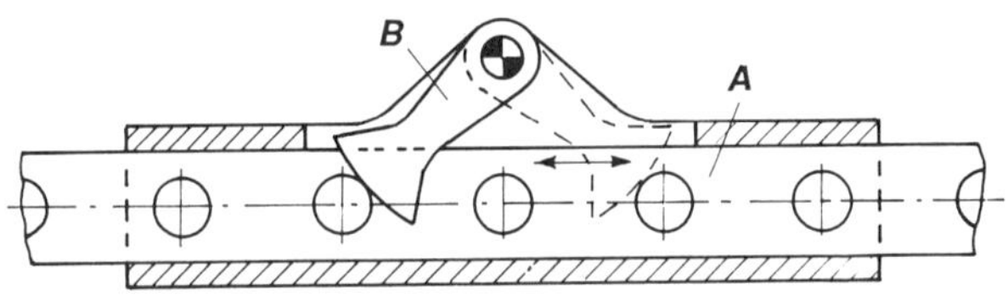

Figure 15.33 Rod A is locked by the action of pawl B. Swinging pawl B to the other side (shown in dashed lines) locks rod A in the opposite direction.

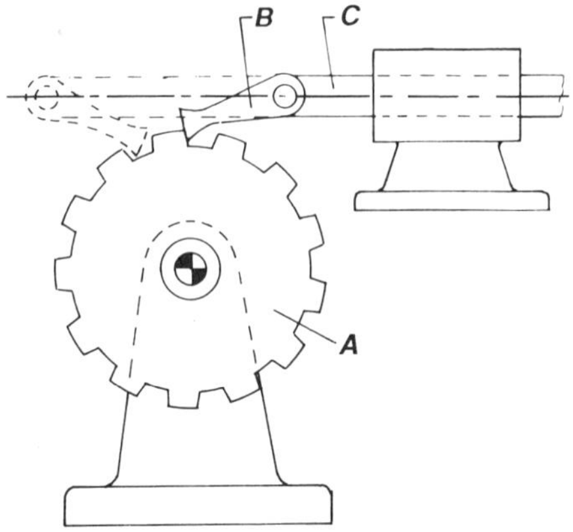

Figure 15.34 When C and B are in the position shown, they prevent ratchet A from rotating CW. The dashed positions of B and C prevent ratchet A from rotating CCW.

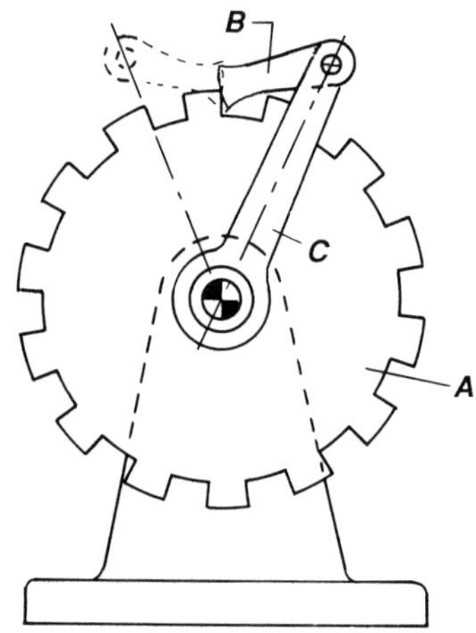

Figure 15.35 Members A, B, and C have same function as in Fig. 15.34.

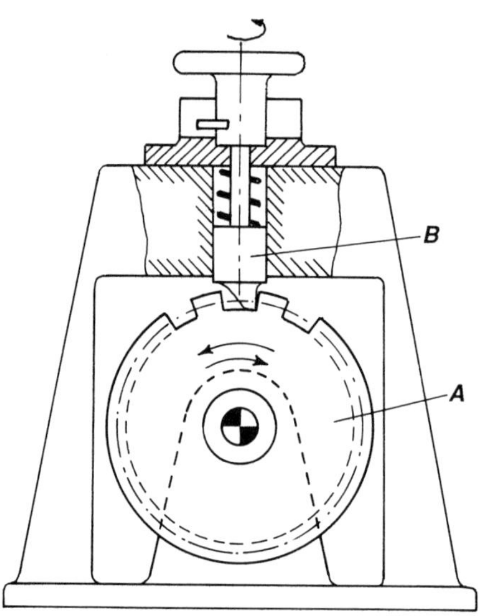

Figure 15.36 Turning pawl B through 180° prevents ratchet A from rotating in the opposite direction.

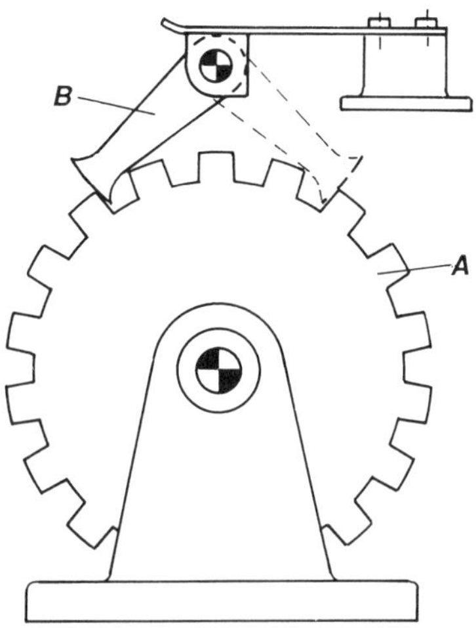

Figure 15.37 Pawl B has three positions. One drawn in solid line, one in dashed line, and one neutral position drawn in between the two other positions in an upwards direction, where no detent action occurs.

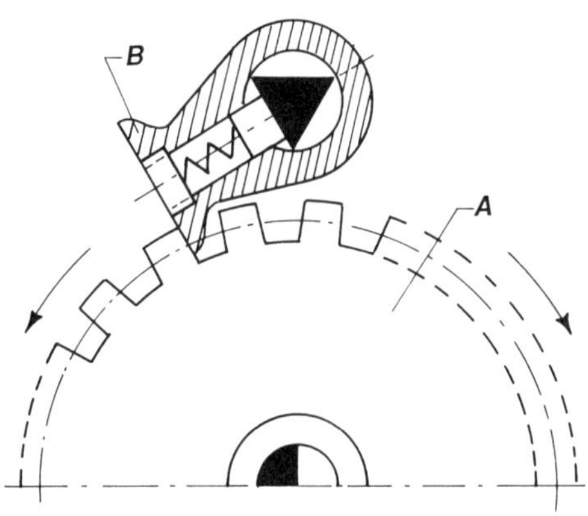

Figure 15.38 As in the previous example, pawl B has three positions: two active positions and one neutral, where it is in a vertical position.

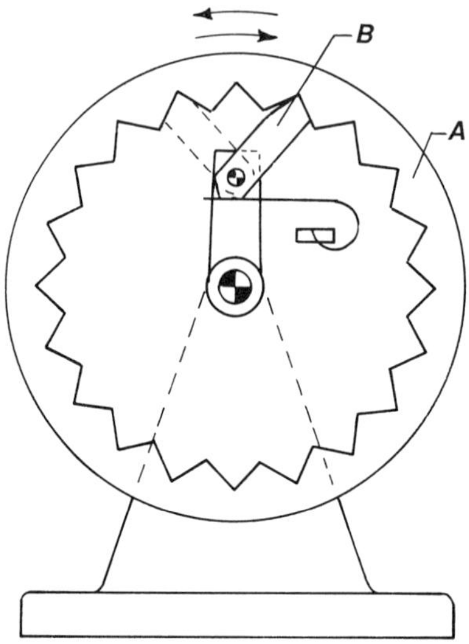

Figure 15.39 An internal ratchet A with a pawl B with two active positions.

INDEXING MECHANISMS

If motion is prevented by form closure (see Fig. 15.9), then the action is often called indexing.

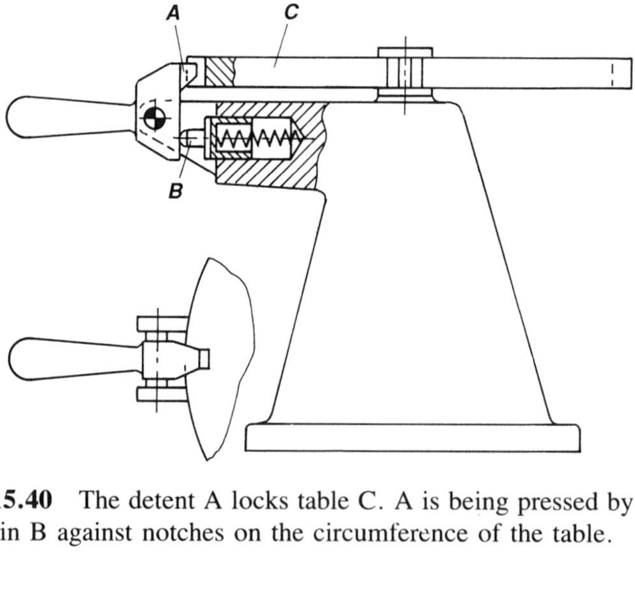

Figure 15.40 The detent A locks table C. A is being pressed by a spring-loaded pin B against notches on the circumference of the table.

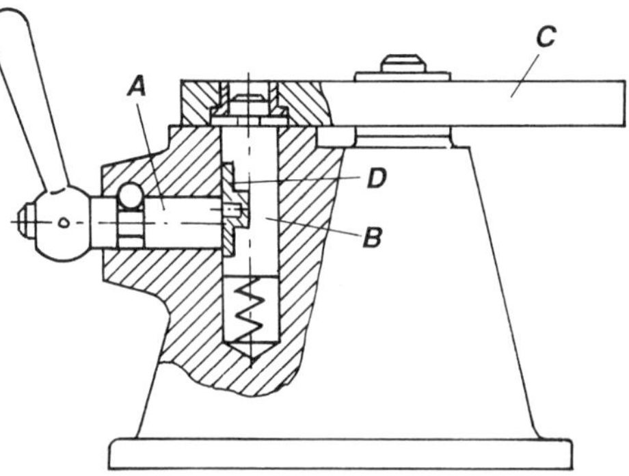

Figure 15.41 An indexing motion is provided to table C by the spring-loaded plunger B, which is moved by the handle A by an eccentric motion D.

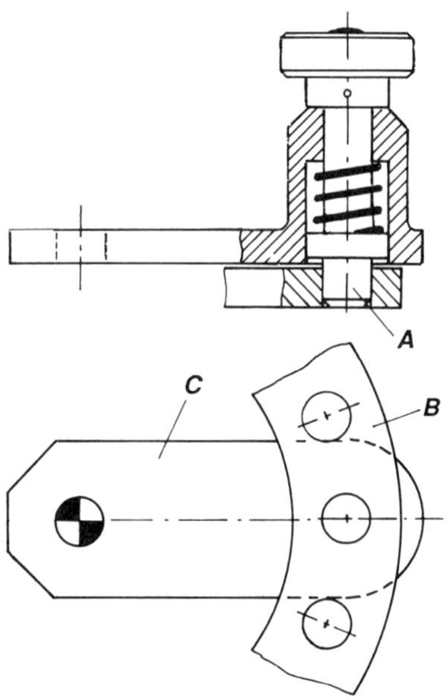

Figure 15.42 Positive indexing motion is provided by the spring-loaded, hand-operated detent A.

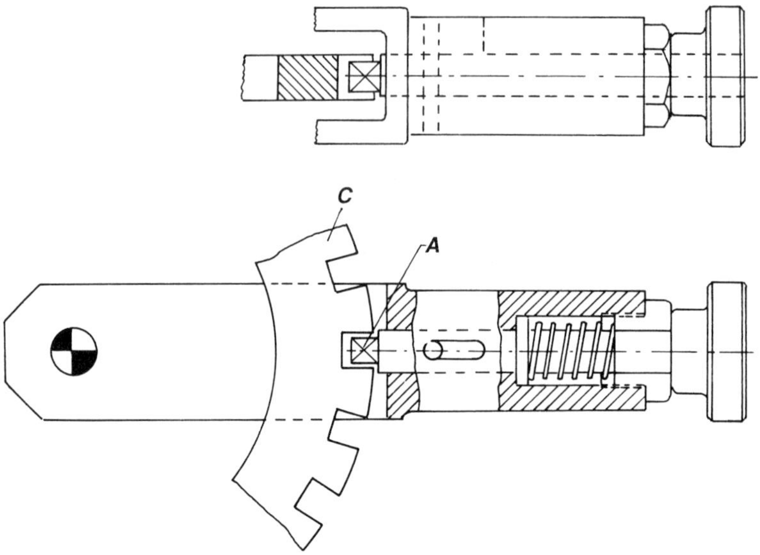

Figure 15.43 Positive indexing motion is provided by the spring-loaded radial detent A.

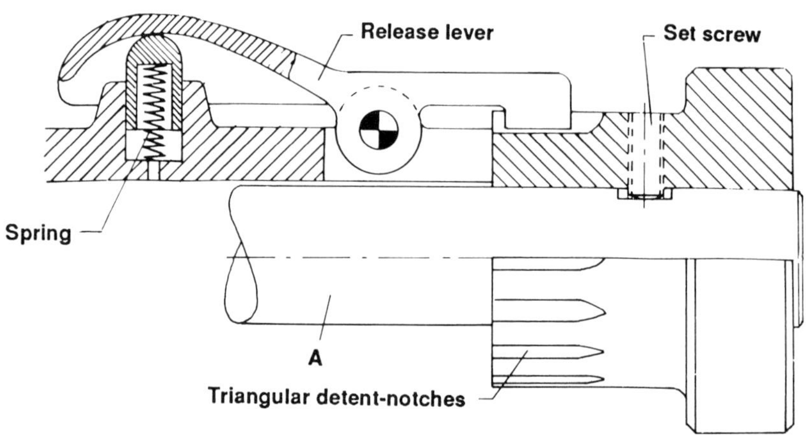

Figure 15.44 Adjustment is provided in radial indexing of shaft A on which is mounted an indexing wheel, which can be adjusted in a radial direction by a set screw. Indexing is neutralized by pressing the release lever.

RATCHET MECHANISMS

Pawl-and-Ratchet Mechanisms

Pawl and ratchet mechanisms are intermittent motion mechanisms. The ratchet motion is often generated by a crank or an eccentric that produce modified simple harmonic motion.

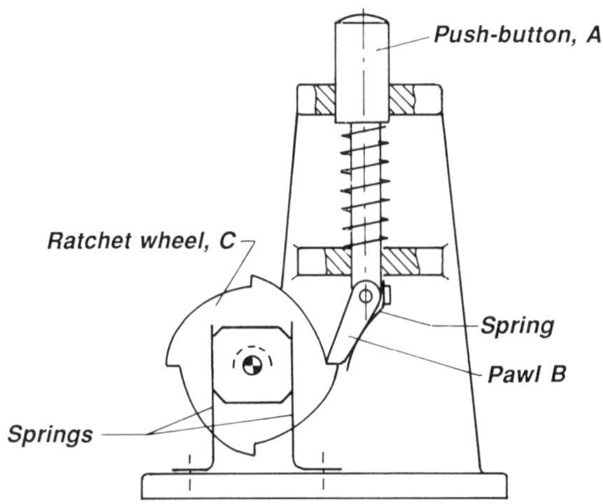

Figure 15.45 Push-button ratchet mechanism. A push on the push-button A indexes ratchet wheel C through the action of spring-loaded pawl B. Ratchet carries square-shaped member so that it is locked in position after indexing.

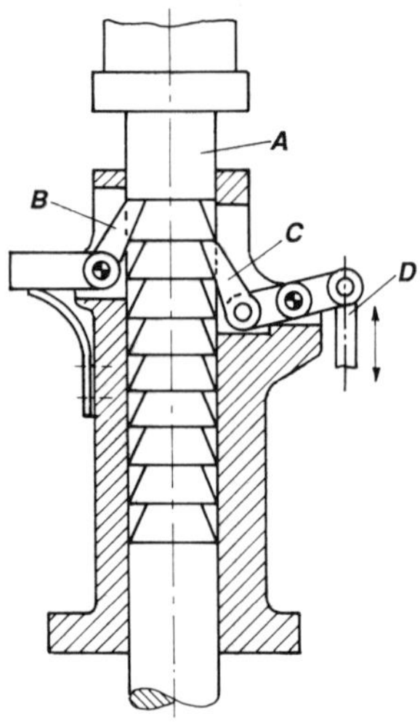

Figure 15.46 Circular rod A is prevented from moving downwards by pawl B. It is being lifted by pawl C, which is activated by the up-and-down motion of D. Making the rod circular allows it to be rotated during motion or during operation.

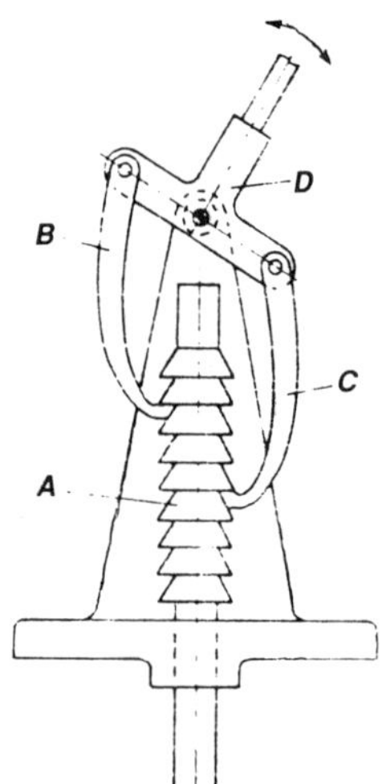

Figure 15.47 The oscillating motion of D is transmitted to the two pawls B and C. While one pawl moves down and out of engagement, the other pawl moves up and takes the circular rod A with it.

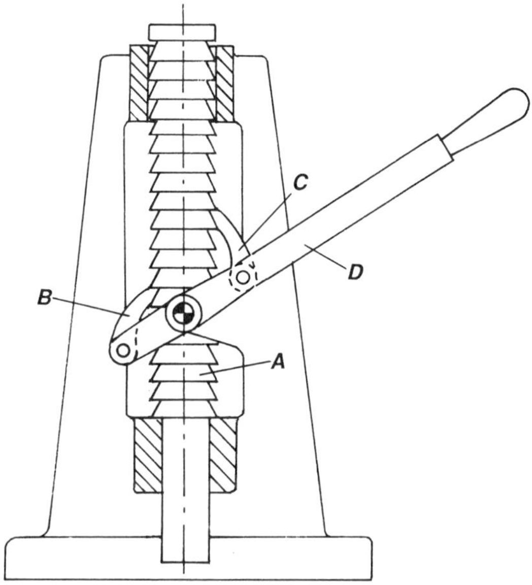

Figure 15.48 An arrangement similar to the mechanism of Fig. 15.47.

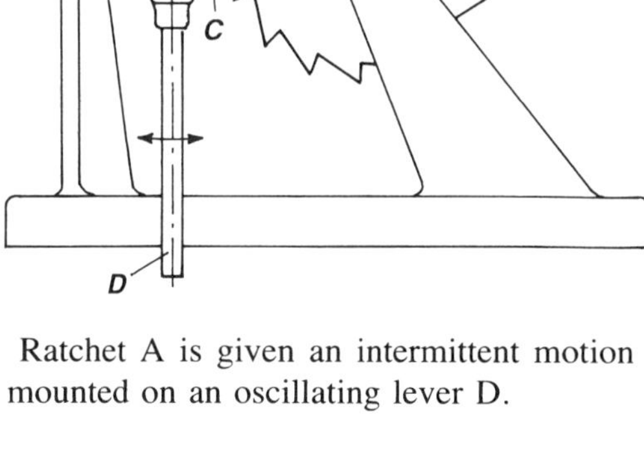

Figure 15.49 Ratchet A is given an intermittent motion by the pawls B and C that are mounted on an oscillating lever D.

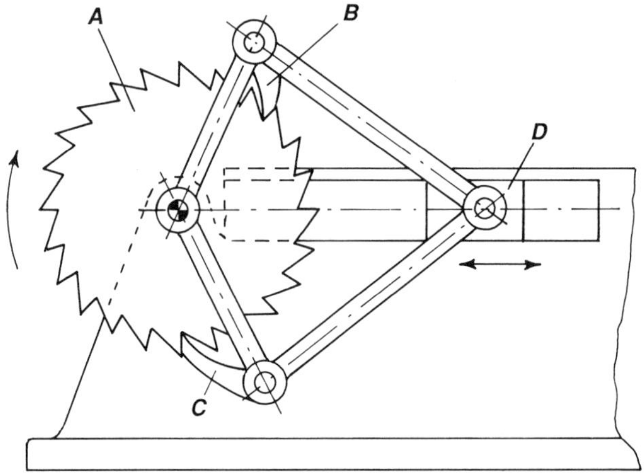

Figure 15.50 The back-and-forth motion of slider D is transformed to the oscillating motion of pawls B and C, thereby imparting an intermittent motion to ratchet A.

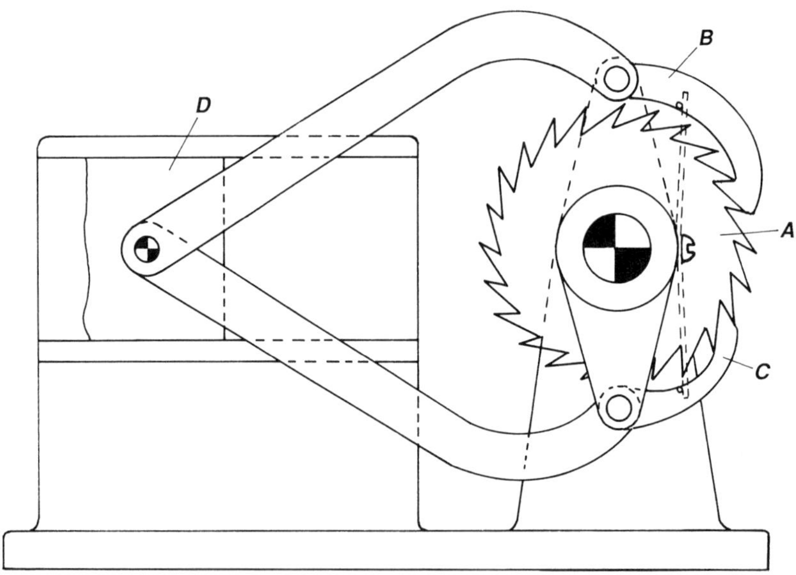

Figure 15.51 A slightly different version of the mechanism of Fig. 15.50.

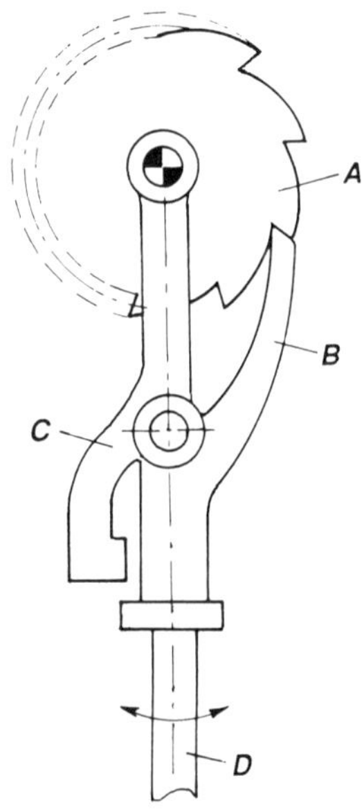

Figure 15.52 Silent ratchet mechanism. Oscillating arm D and pawl B are integral. By a CCW motion, pawl B engages ratchet A, but on the backwards stroke pawl B moves completely out of engagement with ratchet A and does not emit any "clicking" noise.

Mechanism-Operated Pawl-and-Ratchet Mechanisms with Fixed Stroke

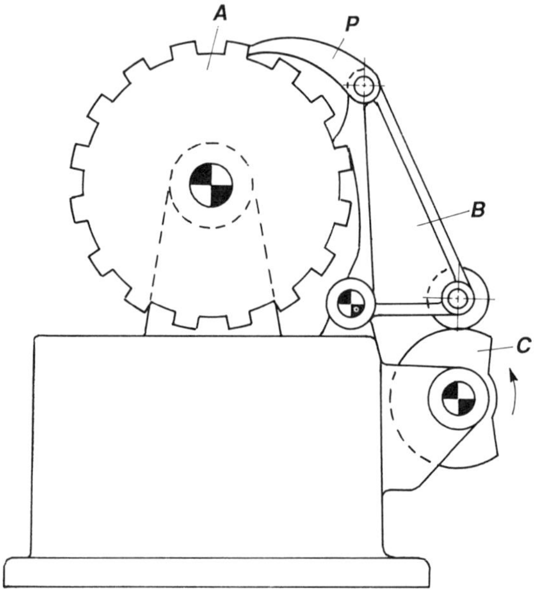

Figure 15.53 The continuous rotation of cam C imparts an oscillating motion to arm B, which carries pawl P, which engages ratchet A. Ratchet A moves with intermittent motion.

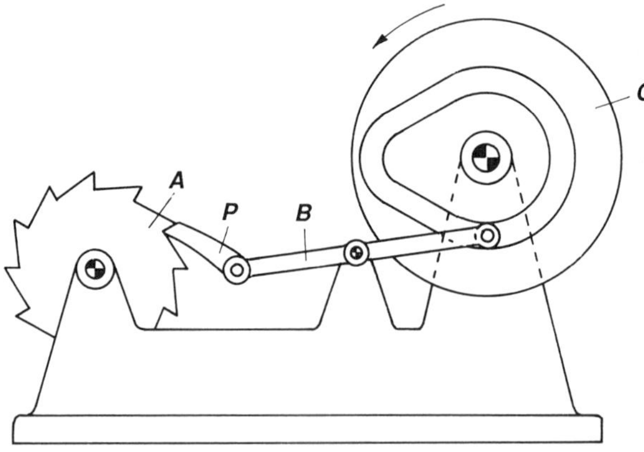

Figure 15.54 A closed-track cam C imparts an oscillating motion to follower B on which is mounted pawl P, which imparts an intermittent motion to ratchet A.

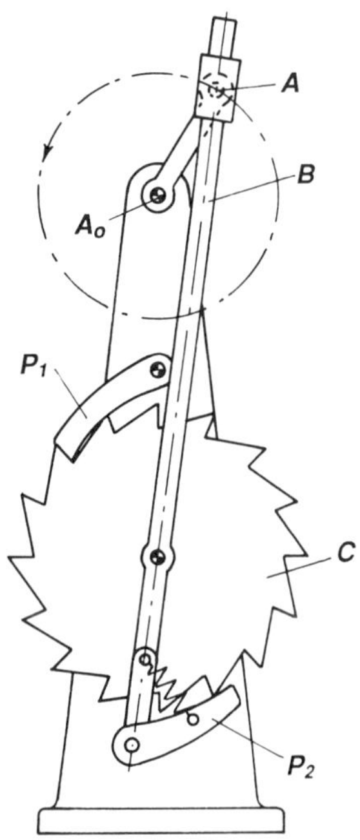

Figure 15.55 The slider crank mechanism with crank A_0A oscillates rod B on which is mounted pawl P_2, which engages ratchet C and causes it to turn. When P_2 is not in engagement, pawl P_1 locks the ratchet in its dwell position.

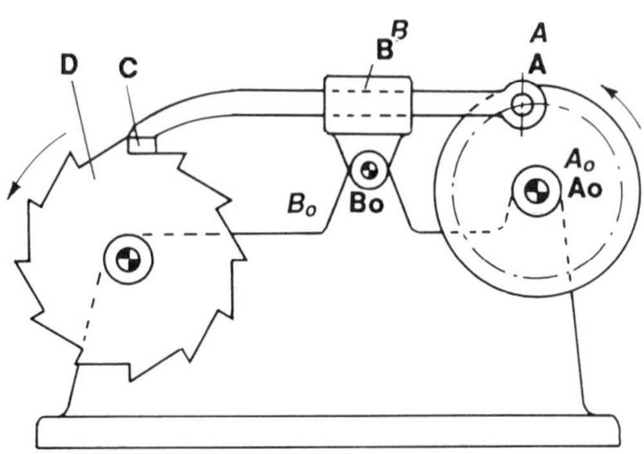

Figure 15.56 An oscillating block mechanism A_0AB_0, where the oscillating block B moves the engaging part C of the oscillating rod into and out of engagement with ratchet D.

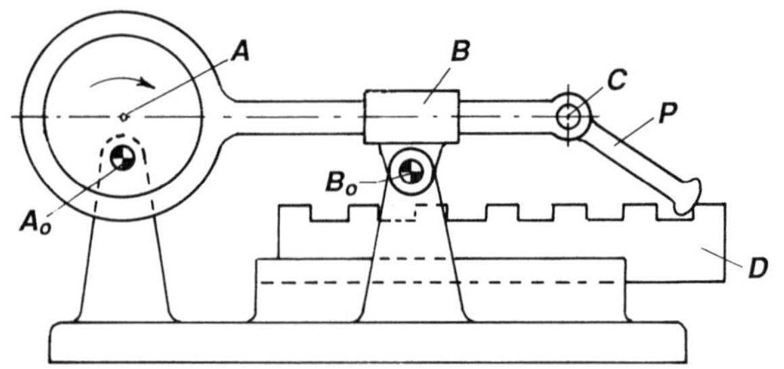

Figure 15.57 An oscillating block mechanism A_0AB_0, where the oscillating block B moves C on a coupler course (not shown) whereby pawl P gets an back and forth horizontal motion. When moving to the right pawl P engages rod D and moves it to the right.

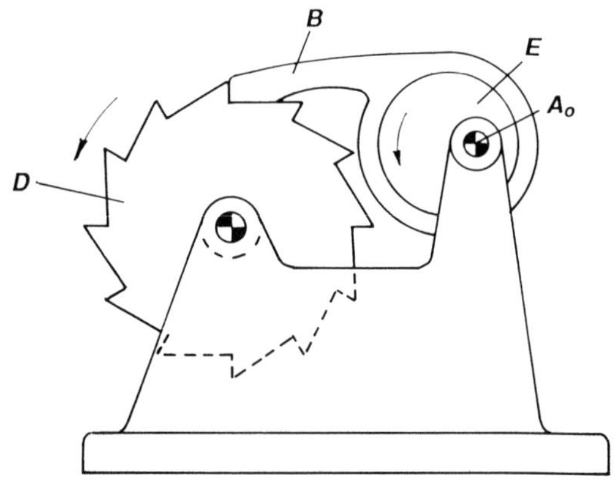

Figure 15.58 The rotary motion of eccentric E around A_0 imparts an oscillating motion to pawl B. Pawl B is kept in engagement with the ratchet partly through gravity and partly because of the rotational direction of the eccentric, which tends to force the pawl downwards, too.

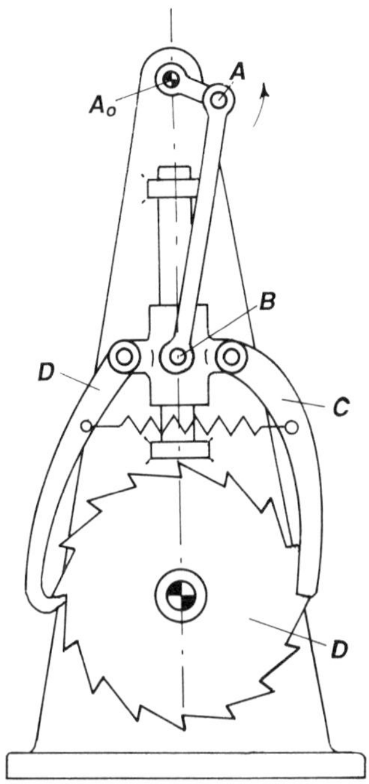

Figure 15.59 The centric slider crank A_0AB imparts an oscillating motion to pawls C and D. (Compare with Fig. 15.47.)

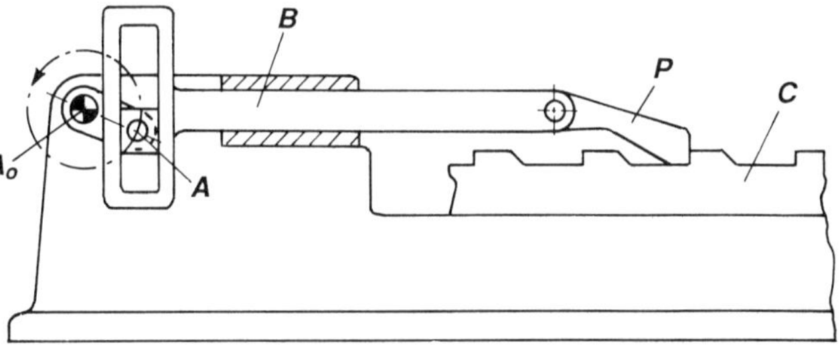

Figure 15.60 The scotch-yoke mechanism A_0AB imparts a back-and-forth motion to pawl P, which moves rod C to the right intermittently. The motion characteristics is 180° dwell and 180° motion.

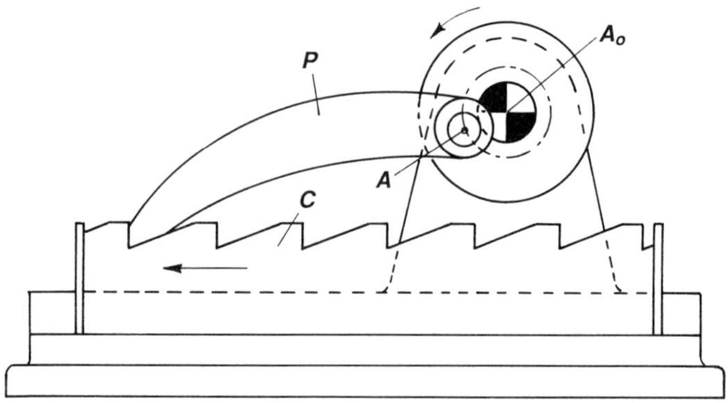

Figure 15.61 Crank A_0A moves pawl P back and forth, thereby imparting an intermittent motion to the left of rack C.

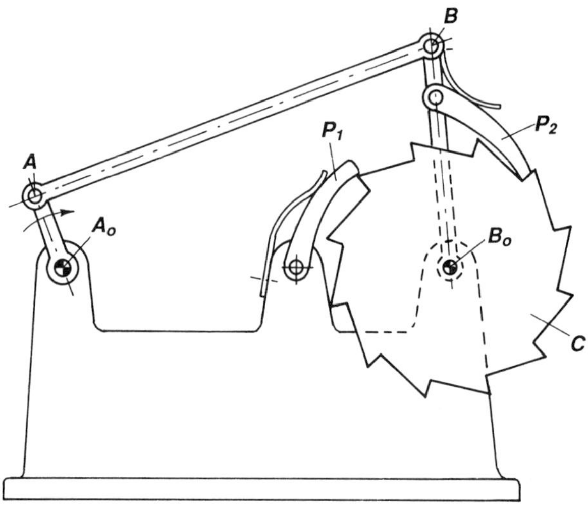

Figure 15.62 A four-bar linkage A_0ABB_0 imparts an intermittent motion to ratchet C through pawl P_1. Pawl P_2, which is spring loaded, keeps ratchet C from moving back.

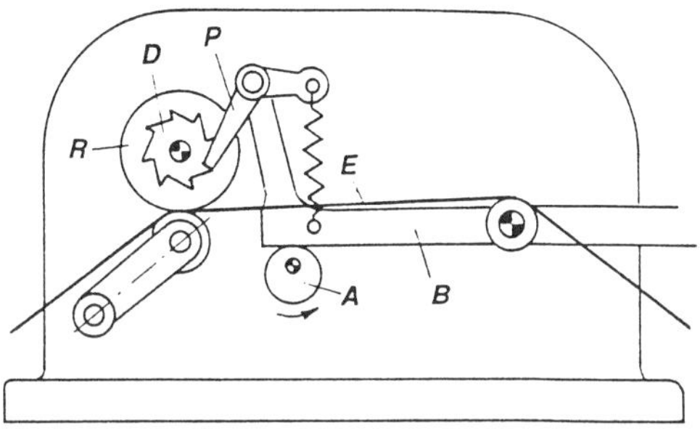

Figure 15.63 Eccentric A oscillates arm B and pawl P, which is mounted on B. The ratchet D gets an intermittent motion and through roller R paper or web E is advanced one step at a time.

Mechanism-Operated Pawl-and-Ratchet Mechanisms with Variable Stroke

If the mechanism that drives the pawl-and-ratchet mechanism is made adjustable, then a variable stroke or rotation is obtained. The adjustment can be made during stopping in some cases and during operation in other cases. If a mechanism is to be adjusted while running, the speed has to be so low that the adjustment can be made by hand, or else the adjustment is made to a link connected to the frame.

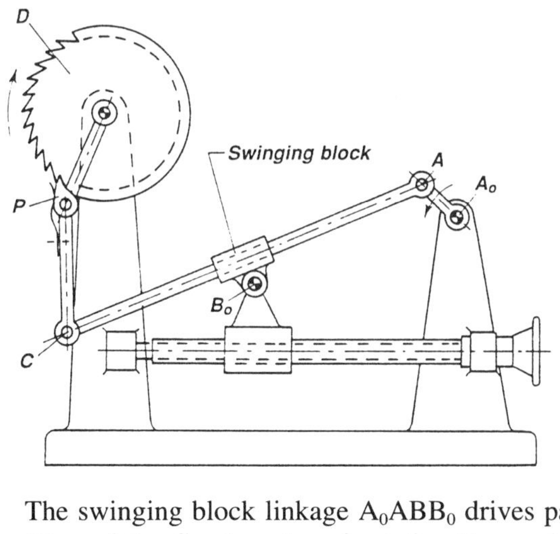

Figure 15.64 The swinging block linkage A_0ABB_0 drives pawl P through link C. The position of the fixed center of rotation B_0 can be adjusted by the screw S, whereby a variable stroke is imparted to ratchet D.

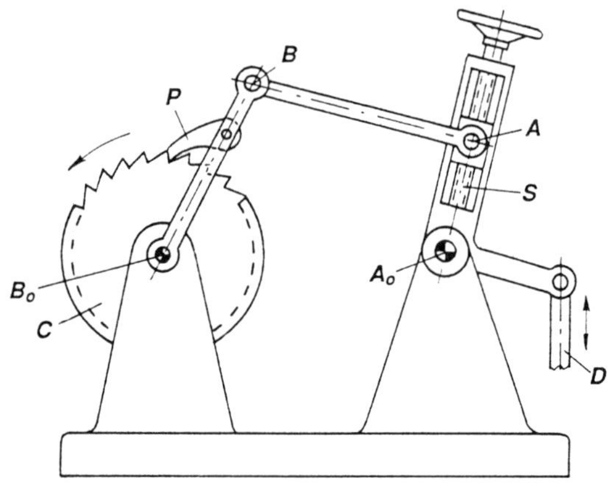

Figure 15.65 Oscillating link D oscillates link A_0A around A_0. This motion is transmitted to rocker B_0B on which pawl P is mounted. The oscillating motion of link B_0B is adjusted through the variable-length crank A_0A by means of a screw S. Ratchet C can index a number of pitches.

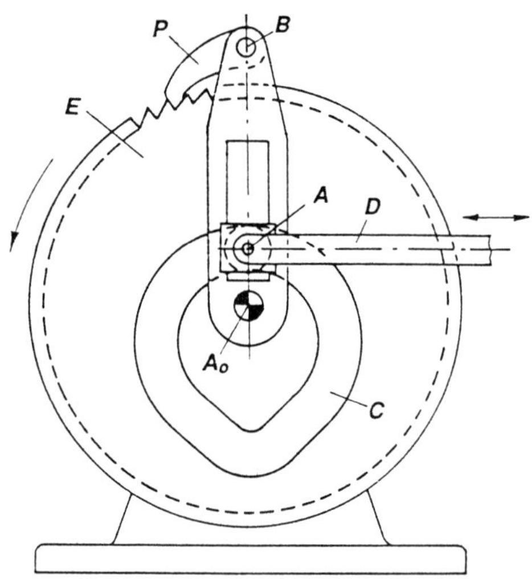

Figure 15.66 Continuously varying–stroke mechanism. The oscillating motion of rod D is transmitted to the oscillating motion of lever A_0B on which pawl P is mounted. Pawl P drives ratchet E. A closed-track cam C is cut in ratchet E, and the constant stroke of rod D is transformed to a varying stroke of the slotted member because the position of A changes with the position of E.

Friction, or Silent, Ratchet Mechanisms

Friction, or silent, ratchet mechanisms are used extensively. Motion is transmitted by friction. They are called silent ratchets because there are only frictional forces to transmit motion and not a pawl to suddenly engage the ratchet wheel. The use of an adjustable stroke does not change the silent feature because the wheel is picked up at the beginning of the stroke. Many of the foregoing pawl and ratchet mechanisms can be converted to friction ratchet mechanisms. Only two examples will be shown of friction ratchet mechanisms.

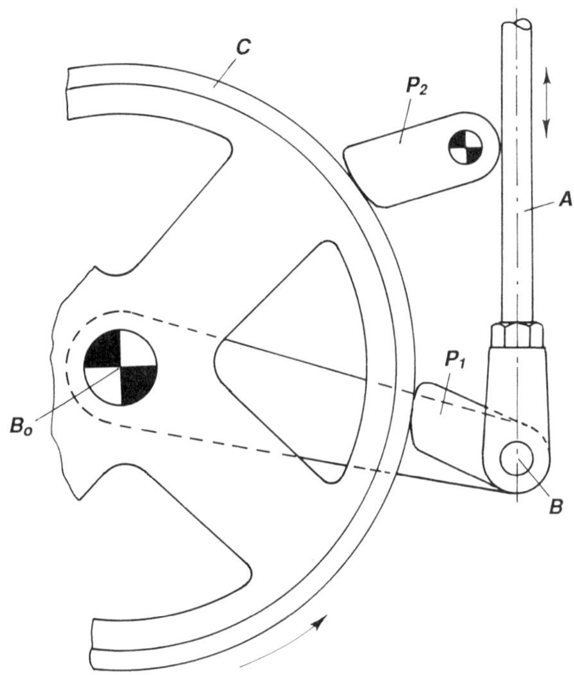

Figure 15.67 The oscillating motion of rod A is converted to a swinging motion of arm B_0B, which carries pawl P_1 and which when moved in a CCW direction around B_0 moves the ratchet CCW, too. Pawl P_2 prevents the ratchet wheel from turning back.

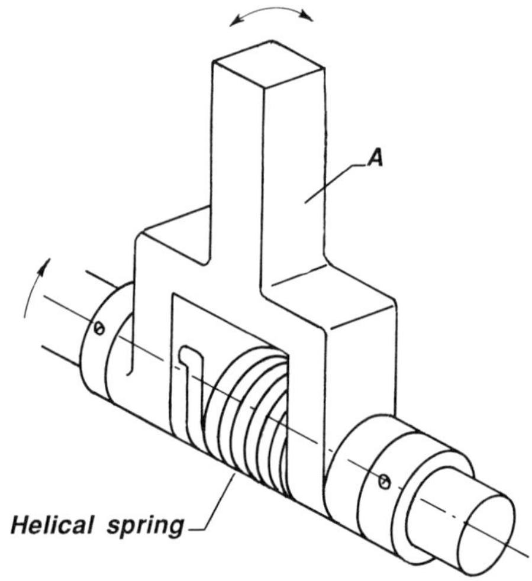

Figure 15.68 Helical spring is attached to arm A and grips around shaft because inner diameter of spring is smaller than shaft diameter. During the forward stroke the spring winds around the shaft and indexes it; during the return stroke it loosens.

16
Overload and Overrunning Clutches

OVERLOAD CLUTCHES

To prevent overload in machines, overload clutches are used. They prevent the torque at a certain point of the machine from exceeding a prescribed level. Certain types of overload clutches when activated require stopping the machine to repair the overload clutch. Other types allow resumption of full operation without having to stop the machine.

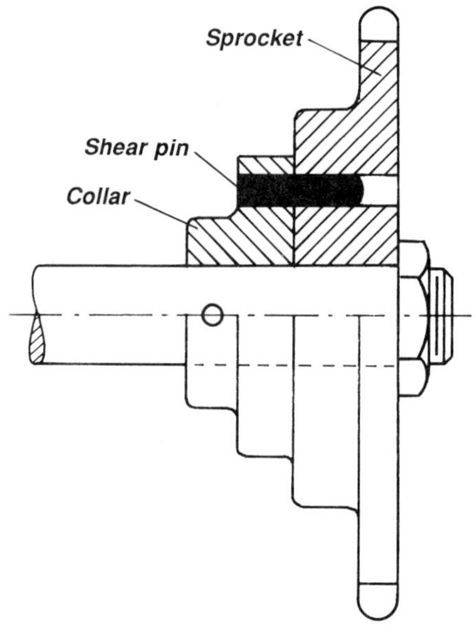

Figure 16.1 If the torque on the sprocket exceeds a certain level, the shear pin is sheared off. The machine must, however, be stopped in order to replace the pin.

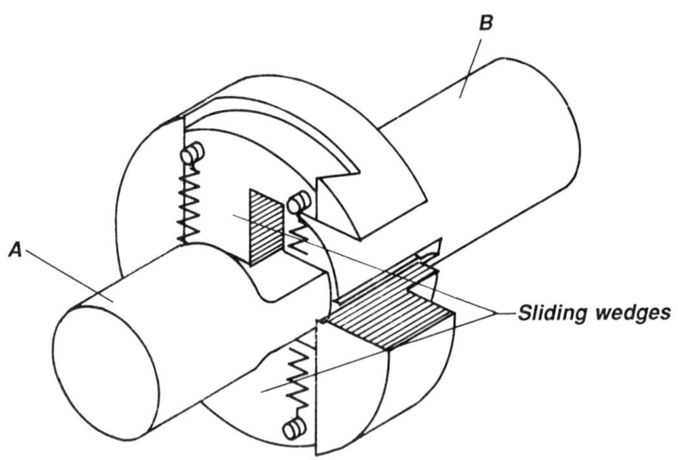

Figure 16.2 Shaft A is flattened at one end. Shaft B carries the clutch with spring-loaded sliding wedges that are forced away from each other if torque exceeds a certain value.

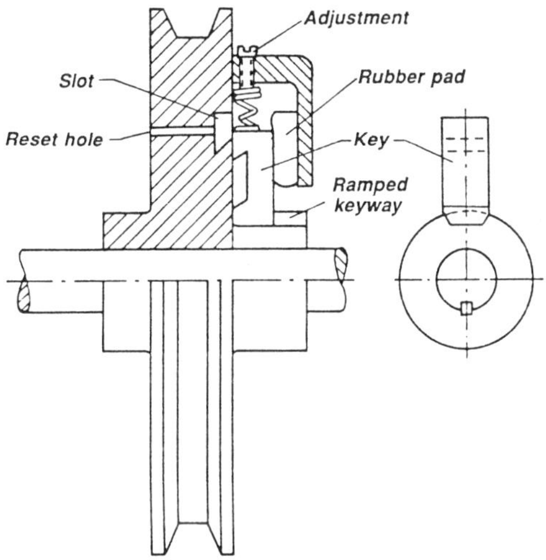

Figure 16.3 A spring-loaded retracting key is forced to move outwards if overload occurs. The rubber pad forces the key into the slot so that clutch is enabled and has to be reset. This is done to prevent wear. The key is reset by pushing on it through a reset hole.

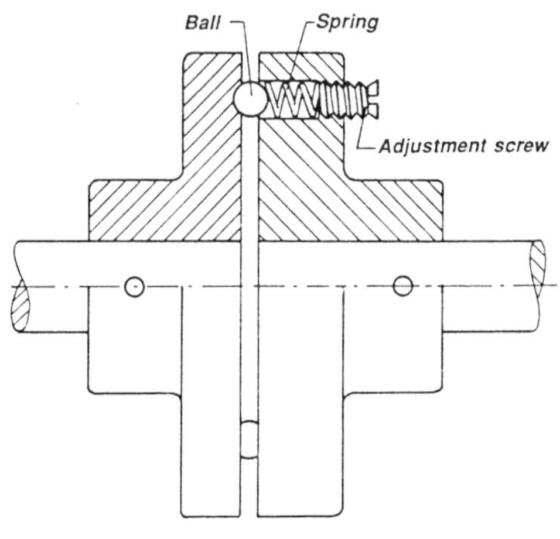

Figure 16.4 A detenting action is used here to couple the two clutch halves together. If overload occurs, the ball or balls are forced out of engagement and the two halves rotate relative to each other. If overload occurs frequently, wear is excessive.

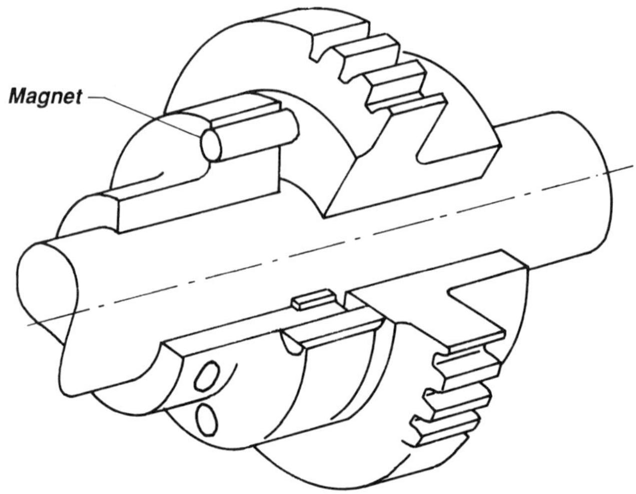

Figure 16.5 With today's powerful magnets, their use as overload protection is improved. The two clutch halves are kept together by the magnets.

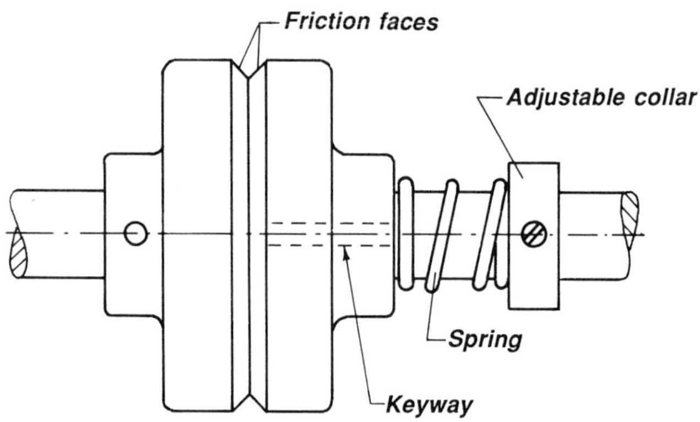

Figure 16.6 Torque can be transmitted by friction between the two clutch halves. The spring force and the friction characteristics determine the torque that can be transmitted before slipping occurs.

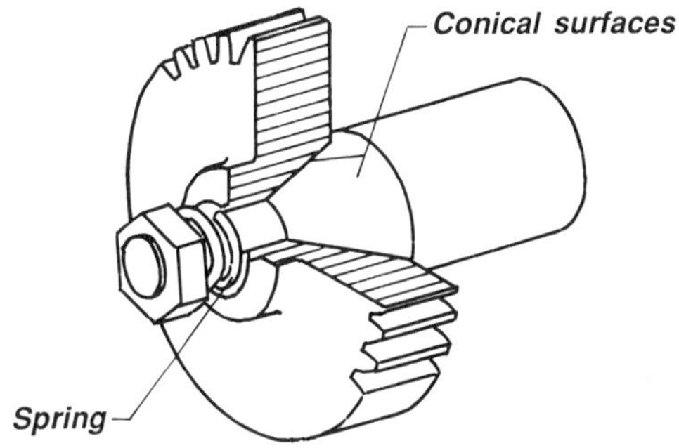

Figure 16.7 Torque transmitted between the two parts can be increased by using coned friction surfaces.

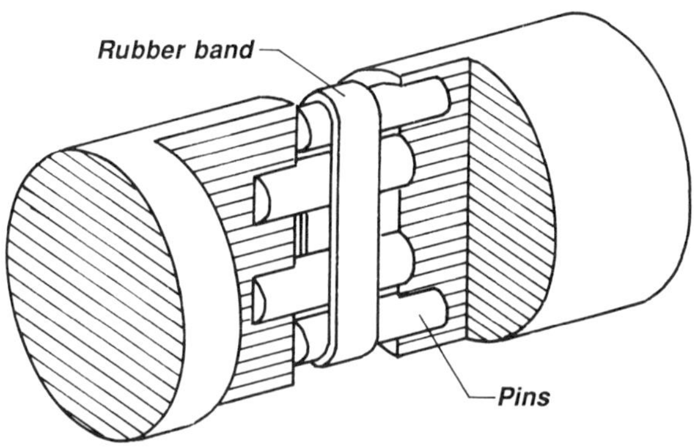

Figure 16.8 Rubber band wrapped around pins provides a cheap overload protection but can be used only for very small torques.

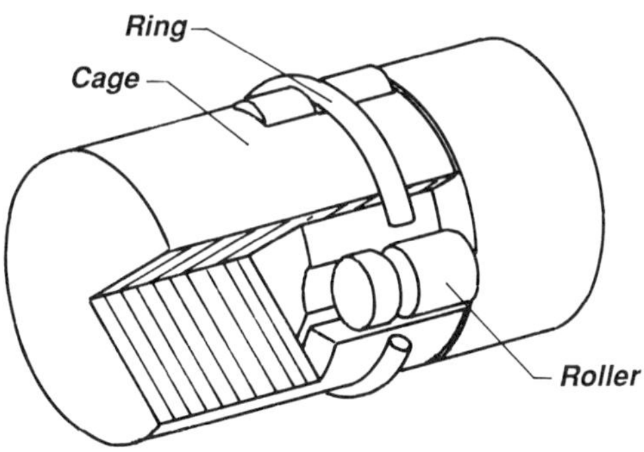

Figure 16.9 Ring presses rollers against grooves in the other clutch half. High torque can be transmitted.

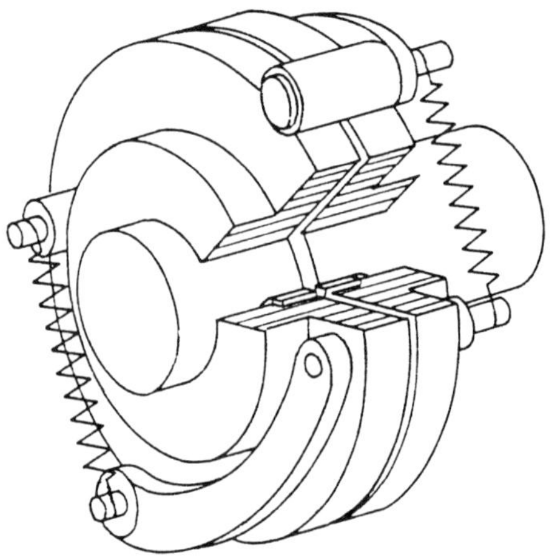

Figure 16.10 Springs keep rollers in slots cut on the circumference of the clutch halves. The rollers are mounted on arms that can pivot.

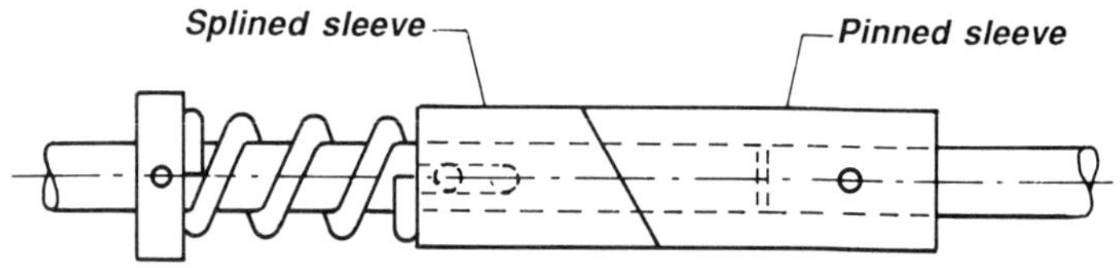

Figure 16.11 Splined sleeve is pressed against a pinned sleeve. Both sleeves are angle-cut. The spring force determines the torque that can be transmitted from the splined to the pinned shaft.

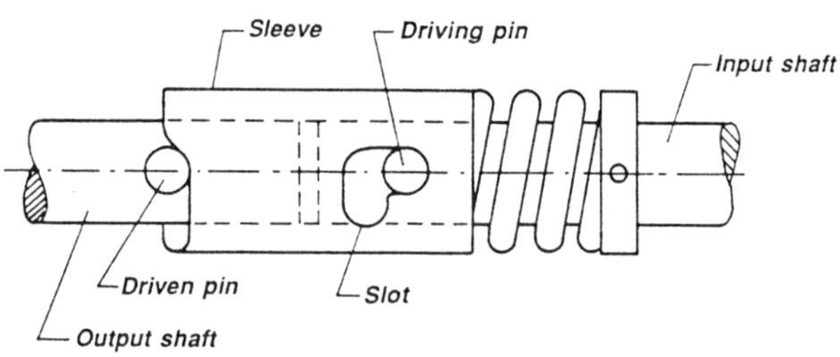

Figure 16.12 Sleeve with cammed surface C connects input and output shafts. Driven pin pushes sleeve to the right against spring. When overload occurs, driving pin drops into slot to keep shaft disengaged. Resetting is done by turning shaft backwards.

OVERRUNNING CLUTCHES

Overrunning clutches allow freewheeling, indexing, and backstopping, which are applicable to many designs.

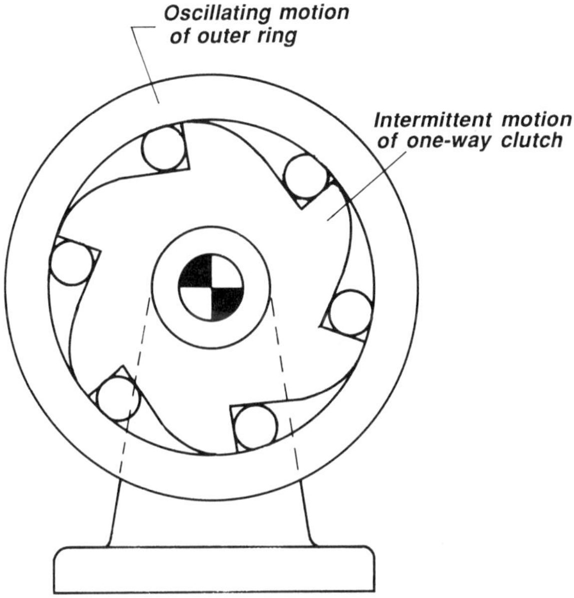

Figure 16.13 Motion of the outer race wedges rollers against the inclined surfaces of the ratchet wheel. Oscillating motion of outer ring results in intermittent motion of inner ring, but inner ring always turns CCW.

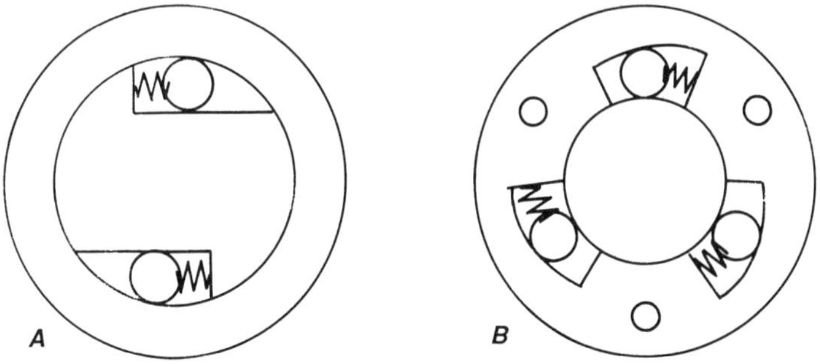

Figure 16.14 To improve wedging action, balls are pressed against the wedges by spring action. Wedges can be on outer or inner ring.

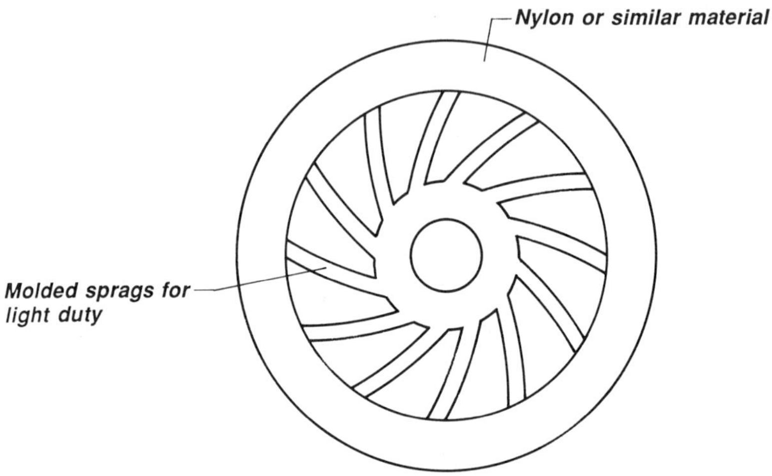

Figure 16.15 Molded sprags of nylon provides springy wedging action.

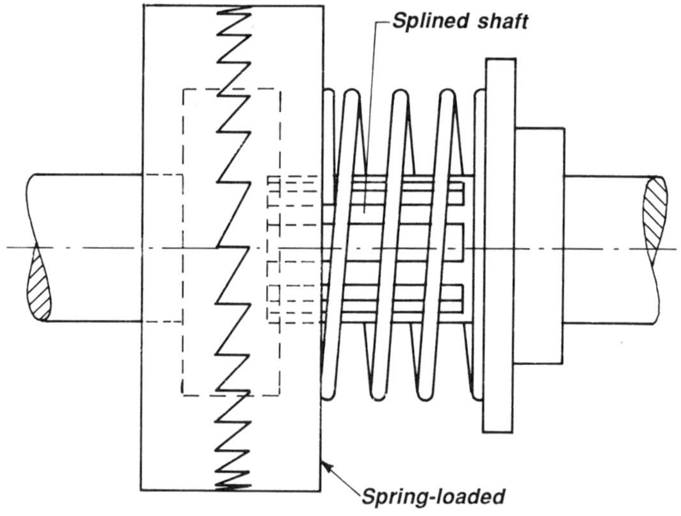

Figure 16.16 The teeth are shaped so that motion can be transferred from one shaft to the other in only one direction.

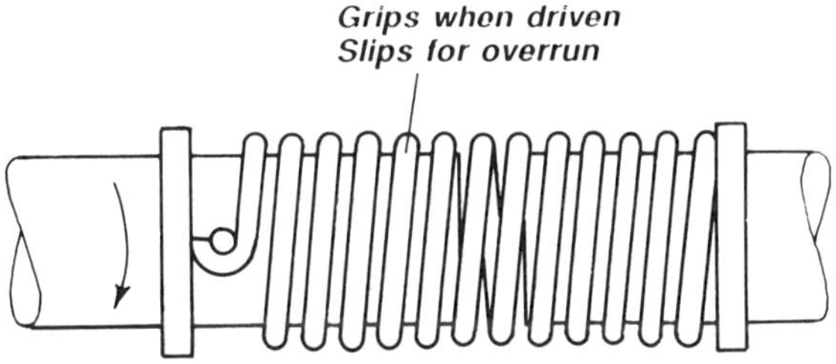

Figure 16.17 Friction between spring and shaft provides overrunning clutch.

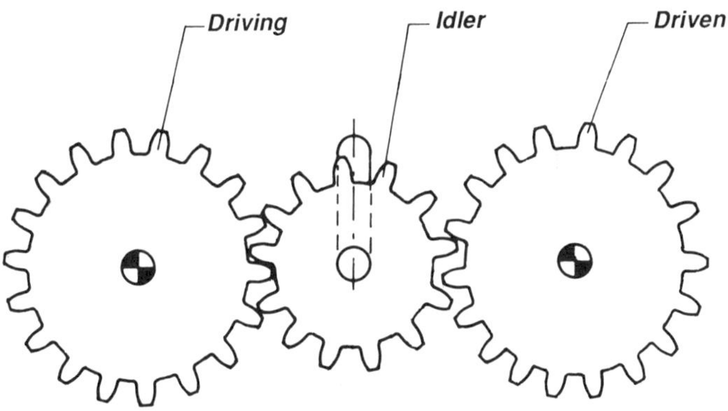

Figure 16.18 An idler between driving and driven gear provides one-way clutch. If drive rotates CCW, it forces the idler gear out of mesh.

Applying Overrunning Clutches

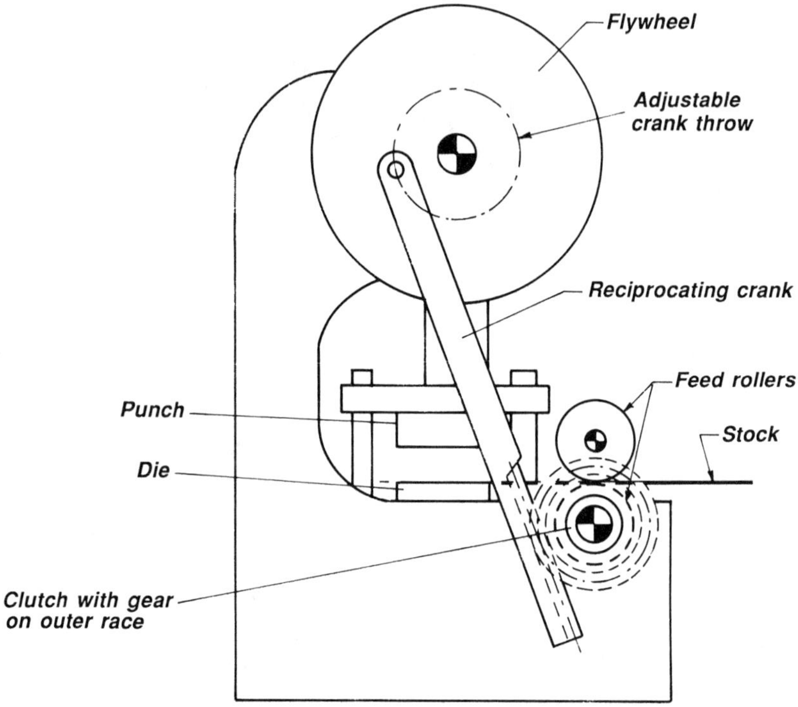

Figure 16.19 Punch press feed is arranged so that strip is not moving on downward stroke because clutch freewheels. Feed occurs by upward motion when clutch transmits torque. Crank length can be adjusted.

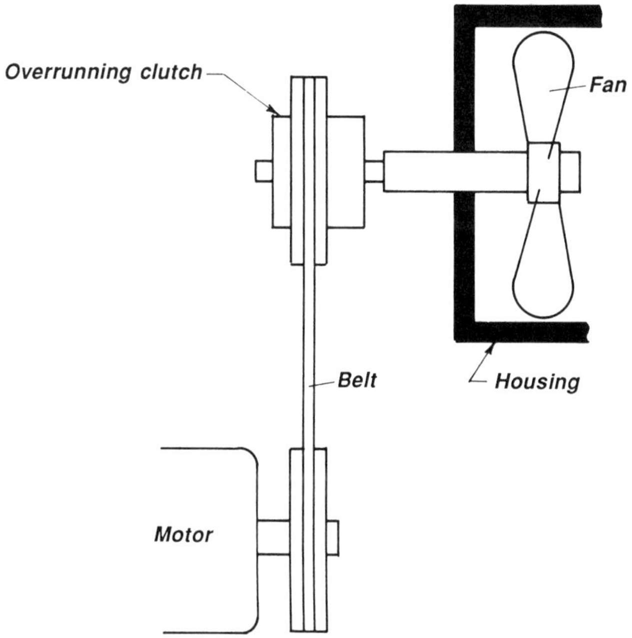

Figure 16.20 Fan freewheels when power is shut off because otherwise momentum of the fan might cause breakage of the belt.

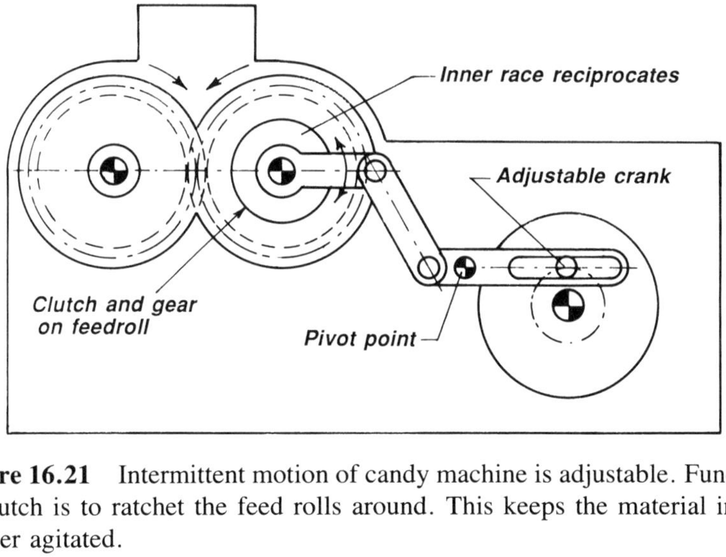

Figure 16.21 Intermittent motion of candy machine is adjustable. Function of clutch is to ratchet the feed rolls around. This keeps the material in the hopper agitated.

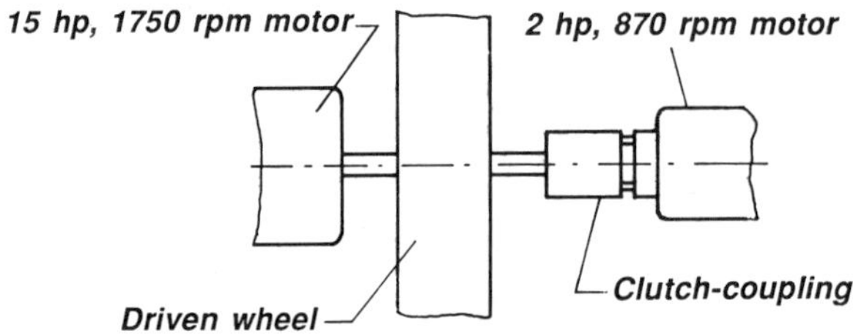

Figure 16.22 Two-speed drive for grinding wheel can be simple, in-line design if overrunning clutch couples two motors. Outer race of clutch is driven by gearmotor; inner race is keyed to grinding-wheel shaft. When gearmotor drives, clutch is engaged; when larger motor drives, inner race overruns.

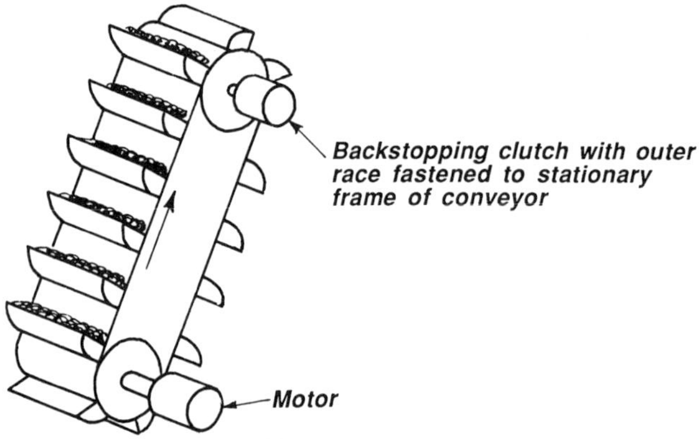

Figure 16.23 Backstopping permits rotation in one direction only. Clutch serves as a counterrotation holding device. An example is a clutch mounted on the headshaft of a conveyor, with the outer race restrained by means of torque-arming to the stationary frame of the conveyor. If for any reason power to the conveyor is interrupted, the backstopping clutch will prevent the buckets from running backwards and dumping the load.

17
Systematic Mechanism Design

In the foregoing chapters many mechanisms have been shown derived from the five basic mechanisms described in Chapter 1. These mechanisms can be incorporated into more complex mechanisms, whereby new and useful mechanisms can be found. To create new mechanisms, structural analysis (also called number synthesis) and type synthesis are applied. The basic building blocks of structural analysis are open and closed kinematic chains.

KINEMATIC CHAINS

The simplest open kinematic chain is two links connected by a turning joint, also called a binary chain. Three links connected in series with two turning joints are called a ternary chain. The degrees of freedom of a chain designate the number of independent input motions that are necessary to determine the motion of the chain, considering one link as the frame.

Kinematic Chains with One Degree of Freedom and a Maximum of Ten Links

Kinematic chains constitute the basis for developing all kinds of mechanisms. M. Gruebler in his classical work *Getriebelehre* (*Kinematics*) found

Designation of Links and Chains

singular link

binary link

ternary link

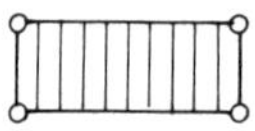
quarternary link

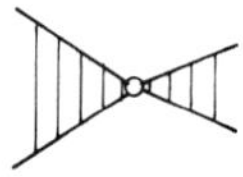
E-chain

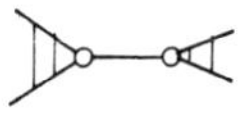
Z-chain

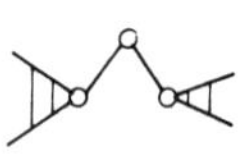
D-chain

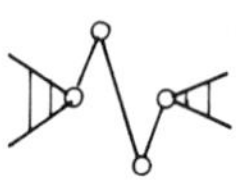
V-chain

a relationship between the number of links (n), the number of joints (g), and the degrees of freedom (f) of a kinematic chain:

$$f = 3(n - 1) - 2g \quad \text{(Gruebler)} \tag{17.1}$$

The equation can easily be derived. Each of the n links has three degrees of freedom when moving in a plane, for a total of 3n degrees of freedom. Each turning joint reduces the degree of freedom of a link by 2, for a total of 2g. If one link is fixed, the degree of freedom of the kinematic chain is reduced by 3, whence

$$f = 3n - 2g - 3 = 3(n - 1) - 2g$$

Table 17.1 shows the various combinations of links and turning joints up to 10 links for chains with one degree of freedom. The number of solutions is indicated in the right column. Gruebler found 12 solutions in class III(1), but R. Franke showed that the correct number is 16. Table 17.2 shows kinematic chains with f = 1, according to Table 17.1. L. S. Woo gave the number of solutions in class IV(1) as 230. This number is correct but the solutions are not and are therefore listed here. Of particular interest is the open kinematic chain in class 0(1). This chain is usually designated a turning joint in the kinematic literature; it is, however, also an open kinematic chain and has found a very interesting application as a high-speed mechanism (see Fig. 1.3.).

Table 17.1 Link Combinations with f = 1

Class		g	n	n_1	n_2	n_3	n_4	n_5	n_6	n_7	n_8	No. of Solutions	
O		1	2	2								1	1
I		4	4		4							1	1
II	a	7	6		4	2						2	2
	b				5	0	1					0	
	a				4	4						9	
	b				5	2	1					5	
III	c	10	8		6	0	2					2	16
	d				6	1	0	1				0	
	e				7	0	0	0	1			0	
	a				4	6						50	
	b				5	4	1					95	
	c				6	3	0	1				15	
	d				6	2	2					57	
IV	e	13	10		7	1	1	1				8	230
	f				7	0	3					3	
	g				7	2	0	0	1			0	
	h				8	0	0	2				2	
	i				8	0	1	0	1			0	
	j				8	1	0	0	0	1		0	
	k				9	0	0	0	0	0	1	0	

Table 17.2 Kinematic Chains with f = 1 According to Table 17.1

Class	Kinematic Chains			
0(1)				
I(1)	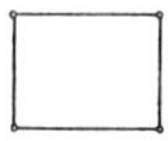			
IIa(1)	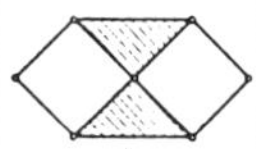 1	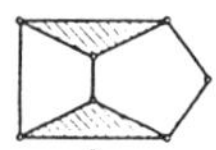 2		
IIIa(1)	 1	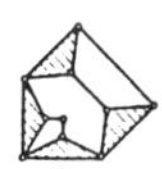 2	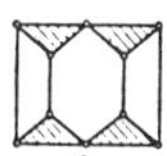 3	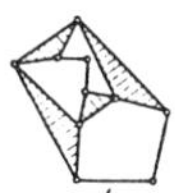 4
	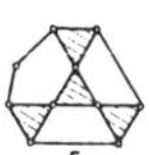 5	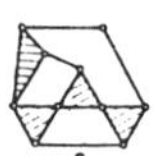 6	 7	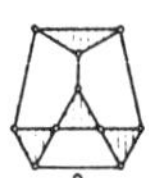 8
	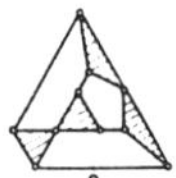 9			
IIIb(1)	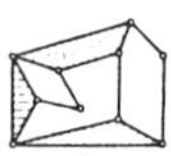 1	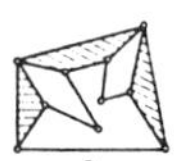 2	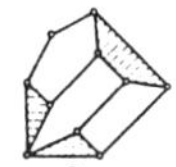 3	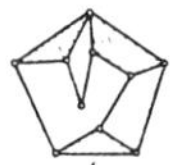 4
	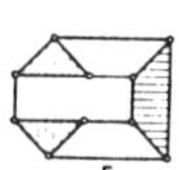 5			
IIIc(1)	 1	 2		

Table 17.2 (Continued)

Class	Kinematic Chains			
IVa(1)	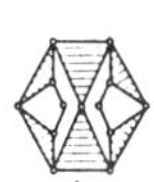 1	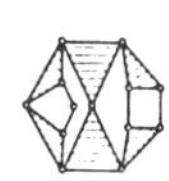 2	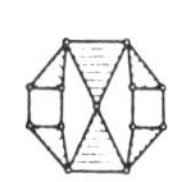 3	 4
	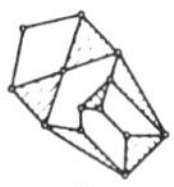 5	 6	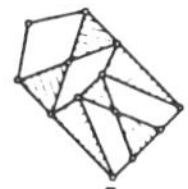 7	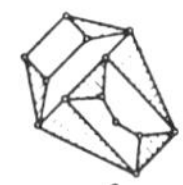 8
	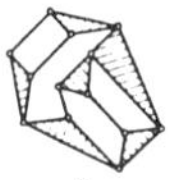 9	 10	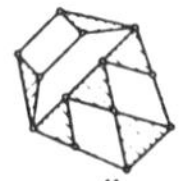 11	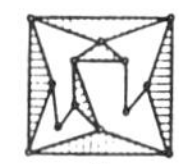 12
	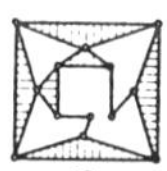 13	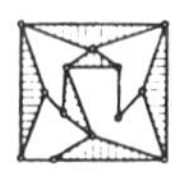 14	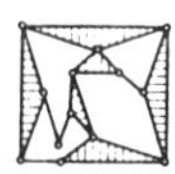 15	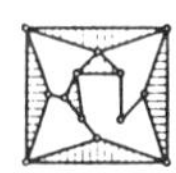 16
	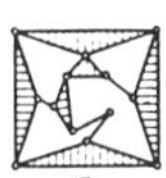 17	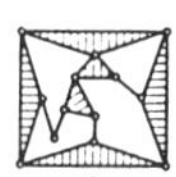 18	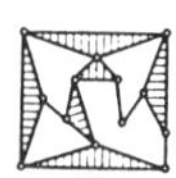 19	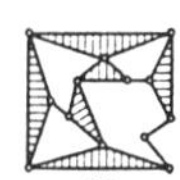 20
	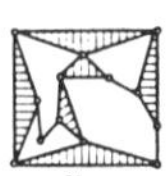 21	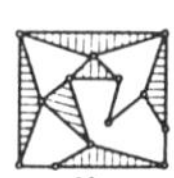 22	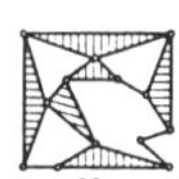 23	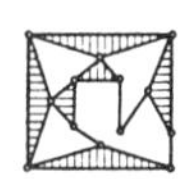 24
	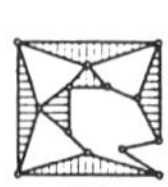 25	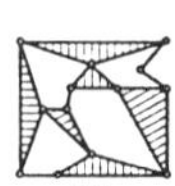 26	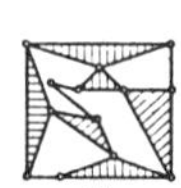 27	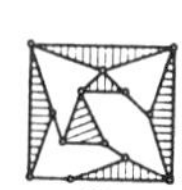 28
	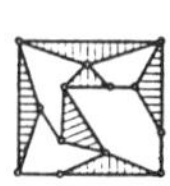 29	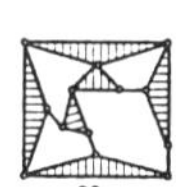 30	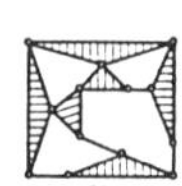 31	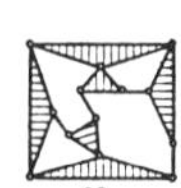 32
	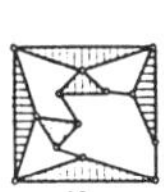 33	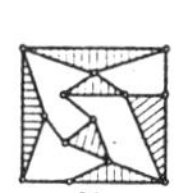 34	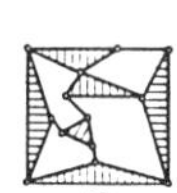 35	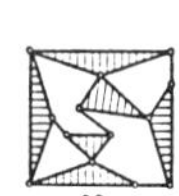 36

Table 17.2 (Continued)

Class	Kinematic Chains			
IVa (1) (cont.)	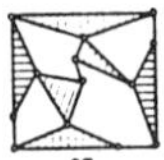 37	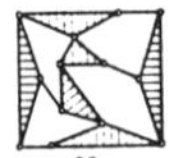 38	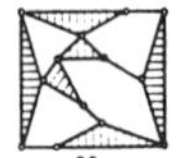 39	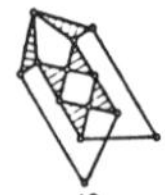 40
	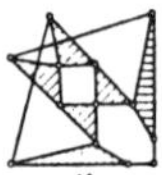 41	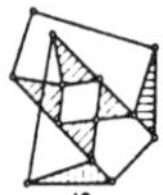 42	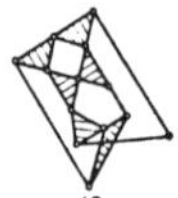 43	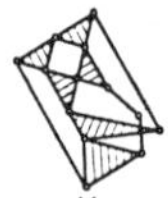 44
	 45	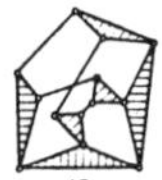 46	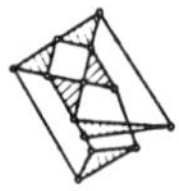 47	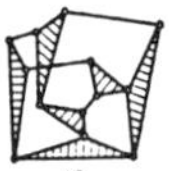 48
	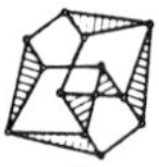 49	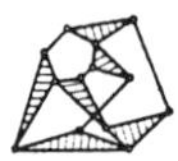 50		
IVb(1)	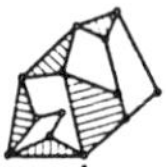 1	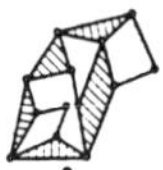 2	 3	 4
	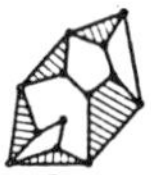 5	 6	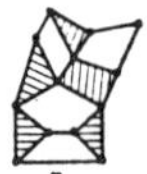 7	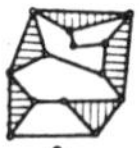 8
	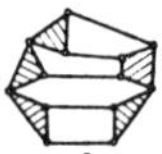 9	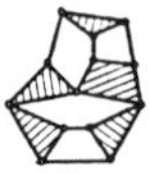 10	 11	 12
	 13	 14	 15	 16
	 17	 18	 19	 20

Table 17.2 (Continued)

Class	Kinematic Chains			
	 21	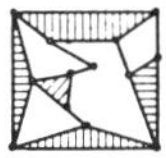 22	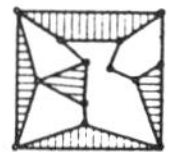 23	 24
	 25	 26	 27	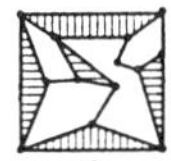 28
	 29	 30	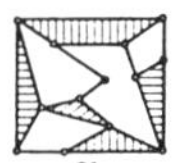 31	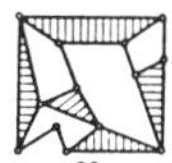 32
	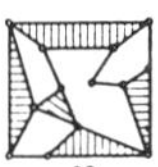 33	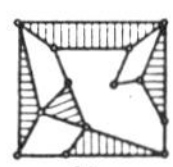 34	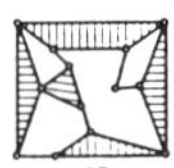 35	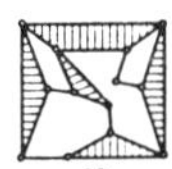 36
	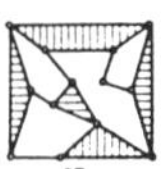 37	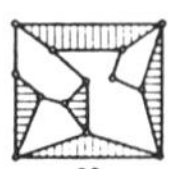 38	 39	 40
	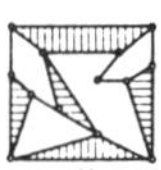 41	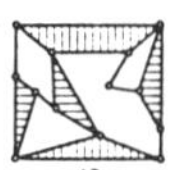 42	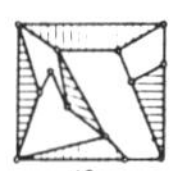 43	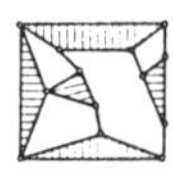 44
	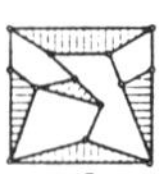 45	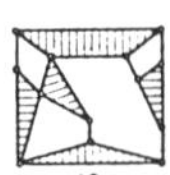 46	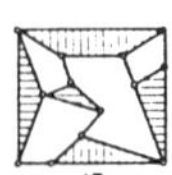 47	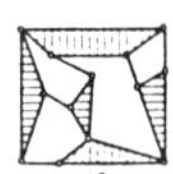 48
	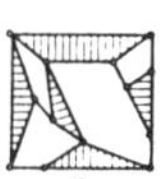 49	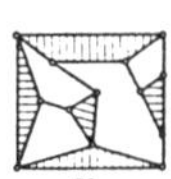 50	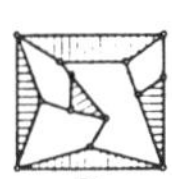 51	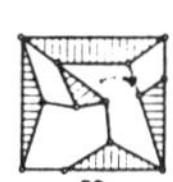 52
	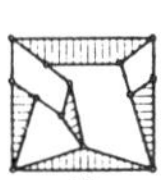 53	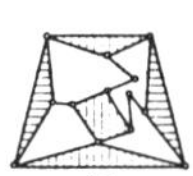 54	 55	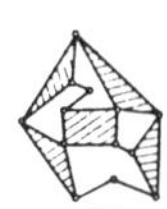 56

Table 17.2 (Continued)

Class	Kinematic Chains			
IVb (1) (cont.)	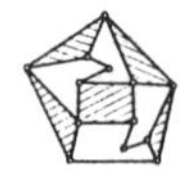 57	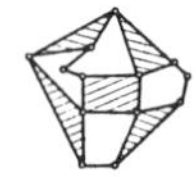 58	 59	 60
	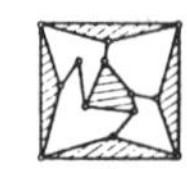 61	 62	 63	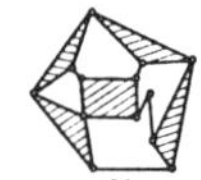 64
	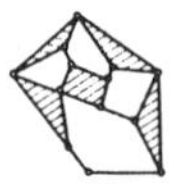 65	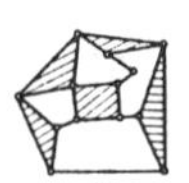 66	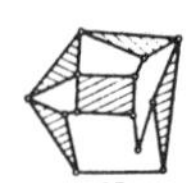 67	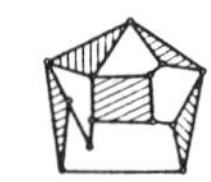 68
	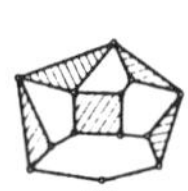 69	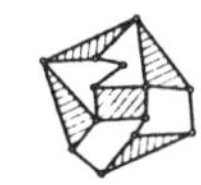 70	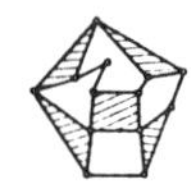 71	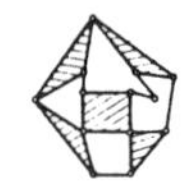 72
	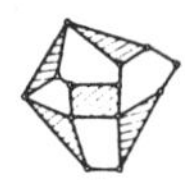 73	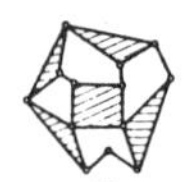 74	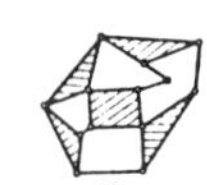 75	 76
	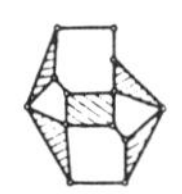 77	 78	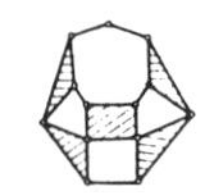 79	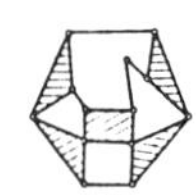 80
	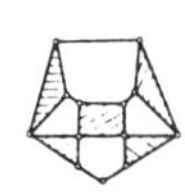 81	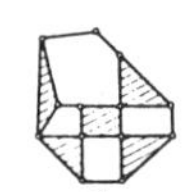 82	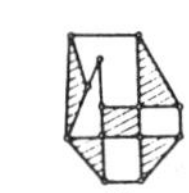 83	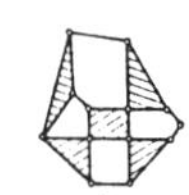 84
	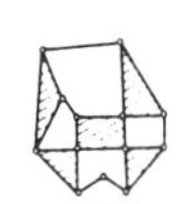 85	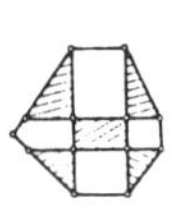 86	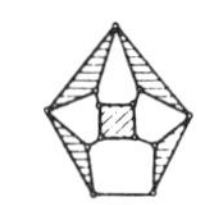 87	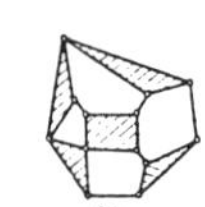 88
	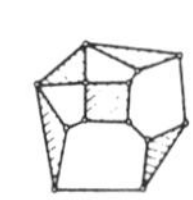 89	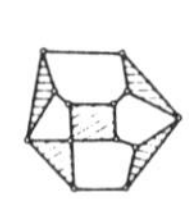 90	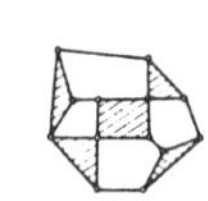 91	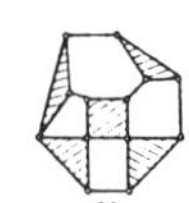 92

Table 17.2 (Continued)

Class	Kinematic Chains
IVb (1) (cont.)	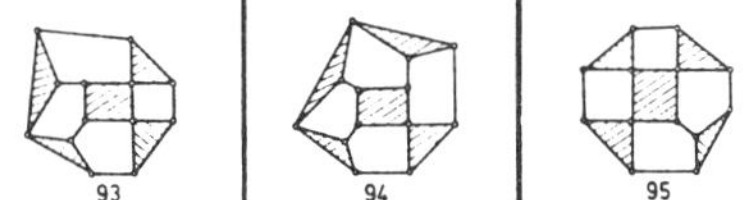 93 94 95
IVc (1)	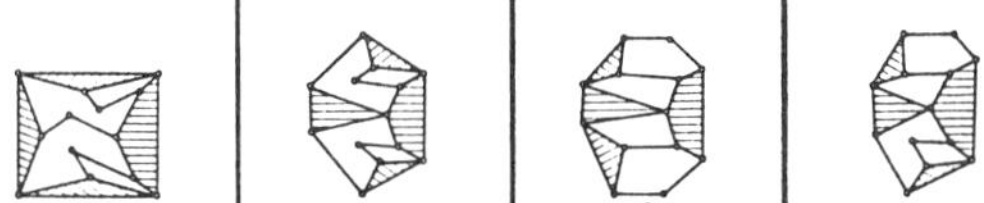 1 2 3 4
	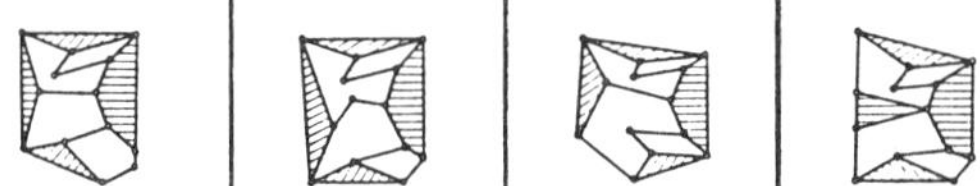 5 6 7 8
	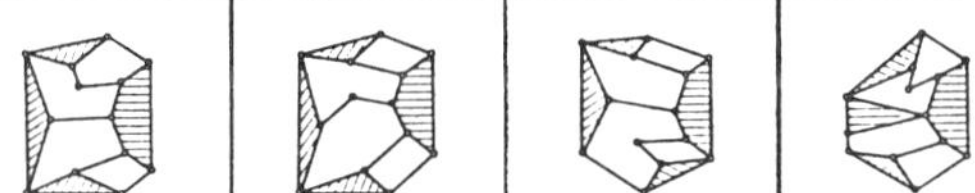 9 10 11 12
	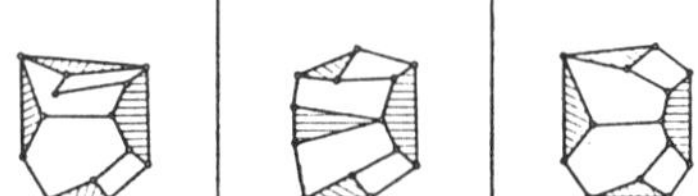 13 14 15
IVd (1)	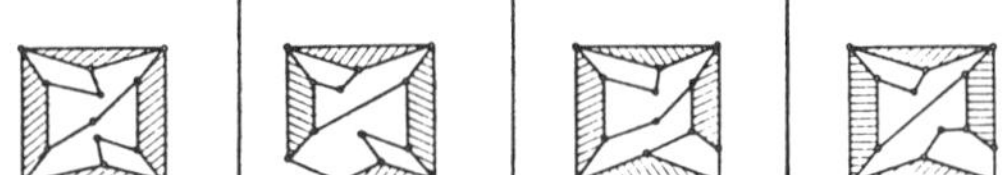 1 2 3 4
	 5 6 7 8
	 9 10 11 12
	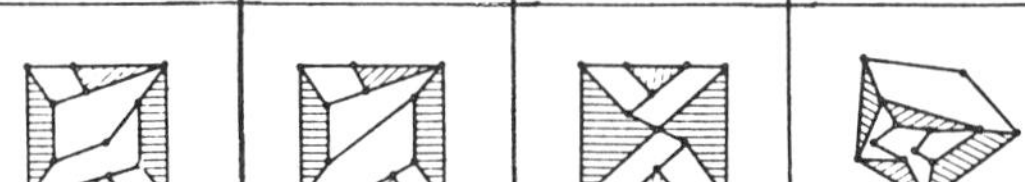 13 14 15 16

Table 17.2 (Continued)

Class	Kinematic Chains			
IVd (1) (cont.)	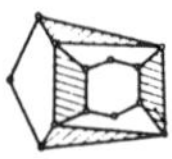 17	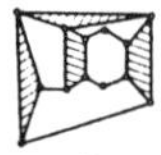 18	 19	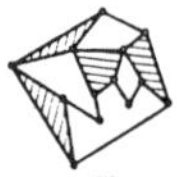 20
	 21	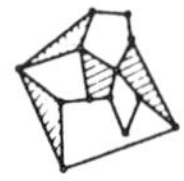 22	 23	 24
	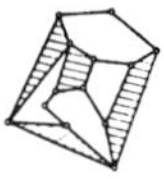 25	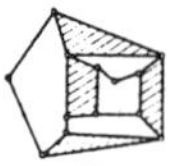 26	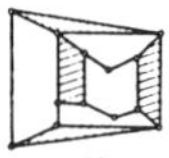 27	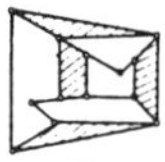 28
	 29	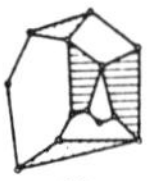 30	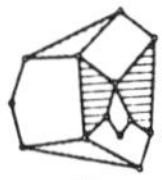 31	 32
	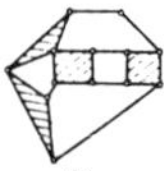 33	 34	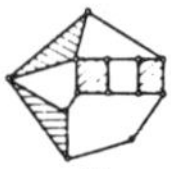 35	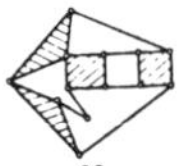 36
	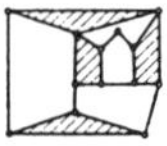 37	 38	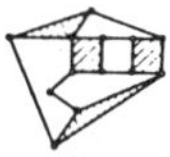 39	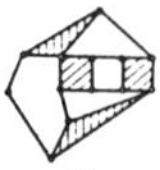 40
	 41	 42	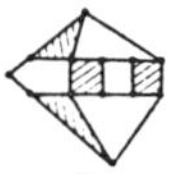 43	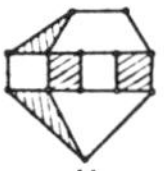 44
	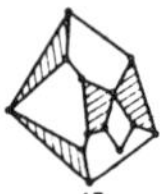 45	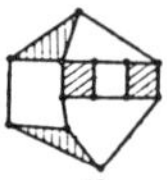 46	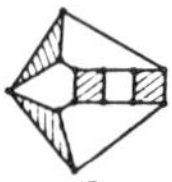 47	 48
	 49	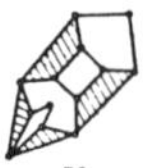 50	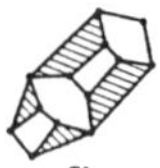 51	 52

Table 17.2 (Continued)

Class	Kinematic Chains			
IVd (1) (cont.)	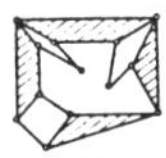 53	 54	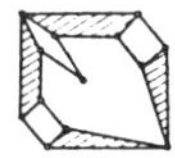 55	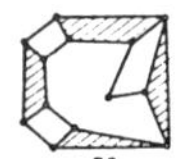 56
	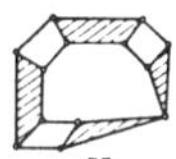 57			
IVe(1)	 1	 2	 3	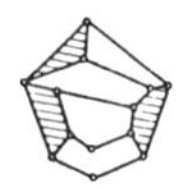 4
	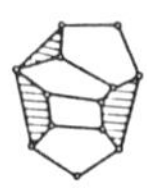 5	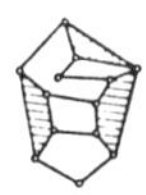 6	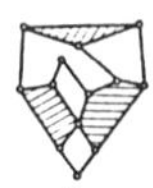 7	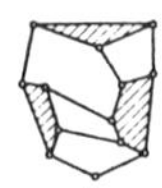 8
IVf(1)	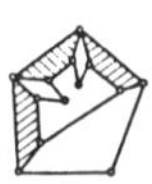 1	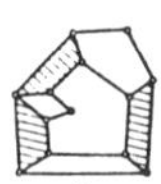 2	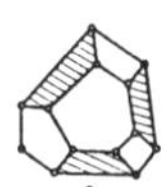 3	
IVh(1)	 1	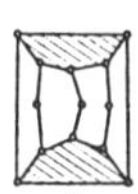 2		

Open and Closed Kinematic Chains with More than One Degree of Freedom

A simple method to find the possible combinations of open and closed kinematic chains is shown in the following. The practical applications of open kinematic chains have already been demonstrated in Chapter 1.

The following two equations expand Gruebler's equation to include type of links and type of joints

$$n = n_1 + n_2 + n_3 + \ldots$$

$$2g = 1n_1 + 2n_2 + 3n_3 + \ldots$$

Possible Combinations of Links and Joints

Tables 17.1, 17.2, 17.3, 17.4, 17.5, and 17.6 show the possible number of joints and links for kinematic chains with one to four degrees of freedom and up to twelve links. To find the possible combinations, a method developed by R. Franke is employed, but his method has been extended further.

A Simple Method of Structural Analysis

A kinematic chain is composed of links and turning joints. The question now is how can all possible solutions be developed from a given assortment of links and joints. To this purpose, each ternary, quarternary, etc., link is represented by a circle surrounding a "3," "4," etc., respectively (Fig. 17.1). As an example, class IV d(1), (Table 17.1) is chosen with $g = 13$, $n = 10$, $n_1 = 0$, $n_2 = 6$, $n_3 = 2$, $n_4 = 2$, and $f = 1$. Consider the two ternary and two quarternary links. They are represented by a circle around a "3" and a "4," respectively. The problem now is to connect the four circles in such a way that three connections radiate from each ternary link and four connections radiate from each quarternary link. The four circles are divided into two groups, one group with $n_3 = 1$ and n_4 2, and the other group with $n_3 = 1$.

Take the first group and try to connect the circles in as many ways as possible. It is possible to connect only two of the circles in two different ways, namely, 34 and 33. The combination 34 can be extended in three ways with one more connection (indicated by a connecting line). The same is valid for the combination 44. The process is continued. As soon as a combination that already exists is found, it is not pursued any further.

If it is possible, through rotation or mirroring of some or all of the circles, to obtain a solution that is identical to the original, the solution is not considered any further. The two solutions (Fig. 17.2) do not change their appearance if mirrored around a horizontal or vertical axis. The process is continued until only three connections are left. Now the second group (in this case a circle with a "3") is connected to the first group in as many ways as possible. Six combinations are possible, namely, A to F. The question is, which of the six combinations can be used? Combination F cannot be used because that would result in two kinematic chains. Combination A cannot be used because the result would be a chain with $f \geq 3$. The rest of the combinations, B, C, D, and E, can all be used to obtain useable solutions.

Permutations of E-, Z-, D-, and V-Chains

The last step in determining possible combinations is to find out what possibilities exist to replace the circle connections with E-, Z-, D-, and V-chains. As an example of how to develop the possible permutations, the

Table 17.3 Link Combinations with f = 2

Class		g	n	n_1	n_2	n_3	n_4	n_5	n_6	n_7	n_8	No. of Solutions	
0		2	3	2	1							1	1
I	a	5	5	1	3	1						1	2
	b				5							1	
	a			1	3	3						3	
II	b	8	7	1	4	1	1					2	9
	c				5	2						3	
	d				6	0	1					1	
	a			1	3	5						17	
	b			1	4	3	1					36	
	c			1	5	1	2					11	
	d			1	5	2	0	1				5	
III	e	11	9	1	6	0	1	1				1	110
	f				5	4						19	
	g				6	2	1					16	
	h				7	0	2					3	
	i				7	1	0	1				2	
	j				8	0	0	0	1			0	
	a			1	3	7						144	
	b			1	4	5	1					575	
	c			1	5	3	2					548	
	d			1	5	4	0	1				165	
	e			1	6	1	3					105	
	f			1	6	2	1	1				120	
	g			1	6	3	0	0	1			15	
	h			1	7	0	2	1				15	
	i			1	7	1	0	2				11	
IV	j	14	11	1	7	1	1	0	1			8	2537
	k			1	8	0	0	1	1			2	
	l				5	6						155	
	m				6	4	1					358	
	n				7	3	0	1				74	
	o				7	2	2					193	
	p				8	1	1	1				24	
	q				8	0	3					13	
	r				8	2	0	0	1			8	
	s				9	0	0	2				2	
	t				9	0	1	0	1			2	
	u				9	1	0	0	0	1		0	
	w				10	0	0	0	0	0	1	0	

Table 17.4 Kinematic Chains with f = 2

Class				
0(2)	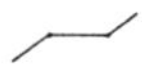 1			
Ia(2)	 1			
Ib(2)	 1			
IIa(2)	 1	 2	 3	
IIb(2)	 1	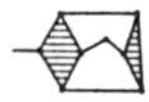 2		
IIc(2)	 1	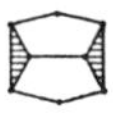 2	 3	
II d(2)	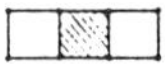 1			
IIIa(2)	 1	 2	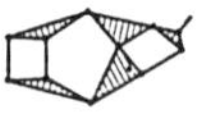 3	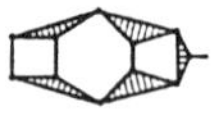 4
	 5	 6	 7	 8

Table 17.4 (Continued)

Class	
IIIa(2) (cont.)	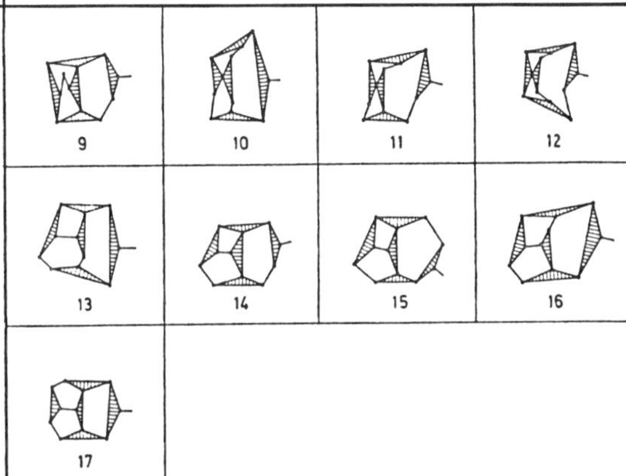 9, 10, 11, 12, 13, 14, 15, 16, 17
IIIb(2)	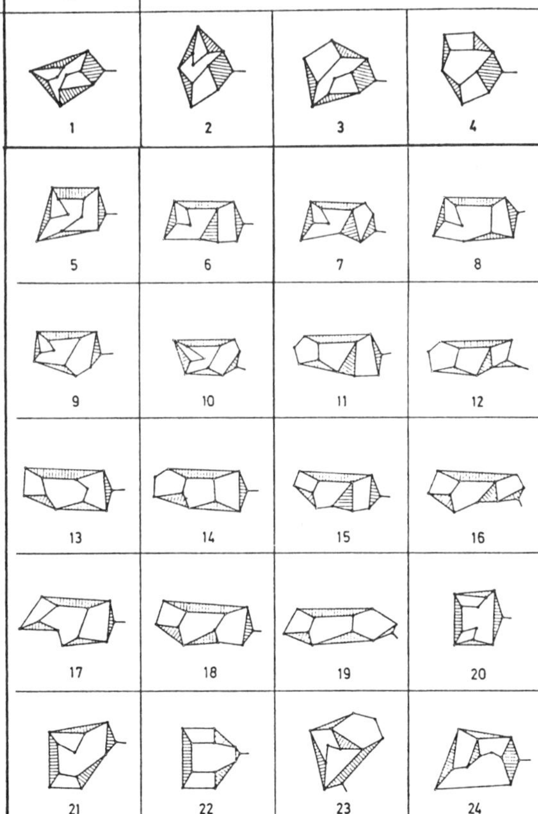 1, 2, 3, 4, 5, 6, 7, 8, 9, 10, 11, 12, 13, 14, 15, 16, 17, 18, 19, 20, 21, 22, 23, 24

Table 17.4 (Continued)

Class				
IIIa(2) (cont.)	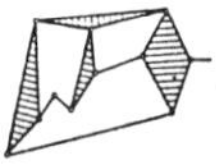 25	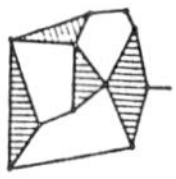 26	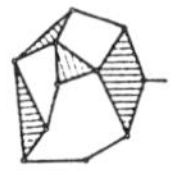 27	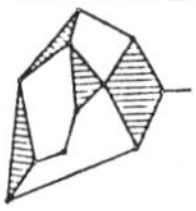 28
	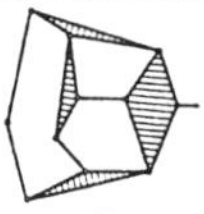 29	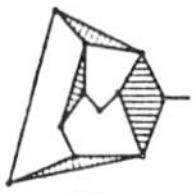 30	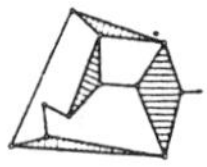 31	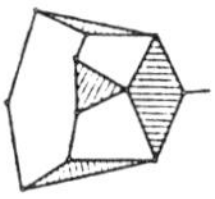 32
	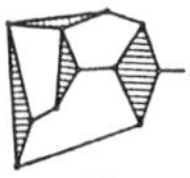 33	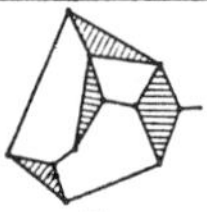 34	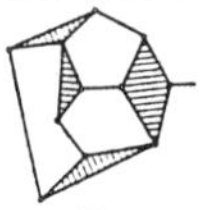 35	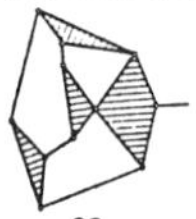 36
IIIc(2)	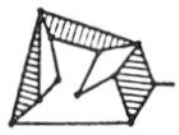 1	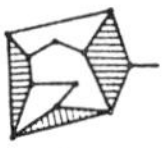 2	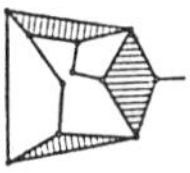 3	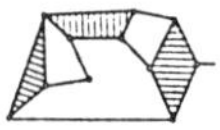 4
	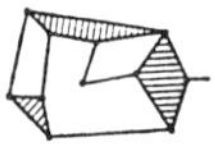 5	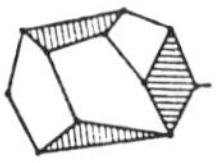 6	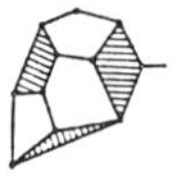 7	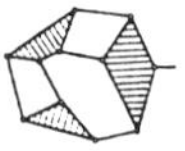 8
	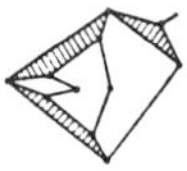 9	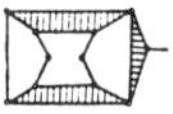 10	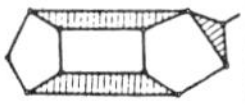 11	
IIId(2)	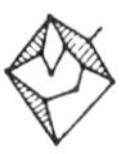 1	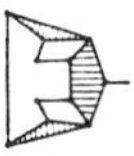 2	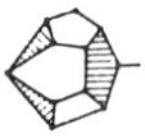 3	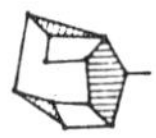 4
	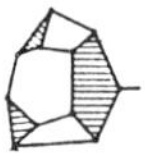 5			
IIIe(2)	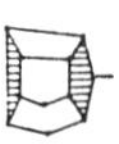 1			

Table 17.4 (Continued)

Class	
IIIf(2)	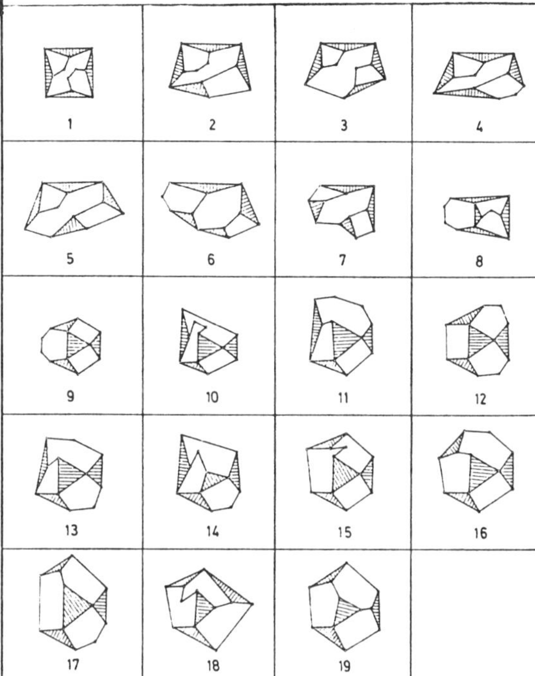
IIIg(2)	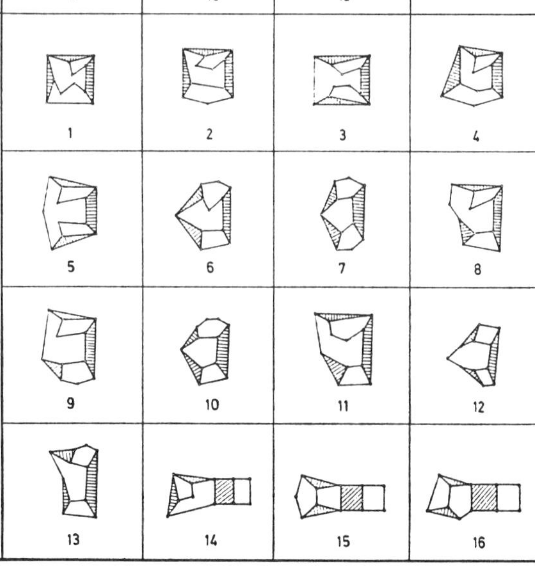

Table 17.4 (Continued)

Class				
IIIh(2)	 1	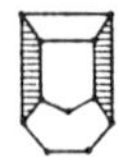 2	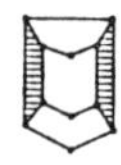 3	
IIIi(2)	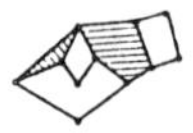 1	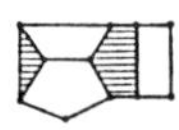 2		

arrangement in Fig. 17.3 is chosen. This combination belongs to class III b (Table 17.2) with g = 11, n = 9, and f = 2. It has 11 joints and 7 connections, and the possible combinations of connections are:

E	Z	D	V
5	1	0	1
5	0	2	
4	2	1	
3	4		

Table 17.5 Link Combinations with f = 3

Class	g	n	n_1	n_2	n_3	n_4	n_5	n_6	n_7	n_8
0	3	4	2	2						
			2	2	2					
			2	3	0	1				
I	6	6								
			1	4	1					
					6					
			2	2	4					
			2	3	2	1				
			2	4	0	2				
II	9	8	2	4	1	0	1			
			2	5	0	0	0	1		
			1	4	3					
			1	5	1	1				
				6	2					
				7	0	1				
			2	2	6					
			2	3	4	1				
			2	4	2	2				
			2	4	3	0	1			
			2	5	0	3				
			2	5	1	1	1			

Table 17.5 (Continued)

Class	g	n	n_1	n_2	n_3	n_4	n_5	n_6	n_7	n_9
III	12	10	2	5	2	0	0	1		
			2	6	0	1	0	1		
			2	6	1	0	0	0	1	
			2	7	0	0	0	0	0	1
			2	6	0	0	2			
			1	7	0	1	1			
			1	7	1	0	0	1		
			1	6	2	0	1			
			1	6	1	2				
			1	5	3	1				
			1	4	5					
				6	4					
				7	2	1				
				8	0	2				
				8	1	0	1			
				9	0	0	0	1		
			2	2	8					
			2	3	6	1				
			2	4	4	2				
			2	4	5	0	1			
			2	5	2	3				
			2	5	3	1	1			
			2	5	4	0	0	1		
			2	6	0	4				
			2	6	1	2	1			
			2	6	2	0	2			
			2	6	2	1	0	1		
			2	6	3	0	0	0	1	
			2	7	0	1	2			
			2	7	0	2	0	1		
			2	7	1	0	1	1		
IV	15	12	2	7	1	1	0	0	1	
			2	8	0	0	0	2		
			2	8	0	0	1	0	1	
			1	4	7					
			1	5	5	1				
			1	6	3	2				
			1	6	4	0	1			
			1	7	1	3				
			1	7	2	1	1			
			1	7	3	0	0	1		
			1	8	0	2	1			
			1	8	1	0	2			
			1	8	1	1	0	1		
			1	9	0	0	1	1		
				6	6					
				7	4	1				
				8	2	2				
				8	3	0	1			
				9	0	3				
				9	1	1	1			
				9	2	0	0	1		
				10	0	1	0	1		
				10	0	0	2			
				10	1	0	0	0	1	
				11	0	0	0	0	0	1

Table 17.6 Link Combinations with f = 4

Class	g	n	n_1	n_2	n_3	n_4	n_5	n_6	n_7	n_8
0	4	5	2	3						
			(3)	1	1					
			(3)	1	3					
			(3)	2	1	1				
I	7	7	2	3	2					
			2	4	0	1				
			1	5	1					
				7						
			(3)	1	5					
			2	3	4					
			2	4	2	1				
			2	5	0	2				
II	10	9	2	5	1	0	1			
			2	6	0	0	0	1		
			1	5	3					
			1	6	1	1				
			1	7	0	0	1			
				7	2					
				8	0	1				
			(3)	1	7					
			2	3	6					
			2	4	4	1				
			2	5	2	2				
			2	5	3	0	1			
			2	6	0	3				
			2	6	1	1	1			
			2	6	2	0	0	1		
			2	7	1	0	0	0	1	
			2	8	0	0	0	0	0	1
			2	7	0	1	0	1		
III	13	11	2	6	0	0	2			
			1	5	5					
			1	6	3	1				
			1	7	1	2				
			1	7	2	0	1			
			1	8	1	0	0	1		
			1	9	0	0	0	0	1	
			1	8	0	1	1			
				7	4					
				8	2	1				
				9	0	2				
				9	1	0	1			
			2	3	8					
			2	4	6	1				
			2	5	4	2				
			2	5	5	0	1			
			2	6	3	1	1			
			2	6	2	3				
			2	6	4	0	0	1		
			2	7	1	2	1			
			2	7	0	4				
			2	7	2	0	2			
			2	7	2	1	0	1		
			2	7	3	0	0	0	1	
			2	8	0	1	2			
			2	8	0	2	0	1		
			2	8	1	0	1	1		
			2	8	1	1	0	0	1	
			2	9	0	0	0	2		
			2	9	0	0	1	0	1	

Table 17.6 (Continued)

Class	g	n	n_1	n_2	n_3	n_4	n_5	n_6	n_7	n_8
IV	16	13	1	5	7					
			1	6	5	1				
			1	7	3	2				
			1	7	4	0	1			
			1	8	1	3				
			1	8	2	1	1			
			1	8	3	0	0	1		
			1	9	0	2	1			
			1	9	1	0	2			
			1	9	1	1	0	1		
			1	10	0	0	1	1		
				7	6					
				8	4	1				
				9	2	2				
				9	3	0	1			
				10	0	3				
				10	1	1	1			
				10	2	0	0	1		
				11	0	0	2			
				11	0	1	0	1		
				11	1	0	0	0	1	
				12	0	0	0	0	0	1

The connections are now numbered as shown in Fig. 17.3. Because the E-chains are permutated first, multiple connections (those connecting the same circles) are given only one number (4 in Fig. 17.3). Multiple connections cannot have more than one E-chain, otherwise relative motion between the two links is not possible (see the next section, "Rigid Structures of E-, Z-, D-, and V-Chains").

One E-chain can be permutated in five different ways. The following combination can be written:

1 2	1 3	1 4	1 5
	2 3	2 4	2 5
		3 4	3 5
			4 5

Two E-chains, therefore, can be permutated 10 different ways.

Three E-chains can be permutated in 10 different ways:

1 2 3	1 2 4	1 2 5*
	1 3 4	1 3 5
		1 4 5
	2 3 4*	2 3 5
		2 4 5
		3 4 5

but only 8 combinations (those not marked with a "*") can be used.

Class IVd	g	n	n_1	n_2	n_3	n_4	
	13	10		6	2	2	$f = 1$

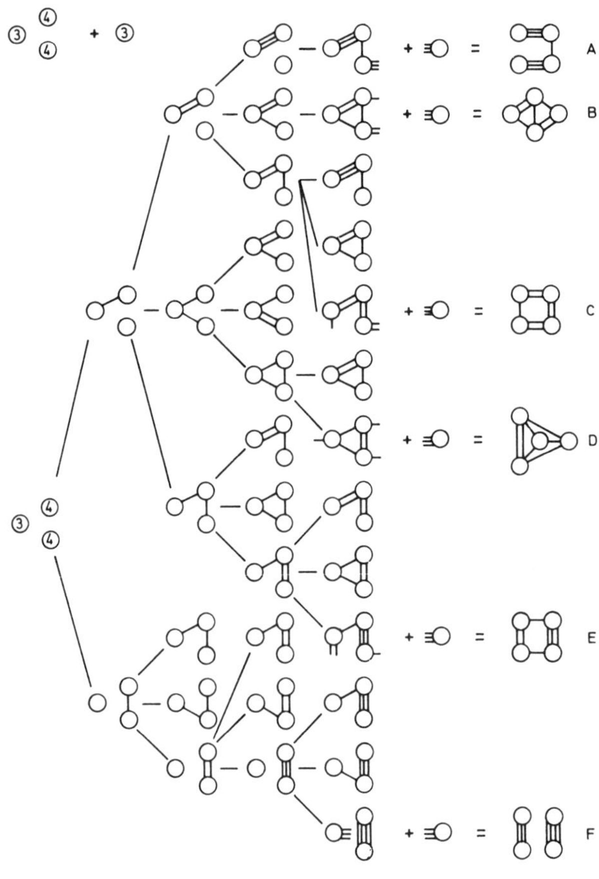

Figure 17.1 Systematic development of skeletons.

Four E-chains can be permutated in 5 different ways:

1 2 3 4*	1 2 3 5*
	1 2 4 5*
	1 3 4 5
	2 3 4 5*

but only one combination is not partly rigid.

Five E-chains yield no solutions at all.

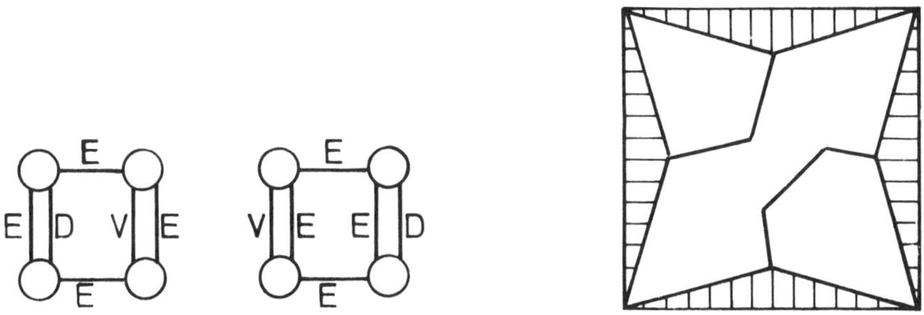

Figure 17.2 Two identical skeletons and the corresponding kinematic chain.

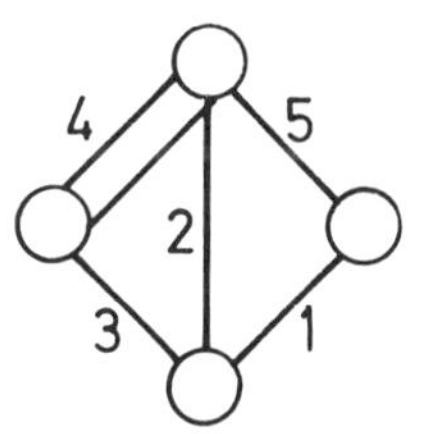

Figure 17.3 Skeleton.

After having found all possible permutations of the E-chains, the Z- and D-chains are permutated. The possible combinations of links and joints for kinematic chains are found here in a systematic way and the method can be programmed on a computer.

Rigid Structures of E-, Z-, D-, and V-Chains

The following combinations of E- and Z-chains (Fig. 17.4) cannot be used because the result is a rigid structure. The combination ZZZ can occur in various ways (Figs. 17.4 and 17.5). The result is a rigid structure. V-chains cannot be used if the corresponding kinematic chain with singular links has f = 2, because the degree of freedom of the kinematic chain would be >2. The only exception is II d(2) (Table 17.2).

Kinematic chains with D- and V-chains are of practical use only if these chains are used as input, output, or frame. It is assumed that it is possible to arrange chains connecting "circle" members to permutate these without creating a new combination (Fig. 17.6).

The number of solutions of open and closed kinematic chains with f = 1 and up to 10 links is 250. The number of solutions of open and closed kinematic chains with f = 2 and up to 11 links is approximately 2660. This

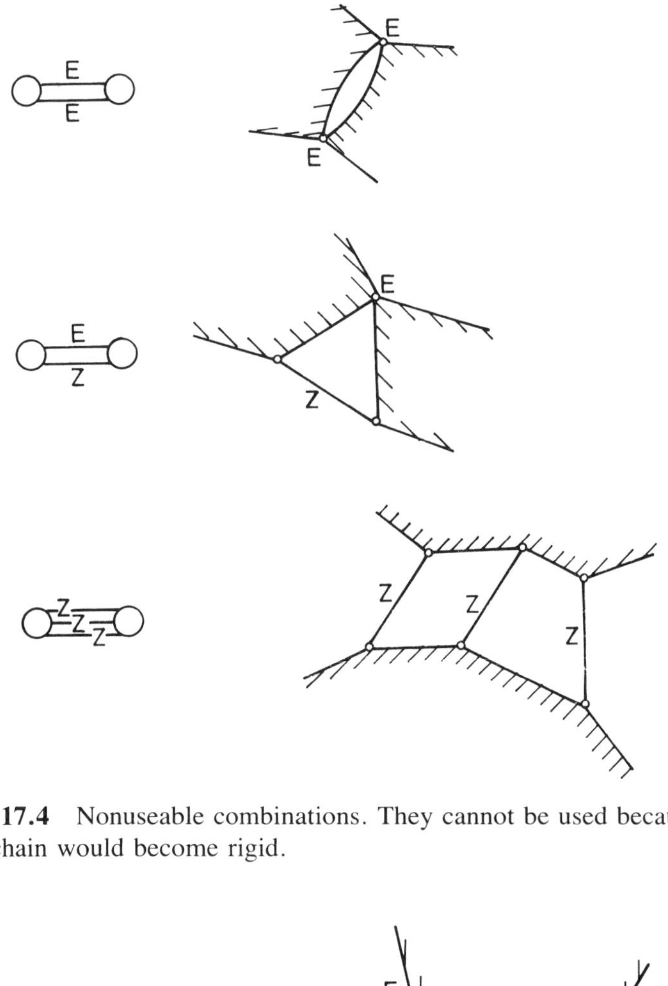

Figure 17.4 Nonuseable combinations. They cannot be used because part of the chain would become rigid.

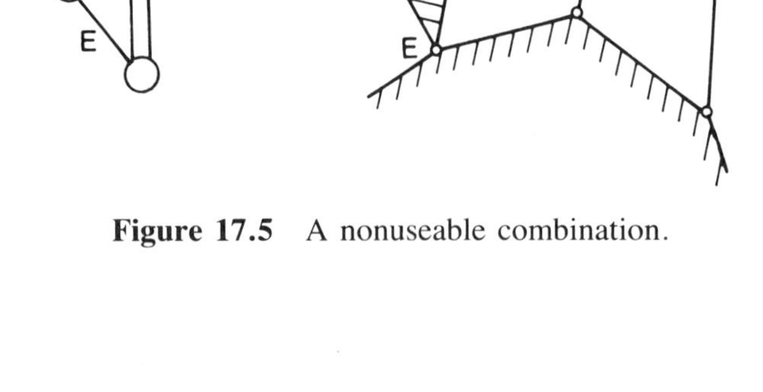

Figure 17.5 A nonuseable combination.

Figure 17.6 Interchanging connections between two links is considered identical solutions.

number was found by me more than two decades ago, but I have never had an opportunity to check the number, so the actual number is probably different. It does indicate that the number of solutions increases disproportionately with the number of links.

The Number of Linkages That Can Be Developed from Kinematic Chains with f = 1 and 6 Links

Recall that in chapter 1 a mechanism was defined in a new way: as an open or closed kinematic chain, where one link is fixed and with given input and output link(s), where the input and output link may be two different links or the same link.

The investigation is started with two kinematic chains, the Watt and the Stephenson chains (Figs. 17.7 and 17.8). Each chain has six links and seven

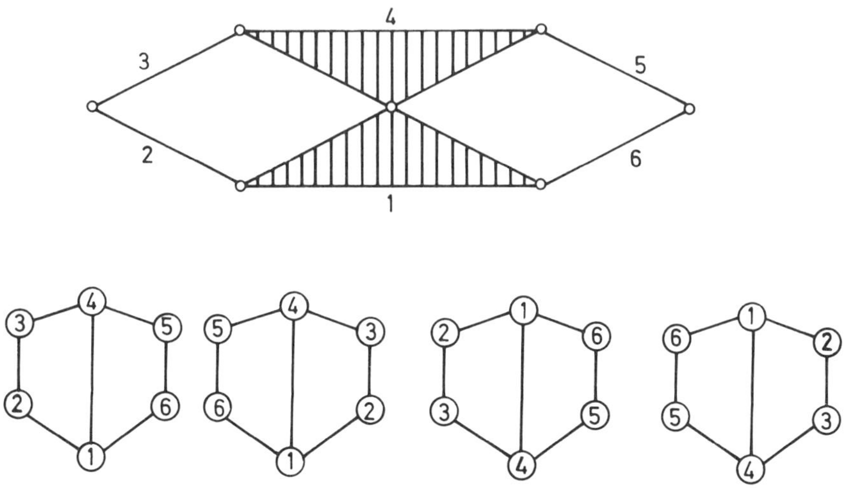

Figure 17.7 Watt's Chain.

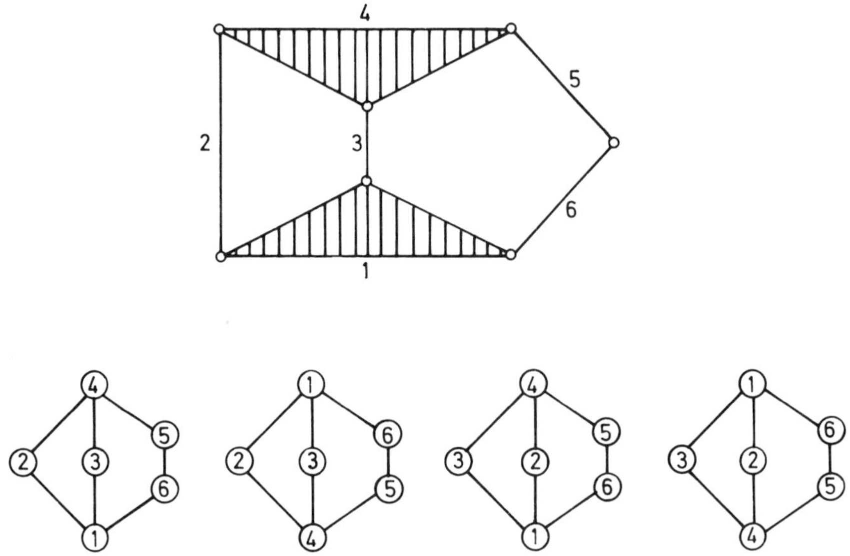

Figure 17.8 Stephenson's chain.

turning joints. It is desired to find how many different linkages can be developed from these chains, provided that one link is input, one link is output, and one link is made a frame link. Beginning with the Watt chain (Fig. 17.7) each link is given a number and represented by a circle as shown in the lower half of the figure. This skeleton is now drawn in as many different positions as possible. Three other configurations are possible. The following combinations are listed:

1	2	3	4*	5*	6*
11	12	13	14	15*	16*
	22	23	24*	25	26
	112	113	114	115*	116*
	122	123	124	125	126
		113	134*	135	1368
			144*	145*	146*
		223	224*	225	226
		233*	234*	235	236*
				255*	256*
					266*

A "*" means that the corresponding number combination is identical to a previous combination and is therefore discarded. For instance, "4" is marked with a "*" because "4" is at the same place as "1" in the first column. It is always the highest number that is discarded.

Useful combinations are to be found in numbers with three digits. It is, however, a requirement that both D-chains, 23 and 56, are necessary for motion transmission. Therefore, only the combinations 125, 126, 135, 226, and 235 can be used. The possible combinations are listed in Table 17.7.

The same process is now applied to the Stephenson chain (Fig. 17.8), and the possible combinations are listed in Table 17.8. The number of solutions for the two kinematic chains is 25 + 34 = 59 (with one output).

Kinematic Chains with Multiple Turning Joints

A multiple joint is a joint that connects more than two links. These joints can be developed by letting the distances between the three separate joints of a ternary link be equal to zero. The same procedure is applied to quar-

Table 17.7 Link Combinations for the Watt Chain

Input	Output	Frame	Solution No.
1	2	5	1
2	1	5	2
1	5	2	3
5	1	2	4
2	5	1	5
5	2	1	6
1	2	6	7
2	1	6	8
1	6	2	9
6	1	2	10
2	6	1	11
1	3	5	12
3	1	5	13
1	5	3	14
5	1	3	15
3	5	1	16
5	3	1	17
2	2	5	18
2	2	6	19
2	3	5	20
3	2	5	21
2	5	3	22
5	2	3	23
3	5	2	24
5	3	2	25

Table 17.8 Link Combinations for the Stephenson Chain

n_3	n_4	g_3	g_4	No. of Solutions		
(2)						
1		1		1		
		2		1		
	(1)					
			1	0		
(4)						
3		1		15		
2		2		16		
1		3		3	47	$f = 1$
		4		1		
(2)	(1)					
1	1	1		5		
	1	2		1		
2			1	1		
1		1	1	1		
		2	1	0		
	(2)					
	1		1	1		
			2	1		

ternary links, and so forth. The entire process can be carried out easily for chains with relatively few links but grows rapidly more complex when the number of links is increased.

To develop chains with multiple joints, let us choose $N_3 = 4$. The ternary links are systematically replaced by multiple joints connecting three links each. The result of this process is shown in Table 17.9. Multiple joints are designated with two concentric circles and are to be considered different from a ternary link. The process is applied to the combinations $n_3 = 3$, $g_3 = 1$, and $f = 1$. By comparison with $n_3 = 4$, it is concluded that there are two solutions (Fig. 17.9). The possible combinations of E-, Z-, and D-chains are

E	Z	D
3	1	2
2	3	1
1	5	0

Table 17.9 Link and Joint Combinations with Multiple Joints and f = 1

Input	Output	Frame	Solution No.
1	1	5	1
1	2	5	2
2	1	5	3
1	5	2	4
5	1	2	5
2	5	1	6
5	2	1	7
1	2	6	8
2	1	6	9
1	6	2	10
6	1	2	11
2	6	1	12
6	2	1	13
1	4	5	14
4	1	5	15
1	5	4	16
5	1	4	17
4	5	1	18
5	4	1	19
5	5	1	20
1	5	6	21
5	1	6	22
1	6	5	23
6	1	5	24
5	6	1	25
6	5	1	26
2	2	5	27
2	3	5	28
2	5	3	29
5	2	3	30
5	5	2	31
2	5	6	32
5	6	6	33
5	6	2	34

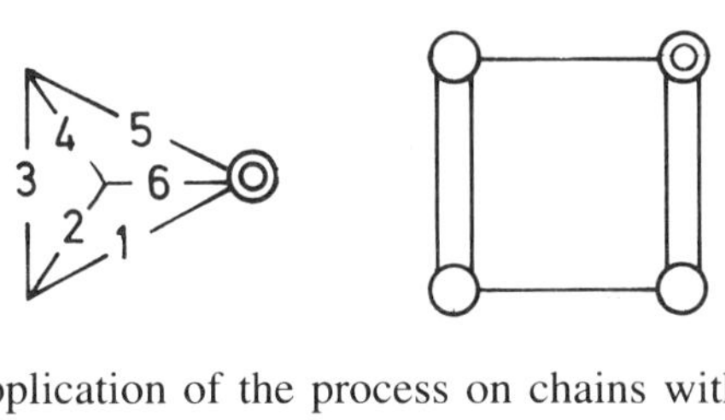

Figure 17.9 Application of the process on chains with multiple joints.

The first selection is shown in Fig. 17.10 in as many positions as possible. The combinations are written as follows:

1	2	3*	4*	5*	6*
	12	12*	14	15	16*
		23	24*	25*	26*
		123	124	125	126*
				145*	146*
					156
			234*	235*	236*
					etc.

The combinations listed in Table 17.9 are drawn in Table 17.10.

TYPE SYNTHESIS (JOINT SUBSTITUTION)

So far, kinematic chains and the linkages (mechanisms) that can be developed from these chains have been considered. However, there are a great many other mechanisms. The solution to finding alternate mechanisms is to apply the method of type synthesis or joint substitution. One type of joint with one degree of freedom can be substituted by another type of joint with one degree of freedom. Two joints with one degree of freedom each can be replaced by one joint with two degrees of freedom. In the following are considered plane mechanisms, that is, where all links move parallel to each other in a plane.

A Complete Collection of Plane Joints

In order to make joint substitutions it is necessary to have a complete collection of plane joints. Such a collection is shown in Fig. 17.11. The upper

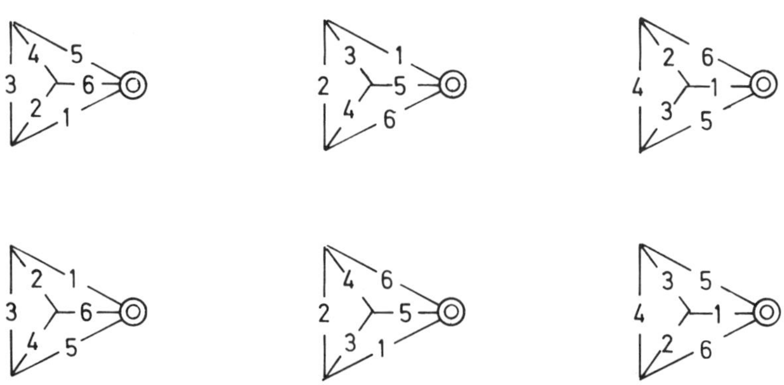

Figure 17.10 Skeleton combinations of the left chain in Fig. 17.9.

Table 17.10 Kinematic Chains with Multiple Joints and f = 1

Para-meter				
$n_3 = 1$ $g_3 = 1$	 1			
$g_3 = 2$	 1			
$n_3 = 3$ $g_3 = 1$	 1	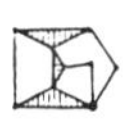 2	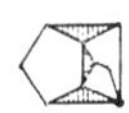 3	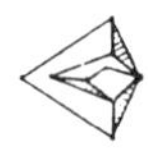 4
	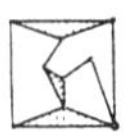 5	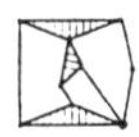 6	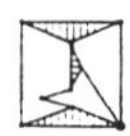 7	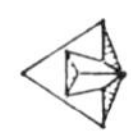 8
	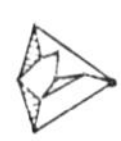 9	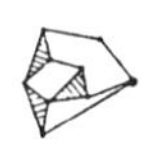 10	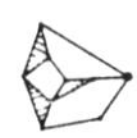 11	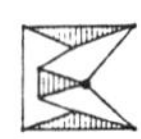 12
	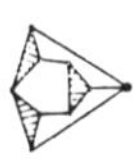 13	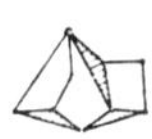 14	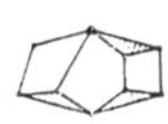 15	
$n_3 = 2$ $g_3 = 2$	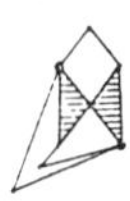 1	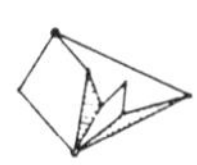 2	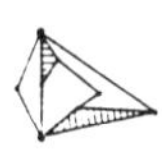 3	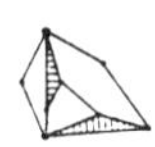 4
	 5	 6	 7	 8
	 9	 10	 11	 12

Table 17.10 (Continued)

Parameter				
$n_3 = 2$ $g_3 = 2$ (cont.)	 13	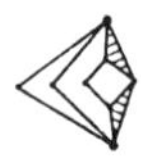 14	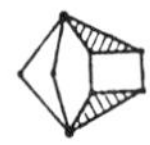 15	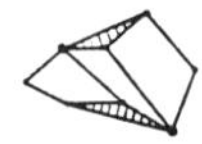 16
$n_3 = 1$ $g_3 = 3$	 1	 2	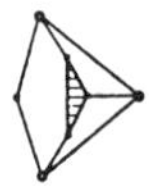 3	
$g_3 = 4$	 1			
$n_3 = 1$ $n_4 = 1$ $g_3 = 1$	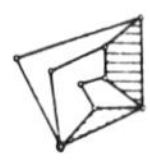 1	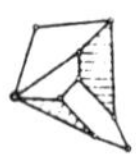 2	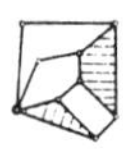 3	 4
	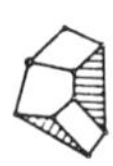 5			
$n_4 = 1$ $g_3 = 2$	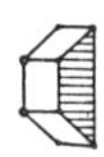 1			
$n_3 = 2$ $g_4 = 1$	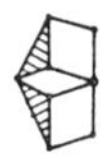 1			
$n_3 = 1$ $g_3 = 1$ $g_4 = 1$	 1			
$n_4 = 1$ $g_4 = 1$	 1			
$g_4 = 2$	 1			

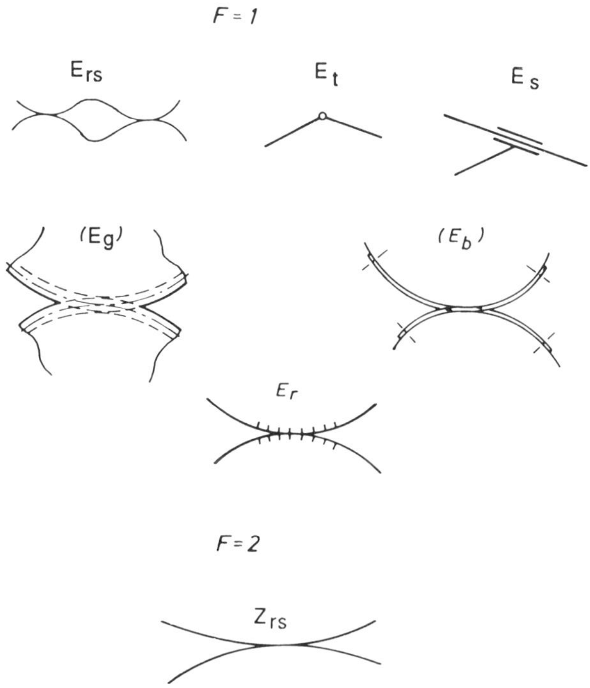

Figure 17.11 The complete collection of all plane joints with one and two degrees of freedom (F = 1 and F = 2).

part shows joints with one degree of freedom (F = 1). These joints are designated with E; the subscript refers to the characteristic motion of joint E. E_t is a turning joint. It has one degree of freedom because the two links connected by the joint can make a rotation only relative to each other. The sliding joint E_s permits sliding only between the two links that the joint connects. The gear joint E_g and the band joint E_b are both designated E_r (rolling joint). The rolling-sliding joint Z_{rs} has two degrees of freedom (F = 2).

Recapitulating, joints with one degree of freedom are as follows:

- E_{rs} is rolling-sliding joint. It has two points of contact. For certain proportions the number of contact points can be increased (Fig. 1.7).
- E_t is a turning joint (revolute pair) and allows pure rotation of one link relative to the other.
- E_s is a sliding joint and allows a translation between the two connected links.

E_r is a rolling joint as known from the pitch circles of two gears in mesh (E_g). The same action can be obtained by the steel band shown (E_b).

There is but one type of joint with two degrees of freedom: Z_{rs}, a rolling-sliding joint.

In plane motion the number of degrees of freedom of a joint is a maximum of two because the purpose of a joint is to limit the motiton of one link relative to the other; the number of degrees of freedom of a link is three, namely, one rotation and two translations. The five joints listed above are considered to be a complete collection of all joints where the links connected move in parallel planes.

Development of Mechanisms from Kinematic Chains

How can mechanisms be developed from the four-bar linkage? A four-bar linkage is shown in Fig. 17.12(a), where I is the frame and II, III, and IV

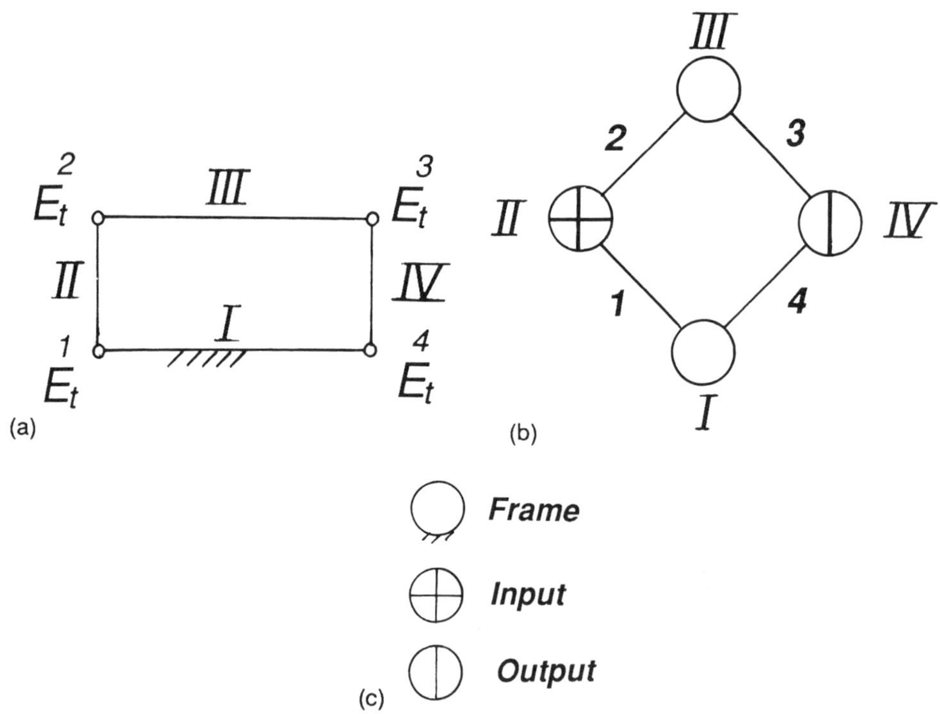

Figure 17.12 (a) Four-bar linkage. (b) Symbolic representation of the four-bar linkage with input and output defined. (c) Explanation of the symbols used in Fig. 17.12(b).

designate the other links. All four links are connected by turning joints E_t. Each joint is numbered as shown in Fig. 17.12(a).

The four-bar linkage can be written symbolically as

1	2	3	4
E_t	E_t	E_t	E_t

A symbolic representation is shown in Fig. 17.12(b). Fig. 17.12(c) explains the symbols. Because input and output are not defined at this point, and input and output members are assumed to be links connected to the frame link, it follows that the arrangement in Fig. 17.12(b) can be mirrored about its vertical axis and therefore the following numbers can be written:

1	2	3	4
4	3	2	1

This is done to decide whether two solutions are identical. For instance, it is seen that

1	2	3	4	=	1	2	3	4
E_g	E_t	E_t	E_t	=	E_t	E_t	E_t	E_g

The following formula can be written:

$$\Sigma E_{rs}, E_t, E_r, E_s, Z_{rs} = 4$$

This means that the summation of the degrees of freedom of the various joints is 4. Continuing systematically,

	1	2	3	4
$4E_{rs}$:	E_{rs}	E_{rs}	E_{rs}	E_{rs}
	1	2	3	4
$3E_{rs}$, $1E_t$:	E_{rs}	E_{rs}	E_{rs}	E_t
	1	2	3	4
	E_{rs}	E_{rs}	E_t	E_{rs}
	1	2	3	4
$3E_{rs}$, $1E_r$:	E_{rs}	E_{rs}	E_{rs}	E_r
	1	2	3	4
	E_{rs}	E_{rs}	E_r	E_{rs}
	1	2	3	4
$3E_{rs}$, $1E_s$:	E_{rs}	E_{rs}	E_{rs}	E_s
	1	2	3	4
	E_{rs}	E_{rs}	E_s	E_{rs}
$2E_{rs}$, $1Z_{rs}$:	No solution			

etc.

The above solutions are shown in Table 17.11. The number of solutions

Table 17.11 Mechanisms Developed from the Four-Bar Linkage by Joint Substitution

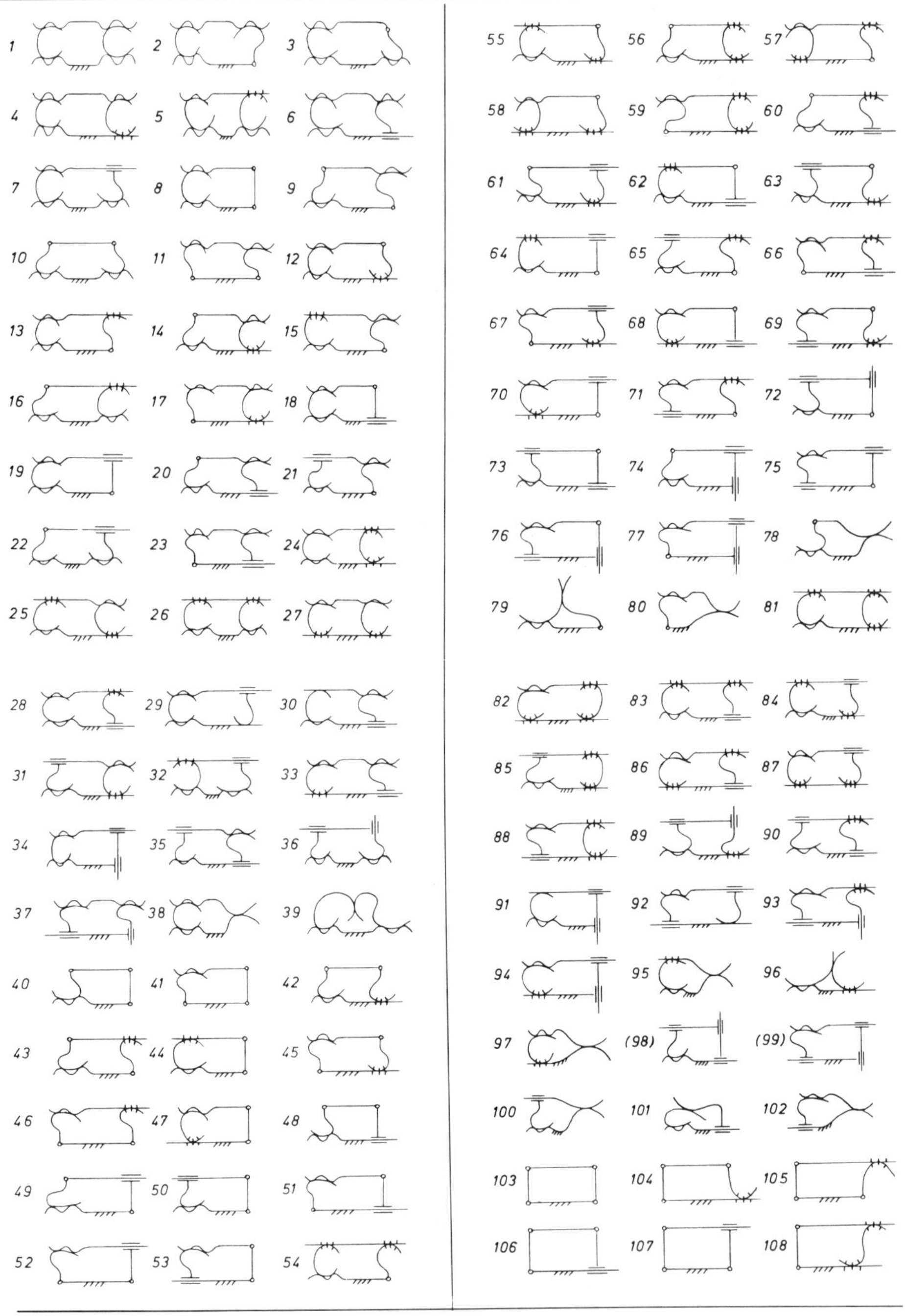

Table 17.11 (Continued)

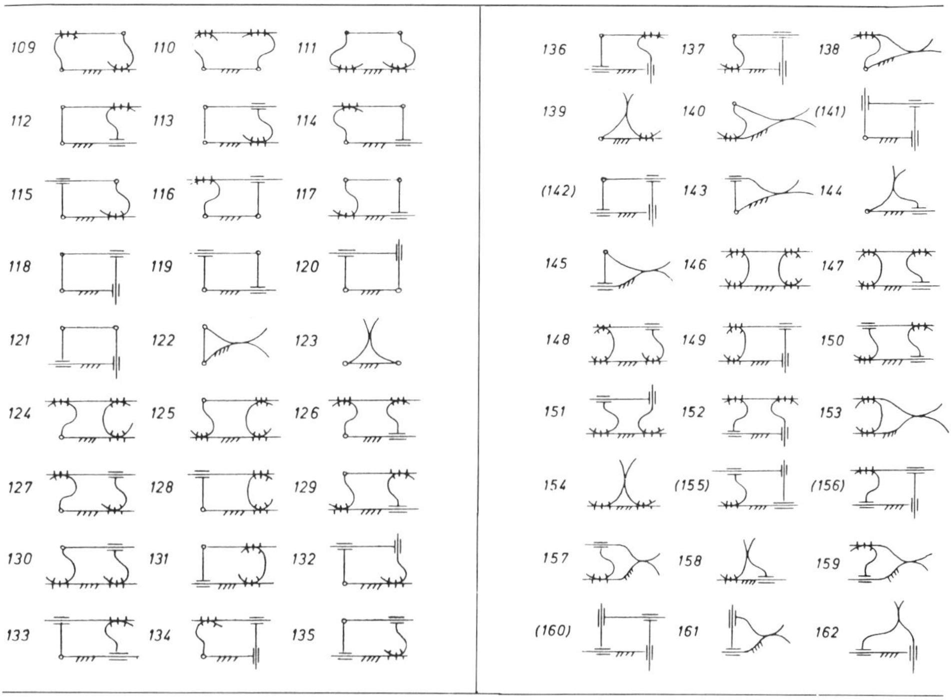

is 162, but from this number must be subtracted solutions that cannot be used as stated by Gruebler. These solutions are shown in parantheses. If input and output are defined and supported directly in the frame, the number of solutions is increased to 272. Most of the solutions are new. An observant reader can use his or her imagination and find applications for many of these new mechanisms.

The above procedure is now applied to a six-bar linkage (Table 17.12, Fig. 1). This is the well-known shaper mechanism, where the rotary input of 12 is converted to the translating output of 67. The corresponding kinematic chain is shown in Fig. 17.13(a). The practical requirements imposed on this mechanism are that the input should be rotary and the output translating (to obtain a plane surface). Joint 4 cannot be replaced by a sliding joint because of the desired motion characteristics of the mechanism. In other words, the turning joints $E_t(1)$ and $E_t(4)$ as well as the sliding joint $E_s(7)$ must remain unchanged. Joint substitution can be made only within a

Table 17.12 Six-Bar Solutions for a Shaper Mechanism, Developed from Watt's Chain by Joint Substitution

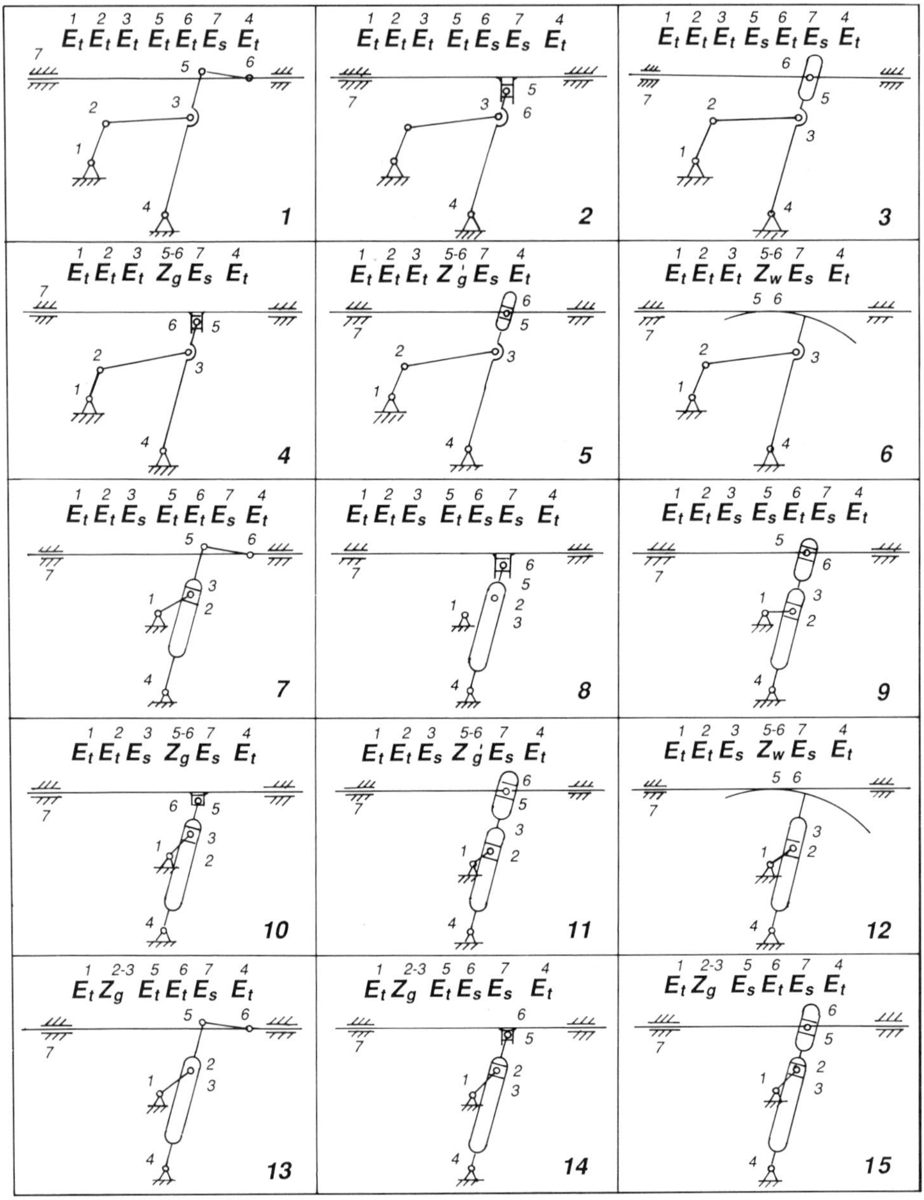

Table 17.12 (Continued)

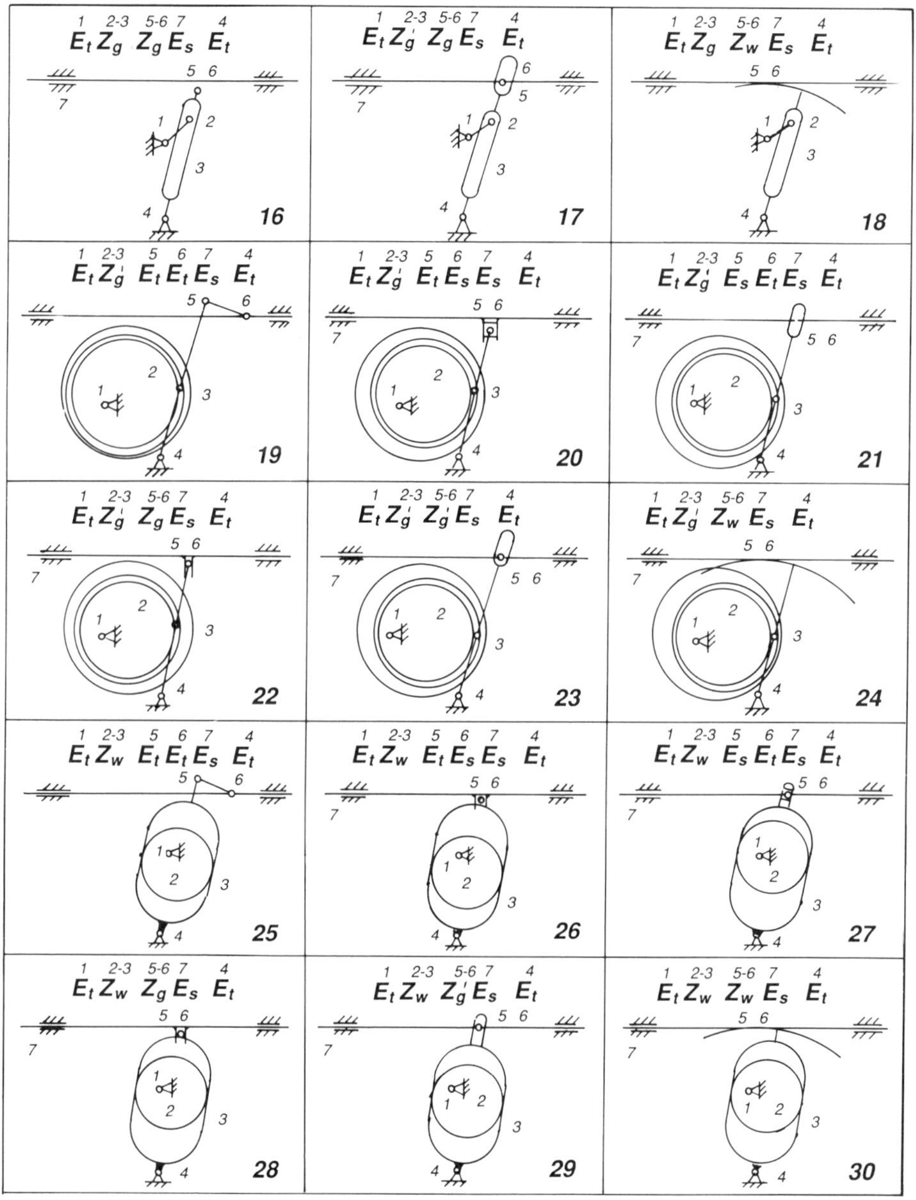

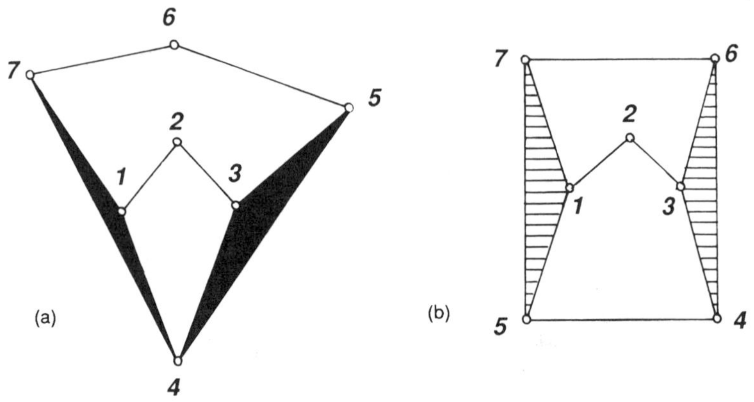

Figure 17.13 Two chains (the Watt and Stephenson chain) from which to develop shaper mechanisms.

chain of binary links, and therefore the formula for this mechanism is written the following way:

$$\overset{1}{E_t}\overset{2}{E_t}\overset{3}{E_t}\quad \overset{5}{E_t}\overset{6}{E_t}\overset{7}{E_s}\quad \overset{4}{E_t}$$

Applying the method of joint substitution, the 30 solutions in Table 17.12 are found.

Another mechanism that will also convert a rotary input to a translating output is shown in Table 17.13, Fig. 31. The corresponding kinematic chain is shown in Fig. 17.13(b). Making a systematic joint substitution yields the 30 more solutions in Table 17.13. The designer will now have to make a choice considering the geometric properties of the 60 solutions together with the three factors of space, speed, and savings.

DEVELOPMENT OF GEAR LINKAGES

It was shown in the section on type synthesis that a four-bar linkage can serve as the basis from which to develop other mechanisms. A symbolic representation is shown in Fig. 17.12(b), and Fig. 17.12(c) explains the symbols. Consider now the mechanism $E_tZ_{rs}E_t$ (Fig. 17.14(b)), which has been developed from Fig. 17.14(a). The two curves roll and slide on each other. If one were to employ E_r (Fig. 17.14(c)) instead of Z_{rs}, the resultant mechanism would be rigid. Assume now that the shapes of the two curves are changed as shown in Fig. 17.15(a), which shows noncircular gears; then motion is possible. The case of circular gears is shown in Fig. 17.15(b).

Table 17.13 Six-Bar Solutions for a Shaper Mechanism, Developed from Stephenson's Chain by Joint Substitution

1 2 3 4 5 6 7 $E_t E_t E_t \; E_t E_t \; E_t E_s$ 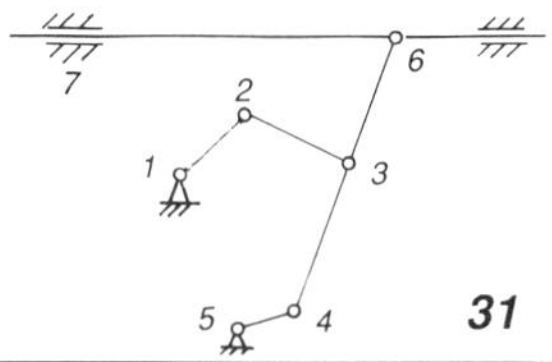 **31**	1 2 3 4 5 6 7 $E_t E_t E_t \; E_t E_s \; E_t E_s$ 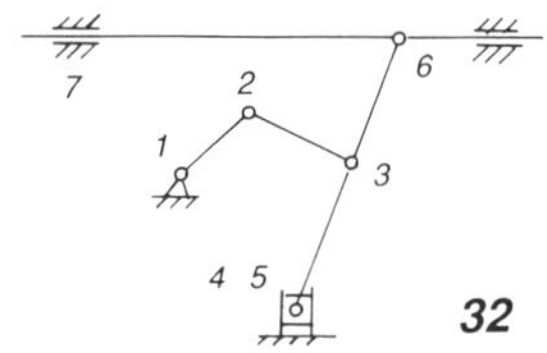 **32**	1 2 3 4 5 6 7 $E_t E_t E_t \; E_s E_t \; E_t E_s$ 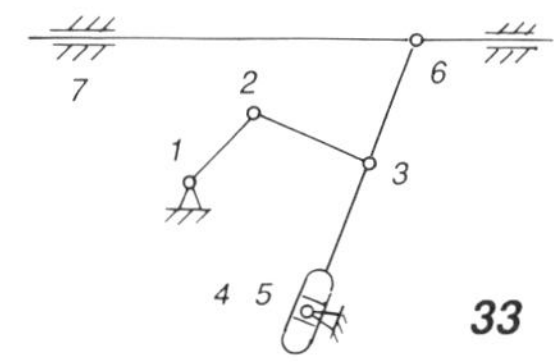 **33**
1 2 3 4-5 6 7 $E_t E_t E_t \; Z_g \; E_t E_s$ 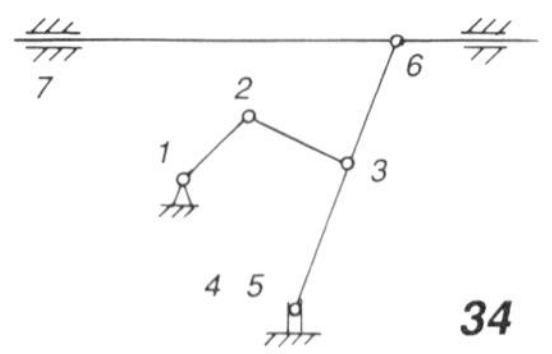 **34**	1 2 3 4-5 6 7 $E_t E_t E_t \; Z_g' \; E_t E_s$ 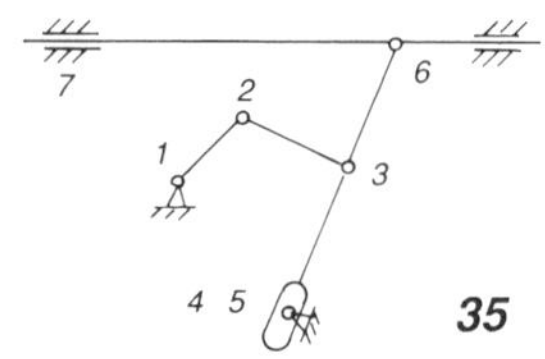 **35**	1 2 3 4-5 6 7 $E_t E_t E_t \; Z_w \; E_t E_s$ 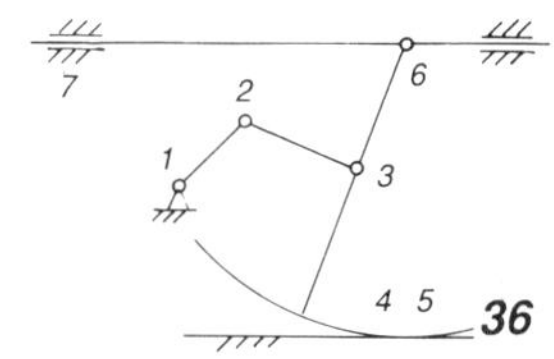 **36**
1 2 3 4 5 6 7 $E_t E_t E_s \; E_t E_t \; E_t E_s$ 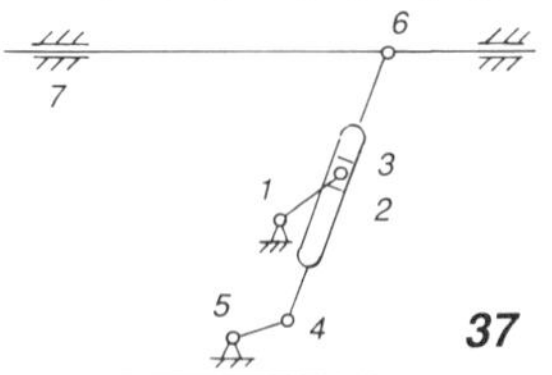 **37**	1 2 3 4 5 6 7 $E_t E_t E_s \; E_t E_s \; E_t E_s$ 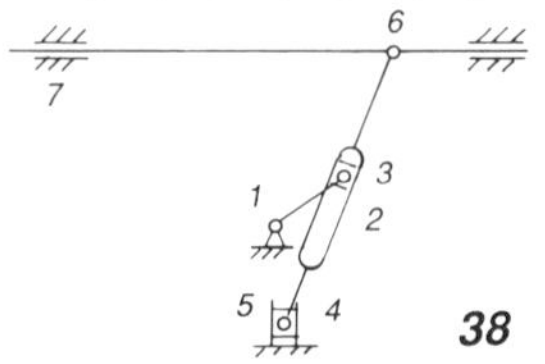 **38**	1 2 3 4 5 6 7 $E_t E_t E_s \; E_s E_t \; E_t E_s$ 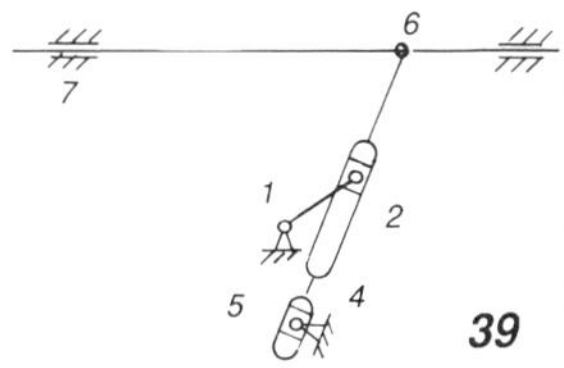 **39**
1 2 3 4-5 6 7 $E_t E_t E_s \; Z_g \; E_t E_s$ 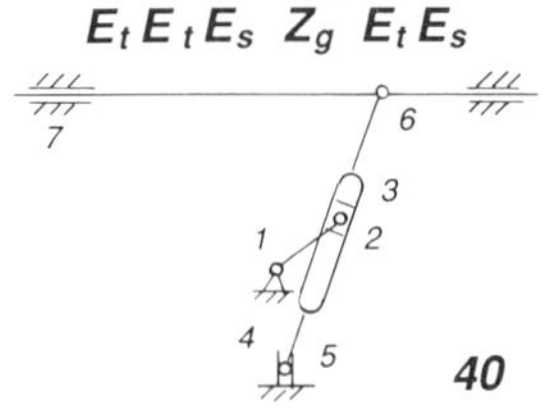 **40**	1 2 3 4-5 6 7 $E_t E_t E_s \; Z_g' \; E_t E_s$ 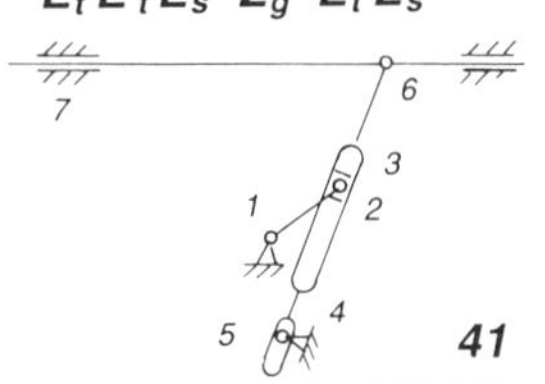 **41**	1 2 3 4-5 6 7 $E_t E_t E_s \; Z_w \; E_t E_s$ 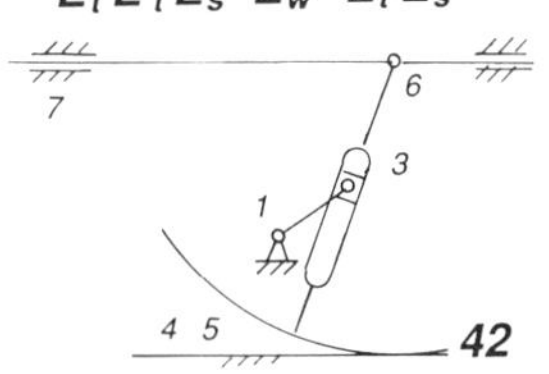 **42**
1 2-3 4 5 6 7 $E_t Z_g \; E_t E_t \; E_t E_s$ 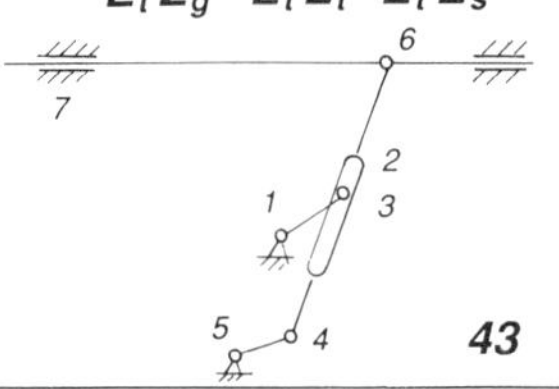 **43**	1 2-3 4 5 6 7 $E_t Z_g \; E_t E_s \; E_t E_s$ 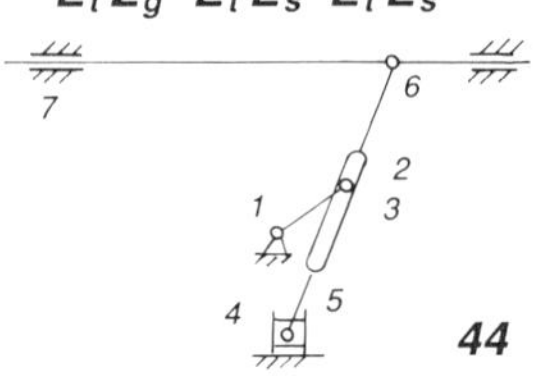 **44**	1 2-3 4 5 6 7 $E_t Z_g \; E_s E_t \; E_t E_s$ 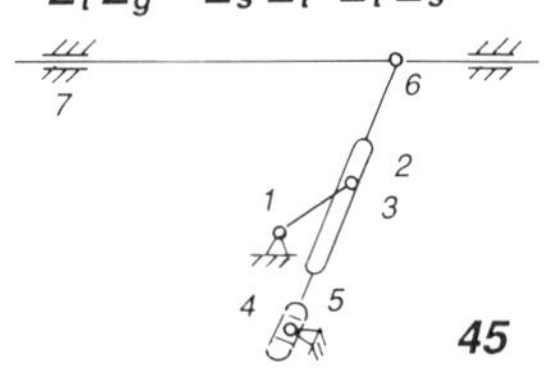 **45**

Table 17.13 (Continued)

1 2-3 4-5 6 7 $E_t Z_g\ Z_g\ E_t E_s$ 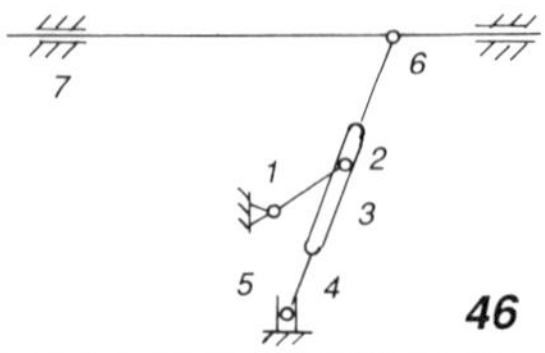 46	1 2-3 4-5 6 7 $E_t Z_g\ Z_g\ E_t E_s$ 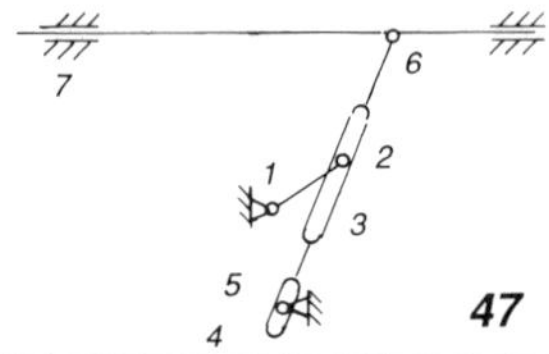 47	1 2-3 4-5 6 7 $E_t Z_g\ Z_w\ E_t E_s$ 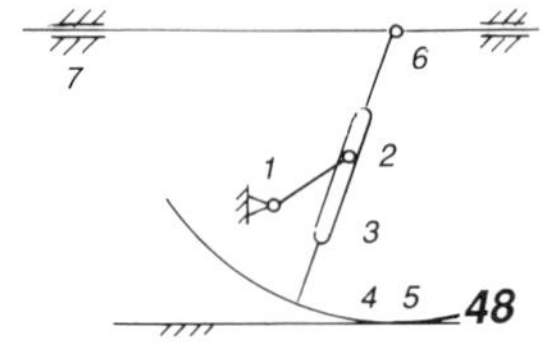 48
1 2-3 4 5 6 7 $E_t Z_g\ E_t E_t\ E_t E_s$ 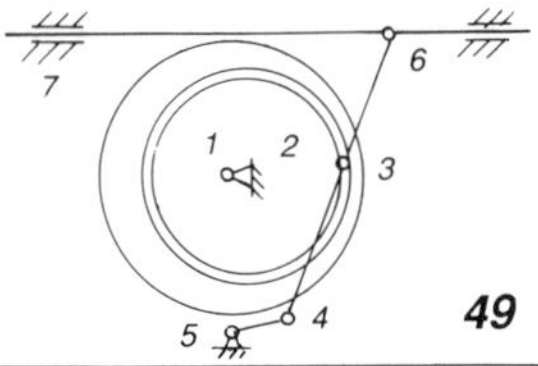 49	1 2-3 4 5 6 7 $E_t Z_g'\ E_t E_s\ E_t E_s$ 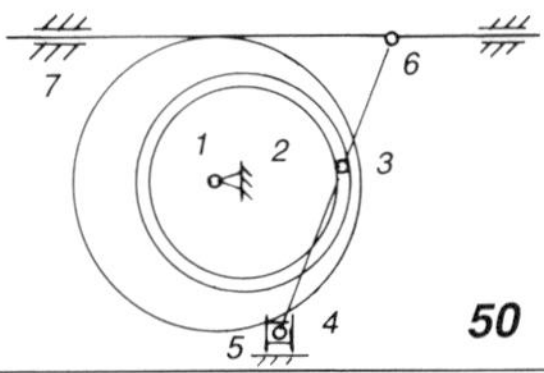 50	1 2-3 4 5 6 7 $E_t Z_g'\ E_s E_t\ E_t E_s$ 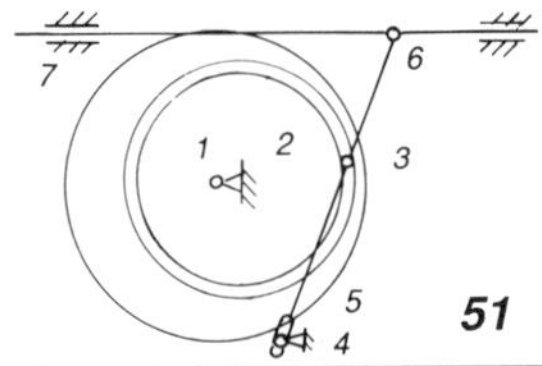 51
1 2-3 4-5 6 7 $E_t Z_g'\ Z_g\ E_t E_s$ 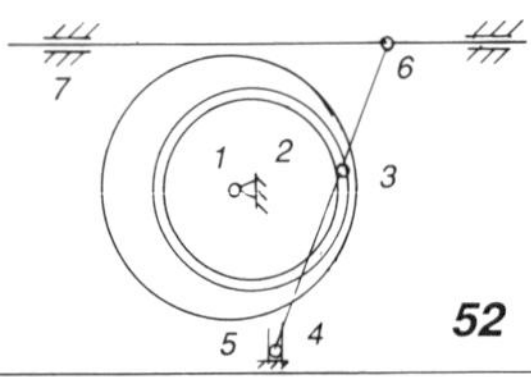 52	1 2-3 4-5 6 7 $E_t Z_g'\ Z_g\ E_t E_s$ 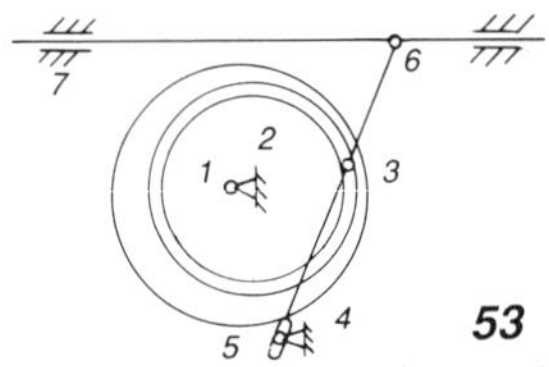 53	1 2-3 4-5 6 7 $E_t Z_g'\ Z_w\ E_t E_s$ 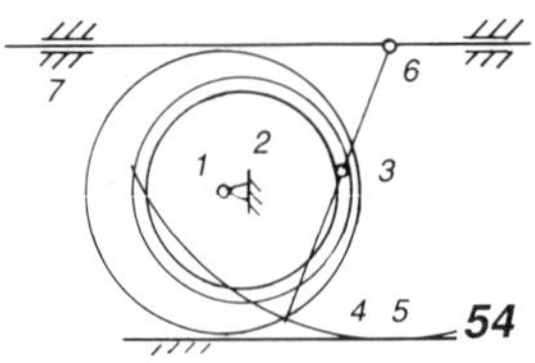 54
1 2-3 4 5 6 7 $E_t Z_w\ E_t E_t\ E_t E_s$ 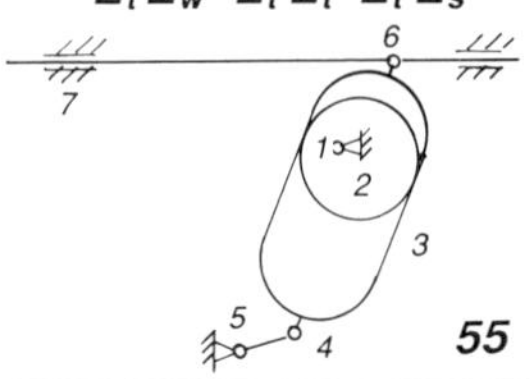 55	1 2-3 4 5 6 7 $E_t Z_w\ E_t E_s\ E_t E_s$ 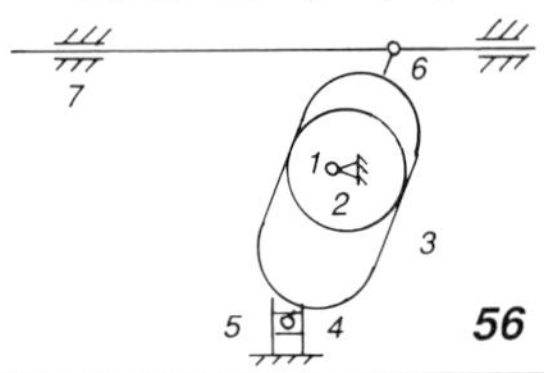 56	1 2-3 4 5 6 7 $E_t Z_w\ E_s E_t\ E_t E_s$ 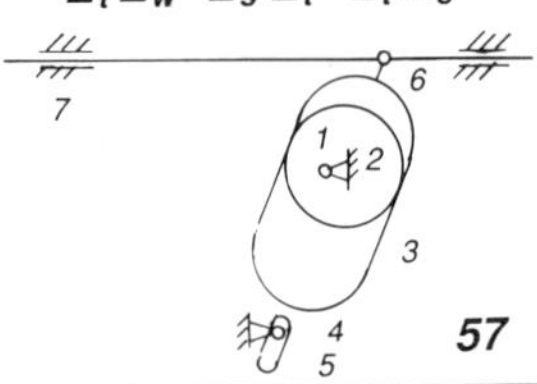 57
1 2-3 4-5 6 7 $E_t Z_w\ Z_g\ E_t E_s$ 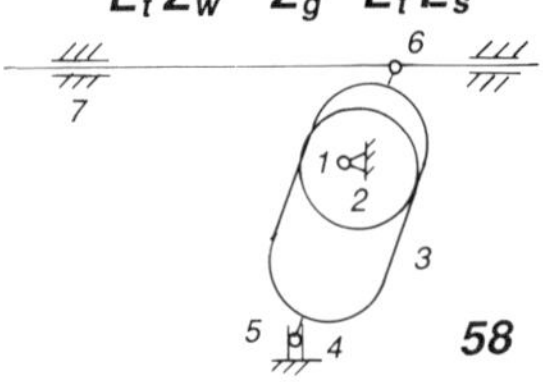 58	1 2-3 4-5 6 7 $E_t Z_w\ Z_g'\ E_t E_s$ 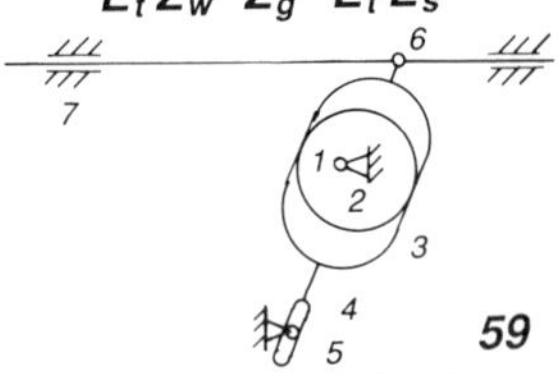 59	1 2-3 4-5 6 7 $E_t Z_w\ Z_w\ E_t E_s$ 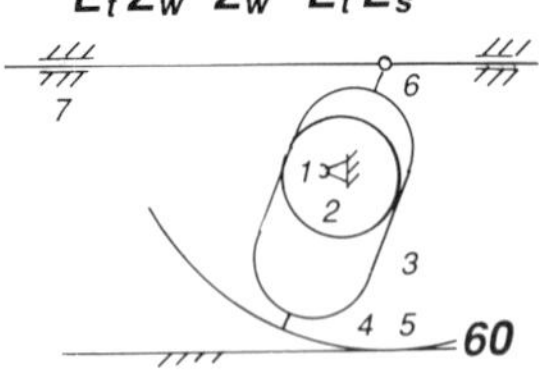 60

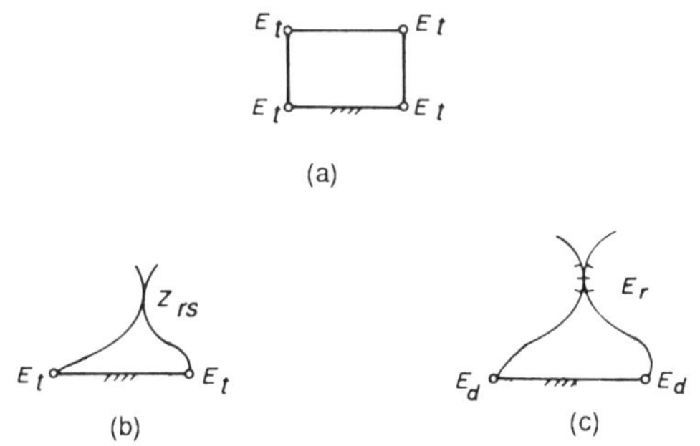

Figure 17.14 (a) A four-bar linkage with designation of joints. (b) Mechanism with a rolling-sliding joint, developed from Figure 17.14(a). (c) Combination of joints resulting in a rigid structure.

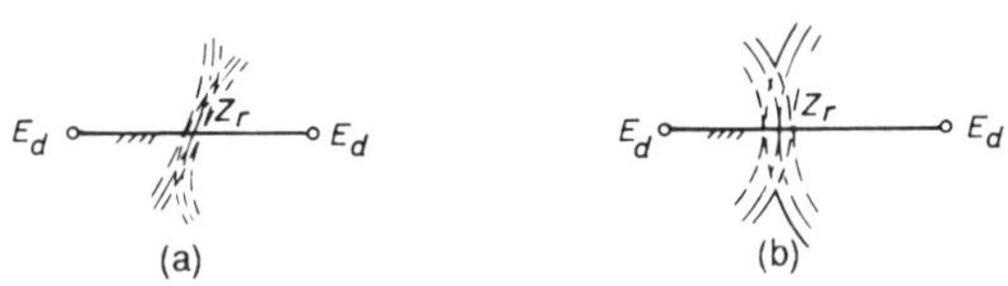

Figure 17.15 (a) Noncircular gear mechanism. (b) Circular gear mechanism.

It is now realized that in any kinematic chain, gear joints can be used, provided that the chain contains one or more four-bar linkages, that is, where there is a constant center distance between the points of rotation of the gears. This procedure has been carried through for classes I, II, and III, with f = 1. As many Z_{rs} joints as possible (see Figs. 17.15(a) and (b)) have been used (Table 17.14). Only external gears are shown, but internal gearing can be used, too.

Development of Gear-Linkage Mechanisms from Kinematic Chains with Multiple Joints

In the foregoing 47 kinematic chains with multiple joints with f = 1 were developed (Tables 17.9 and 17.10). Table 17.14 shows the corresponding gear-linkage mechanisms where as many Z-chains as possible have been replaced by gears.

Table 17.14 Development of Gear Linkages from Kinematic Chains with Singular Joints

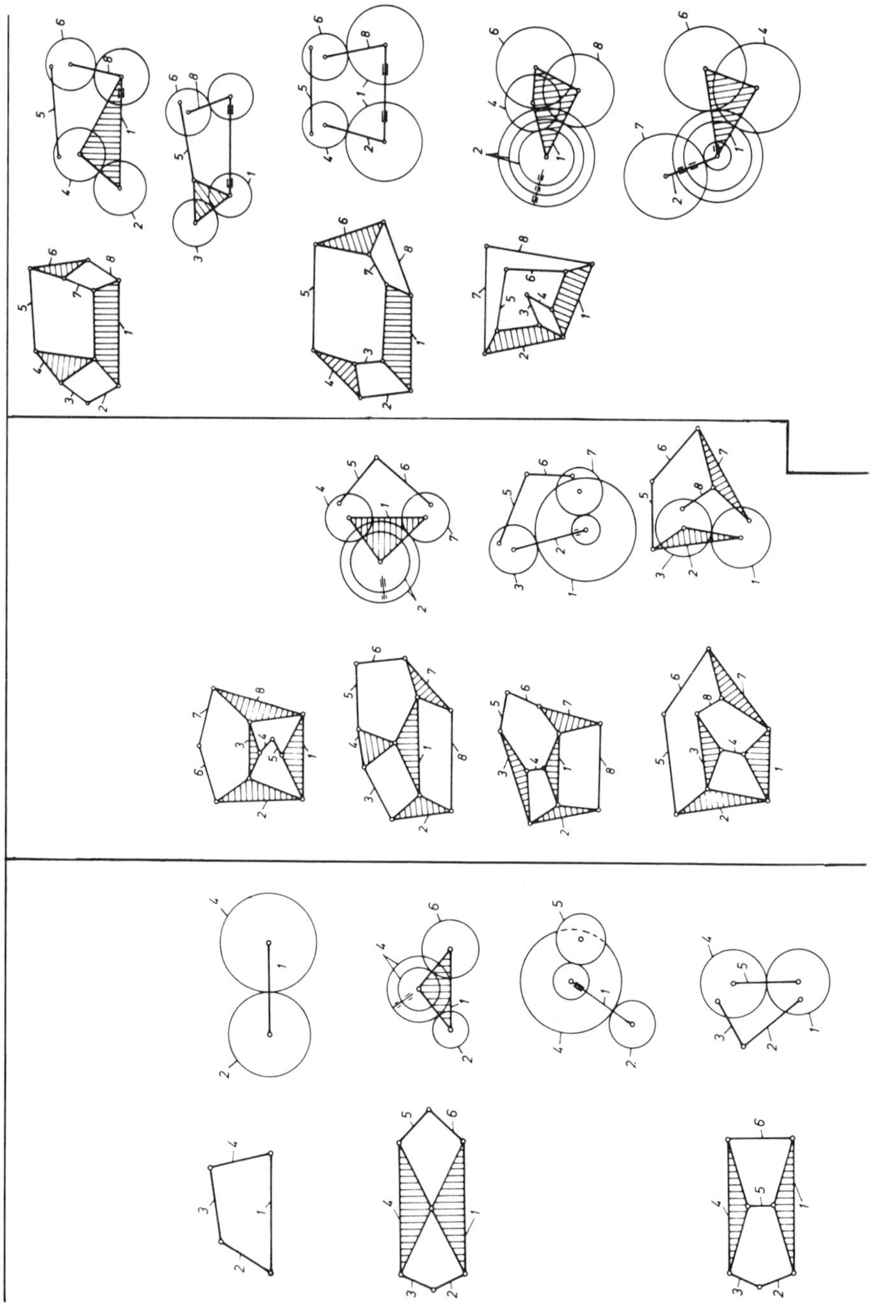

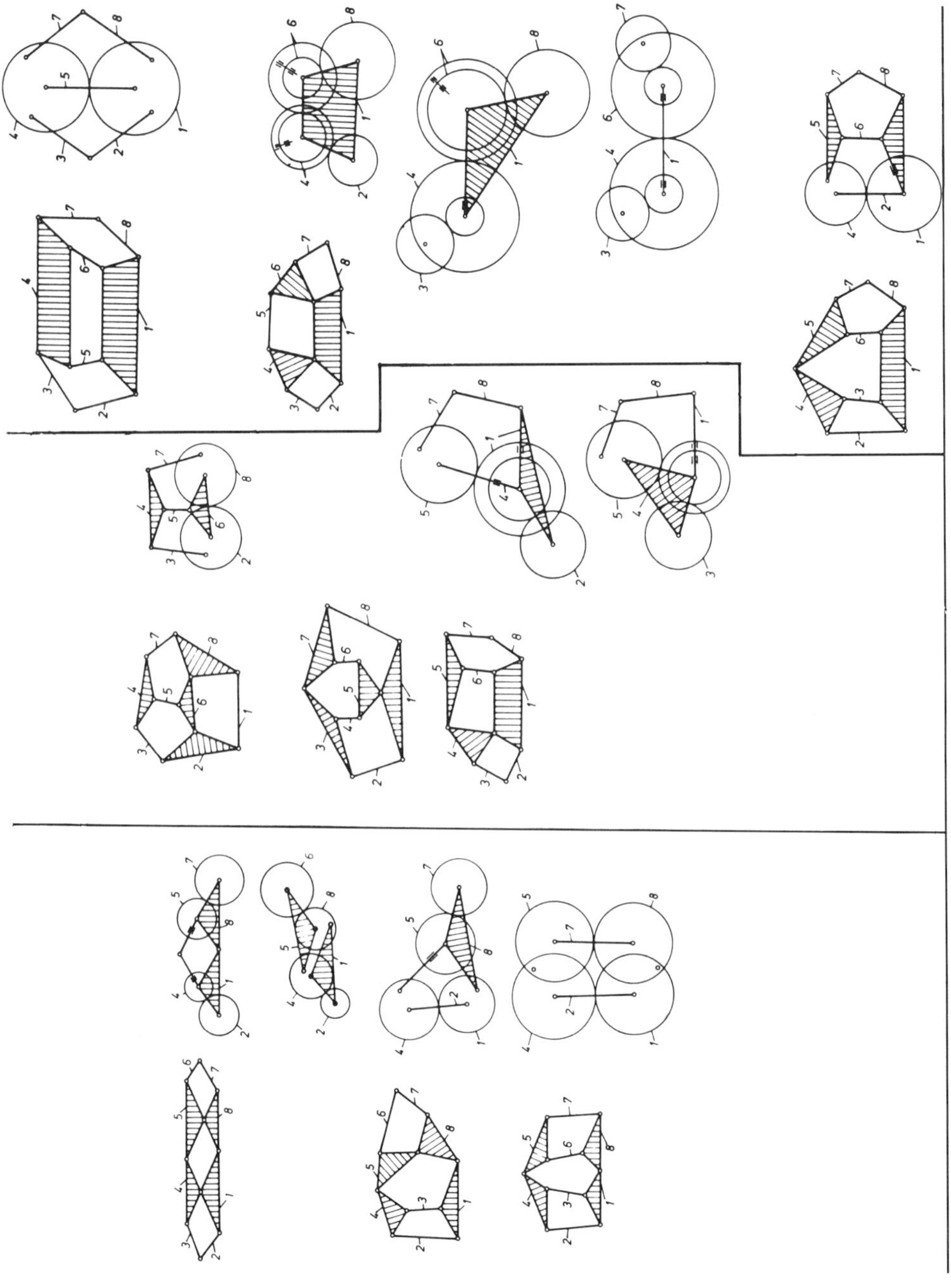

Table 17.15 Development of Gear Linkages from Kinematic Chains with Multiple Joints

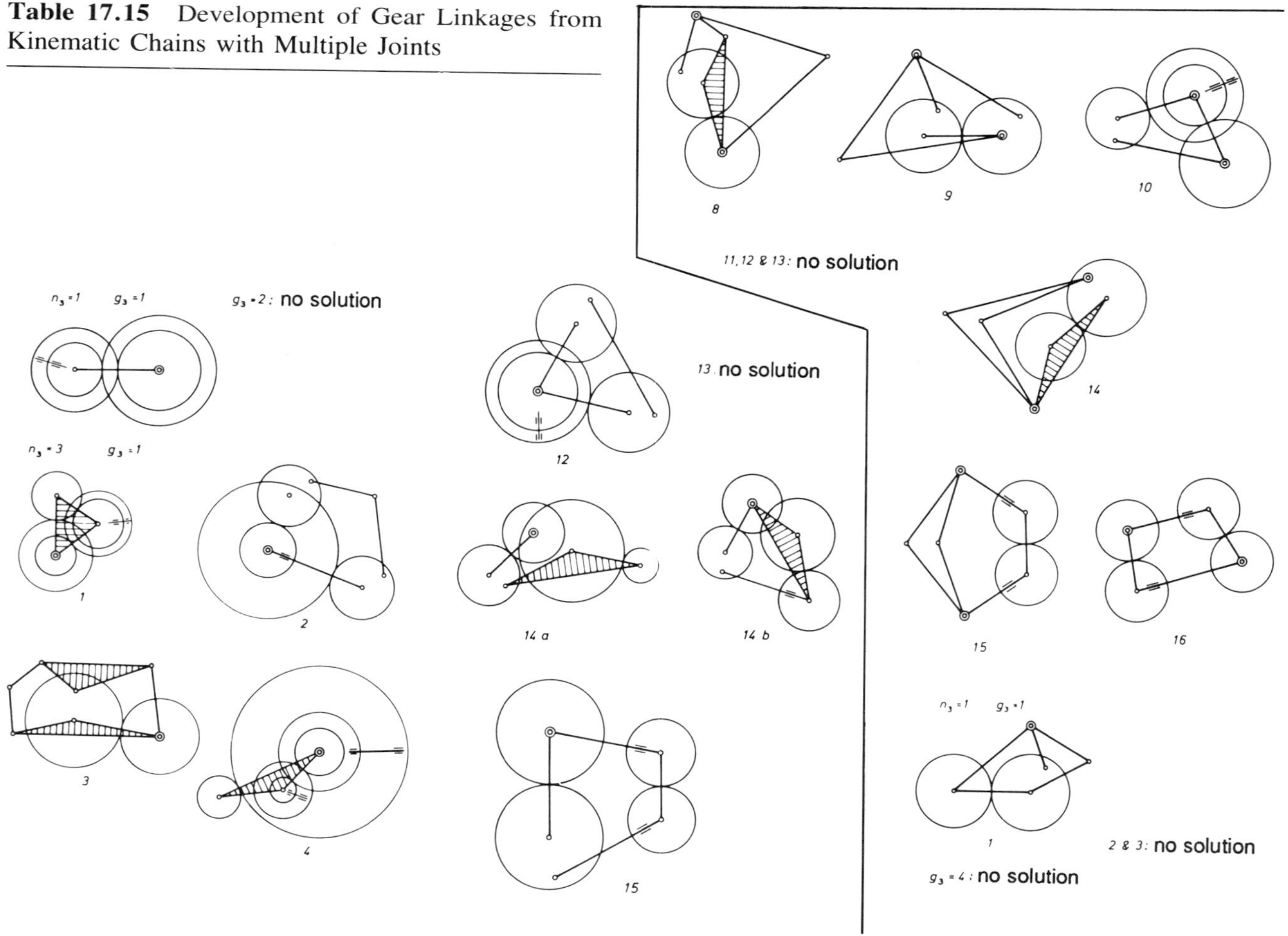

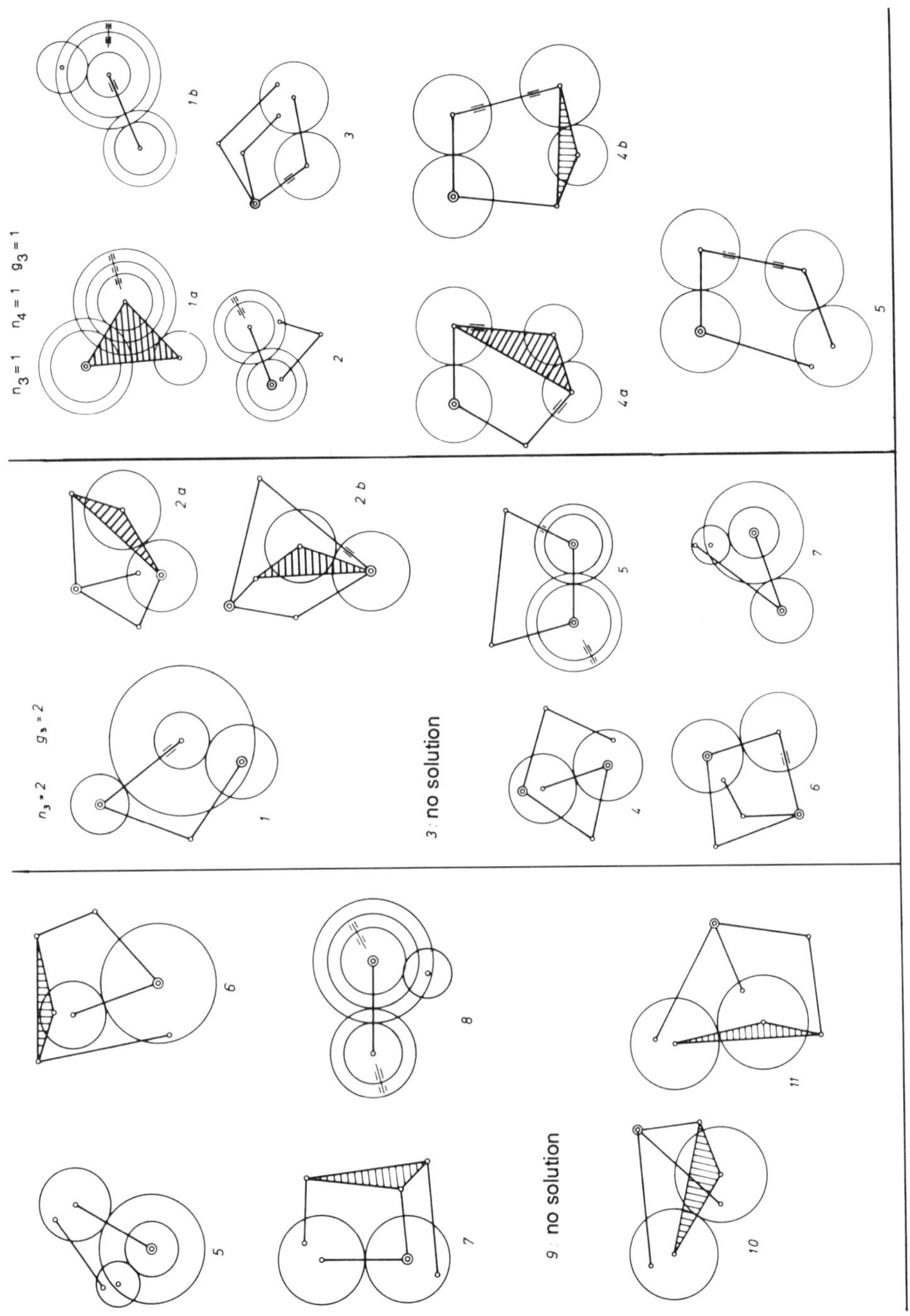
$n_3 = 1$ $n_4 = 1$ $g_3 = 1$
1a
1b
2
3
4a
4b
5
$n_3 = 2$ $g_3 = 2$
1
2a
2b
3 : no solution
4
5
6
7
5
6
7
8
9 : no solution
10
11

Table 17.15 (Continued)

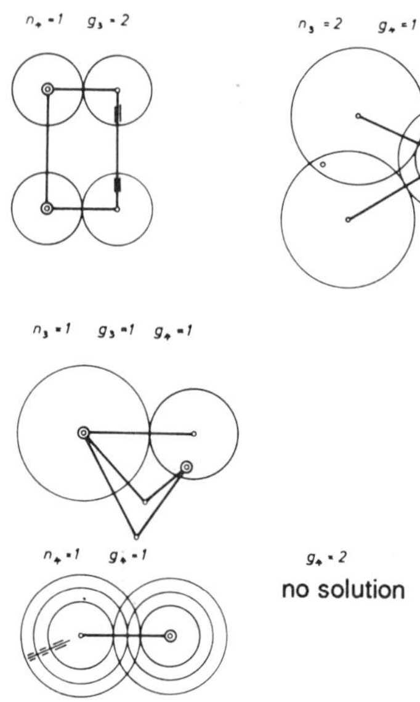

Development of Planetary Gears from Chains with Multiple Joints

Planetary gears can be developed from kinematic chains with multiple joints. Two planetary gear systems are found in Table 17.15 and are shown in Fig. 17.16.

A planetary gear system is characterized by colinearity of the axes of the input and output shafts. In many cases they are also characterized by the fact that one or more gears can be blocked to provide different output speeds (for instance in automatic automobile transmissions). The latter distinction kinematically characterizes planetary gear systems with two or more degrees of freedom. They can be developed from kinematic chains with two or more degrees of freedom.

Complex Planetary Gear Systems

Complex planetary gear systems are developed from kinematic chains with multiple joints, and the combination $n_3 = 4$, $g_4 = 1$, and $f = 1$ is chosen. The investigation is started by findig the corresponding kinematic chains and then reducing the chains to multiple-joint chains.

Four chains I-IV are found (Figs. 17.17–17.20). From each of these

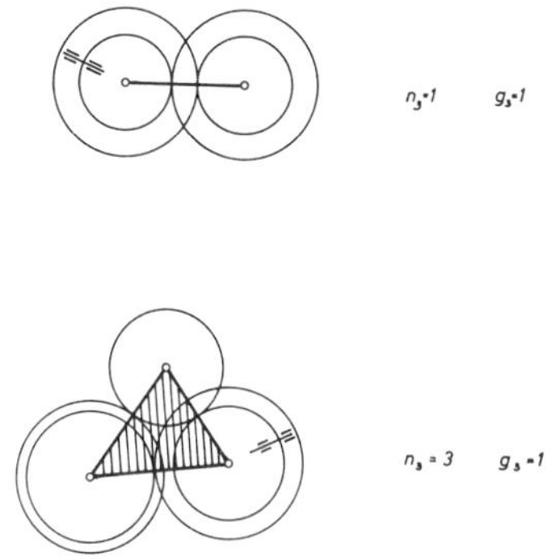

Figure 17.16 Two planetary gear systems developed from Table 17.15.

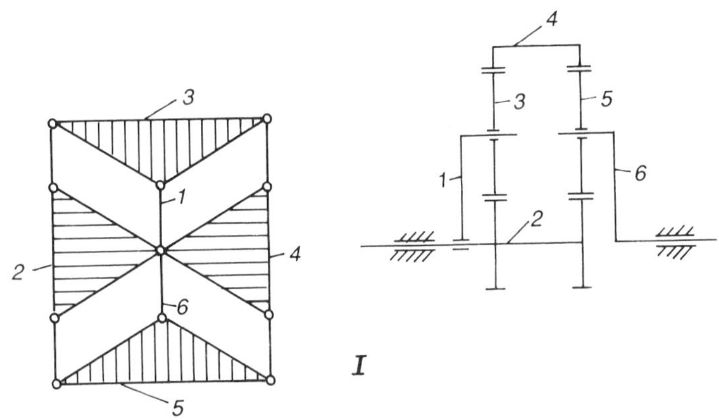

Figure 17.17 Kinematic chain and a corresponding planetary gear system. (See also Figs. 17.18–17.28.)

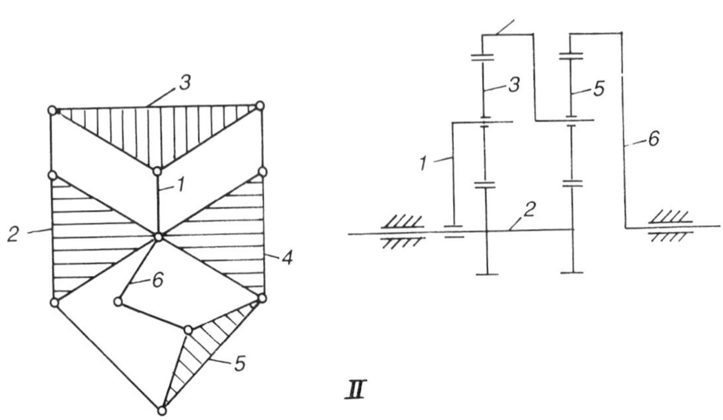

Figure 17.18

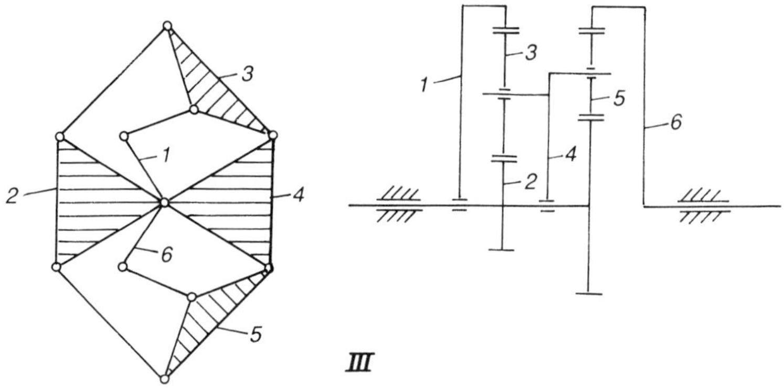

III

Figure 17.19

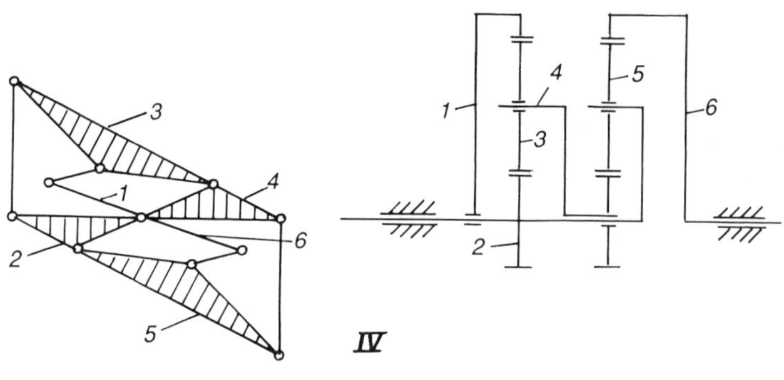

IV

Figure 17.20

chains a number of different planetary gear systems can be found by assigning input, output, and frame values to various members. The solutions are listed in Table 17.16. All these systems have been investigated with respect to transmission ratio, space requirements, and efficiency (see Chapter 5).

A number of more complex systems are shown in Figs. 17.21–17.28.

Table 17.16 Solutions Based on the Kinematic Chains I–IV

Input	Output	Frame	System No.
1	2	6	I
2	1	6	
1	6	2	
1	2	6	II
2	1	6	
1	6	2	
6	1	2	
1	4	6	
4	1	6	
1	6	4	
6	1	4	
2	6	1	
6	2	1	
4	6	1	
6	4	1	
1	2	6	III
2	1	6	
1	4	6	
4	1	6	
1	6	2	
1	2	6	IV
1	6	2	
1	4	6	
1	6	4	
2	1	6	
2	6	1	

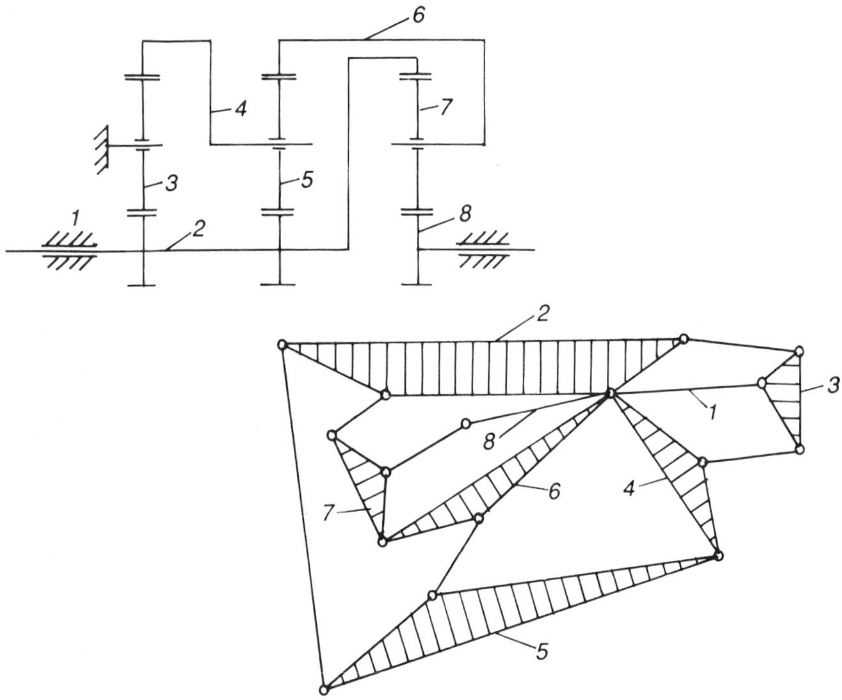

Figure 17.21

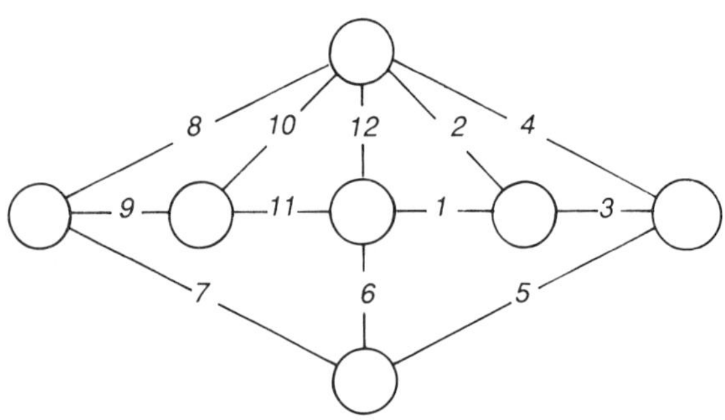

E	Z	D
8	0	4
7	2	3
6	4	2
5	6	1
4	8	0

Planetary gear systems can be found only from these combinations (rows 7 2 3 through 4 8 0)

Figure 17.22

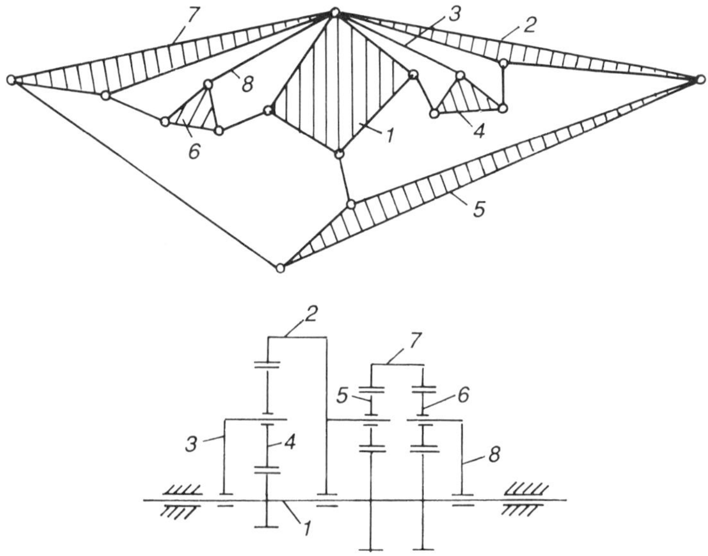

Figure 17.23

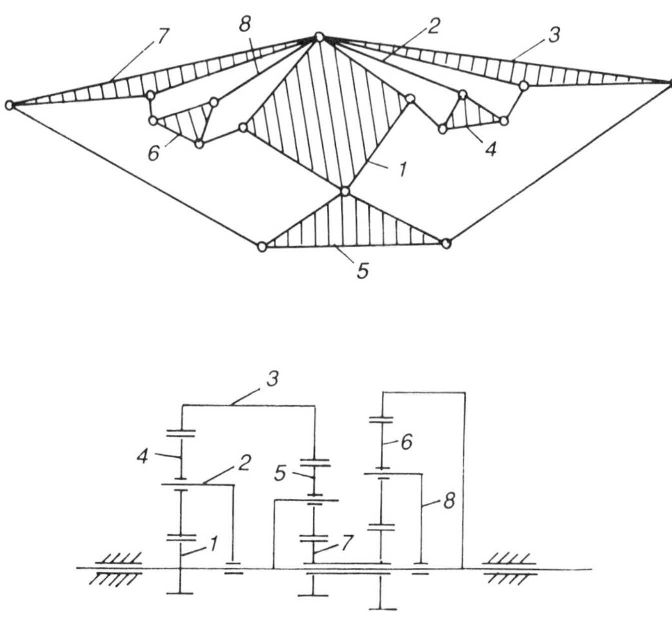

Figure 17.24

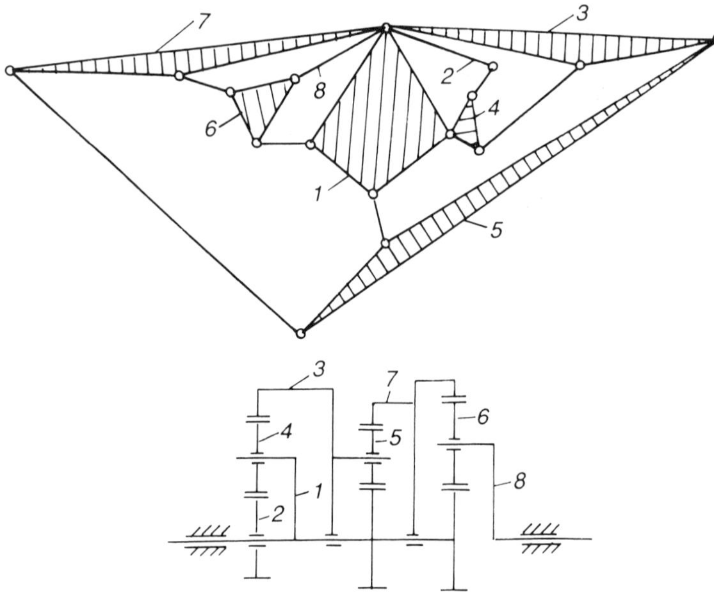

Figure 17.25

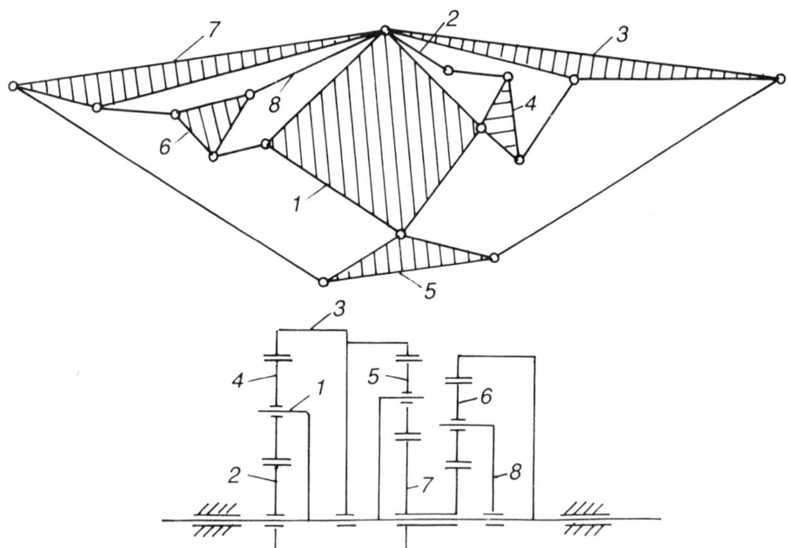

Figure 17.26

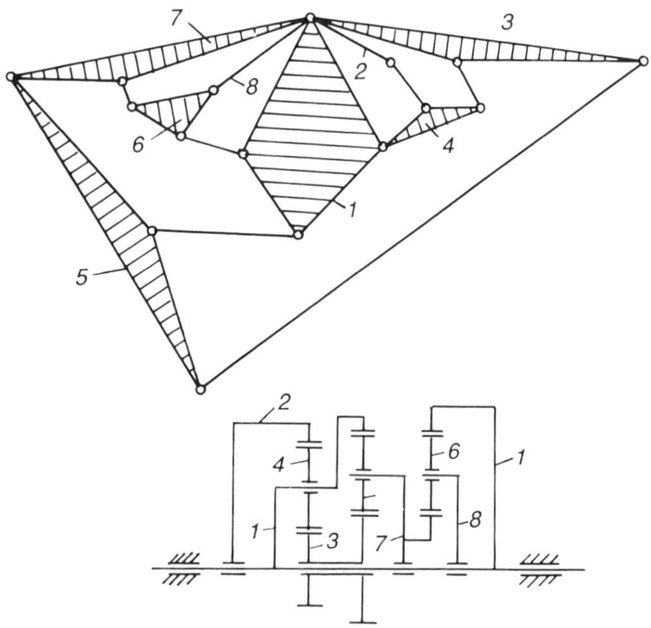

Figure 17.27

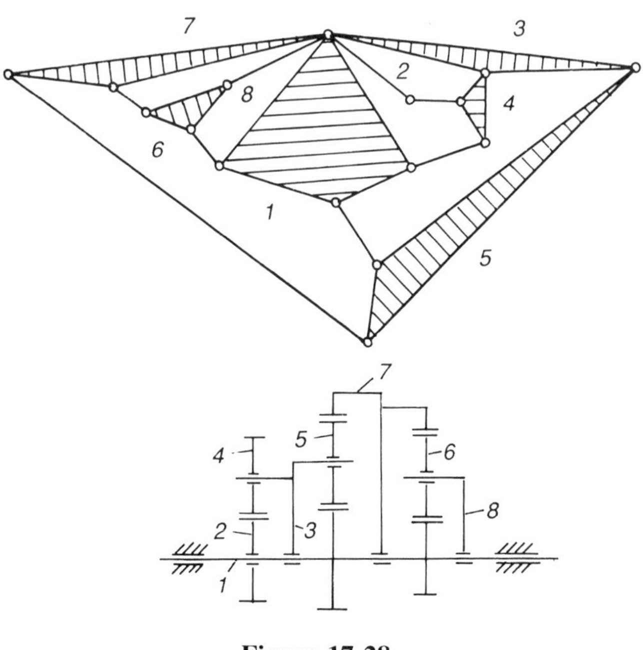

Figure 17.28

In this book, a great number of planetary gear systems and other mechanisms have been described. Planetary gear systems were treated in Chapter 5 with respect to transmission ratio, space requirements, and efficiency. A designer in general will never have the time to investigate every solution; thus design charts have been provided so that for a given transmission ratio a limited number of systems can be found. The design charts enable the designer to make an intelligent choice.

The author is now working in the area of linkage synthesis, and it is his intention to provide the designer with a number of advanced methods that will be programmed for interactive synthesis. I gladly accept challenges from industry concerning difficult motion problems. My programs and methods will enable you to handily solve problems that offer great difficulty for the time being.

SYMBOLS

F	degree of freedom of a joint
f	degree of freedom of chain with one link fixed
g	number of joints
n	number of links
n_1	number of singular links
n_2	number of binary links
n_3	number of ternary links
n_4	number of quarternary links
g_2	number of joints connecting two links
g_3	number of joints connecting three links
g_4	number of joints connecting four links

A FINAL WORD TO MY READERS

It is with a certain sadness I leave you, the reader, at the end of my book. I hope it has been an interesting journey for you and I do hope it will be an inspiration to you, too.

General Bibliography

Altmann, F. G. *Schraubgetriebe, ihre moegliche und zweckmaessige Ausbildung*, VDI-Verlag, 1932.

Altmann, F. G. Ausgewaehlte Raumgetriebe, *VDI-Berichte* 12, 1956, pp. 69–76.

Altmann, F. G. Koppelgetriebe fuer gleichfoermige Uebersetzung, Z.*VDI* 92, no. 33, 1950, pp. 909–916.

Arnesen, L. Planetary Gears, *Machine Design* 31, September 3, 1959, pp. 135–139.

Bareiss, R. A., and Brand, P. A. Which Type of Torque-Limiting Device, *Product Engineering*, 29, August 4, 1958, pp. 50–53.

Beggs, J. S. *Mechanisms*, New York, McGraw-Hill Book Company, 1955.

Beggs, J. S. Cams and Gears Join to Stop Shock Loads, *Product Engineering* 28, September 16, 1957, pp. 84–85.

Berthold, H. *Stufenlos verstellbare mechanische Getriebe*, Berlin, Verlag Technik, 1954.

Beyer, R. *Technische Kinematik*, Leipzig, Barth Verlag, 1931.

Beyer, R. *Kinematische Getriebesynthese*, Berlin, Springer Verlag, 1953.

Beyer, R. Raeumliche Malteserkreuzgetriebe, *VDI-Forschungsheft* 461, VDI-Verlag, 1957, pp. 32–36.

Bloomfield, B. Non-Circular Gears, *Product Engineering* 31, March 14, 1960, pp. 55–66.

Bogardus, F. J. A Survey of Intermittent-Motion Mechanisms, *Machine Design* 28, September 20, 1956, pp. 124–131.

Botstiber, D. W. and Kingston, L. Cycloid Speed Reducer, *Machine Design* 28, June 28, 1956, pp. 65–69.

Brandenberger, H. *Kinematische Getriebemodelle,* Zuerich, Schweizer Druck and Verlagshaus AG, 1955.

Brunell, K. Constant Diameter Cams, ASME Paper 61SA-2, June, 1961.

Burmester, L. *Lehrbuch der Kinematik,* Leipzig, Felix Verlag, 1888.

Conway, H. G. Straight-Line Linkages, *Machine Design* 22, 1950, p. 90.

Creech, M. D. Dynamic Analysis of Slider-Crank Mechanisms, *Product Engineering* 33, October 29, 1962, pp. 58–65.

Crossley, F. The Permutations of Ten Link Plane Kinematic Chains, *Antriebstechnik* 3, 1964, pp. 181–185.

Cunningham, F. W. Non-Circular Gears, *Machine Design* 31, February 19, 1959, pp. 161–164.

Doughtie, V. L. and James, W. H. *Elements of Mechanism,* New York, John Wiley and Sons, Inc., 1962.

Franke, R. *Vom Aufbau der Getriebe,* Vols. I-II, Duesseldorf, VDI-Verlag, 1985 and 1951.

Fry, M. When Will a Toggle Snap Open, *Machine Design* 21, August 1949, p. 126.

Gagne, A. F. One-Way Clutches, *Machine Design* 22, April 1950, pp. 120–128.

Grodzinski-Polster. *Getriebelehre,* Sammlung Goeschen no. 1061, Volume 1-2, Berlin and Leipzig, 1933.

Grodzinski, P. Eccentric Gear-Mechanisms, *Machine Design* 25, May, 1953, pp. 141–150.

Grodzinski, P. Straight-Line Motion, *Machine Design* 23, June, 1951, pp. 125–127.

Grodzinski, P. and M'Ewen, E. Link Mechanisms in Modern Engineering, *Proceedings of the Institute of Mechanical Engineers* (London) 168, no. 37, 1954, pp. 877–896.

Grodzinski, P. Applying Eccentric Gearing, *Machine Design* 26, July, 1954, pp. 147–151.

Grodzinski, P. *A Practical Theory of Mechanisms,* Manchester, England, Emmett & Co., Ltd., 1947.

Gruebler, M. *Getriebelehre, eine Theorie des Zwanglaufes und der ebenen Mechanismen,* Berlin, Springer Verlag, 1917.

Haenchen, R. *Sperrwerke und Bremsen,* Berlin, Springer Verlag, 1930.

Hannula, F. W. Designing Non-Circular Surfaces, *Machine Design* 23, July, 1951, pp. 111–114, 190.

Harrisberger, L. *Mechanization of Motion,* New York, John Wiley and Sons, 1961.

Hartmann, W. *Die Maschinengetriebe,* Stuttgart-Berlin, 1913.

Heidler, G. R. Spring-Loaded Differential Drive, *Machine Design* 30, April 3, 1958, p. 40.

Hekeler, C. B. Flexible Metal Tapes, *Product Engineering* 32, February 20, 1961, pp. 65–69.

Hinkle, R. T. *Kinematics of Machines,* 2nd Edition, Englewood Cliffs, New Jersey, Prentice-Hall, Inc., 1960.

Hornsteiner, M. Getriebetechnische Fragen bei der Faltschachtel-Fertigung, Getriebetechnik, *VDI-Berichte* 5, 1955, pp. 66–74.

Jahr, W. and Knechtel, P. *Grundzuege der Getriebelehre*, vol. 1–2, Leipzig, Fachbuch Verlag, 1955, 1956.

Jensen, P. W. Retliniet foering af krankrogen ved vippekraner, *Ingenioer-og Bygningsvaesen*, February 10, 1957, pp. 52–53.

Jensen, P. W. Prize winner, *Machinery's* Ingenious Mechanisms Competition, February 1960.

Jensen, P. W. Four-Bar Linkage Proportions, *Product Engineering*, September 19, 1960, pp. 76–80.

Jensen, P. W. Intermittent Motion from Two Synchronized Cams, *Machinery*, November 1960, pp. 134–135.

Jensen, P. W. Mouvement intermittent communique par deux cams synchronisees, *Machine Moderne*, April 1961, pp. 47–48.

Jensen, P. W. Slide Motion Differential, *Machinery*, December 1961, p. 135.

Jensen, P. W. Rotary Work-Transfer Device, *Machinery*, March 1962, pp. 138–139.

Jensen, P. W. Differential a mouvement de coulisse, *Machine Moderne*, April 1962, p. 49.

Jensen, P. W. Four-Bar Power Linkages, *Product Engineering*, November 12, 1962, pp. 71–76 (original work by J. Volmer).

Jensen, P. W. Single Closed-Track Cam Drives Glue-Transfer Mechanism, *Machinery*, February, 1963, pp. 119–120.

Jensen, P. W. Cardano's Cirkler, *Ingenioer-og Bygningsvaesen*, May 25, 1963, pp. 284–285.

Jensen, P. W. Concave and Convex Machining on a Shaper, *Machinery*, July 1963, p. 114.

Jensen, P. W. Adjustable Indexing Mechanism with 180-Degree Dwell, *Machinery*, July 1963, p. 120.

Jensen, P. W. Simple Linkage Replaces Three Gears, *Machinery*, August 1963, p. 107.

Jensen, P. W. Mechanisms, *Product Engineering*, September 2, 1963, p. 67.

Jensen, P. W. Geneva Drive in Which Gear Ratios Control Motion Time, *Machinery*, September 1963, pp. 136–137.

Jensen, P. W. Cam unique a chemin de roulement ferme command un mecanisme transporteur de colle, *Machine Moderne*, September 1963, pp. 45–46.

Jensen, P. W. Mechanisms, *Product Engineering*, September 2, 1963, p. 67.

Jensen, P. W. Tringlerie a engrenages a cinq barres pour mouvement rectiligne, *Machine Moderne*, October 1963, p. 48.

Jensen, P. W. Zur Analyse und Synthese der dreigliedrigen Kurvengetriebe, Getriebetechnik, *VDI-Berichte* 77, October 1963, pp. 75–82.

Jensen, P. W. Kobbelkurver, *Ingenioer-og Bygningsvaesen*, December 10, 1963, p. 610.

Jensen, P. W. Indexing Mechanism with 180-Deg. Dwell, *Machinery*, (London), December 18, 1963, p. 139.

Jensen, P. W. Mecanisme de division reglable avec periode d'arret de 180 degres, *Machine Moderne,* December 1963, p. 47.

Jensen, P. W. Une tringlerie simple remplace trois engrenages, *Machine Moderne,* December 1963, p. 48.

Jensen, P. W. Oscillating Shaft Driven with Simple Hermonic Motion, *Machinery,* February 1964, p. 142.

Jensen, P. W. Usinage concave et convexe sur etau-limeur, *Machine Moderne,* February 1964, p. 48.

Jensen, P. W. Commande type Croix de Malte dans lequelle les rapports d'engrenages regissent le temps de mouvement, *Machine Moderne,* February 1964, p. 44.

Jensen, P. W. Bevaegelsesmekanisme, *Ingenioer-og Bygningsvaesen,* May 10, 1964, p. 157.

Jensen, P. W. Arbre oscillant entraine avec mouvement harmonique simple, *Machine Moderne,* May 1964, p. 47.

Jensen, P. W. Machinery Mechanisms, *Product Engineering,* June 8, 1964, pp. 66–74 (also reprinted).

Jensen, P. W. Constant Pivot Linkages Replace Ball Bearings, *Machinery,* June 1964, p. 157.

Jensen, P. W. Prize winner, Machinery's Ingenious Mechanisms *Competition,* July 1964.

Jensen, P. W. Sprocket-Operated Geneva Drive Provides Wide Design Possibilities, *Machinery,* July 1964, p. 125.

Jensen, P. W. Mekanismehjoernet, *Maskinindustrien,* September 7, 1964, p. 647.

Jensen, P. W. Mekanismehjoernet, *Maskinindustrien,* September 2, 1964, p. 688.

Jensen, P. W. Cam-Controlled Differential Mechanism, *Machinery,* October 1964, p. 156.

Jensen, P. W. Machinery Mechanisms, *Product Engineering*, October 26, 1964, pp. 108–115 (also reprinted).

Jensen, P. W. Mekanismehjoernet, *Maskinindustrien,* October, Nr. 1, 1964, p. 738.

Jensen, P. W. Cam-Controlled Differential Mechanism, *Machinery,* October 1964, pp. 156–157.

Jensen, P. W. Mekanismehjoernet, *Maskinindustrien,* October Nr. 2, 1964, p. 784.

Jensen, P. W. Analysis and Synthesis of Some Gear Drives, Having as a Basic Mechanism the Centric Slider-Crank or the Turning-Block Linkage, ASME-Paper 64-Mech-25, October 19, 1964.

Jensen, P. W. Machinery Mechanisms, *Product Engineering,* October 26, 1964, pp. 108–116.

Jensen, P. W. Mekanismehjoernet, *Maskinindustrien,* November, Nr. 1, 1964, p. 825.

Jensen, P. W. Mekanismehjoernet, *Maskinindustrien,* November, Nr. 2, 1964, p. 873.

Jensen, P. W. Mekanismehjoernet, *Maskinindustrien,* December, Nr. 1, 1964, p. 647.

Jensen, P. W. Mekanismehjoernet, *Maskinindustrien,* January, Nr. 1, 1965, p. 158.

Jensen, P. W. Mekanismehjoernet, *Maskinindustrien,* January, Nr. 2, 1965, p. 55.
Jensen, P. W. Gear-Slider Mechanisms, *Product Engineering,* February 1, 1965, pp. 757–760.
Jensen, P. W. Mekanismehjoernet, *Maskinindustrien,* February, Nr. 3, 1965, p. 33.
Jensen, P. W. Mekanismehjoernet, *Maskinindustrien,* February, Nr. 4, 1965, p. 10.
Jensen, P. W. Mekanismehjoernet, *Maskinindustrien,* March, Nr. 5, 1965, p. 11.
Jensen, P. W. Mekanismehjoernet, *Maskinindustrien,* April Nr. 6, 1965, p. 12.
Jensen, P. W. Mekanismehjoernet, *Maskinindustrien,* April, Nr. 8, 1965, p. 271.
Jensen, P. W. Mekanismehjoernet, *Maskinindustrien,* May, Nr. 9, 1965, p. 309.
Jensen, P. W. Mekanismehjoernet, *Maskinindustrien,* May, Nr. 10, 1965, p. 347.
Jensen, P. W. Apparatus for Separating Lengths of Can Stock, US Patent 3,183,754, May 18, 1965.
Jensen, P. W. Systematic Mechanism Design, ASME-Paper 65-MD-37, 1965.
Jensen, P. W. Mekanismehjoernet, *Maskinindustrien,* June, Nr. 11, 1965, p. 385.
Jensen, P. W. Mekanismehjoernet, *Maskinindustrien,* June, Nr. 12, 1965, p. 417.
Jensen, P. W. Two-gear Drive Produces Variable Output Motions, *Machinery,* June 1965, pp. 104–105.
Jensen, P. W. Cycloid Gear Mechanisms, *Product Engineering,* June 21, 1965, pp. 82–88 (also reprinted).
Jensen, P. W. Mekanismehjoernet, *Maskinindustrien,* July, Nr. 13, 1965, p. 453.
Jensen, P. W. Mekanismehjoernet, *Maskinindustrien,* August Nr. 15, 1965, p. 493.
Jensen, P. W. Mekanismehjoernet, *Maskinindustrien,* August Nr. 16, 1965, p. 533.
Jensen, P. W. Cykloide mekanismer, *Maskinindustrien,* May, Nr. 16, 1965, p. 518–536.
Jensen, P. W. Mekanismehjoernet, *Maskinindustrien,* September, Nr. 17, 1965, p. 569.
Jensen, P. W. Mekanismehjoernet, *Maskinindustrien,* September, Nr. 18, 1965, p. 609.
Jensen, P. W. Geneva Mechanisms, *Design News,* September 29, 1965 (also reprinted).
Jensen, P. W. Mecanismes a courbes cycloidale, *Machine Moderne,* September 1965, pp. 39–44.
Jensen, P. W. Mekanismehjoernet, *Maskinindustrien,* September Nr. 19, 1965, p. 653.
Jensen, P. W. Mekanismehjoernet, *Maskinindustrien,* October, Nr. 20, 1965, p. 693.
Jensen, P. W. High-Speed Cutoff Device, *Machinery,* October, 1965, pp. 114–115.
Jensen, P. W. Mekanismehjoernet, *Maskinindustrien,* November, Nr. 21, 1965, p. 729.
Jensen, P. W. Mekanismehjoernet, *Maskinindustrien,* November, Nr. 22, 1965, p. 769.
Jensen, P. W. Mekanismehjoernet, *Maskinindustrien,* December, Nr. 23, 1965, p. 809.
Jensen, P. W. Mekanismehjoernet, *Maskinindustrien,* December, Nr. 24, 1965, p. 845.

Jensen, P. W. *Cam Design and Manufacture,* New York, The Industrial Press, 1965.

Jensen, P. W. Mekanismehjoernet, *Maskinindustrien,* January, Nr. 1, 1966, p. 20.

Jensen, P. W. Mekanismehjoernet, *Maskinindustrien,* January, Nr. 2, 1966, p. 249.

Jensen, P. W. High-Speed Cutoff Mechanism, *Machinery,* February 1966, pp. 88–89.

Jensen, P. W. Getriebe und Getriebetechnik fuer Konstrukteure, *Maschinenmarkt* 72, 1966, pp. 19–26.

Jensen, P. W. The Application of the Method of Turned Velocities and the Acceleration of the Instant Center to Determine Velocities and Accelerations in Complex Mechanisms, ASME-Paper 66-Mech-33, 1966.

Jensen, P. W. Mekanismehjoernet, *Maskinindustrien,* January, Nr. 1, 1967, p. 49.

Jensen, P. W. Mekanismehjoernet, *Maskinindustrien,* January, Nr. 2, 1967, p. 57.

Jensen, P. W. Mekanismehjoernet, *Maskinindustrien,* February, Nr. 4, 1967, p. 125.

Jensen, P. W. Mekanismehjoernet, *Maskinindustrien,* March, Nr. 6, 1967, p. 212.

Jensen, P. W. Mekanismehjoernet, *Maskinindustrien,* April, Nr. 8, 1967, p. 281.

Jensen, P. W. Mekanismehjoernet, *Maskinindustrien,* May, Nr. 9, 1967, p. 333.

Jensen, P. W. Bewegungsverhaeltnisse an Schubkurbeln, *Maschinenmarkt,* May 5, 1967, pp. 776–779.

Jensen, P. W. Geschwindigkeiten und Beschleunigungen am Gelenkviereck, *Maschinenmarkt* 73, 1967, pp. 1675–1676.

Jensen, P. W. Krumtapmekanismens kinematik, *Maskinindustrien,* May, Nr. 10, 1967, p. 355–362.

Jensen, P. W. Mekanismehjoernet, *Maskinindustrien,* June, Nr. 12, 1967, p. 455.

Jensen, P. W. Mekanismehjoernet, *Maskinindustrien,* October, Nr. 19, 1967, p. 681.

Jensen, P. W. Mekanismehjoernet, *Maskinindustrien,* October, Nr. 21, 1967, p. 781.

Jensen, P. W. Mekanismehjoernet, *Maskinindustrien,* October, Nr. 22, 1967, p. 825.

Jensen, P. W. Mekanismehjoernet, *Maskinindustrien,* December, Nr. 23, 1967, p. 877.

Jensen, P. W. Planethjulsmekanismer I, *Maskinindustrien,* January, Nr. 1, 1968, pp. 23–24, 27.

Jensen, P. W. Planethjulsmekanismer II, *Maskinindustrien,* January, Nr. 2, 1968, pp. 13, 15, 17–19, 21–22.

Jensen, P. W. Kinematic Space Requirements and Efficiency of Coupled Planetary Gear Systems, ASME-Paper 68-Mech-45 (10th U.S. Conference on Mechanisms).

Jensen, P. W. Raumbedarf und Wirkungsgrad zusammengesetzter Planetenraedergetriebe mit einstufigem Planetenrad, *Konstruktion,* no. 5, pp. 178–184.

Jensen, P. W. Offene und geschlossene kinematische Ketten, *Maschinenmarkt* 76, 1970, pp. 2336–3444.

Jensen, P. W. Kinematische Ketten mit einem Freiheitsgrad und bis zu 10 Gliedern, *Maschinenmarkt* 76, 1970, pp. 635–638.

Jensen, P. W. The Missing Link: An Extension of Gruebler's Theory, ASME-Paper 70-Mech-70, (11th U.S. Conference on Mechanisms).

Jensen, P. W. *Kinematics and Dynamics of Mechanisms* (in Danish), Teknisk Forlag, Copenhagen, 1970.

Jensen, P. W. *Mekanismesystematik* (Systematic Design of Mechanisms), Transactions of the Danish Society of Mechanical Engineers, DIF, 1973.

Jensen, P. W. Converting Rotary Motion to Indexing Motion with Geneva Mechanisms, *Project* (Copenhagen), no. 6, December 1982, pp. 19–32.

Jensen, P. W. From Rotating to Indexing Motions with Geneva Mechanisms, *8th Applied Mechanisms Conference*, 1983.

Jensen, P. W. Burmester—Up Two Points (This Is Not the Dow Jones Index), *9th Applied Mechanisms Conference*, Kansas City, Missouri, 1985.

Jensen, P. W. Synthesis of Slider-Cranks, Turning-Block Linkages and Geared Link Mechanisms, *9th Applied Mechanisms Conference*, Kansas City, Missouri, 1985.

Jensen, P. W. The Inverted Coupler Curve method for Path Generation, *10th Applied Mechanisms Conference*, 1987.

Jensen, P. W. Path Generation by the Slider-Crank Exemplified by Four-Point Straight-Line Mechanisms, *10th Applied Mechanisms Conference*, 1987.

Jensen, P. W. Synthesis of Geared Double-Crank Mechanisms for Four Precision Points and Three Crank Angles for Any Positive or Negative Transmission Ratio, *10th Applied Mechanisms Conference*, 1987.

Jensen, P. W. Using Dead-Center Positions in Kinematic Synthesis (Path Generation), *Proceedings of the 7th World Congress on Mechanisms*, September 17–22, Sevilla, Spain, vol. 1, New York, Pergamon Press.

Jensen, P. W. *Cam Design and Manufacture*, 2nd Edition, New York, Marcel Dekker, Inc., 1987.

Jensen, P. W. The Polode Synthesis Method, *1st National Applied Mechanisms and Robotics Conference*, Cincinnati, Ohio, 1989.

Jensen, P. W. Die Pseudopol-Methode—ein neues Verfahren der Getriebesynthese, *VDI-Forschungsheft* 660, July 1990, Berlin, VDI-Verlag.

Jensen, P. W. Streamlined Linkage Design, *Machine Design*, March 21, 1991, pp. 169–172.

Johnson, R. C. Geneva Mechanisms, *Machine Design* 28, March 22, 1956, pp. 107–111.

Johnson, R. C. Development of a High-Speed Indexing Mechanism, *Machine Design* 30, September 4, 1958, pp. 134–138.

Jones, F. D. *Ingenious Mechanisms for Designers and Inventors*, vol. 1–4, New York, The Industrial Press.

Kaplan, J., and North, H. Cyclic Three-Gear Drives, *Machine Design* 31, March 19, 1959, pp. 185–188.

Kearny, W. R., and Wright, M. G. Straight-Line Mechanisms, *Transactions of the Second Conference on Mechanisms*, Purdue University, West Lafayette, Indiana, October 1954, pp. 209–216.

Kiper, G. Kurvendifferentialgetriebe, *Konstruktion*, February 1968, pp. 41–48.

Kist, K. E. Modified Starwheels, *Transactions of the Third Conference on Mechanisms*, Purdue University, West Lafayette, Indiana, May 1956, pp. 16–20.

Kraus, R. *Grundlagen des systematischen Getriebeaufbaus,* Verlag Technik, Berlin, 1952.

Kraus, R. *Geradfuehrungen durch das Gelenkviereck,* VDI-Verlag, 1955.

Kraus, R. *Getriebelehre,* vol. 1, Berlin, Verlag Technik, 1954.

Kuhlenkamp, A. Linkage Layouts by Mathematical Analysis, *Product Engineering,* August 1955, pp. 165–170.

Lanz, M. M. and Betancourt. *l'Essay sur la Composition des Machines,* 1810.

Lichtwitz, O. Mechanisms for Intermittent Motion, *Machine Design* 23, December 1951, pp. 134–148.

Lichtwitz, O. Mechanisms for Intermittent Motion, *Machine Design* 24, January 1952, pp. 127–141.

Lichtwitz, O. Mechanisms for Intermittent Motion, *Machine Design* 24, February 1952, pp. 146–155.

Lichtwitz, O. Mechanisms for Intermittent Motion, *Machine Design* 24, March 1952, pp. 147–155.

Lichtwitz, O. *Getriebe fuer aussetzende Bewegung,* Berlin, Springer Verlag, 1953.

Lockenvitz, A. E. Non-Circular Cams and Gears, *Machine Design* 24, May, 1952, pp. 141–145.

Morgan, P. Mechanisms for Puppets, *Machine Design* 34, December 20, 1962, pp. 105–109.

Morrison, R. A. Rolling-surface Mechanisms, *Machine Design* 30, December 11, 1958, pp. 119–123.

Peyrebrune, H. E. Application and Design of Non-circular Gears, *Transactions of the First Conference on Mechanisms,* Purdue University, West Lafayette, Indiana, 1953, pp. 13–21.

Pollitt, E. P. High-speed Web-cutting, *Machine Design* 27, December 1955, pp. 155–160.

Rappaport, S. Kinematics of Intermittent Mechanisms, *Product Engineering* 20, July 1949, pp. 110–112.

Rappaport, S. Kinematics of Intermittent Mechanisms, *Product Engineering* 20, August 1949, pp. 109–112.

Rappaport, S. Kinematics of Intermittent Mechanisms, *Product Engineering* 20, October 1949, pp. 137–139.

Rappaport, S. Kinematics of Intermittent Mechanisms, *Product Engineering* 21, January 1950, pp. 133–134.

Rappaport, S. Crank-and-Slot Drive, *Product Engineering* 21, July 1950, pp. 136–138.

Rappaport, S. Shearing Moving Webs, *Machine Design* 28, May 3, 1956, pp. 101–104.

Rappaport, S. Elliptical Gears for Cyclic Speed Variation, *Product Engineering* 31, March 28, 1960, pp. 68–70.

Rasche, W. H. Design Formulas for Three-link Screw Mechanisms, *Machine Design* 17, August 1945, pp. 147–149.

Rauh, K. *Aufbaulehre der Verarbeitungsmaschinen,* Stuttgart, Francksche Verlagshandlung, 1949.

Rauh, K. *Entwicklungslinien im Landmaschinenbau,* W. Girardet, Essen, 1949.
Rauh, K. *Praktische Getriebelehre,* vols. I–II, Berlin, Springer Verlag, 1951, 1954.
Reinecke, H. Systematik der Spreizmechanismen by Photoapparaten, *Feinwerktechnik,* 1951, no. 5, pp. 101–110.
Reuleaux, F. *Theoretische Kinematik,* part 1, Braunschweig, Verlag Vieweg und Sohn, 1875.
Reuleaux, F. *Die praktischen Beziehungen der Kinematik zur Geometrie und Mechanik,* part 2, Braunschweig, Verlag Vieweg und Sohn, 1900.
Reuleaux, F. *Der Konstrukteur,* Braunschweig, Verlag Vieweg und Sohn, 1899.
Reuleaux, F. *The Constructor* (translated by H. H. Supplee), Princeton, New Jersey, D. Van Nostrand Company, Inc., 1893.
Reuleaux, F. *The Kinematics of Machinery,* New York, Dower Publications, 1965.
Ring, F. Remote Control Handling Devices, *Mechanical Engineering* 78, September 1956, pp. 828–831.
Roemer, R. L. Flight-control Linkages, *Mechanical Engineering 80,* June 1958, pp. 55–60.
Roessner, E. E. Ratchet Layout, *Product Engineering* 29, January 20, 1958, pp. 89–991.
Rumsey, R. D. Redesigned for Higher Speed, *Machine Design* 23, April 1951, pp. 123–129.
Schmidt, E. H. Cyclic Variations in Speed, *Machine Design* 19, March 1947, pp. 108–111.
Schmidt, E. H. Cycloidal-crank Mechanisms, *Machine Design* 31, April 2, 1959, pp. 111–114.
Schulze, E. F. C. Designing Snap-action Toggles, *Product Engineering* 26, November 1955, pp. 168–170.
Sheppard, W. H. Rolling Curves and Non-circular Gears, *Mechanical World,* January 1960, pp. 5–11.
Sieker, K.-H. *Einfache Getriebe,* Leipzig, Akademischer Verlagsanstalt, 1950.
Sieker, K.-H. *Getriebe mit Energiespeichern,* Verlagshandlung, Fuessen, C. F. Wintersche 1954.
Simonis, F. W. *Stufenlos verstellbare Getriebe,* Berlin, Springer Verlag, 1949.
Spector, L. F. Mechanical Adjustable-speed Drives, *Machine Design* 27, June, 1955, pp. 178–189.
Spector, L. F. Mechanical Adjustable-speed Drives, *Machine Design* 27, April, 1955, pp. 163–196.
Spotts, M. F. Kinematic Properties of the Three-gear Drive, *Journal of the Franklin Institute* 268, no. 6, December 1959, pp. 464–473.
Strasser, F. Ten Universal Shaft-couplings, *Product Engineering* 29, August 18, 1958, pp. 80–81.
Talbourdet, G. J. Intermittent Mechanisms, *Machine Design* 20, September 1948, pp. 159–162.
Talbourdet, G. J. Intermittent Mechanisms, *Machine Design* 20, October 1948, pp. 135–138.

Talbourdet, G. J. Motion Analysis of Several Intermittent Variable-speed Drives, *Transactions of the ASME* 71, 1949, pp. 83–96.

Talbourdet, G. J. Intermittent Mechanisms, *Machine Design* 22, September 1950, pp. 141–146.

Talbourdet, G. J. Intermittent Mechanisms, *Machine Design* 22, October 1950, pp. 121–125.

Vandeman, J. E. and Wood, J. R. Modifying Starwheel Mechanisms, *Machine Design* 25, April 1953, pp. 255–261.

Vogel, W. F. Crank Mechanism Motions, *Product Engineering* 12, June 1941, pp. 301–305.

Vogel, W. F. Crank Mechanism Motions, *Product Engineering* 12, July 1941, pp. 374–379.

Vogel, W. F. Crank Mechanism Motions, *Product Engineering* 12, August 1941, pp. 423–428.

Vogel, W. F. Crank Mechanism Motions, *Product Engineering* 12, September 1941, pp. 490–493.

Volmer, J. Konstruktion eines Gelenkgetriebes fuer eine Geradfuehrung, *VDI-Berichte* 12, 1956, pp. 175–183.

Volmer, J. Systematik, Kinematik des Zweiradgetriebes, *Maschinenbautechnik* 5, no. 11, 1956, pp. 583–589.

Volmer, J. Gelenkgetriebe zur Geradfuehrung einer Ebene, *Zeitschrift praktischer Metallbearbeitung* 53, no. 5, 1959, pp. 169–174.

Weise, H. Bewegungsverhaeltnisse an Filmschaltgetriebe, *VDI-Berichte* 5, 1955, pp. 99–106.

Weise, H. Getriebe in photographischen und kinematographischen Geraete, *VDI-Berichte* 12, 1956, pp. 131–137.

Weyth, N. C., and Gagne, A. F. Mechanical Torque-limiting Devices, *Machine Design* 18, May, 1946, pp. 127–130.

Willis, R. *Principles of Mechanism,* 2nd Edition, Cambridge, England, 1870.

Woo, L. S. Type Synthesis of Plane Linkages, *ASME-Transactions, Journal of Engineering for Industry,* February, 1967, pp. 159–172.

Ziebolz, H. A New Approach to Design, *Machine Design,* June 1947.

Ziebolz, H. Systematic Design of Mechanisms, *Machine Design,* December 1950, pp. 126–136.

Bibliography: Planetary Gear Systems

Abild, R. N. Epicyclic Gear Systems, *Machine Design* 27, no. 8, August 1955, p. 171.

Allen-Stoeckicht High-Speed Epicyclic Gearing. *Engineering* 171, no. 4436, February 2, 1951, pp. 117–119.

Allen, H. N. G. and Jones, T. P. Epicyclic Gears, *Inst. Mar. Engr., Tran* 64, no. 5, May, 1952, pp. 79. (See also *Engineering* 193, no. 5016, 5017, March 14, 1952, p. 373.)

Allen, H. N. G. and Jones, T. P. Marine and Industrial High Powered Epicyclic Gearing, *Eng.-J.* 43, no. 9, September 1960, pp. 83–92.

All Epicyclic Transmission in the German Escort Vessels, *Marine Engineering,* June, 1960.

Alt, E. and Netze, G. Krupp-Stoeckicht-Planetengetriebe, *Techn. Mitt. Krupp,* 1954, no. 6 (Sonderdruck).

Alt, E. Leistungsgrenzen von Planetengetrieben, *Z. VDI* 97, 1955, no. 7, pp. 214–215.

Altmann, F. G. Die Bauformen gleichachsiger Stirnradumformer, *Masch.-Bau* 6, 1927, pp. 1083–1087.

Altmann, F. G. Antrieb von Hebezeugen durch hochuebersetzende, raumsparende Stirnradgetriebe, *Masch.-Bau* 6, 1927, pp. 1091–1094.

Altmann, F. G. Zwei neue Getriebe fuer gleichfoermige Uebersetzung, *Masch.-Bau* 8, 1929, pp. 721–723.

Altmann, F. G. Bauformen der Getriebe bei neueren Getriebemotoren, *Z. VDI* 76, 1932, pp. 235–240.

Altmann, F. G. Stufenlos regelbare Schaltwerkgetriebe, *Z. VDI* 84, 1940, pp. 333–342.

Altmann, F. G. Ausgleichgetriebe fuer Kraftfahrzeuge, *Z. VDI* 84, 1940, pp. 545–551.

Altmann, F. G. Koppelgetriebe fuer gleichfoermige Uebersetzung, *Z. VDI* 92, 1950, pp. 909–916.

Altmann, F. G. Zur Zahlsynthese der raeumlichen Koppelgetriebe, *Z. VDI*, Duesseldorf, March 21, 1951, pp. 205–208.

Altmann, F. G. and Heimann, G. Pruefstandversuche an Sperr-Ausgleichsgetrieben fuer Kraftfahrzeuge, *Z. VDI* 85, 1941, pp. 749–754.

Andersson, S. On Hydrodynamic Torque Converters, Dissertation, Lund Technical University, Lund, Sweden, 1982.

Andersson, S. Analysis of Multi-Element Torque Converter Transmissions, *Int. J. Mech. Sci.* 28, no. 7. 1086, pp. 431–441.

Arlt, W. Uebersetzungsverhaeltnisse bei Planetenradgetrieben und umlaufenden Raederwerken, *Reuleaux-Mitt.* 2, 1934, pp. 79–89.

Arlt, W. Zusammenhaenge und Gesetze der Drehzahlverhaeltnisse ebener Raederwerke, *Wiss. Arb. d. Staatl. Akad. f. Technik,* Chemnitz, Mitte, Oktober 1936, pp. 3–33.

Arnaudow, K. *Untersuchung des Lastausgleiches in Planetengetrieben,* Dissertation Technical University, Dresden, 1968.

Arnaudow, K. Ueber einige Grundfragen des Lastaugleiches bei Planetengetrieben, *Maschinenbautechnik* 12, 1963, no. 8, pp. 433–437.

ATZ: Die automatischen Getriebe der amerikanischen, "Compact Cars," *ATZ* 63, no. 8, pp. 227–235.

Autenrieth, E. and Ensslin, M. Selflocking, in *Technische Mechanik,* Berlin, J. Springer, 1922, pp. 190–192.

Autorenkollektiv: *Ausruestungen fuer Bergbau und Schwerindustrie,* Vol. 3, Getriebe und Triebwerksteile, Fachbuchverlag, Leipzig, 1955.

Bachman, G. Flanschmotor-Gelenk-Planetengetriebe, *AEG-Mitt.,* 1935, p. 104.

Balogh, A. Epicyclic Gears, *Automotive Engineering* 40, no. 524, February 1950, pp. 73–74.

Balogh, A. Berechnung der Geschwindigkeit der Planetengetriebe auf einer neuen Grundlage. *Maschinenbautechnik* 3, no. 12, 1954, pp. 618–620.

Barnwell, J. H. The Double-Eccentric Speed Reducer, *Machine Design,* August 17, 1961, pp. 135–137.

Bastert, C.-CH. Die Verlagerung der Zentralraeder in Planetengetrieben. *Forschung im Ingenieurwesen* 37, VDI-Verlag, no. 1, 1971.

Beck, G. Massnahmen zur Steigerung der Gelaendegaengigkeit bei Sonderfahrzeugen. Das Wenden mit Gleiskettenfahrzeugen, *ATZ,* 1939, pp. 413–419.

Beckmann, K. Wirkungsgrad und Stellbereich von Planetengetrieben mit Ueberlagerungszweig in Abhaengigkeit vom Verhaeltniss der Ueberlagerungsleistung zur Gesamtleistung, *Konstruktion* 20, no. 9, 1968, pp. 358–364.

Bedaux, M. F. Considerations analytiques et geometriques sur les trains epicycloidaux, *Société des Ingenieurs de l'automobile* 30, Feb. 1965, p. 77.

Beggs, J. S. and Beggs, R. S. Planeten-Kurven-Getriebe, *Z. VDI* 99, no. 19, July 1, 1957, pp. 839–840.

Benford, R. L. Numerical Rules for Designing Planetary Gears, *Machine Design,* August 21, 1958, pp. 129–135.

Berger, A. Zur Kinematik des Umlaufgetriebes, *MTZ* 6, 1944, pp. 67–70.

Berger, G. Automatisches, stufenlos wirkendes hydrostatisches Lastschaltgetribe fuer Kraftfahrzeuge, Report no. 86.6, Institut fuer Konstruktionstechnik, Ruhr-Universitaet Bochum, Federal Republic of Germany, 1986.

Bergstrasser, M. Planetengetriebe mit auswechselbarer Uebersetzung, *Maschinenbautechnik* 10, 1961, pp. 209–211.

Bergstrasser, M. Stirnradpaarungen mit Profilverschiebung. Technische Akademie-Wuppertal. *Berichte* no. 1.

Bernhard, J. M. Die Planetenraedergetriebe und die graphische Darstellung der Bewegungsvorgaenge durch Diagramme etc., *Glasers Annalen* 65, 1941, pp. 35–43, 112–117.

Bernhard, J. M. Die Planetenraedergetriebe. *Glasers' Annalen* 65, no. 4, 1941, pp. 35–43, and 66, no. 12, 1942, pp. 125–132.

Beyer, R. Graphisch-kinematische Methoden an Umlaufgetrieben und am Raederknie. *Z. Maschinenkonstrukteur,* no. 2/3, 1929.

Beyer, R. Graphisch-kinematische Grundlagen der Radgetriebe mit sich schneidenden Achsen, *Masch.-Bau,* 1929, pp. 718–720.

Beyer, R. *Technische Kinematik,* Leipzig, J. A. Barth, 1931.

Beyer, R. Drehzahlvektorenplane ebener Getriebe. *Z. f. Instru.-Kunde,* 1933, pp. 164–172.

Beyer, R. Neue Winkelgeschwindigkeitsplane ebener Getriebe. *Reuleaux-Mitteilungen,* 1934, pp. 90–91.

Beyer, R. Drehzahlberechnung von Umlaufraedergetrieben. *Masch.-Bau* 14, 1935, pp. 109–110.

Beyer, R. Getriebetechnik der Schnellhobler. *Masch.-Bau* 16, 1937, pp. 161–164.

Beyer, R. Uebersetzungsverhaeltnisse von Raederkurbelgetrieben in neuer Behandlungsweise. *Werkstatt und Betrieb* 83, 1950, pp. 389–393.

Birkle, H. G. *Das Betriebsverhalten der stufenlos einstellbaren Koppelgetriebe.* Diss. T. H. Darmstadt, 1968.

Blaise, H. Die relative Verstellung zweier Wellen waehrend des Betriebes und ihr Aufbau, *Reuleaux Mitteilungen,* 1934, pp. 2–5.

Bloch, P. and Schneider, R. C. Hydrodynamic Split Torque Transmissions, *SAE Trans.* 68, 1960, pp. 257–273.

Bock, F. Ueber Differentialraederwerke, *VDI-Z,* 1879, p. 411.

Boege, A. *Die Mechanik der Planetengetriebe,* Braunschweig, F. Vieweg und Sohn, 1980.

Boracci, P. Il cambio automatico Pavesi, *Auto Italiana,* 1938, pp. 19–22.

Bosch, M. *Vorlesungen ueber Maschinenelemente,* Berlin, J. Springer, 1940, pp. 383–391.

Bosch, M. *Berechnung der Maschinenelemente,* Berlin, J. Springer, 1951.

Brand, A. Die Kraftuebertragungsanlagen fuer Verbrennungstriebwagen, *MTZ,* 1940, pp. 156–176.

Braennare, G. On Rolling Variators, Dissertation, Division of Machine Elements, Chalmers University of Technology, Gothenburg, Sweden, 1986.

Brandenberger, H. Wirkungsgrad und Aufbau einfacher und zusammengesetzer Umlaufraedergetriebe. *Masch. Bau-Betrieb* 8, 1929, pp. 249–253, 290–294.

Brandenberger, H. Die Berechnung des Reibungsverlustes und des Wirkungsgrades von Umlaufgetrieben, *Schweizerische Bauzeitung* 70, no. 39, Sept. 27, 1952, p. 559.

Brandenberger, H. Der Wirkungsgrad selbshemmender Umlaufgetriebe bei Umkehrung der Bewegungsrichtung, *Getriebetechnik I,* VDI-Verlag, 1953, pp. 69–76.

Brass, E. A. Two Stage Planetary Arrangements for the 15:1 Turboprop Reduction Gear, *ASME* Paper no. 60-SA-1, 1960.

Breier, K. Planetengetriebe im Huettenbau. *DEMAG-Informationen* Za 109.

Breuer, K. Planetengetriebe fuer grosse Drehmomente und grosse Leistungen. *Antriebstechnik* 1, 1962, pp. 17–24.

Breuer, K. Planetengetriebe fuer grosse Drehmomente und grosse Leistungen, Part I, *Antriebstechnik,* April 1962, pp. 17–29.

Breuer, K. Planetengetriebe fuer grosse Drehmomente und grosse Leistungen, Part II, *Antriebstechnik,* November 1962, pp. 146–157.

Breuer, K. Neue Entwicklung von hydrostatischen Getrieben, *VDI-Z* 106, no. 6, 1964, pp. 193–195.

Breuer, K. Planetengetriebe fur grosse Drehmomente und grosse Leistungen, *Deutsche Hebe und Foerdertechnik* 11, no. 2, 1965, pp. 69–71, no. 3, pp. 109–111.

Breuer, K. Zur Konstruktion von Grossgetrieben, *DEMAG* Nachrichten, no. 163.

Breuer, K. and Schultz, M. Planetengetriebe mit oelhydraulischer Drehzahlregelung, *VDI-Nachrichten,* October 11, 1961.

Brewer, R. C. Synthesis of Epicyclic Gear Trains Using the Velocity Ratio Spectrum, *Trans. of the ASME,* August 1960, pp. 173–178.

Brode, K. Drei neue Motoren aus der Musterreihe, *Luftwissen,* 1941, pp. 57–59.

Brode, K. Die Baumuster des Einspritzhochleistungsmotor, Bramo Fafnir 23, *Luftwissen,* 1941, pp. 189–193.

Brosamler, H. Ein Beitrag zur Berechnung von Huellelementen, insbesondere Rollenketten in Umlaufgetrieben, *Wissenschaftliche Zeitschrift der T. H. Dresden* 8, no. 4, 1958–1959, pp. 753–762.

Brown, W. Compound Epicyclic Gearing, *Engineering Materials and Design* 6, 1963, no. 6, pp. 438–439, 441.

Brown, W. *Epicyclic Gearing,* London and New York, The Machinery Publishing Co.

Buckingham, A. *Manual of Gear Design,* Section 2, Machinery, New York, 1935, pp. 129–136.

Budnick, A. Planeten-Wendegetriebe der Ljungstroem Schiffsturbinen, *Z. VDI,* 1924, p. 1191.

Budnick, A. Zeichnerische Behandlung von Kraeften und Momenten in Koppel- und Raedergetrieben, *VDI-Forschungsheft* 288, Berlin, 1938.

Buetner, P. Planetengetriebe, Konstruktion und Herstellung, *Verein Ind. Kraftwirtschaft-Berichte* 137, Essen, 1963, pp. 13–19.

Bunke, G. T. Turbo-Umlaufraedergetriebe, *ASUG-Mitteilungen* 6, no. 3, 1969.

Burgess, E. G. Minimization of Gear Train Inertia, *Transactions of the ASME,* April 1954, pp. 493–496.

Burmester, L. *Lehrbuch der Kinematik,* Leipzig, Felix, 1888, p. 478.

Bussien, R. *Automobiltechnisches Handbuch,* Berlin, Drayn Verlag, 1941.

Carlson, B. and Larsson, H. Wirkungsgrad und Selbsthemmung einfacher Umlaufgetriebe, *Report* no. 1 from Inst. for Machine Elements, Chalmers Technical University, Gothenburg, 1957.

Carmichael, C. Planetary Gear Calculations, *Machine Design* 20, no. 3, March 1948, p. 165.

Chabert, G. Montage des satellites doubles dans les trains planetaires, Société d'études de l'industrie de l'engrenage, *SEIE,* no. 41, 1963, pp. 1–13.

Chang, Ch.-T. Multi-Stage Planetary Gear Trans, *Design News,* June 23, 1965, pp. 138–145.

Chilton, R. Aircraft-Engine Reductions Gears and Torque Meters, *Automotive Industries,* 1940, pp. 526–529.

Chironis, N. P. Rearranged Linkages Create New Planetary Gear Systems, *Product Engineering,* February 13, 1967, pp. 70–73.

Chonik, S. Embrayages hypocycloidaux, *Mecanique-Electricité* 50, no. 197, 1966, pp. 26–31.

Clausen, H. Efficiency of Gear Trains, *Engineering* 177, no. 4599, March 19, 1954, p. 366.

Coenen, M. *Elemente des Werkzeugmaschinenbaus,* Berlin, 1927, pp. 61–67.

Conklin, R. M. and Ballmer, J. Epicyclic Gear Train Calculations, *Product Engineering,* April 1957, pp. 186–187.

Cotal, I. Changement de vitesse éléctromecanique "Cotal," *J. Soc. Ingrs. Automobile,* 1940, pp. 27–31.

Cowie, A. Nomograph Simplifies Planetary Gear Calculations, *Machine Design* 19, no. 9, September 1947, p. 155.

Cowie, A. Another Look at Planetary Gears, *Design News,* December 7, 1959, pp. 70–73.

Crown, W. R. Positions of Differential Gears, *Design News* November 23, 1959, pp. 56–57.

Dahl, A. Selbstspannende Riementriebe. *Konstruktion* 6, no. 8, 1954.

Danke, P. Aufsteckgetriebe, *Antriebstechnik* 4, 1965, no. 8, p. 310.

Danke, P. Normalgetriebe, Planetengetriebe, Schiffs- und Fahrzeuggetriebe auf der Hannover-Messe 1965, *Antriebstechnik* 4, 1965, no. 6, pp. 209–214.

Delanghe, G. Les changements de vitesse epicycloidaux. *Gvn. civ.,* 1937, pp. 228–230.

Denham, A. F., and Deschamps, H., and Kutzbach, K. Torque Reversal Shifts Gears, *ATZ,* 1933, p. 289. *Pruefung, Wertung und Weiterentwicklung von Flugmotoren,* Verlag C. Schmidt & Co., Berlin, 1921.

Dietrich, G. Reibungskraefte, Laufunruhe und Geraeuschbildung an Zahnraedern, *Deutsche Kraftfahrtforschung,* no. 25, Berlin, 1949.

Dietrich, G. *Berechung von Stirnraedern,* Duesseldorf, VDI-Verlag GmbH, 1952.

Dobbeler, C. V. Umlaufgetriebe, *Betrieb* 1, 1918–19, pp. 171–179.

Dobbeler, C. V. Differentialgetriebe, *Masch. Bau/Gestaltung* 3, 1923, pp. 175–177.

Doerpmund, H. Kickdown verzoegert Hochschalten. *VDI-Nachrichten*, March 27, 1968, p. 14.

Doerpmund, H. Das automatisch schaltende Planetengetriebe des VW 411, sein Wandler und sein Fluid. Sonderdruck aus *Oelhydraulik und Pneumatik* 13, no. 2, 1969.

Dorey, R. E. and McCandlish. The modelling of losses in mechanical gear trains for the computer simulation of heavy vehicle transmission systems, *Proceedings, International Conference on Integrated Engine Transmission Systems*, Bath, IMechE Conference Publications, 1986–87, pp. 69–82.

Dorsch, S. Umlaufraedergetriebe zur Drehzahlregelung von Stroemungsmaschinen, *Die Maschine*—Antriebe & Getriebetechnik, 20, 1966, pp. 51–55.

Droscha, H. Erfolgreiche Umkonstruktion von Antriebselementen. *Antriebstechnik* 5, no. 8, 1966, pp. 274–277.

Dubbel, H. *Taschenbuch fuer den Maschinenbau*, Vol. 1, 1951, pp. 586, 685.

Duda, M. Der geometrische Verlustbeiwert und die Verlustunsymmetrie bei geradverzahnten Stirnradgetrieben, *Forschung im Ingenieurwesen* 37, no. 1, VDI-Verlag, 1971.

Duditza, F. and Miloiu, C. Planetengetriebe mit zwei Zahnraedern. *Antriebstecknik* 8, no. 6, 1969, pp. 219–223.

Dudley, D. W. *Gear Handbook*, New York, McGraw-Hill, 1962, Chapter 3.

Eckert, B. Bauliche Gestaltung von Kraftfahrzeuggetrieben, *ATZ*, 1941, pp. 362–366, 381–385.

Eckhardt, H.-G. Hydrostatische Getriebe fuer Ueberlagerungsgetriebe, *Antriebstechnik* 7, no. 22, 1968, pp. 406–411.

Edgemond, F. W., Jr. Epicyclic Gears for Control Mechanisms, *Product Engineering*, Feb. 1957, pp. 194–198.

Ehrlenspiel, K. Untersuchung eines drehzahlregelbaren Planetengetriebes (Ueberlagerungsgetriebe), *Energie* 18, no. 1, 1966, pp. 25–28.

Ehrlenspiel, K. Zahnraeder und Zahnradgetriebe, *Konstruktion* 19, no. 9, VDI-Tagung, 1966, p. 385.

Ehrlenspiel, K. Betriebserfahrungen mit Stirnrad- und Planetengetrieben—Massnahmen zur Schadenverhuetung. *Der Maschinenschaden* 45, no. 5, 1972, pp. 133–144.

Ehrlenspiel, K. and Dehner, H. Planetengetriebe fuer Schiffsantriebe, Fachzeitschrift and *HANSA* 110, no. 4, February, 1973.

Ehrlenspiel, K. Planetengetriebe—Lastausgleich und konstruktive Entwicklung, *VDI-Berichte* 105, VDI-Verlag, 1976.

Elwell, J. Development des changements de vitesse éléctromatique éléctromecaniques cotal, *J. Soc. ingrs. autom.*, 1940, pp. 32–33.

Enderlen, H. Turboplanetengetriebe fuer grosse Leistungen und hohe Drehzahlen. *VDI-Z* 109, no. 6, 1967, pp. 225–229.

Englert, O. Ovalzahnraeder fuer die Automatisierung, *Oelhydraulik und Pneumatik* 11, no. 9, 1967, pp. 346–357.

Epicyclic Gearing for Lower Powers, *Mech. World* 124, no. 3215, August 27,

1948, p. 231; *Engr. Digest, Amer. Ed.,* Vol. 5, no. 4, May–June 1948, p. 179.

Epicyclic Gears for Control Mechanisms, *Product Engineering,* February 1957, p. 194.

Epicyclic Gears for Many Purposes, *Engineering,* November 28, 1958.

Epicyclic Gears for Marine Propulsion, *Shipbuilding and Ship Rec.* 83, no. 4, Jan. 1954, p. 116.

Epicyclic Gears for Merchant Ship Propulsion, *Engineering,* Oct. 27, 1961.

Epicyclic Transmission for Geared Diesel Engines, *The Marine Engineer and Naval Architect,* Nov. 1961.

Ernst, H. Verladebruecken mit Verschiebetraeger, *Z VDI* 84, 1940, pp. 959–961.

Ernst, H. Lastsenkvorrichtung fuer Krane mit mechanischer Bremse und Hilfsmotor, *Z VDI* 84, 1940, pp. 157–159.

Ernst, H. *Die Hebezeuge,* Vol. 1, Braunschweig, Vieweg & Sohn, 1952.

Faires, V. M. *Design of Machine Elements,* 3rd edition, New York, MacMillan, 1955.

Faroux, R. C. Comment est fabriquée la boite de vitesses cotal, *La Vie Automobile* 35, 1939, pp. 228–230.

Findeisen, D. Gleichfoermig uebersetzende Getriebe stufenloser Uebersetzungsaenderung: Gegenueberstellung von mechanischer und fluidtechnischer Energieuebertragung, Teil 2, *Konstruktion* 33, no. 1, 1981, pp. 15–24.

Fisher, E. Planetengetriebe mit elektrischer Schaltung, *Der Motorwagen* 29, 1926, pp. 772–774.

Fluecht, H. Die endlose Kette, Gleiskettenfahrzeuge im Betrieb, *Deutsche Kraftfahrt,* January, 1939, pp. 43–46.

Foerster, H. J. *Foettinger-Getriebe in Leistungsverzweigungen,* VDI-Forschungsheft no. 444. Ausg. B, 20, 1954, excerpt in ATZ 58, 1956, no. 9, p. 258.

Foerster, H. J. Zur Berechnung des Wirkungsgrades von Planetengetrieben. *Konstruktion* 21, no. 5, 1969, p. 169.

Formula for Epicyclic Gear Train of Large Reduction, *Product Engineering,* December 1954, p. 213; 1954, p. 192.

Formula for Epicyclic Gear Train of Large Reduction, *Product Engineering,* Mid-Oct., 1957, pp. E-12.

Francis, V. Spacing Planetary Gears, *Machinery* (London) 74, no. 1898, March 10, 1949, p. 302.

Frey, E. Der Wendelhalbmesser von Raupenfahrzeugen, *ATZ* 45, 1942, pp. 129–131.

Frey, E. Lenkgetriebe fuer Gleiskettenwagen, *ATZ* 45, 1942, pp. 515–523.

Friedemann, P. Das Maybach-Fahrgestell mit Siebenlitermotor, *Z. VDI* 70, 1926, pp. 1441–1448.

Fritsch, F. Das Zwillingsgetriebe—Eine Neukonstruktion auf dem Gebiet der Planetengetriebe, *Oesterreichische Ingenieur-Zeitschrift* 3, March 1960, pp. 103–105.

Fritsch, F. Der Lastausgleich im Stirnrad- und Planetengetriebe, *VDI-Berichte* 167, VDI-Verlag, 1971, pp. 199–207.

Fritzson, D. Numerical Methods to Determine the Forces in Traction Drives Assuming Hetzian Pressure Distribution, Report no. 1983-12-19, Division of Machine Elements, Chalmers University of Technology, Gothenburg, 1983.

Gackstetter, G. Auswahl von Planetengetriebe zur Leistungsverzweigung fuer Regelgetriebe, *Konstruktion* 17, no. 9, 1965, pp. 349–355.

Gackstetter, G. Leistungsverzweigung bei der stufenlose Drehzahlregelung mit vierwelligen Planetengetriebe, *VDI-Z* 108, no. 6, 1966, pp. 210–214.

Gastrow, H. Selbsttaetige Spritzgussmaschinen mit Umlaufgetriebe fuer Kunststoffe, *Z. VDI* 86, 1942, pp. 151–152.

Gatcombe, E. K. and Prowell, R. W. Rocket Motor Gear Tooth Analysis, *Trans. ASME,* Series B, Vol. 82, August 1960, pp. 223–230.

Gaunitz, A. Planetengetriebe mit grosser Uebersetzung, *VDI-Z* 92, no. 33, November 21, 1950, p. 956.

Geschelin, J. Making the Hydra-matic Precision Drive, *Automotive Industries* 134, no. 11, 1966, pp. 74–75.

Gelbhaar, P. Beitraege zur Berechnung und Durchbildung der Werkzeugmaschinen II, Umlaufraederwerke, *Werkstattstechnik,* May 1910, pp. 270–278, 390–396, 547–563.

Gelbharr, P. Graphische Tabellen zur Berechnung von Planetengetrieben unter Beruecksichtigung der Verwendung von Normalraedern, *Werkstattstechnik,* Sept. 15, 1912, pp. 476–479, 506–509, 533–536.

General Equations for Solving All Planetary Gear Ratios, *Product Engineering,* 4th Product Design Digest Issue, p. E-12a.

Getriebeblatt 601, Das Kardankreispaar.

Getriebeblatt 605, Umlaufraedergetriebe.

Getriebeblatt 606, Rueckkenhrende Umlaufraedergetriebe.

Getriebeblatt 607, Kegelraederumlaufgetriebe.

Getriebeblatt 652/653 and 654/655 Zahnradkurbel-Getriebe, Part I and II.

Getriebeblatt 661/662 Zahnrad-Kurvengetriebe. Zahnrad-Zugorgangetriebe.

Gloede, V. and Krause, W. Leistungsverhaeltnisse in Zahnradumlaufgetrieben, *Feingeraetetechnik,* no. 8, 1963, pp. 357–359.

Glover, J. H. Planetary Gear Trains Ratios, *Machine Design,* August 3, 1961, pp. 125–127.

Glover, J. H. R-factors Quickly Give Speed Ratios of Planetary Gear Systems, *Product Engineering,* January 6, 1964, pp. 59–68.

Glover, J. H. Efficiency and Speed-ratio Formulas for Planetary Gear Systems, *Product Engineering,* September 27, 1965, pp. 72–79.

Goblau, F. Die internationale Luftfahrtausstellung in Paris, *Z. VDI* 75, 1931, pp. 473–480.

Gosslau, F. Flugmotorenbau, *Z. VDI* 79, 1935, pp. 569–576.

Green, G. A. Power Transmissions for Buses, *SAE Transactions* 46, 1940, pp. 1–16.

Grodzinski, P. Neuere Kraftwagenumlaufgetriebe, *Reuleaux Mitteilungen* 1, 1933, p. 84.

Grodzinski, P. Verbesserungen an Werkzeugmaschinengetrieben, *Reuleaux-Mitteilungen* 1, 1933, p. 80.

Grodzinski, P. Stufenlose Umformer in Verbindung mit Umlaufraedergetrieben zur Veraenderung des Drehzahlbereichs, *AWF-Mitteilungen* 17, 1935, pp. 59–62, 96–97.

Grodzinski, P. Planetary Gear Mechanisms, with Eccentric Gears, *Machinery (London)* 84, no. 2158, March 26, 1954, p. 645.

Groeber, S. F. Leistungsverluste in Umlaufraedergetrieben, *Industrieanzeiger,* no. 48, 1968.

Groebner, S. F. Experimentelle Ermittlung des Wirkunsgrades von Planetengetriebe mit Innen- und Aussenverzahnung, *Z. VDI* 102, November 28, 1960.

Gruebler, M. *Getriebelehre,* Berlin, Springer-Verlag, 1917.

Gsching, W. Diwabus-Getriebe fuer Stadtomnibusse, *ATZ* 61, no. 11, 1959, pp. 329–333.

Gsching, W. Das Voith-Diwabus-Getriebe. *ATZ* 55, no. 3, 1963, pp. 53–60.

Halling, J. and Sutcliffe, W. J. Design Charts for Epicyclic Gear Systems, *Design News,* August 17, 1959, pp. 56–59.

Hardy, H. W. *Planetary Gearing,* The Machinery Publishing Co., London and New York.

Hartmann, H. *Die Maschinengetriebe,* Vol. 1, 1910.

Hartmann, H. Stufenlose Feinregelgetriebe und deren Einsatz, *Industriekurier,* Tech. & Forsch. 19, no. 9, 1966, pp. 17–18.

Havemann, H. Beitrag zur Berechnung von Flugmotorengetrieben, *Motor-technische Zeitschrift* 3, 1941, pp. 315–318, 317–318.

Hedman, A. A Method To Analyze Mechanical Transmission Systems, *Report* no. 1985-11-08, Division of Machine Elements, Chalmers University of Technology, Gothenburg, Sweden, 1985.

Hedman, A. A matrix-based method to analyze complex mechanical transmissions. *Proceedings, Seventh World Congress on the Theory of Machines and Mechanisms,* Seville, Spain, 1987, pp. 1415–1420.

Hedman, A. Mechanical Transmission Systems—A General Computer-Based Method of Analysis, Dissertation, Chalmers University of Technology, Gothenburg, Sweden, 1988.

Hedman, A. Mechanical Transmission Systems with Several Degrees of Freedom—A Method of Analysis, *Proc., International Conference on Gearing,* Zhengzhou, China, 1988.

Helfer, F. Eine Analogie zur Untersuchung von Planetengetrieben, *ATZ* 69, no. 5, 1967, pp. 149–152 and *Forschung und Konstruktion,* Voith, Heidenheim, no. 14.

Helfer, F. Hydrodynamische Koppelgetriebe, VDI-Berichte 228, VDI-Verlag, 1975, pp. 59–67.

Hicks, R. J. Variable-ratio transmission of high power using epicyclic gear trains, *Allen Engineering Review* 48, 1962, pp. 19–26.

Hill, F. Einbaubedingungen bei Planetengetrieben. *Konstruktion* 19, no. 10, 1967, pp. 393–394.

Hock, J. Beitrag zur Ermittlung des Wirkungsgrades einfacher und gekoppelter Umlaufgetriebe. *Fortschr.-Ber.* VDI-Z. Reihe 1, no. 3.

Jakobsson, B. Torque distribution, power flow and zero output conditions of epicyclic gear trains. *Transactions of Chalmers University of Technology* 224, Gothenburg, 1960, pp. 1–55.

Jakobsson, B. Efficiency of epicyclic gears considering the influence of the number of teeth. *Transactions of Chalmers University of Technology,* Gothenburg 309, 1966, pp. 1–36.

Jarchow, F. Stirnraeder-Planetengetriebe mit selbstaetigem Belastungsausgleich. *Z. VDI* 104, no. 12, 1962, pp. 509–512.

Jarchow, F. Leistungsverzweigte Getriebe, Stirnrad-Planetengetriebe mit parallel liegenden Axialkolbengetrieben, *VDI-Z* 106, no. 6, 1964, pp. 196–205.

Jarchow, F. Leistungsverzweigung im Getriebe. *VDI-Nachrichten* 49, 1967.

Jarchow, F. Ueberlagerungsgetriebe fuer stufenlose Drehzahl- und Drehmomentwandlung in Kraftfahrzeugen. *VDI-Berichte* 105, 1967, pp. 69–78.

Jarchov. F. Stufenlose hydrostatische Umlaufgetriebe, *VDI-Berichte* 196, VDI-Verlag, 1973, pp. 59–64.

Jarchov, F. Auslegung und Gestaltung hydrostatischer Koppelgetriebe part 1–2, *Antriebstechnik* 16, no. 1, 1977, pp. 27–30; no. 2, pp. 72–75.

Jarchov, F. and Benthake, K. Hydrostatische Koppelgetribe, *VDI-Berichte* 228, VDI-Verlag, 1975, pp. 69–79.

Jarchow, F., Langenbeck, K., and Benthake, H. Planeten- und Ueberlagerungsgetriebe, *Antriebstechnik* 16, no. 10, 1967, pp. 343–348; no. 11, pp. 402–406; no. 12, pp. 432–440.

Jarchow, F. and Shaker, A. Stufenloses hydrostatisches Koppelgetriebe fuer Fahrzeuge, *Konstruktion* 29, no. 12, 1977.

Jensen, P. W. Planetary Gears (in Danish), *Maskinindustrien,* no. 1, 1968, pp. 23–24; no. 2, pp. 13, 15, 17–19, 21–22.

Jensen, P. W. Raumbedarf und Wirkungsgrad von Umlaufgetrieben, *Maschinenmarkt* 75, 1969, pp. 1909–1918.

Jensen, P. W. Raumbedarf und Wirkungsgrad zusammengesetzter Planetenraedergetriebe, *Konstruktion* 21, 1969, pp. 178–184.

Jokl, W. Planeten-Mehrgang-Schiffsgetriebe, Baureihe MS. Magdeburg; *ASUG-Mitteilungen* 4, no. 1, 1967.

Klein, H. *Planetenrad-Umlaufraedergetriebe,* Munich, Carl Hanser Verlag, 1962.

Klein, H. Aufbau ein- und mehrstufiger Zahnradgetriebe, *Schweizer Maschinenmarkt* 64, no. 46, 1964, pp. 51–54.

Klein, H. Auslegen von Umlaufraedergetrieben im Dreiwellenbetrieb (Ueberlagerungsbetrieb) fuer optimale Uebertragungsgsverhaeltnisse. *Konstruktion* 17, no. 9, 1965, pp. 361–365.

Klein, H. Die Zahnuebertragungsleistung an Umlaufraedergetrieben, *Deutsche Hebe und Foerdertechnik* 11, no. 6, 1965, pp. 342–349.

Klein, H. Das Bemessen von Umlaufraedergetriebe-Uebersetzungen unter Beruecksichtigung des Ueberlagerungsgrades fuer stufenlose Drehzahleinstellung und -aenderung und relativen Gleichlauf, *die Maschine* 19, no. 11, 1965, pp. 40–42, no. 12, pp. 35–38.

Klein, H. Der Beschleunigungsvorgang an einen Umlaufraedergetriebe, *Schweizer Maschinenmarkt* 65, no. 13, 1965, pp. 45–50.

Klein, H. Das Bemessen von Umlaufraedergetriebe Uebersetzungen unter Beruecksichtigung des Ueberlagerungsgrads fuer stufenlose Drehzahleinstellung und-aenderung und relativen Gleichlauf-III, *die Maschine*—Antriebs & Getriebetechnik, 20, no. 1, 1966, pp. 41–45.

Klein, H. Das Umlaufraedergetriebe als Ausgleichselement in Gleichlaufantrieben von Transportwalzen und Transportbaendern, *Betriebs-techn.* 7, no. 1, 1966, pp. 28–29.

Klein, H. *Die Planetenrad-Umlaufraedergetriebe,* Munich, Carl Hanser Verlag, 1962.

Klein, H. Umlaufraedergetriebe fuer die Drehzahlregelung von elektrischen und mechanischen Antrieben, *Automatik* 11, no. 5, 1966, pp. 179–185.

Kluge, H. and Boellinger, H. Wirkungsgradmessungen an Zahradwechselgetrieben, *ATZ* 37, 1934, pp. 3–8.

Kluge, H. and Weiss, W. Wirkungsgrade von Zahnrad- und Kettenwechselgetrieben fuer Motorraeder, *Dtsch. Kraftf.-Forschung,* 10, Berlin, VDI-Verlag, 1938.

Knab, H. *Getriebelehre,* Nuremberg, Selbstverlag, 1928.

Knlerim, F. Differentiale im Kraftfahrzeug- und Maschinenbaueinsatz. *Antriebstechnik* 8, no. 9, 1969, pp. 337–339.

Koechling, P. Das Rieseler-Getriebe, *Motorwagen,* 1928, pp. 142–145, 173, 176.

Kraus, R. A. Zur graphisch-rechnerischen Bestimmung des Wirkungsgrades von Getrieben, *Reuleaux-Mitt.* 2, 1934, pp. 57–58.

Kraus, R. A. Zur Systematik der gleichachsigen Drehzahlwechselgetriebe als Stirnrad-und Umlaufgetriebe *ATZ* 54, no. 4, 1952, pp. 82–86; no. 5, pp. 115–121; no. 6, pp. 137–139.

Kraus, R. A. *Getriebeaufbau,* Verlag Technik, 1952.

Kraus, R. A. *Stirnradwechselgetriebe,* Munich, Carl Hanser Verlag, 1955.

Kraus, R. A. Kritische Untersuchung der Wirkungsgrad-Berechnung einfacher Planetengetriebe, *Maschinenmarkt* 90, 1961, pp. 25–28.

Krejnes, M. A. et al. Selecting reduction-gear systems consisting of three differential three-member mechanisms, *Russian Engineering J.,* no. 11, 1962, pp. 19–23.

Kuehner, K. Der neuzeitliche Getriebebau und das ZF-Media-Getriebe, *ATZ* 55, 1953, p. 63.

Krumme, W. Werkstoffpruefgeraet fuer die Einflanken-Abrollpruefung von Zahnraedern, *Z. VDI* 87, 1943, p. 816.

Kutzbach, K. Reibung und Abnutzung von Zahnraedern, *Z. VDI* 70, 1926, p. 999.

Kutzbach, K. Mehrgliedrige Raederbetriebe und ihre Gesetze, *Masch.-Bau* 6, 1927, pp. 108–185, 1080–1083, 1087.

Kutzbach, K. Die Regelung bei stufenlosen Umformern, *Sonderheft Getriebe,* VDI-Verlag, 1928, pp. 36–38.

Kutzbach, K. Mechanische Leistungsverzweigung, *Masch.-Bau* 8, 1929, pp. 710–716.

Kutzbach, K. Neuere amerikanische Versuche ueber Abnutzung und Wirkungsgrad von Zahnraedern, *Masch.-Bau/Betrieb* 5, 1962, pp. 223–224.

Kuzin, B. V. et al. Load Distribution in Multi-Path Power Transmissions, *Russian Engineering J.,* no. 8, 1965, pp. 3–8.

Lancaster, F. Determination of Change Gears, *American Machinist,* Nov. 7, 1934, pp. 766–767.

Langdon, R. H. Planetary Gear Ratios, *Machine Design,* April 28, 1960, pp. 161–170.

Langenbeck, K. Ueberlagerungsgetriebe fuer den Antrieb von Erzfoerderpumpen, *Antriebstechnik* 8, no. 10, 1969, pp. 380–382.

Langenbeck, K. Antrieb von Metallfolienreckanlagen mit Ueberlagerungsgetrieben. *Antriebstechnik* 9, no. 2, 1970, pp. 44–48.

Langenbeck, K. Planetengetriebe und ihr Einsatz, *Tagungsheft Antriebstechnik,* Hannover Messe, 1974, pp. 39–43.

Laughlin, Holowenko, and Hall. How To Determine Circulating Power in Controlled Epicyclic Gear Systems, *Machine Design,* March 22, 1956, pp. 132–136.

Laughlin, Holowenko, and Hall. Design Charts and Equations for Controlled Epicyclic Gear Systems, *Machine Design,* Nov. 29, 1956, pp. 129–138.

Leboime, P. Les transmissions mecaniques de forte puissance et leur fabrication (Mechanische Getriebe fuer die Uebertragung grosser Leistungen und ihre Herstellung), *Ingenieurs de L'Automobile* 39, 1966, no. 4, pp. 201–209.

Leistner, F. Die Masse als Parameter des Planetengetriebes, *Wissenschaftliche Zeitschrift der TH Magdeburg* 13, 1969, p. 707.

Leunig, G. Differential-Aufladung von Fahrzeug-Dieselmotoren, *ATZ* 68, no. 5, 1966, pp. 165–170.

Levai, Z. Analytische Untersuchung elementarer Planetengetriebe, *Acta Technica Academiae Scientiarum, Hungaricae* 49, nos. 3–4, 1964, pp. 357–371.

Levai, Z. Analytische Untersuchung komplexer Planetengetriebe, *Acta Technica Academiae Scientiarum Hungaricae* 53, nos. 1–2, 1966, pp. 17–58.

Levai, Z. Structure and Analysis of Planetary Gear Trains, *Journal of Mechanisms* 3, pp. 131–148.

Lewis, D. E. Epicyclic Gear Train, *Machine Design* 36, no. 17, 1964, pp. 175–179.

Loersch, G. Umlaufraedergetriebe fuer hohe Uebersetzungen. *Maschinenbautechnik* 16, no. 1, 1967, pp. 15–18.

Loersch, G. Aufbau und Eigenschaften von Getriebekombinationen stufenlos verstellbarer Kettengetriebe mit Umlaufraedergetrieben. Dissertation, TH Karl-Marx Stadt, 1971.

Looman, J. Leistungsverzweigung-positiv oder negativ, *VDI-Z* 108, no. 6, 1966, pp. 221–226.

Looman, J. Selbstsperrdifferentiale in Kraftfahrzeugen, *VDI-Z* 110, no. 6, 1968, pp. 209–216.

Looman, J. *Zahnradgetriebe,* Berlin, New York, Springer-Verlag, 1970.

Lueer, E. Leistungsverzweigung in stufenlosen Getrieben, *Konstruktion* 15, no. 3, 1963, pp. 109–110.

MacClain, A. L. and Buck, R. S. Flight testing with an engine torque indicator, *SAE J.* 42, 1938, pp. 49–52.

Machintosch, D. C. Epicyclic Gears: The mathematics Involved in Calculating Compound Ratios, *Autom. Engr.* 25, 1935, pp. 345–346.

MacMillan, R. H. Epicyclic Gear Trains, *Engineer* 187, no. 4861, March 25, 1949, pp. 318–320.

MacMillan, R. H. Epicyclic Gear Efficiencies, *Engineer* 188, no. 4900, December 23, 1949, p. 727.

Maegi, M. On Efficiencies of Mechanical Coplanar Shaft Power Transmissions, Dissertation, Division of Machine Elements, Chalmers University of Technology, Gothenburg, Sweden, 1974.

Maier, A. Konstruktion und Entwicklung der Kraftfahrzeug-Stufengetriebe, *ATZ* 44, 1941, pp. 415–428.

Maier, A. Neuere Entwicklungen im Personenwagengetriebebau, *ATZ* 55, 1953, p. 205.

Malmborg, B. D. 10-Speed Transmission Combines Conventional and Planetary Gear Units, *Design News* 19, no. 24, 1964, pp. 22–23.

Mann, W. H. Differentialgetriebe, *Spinner und Weber* 53, 1935, p. 8.

Mann, W. H. Graphical Solution of Epicyclic Gear Trains Problems, *Engineering* 171, no. 4442, March 16, 1951, p. 312.

Martin-Prevel, M. Sedution, traitise et utilite des trains epicycloidaux, *J. Soc. Ingrs.* 11, 1938, pp. 337–355.

Massot, M. P. Influence du frottement sur la functionnement des trains epicycloidaux, *Société des Ingenieurs Civils de France,* 1936, pp. 576–593.

Maul, F. Die Zahnrad-Umlaufgetriebe, *Konstruktion*, May, 1952, p. 139.

Mayatt, D. J. Designing Planetary Gear Trains, *Machine Design,* January 16, 1964, pp. 203–204.

Merritt, H. E. Some Considerations Influencing the Design of High-Speed Track-vehicles, *J. Instin. Autom. Engrs.* 7, January 1939, pp. 11–32.

Merritt, H. E. Epicyclic Gears Trains No. II, *The Engineer* (England), March 28, 1941.

Merritt, H. E. *Gear Trains,* Pitman & Sons, Ltd., London, 1947.

Merritt, H. E. *Gears,* 3rd edition, London, Pitman & Sons, Ltd., 1954.

Mertz, R. Neues hydraulisches "Getriebe" *ATZ* 43, 1940, pp. 522–523.

Methods of Specifying Precision of Spur Gears, *Product Engineering,* November 1956, p. 135.

Mettler, E. Das mechanische Kleingetriebe, *Schweizerische Technische Export-Zeitung,* Bern, 1938; Januar., no. 8/I, no. 1–2; March, no. 9/I, pp. 3–6; April, no. 10, pp. 7–8; May, no. 11/I, pp. 3–6; July, no. 12I, pp. 4–6.

Meyer zur Capellen, W. *Geschwindigkeitken und Beschleunigungen an Umlaufraedergetrieben,* Ma-Bau, Getriebetechnik, pp. 577–580.

Meyer zur Capellen, W. Das Ueberlagerungsprinzip bei ebenen und sphaerischen Umlaufraedertrieben, *Industrie-Anzeiger* 87, no. 70, 1965, pp. 1665–1674.

Michalec, G. W. Gear Differentials, *Machine Design,* October and December 1955, pp. 178–188.

Mildenberger, D. Planetengetriebe als Differentialgetriebe, *Maschinenmarkt* 70, no. 46, 1964, pp. 24–25.

Monath, L. Elektrische und elektromechanische Zwischengetriebe fuer Fahrzeuge, *ETZ* 65, 1944, pp. 365–369, 397–401.

Morland, W. V. Diesel engine and transmissions progress in English buses, *Diesel Power* 17, 1939, pp. 642–646.

MP Umlaufgetriebe, *Antriebstechnik* 6, no. 7, 1967, p. 259.

Muir, N. S. On the Art of Dynamometry with Particular Reference to the Measurement of the Engine Power in Flight, *Journ. Roy. Aeronautical Society* 41, 1937, p. 864.

Mueller, M. A. Stufenlose Antriebe als Umlaufraedergetriebe mit hydrostatischem Nebenzweig, *Antriebstechnik* 4, no. 1, 1965, pp. 3–8.

Mueller, H. W. Hydrostatische Umlaufgetriebe. *Oelhydraulik und Pneumatik* 14, no. 11, 1970, pp. 513–516.

Mueller, H. W. *Die Umlaufgetriebe,* Berlin, Springer-Verlag, 1971.

Mueller, H. W. Adaption of Continuously Variable Transmissions to the Characteristics of a Driven Machine by Bicoupler Planetary Transmissions, *Journal of Mechanical Design* 103, no. 1, 1981, pp. 41–47.

Mueller, H. W. and Schaefer, W. F. Geometrische Voraussetzungen bei innenverzahnten Getriebestufen kleinster Zaehnezahldifferenzen, *Forschung im Ingenieurwesen* 36, no. 5, VDI-Verlag, 1970.

Nakamura, K. et al. A Development of a Traction Roller System for a Gas Turbine Driven APU, *SAE-Paper* no. 790106, February, 1979.

Neufeld, K. Getriebeturbinen fuer Dampfkraftwerke, *VDI-Nachrichten,* October 25, 1961.

Neumann, R. Umlaufraedergetriebe in Textilmaschinen, *Deutsche Textiltechnik* 15, no. 4, 1965, pp. 175–182.

Neumann R. Drehzahlberechnung von Umlaufraedergetrieben, *Deutsche Textiltechnik* 17, no. 2, 1967.

Neussel, P. *Untersuchung von rueckkehrenden Umlaufgetrieben mit und ohne Selbsthemmung unter besonderer Beruecksichtigung von Koppelgetrieben,* Dissertation, T. H. Darmstadt, 1962.

New AGMA Classification, *Product Engineering,* 32, pp. 68–74.

Niemann, G. *Maschinenelemente,* Vol. 2, Getriebe, Springer-Verlag, Berlin, 1960.

Niemann, G. Stirnrad-Planetengetriebe mit Einfach-Schraegverzahnung und Lastausgleich, *VDI-Z* 106, no. 6, 1964, pp. 211–212.

Nikolaus, H. Graphische Darstellung der Drehzahl-, Momenten- und Leistungsverteilung in einfach rueckkehrenden Planetengetriebe, *Maschinenmarkt* 90, 1961, pp. 29–30, 35–39; no. 12, pp. 33–36, 39.

Noock, W. G. Getriebe, *Z VDI* 64, 1920, pp. 317–322, 346–350, 377–381.

Nuebling, O. Die Zahnradbeanspruchung und ihre Berechnungsweise bei Flugmotorengetrieben, *Luftf.-Forschg.* 17, 1940, pp. 145–153.

Nuell, W. v. d. and Pfau, H. Antrieb und Reglung der Motorenlader, *Z VDI* 85, 1941, pp. 749–754, 981–989.

Obert, E. F. Speed Ratios and Torque Ratios in Epicyclic Gear Trains, *Product Engineering,* April 1945, pp. 270–271.

Oernhagen, L. On Self-Locking Transmissions, Report no. 24 from the Institute of

Machine Elements, Chalmers University of Technology, Gothenburg, Sweden, 1963.

Ohlendorf, H. *Verlustleistung und Erwaermung von Stirnraedern,* Dissertation, T. H. Muenchen, 1958.

Opitz. Lebensdauerpruefung von Zahnradgetrieben, *Forschungsberichte des Landes Nordrhein-Westfalen,* no. 901.

Perrett, W. Wechselgetriebe mit Umlaufraedern, *Forsch. Ing.-Wes.* 23, 1957, p. 102.

Petit, H. La boite de viteses semiautomatique "Oldsmobile," *Vie Autom.* 34, no. 1129, 1938, p. 1922.

Petit, H. Essai d'une Voiture Peugeot 402B normale avec boite Cotal multiplie, *Vie Autom.* 35, no. 1154, 1939, pp. 27–29.

Petit H. De quelques ments des changements de vitesse automatique, *Vie Autom.* 36, no. 1178, 1940, pp. 21–24.

Pfahl, Die Vorteile des Reihenbaus elektrischer Hubwerke, *Z. VDI* 69, 1925, pp. 823–825.

Pickard, J. Berechnung und Konstruktion einfacher und zusammengesetzter Planetengetriebe, *Survey,* Ingenieur Digest Verlag, 1970.

Pickard, J. and Besson, B. Proposition d'un modèle général representatif des transmissions de puissance. Examples d'applications, *Ingenieur de l'Automobile,* no. 4, 1977, pp. 183–191.

Pickard, J. and Besson, B. Hydrostatic Power Splitting Transmissions Design and Application Examples, *ASME J. of Eng. for Power* 103, no. 1, January 1981, 168–173.

Pickard, J. et al. Planetengetriebe in der Praxis, Wuerttemberg, Expert Verlag, 1981.

Plint, M. A. Designing Epicyclic Gear Trains for Equal Planet Pinion Spacing, *Machinery (London)* 79, pp. 727–728.

Plump, Development of the Reduction Gears for a Large Propeller Turbine, *The Institution of Mechanical Engineers,* London, International Conference, September 1958.

Poincare, L. La mesure des couples en vol, *J. Soc. de L'Automobile* 12, 1939, pp. 227–234.

Polder, J. W. A network theory for variable epicyclic gear trains, *Dissertation,* University of Technology, Eindhoven, 1969.

Pomeroy, L. H. Transmission, *Auto. Eng.* 21, 1951, pp. 596–600.

Poocza, A. Planeten-Nockengetriebe zum Ausgleich der Ungleichfoermigkeit in Kettengetrieben. *VDI-Z* 101, 1959, 24, pp. 1130–1134.

Poocza, A. Eine Moeglichkeit zum Erhoehen der Uebersetzung ins Langsame bei mit Antriebsmotor zusammengebauten Planetengetrieben, *VDI Z* 7, March 1, 1959, pp. 274–276.

Poocza, A. Planentenrad-Nockengetriebe zum Ausgleich der Ungleichfoermigkeit in Kettentrieben, *VDI Z* 101, no. 24, August 1959, pp. 1130–1134.

Poppinga, R. Die Berechnung von Stirnrad Planetengetrieben, Habilitationsschrift der T. H., Berlin, 1944.

Poppinga, R. *Stirnrad-Planetengetriebe,* Stuttgart, Franckh'scher Verlag, 1949.

Poppinga, R. Die Zahnwaelzleistung in Planetengetrieben, *Die Technik* 4, no. 6, June 1949, pp. 257–264.

Poppinga, R. Der Wirkungsgrad der Planetengetriebe, *Ing. Archiv* 18, no. 1, 1950, pp. 39–52.

Poppinga, R. Berechnung und Gestaltung von Zahnrad Planetengetriebe, *Konstruktion* 2, 1950, pp. 33–41.

Poppinga, R. Efficiency of Planetary Gear Trains, *Engineering Digest* 11, December 4 and 21, 1950.

Poppinga, R. Der Wirkunsgrad der Planetengetriebe, *VDI Z*, March 21, 1951, pp. 250–252.

Proeger, F. *Die Getriebekinematik als Ruestzeug der Getriebedynamik*, VDI-Verlag, Berlin, 1926.

Prowell, R. W. A Study of Friction Loss for Spur Gear Teeth, *ASME* 61, WA-85.

Pueschl, T. Dynamik des Differentialgetriebes, *Z VDI* 78, 1934, pp. 799–800.

Puls, E. F. Ein neues selbstschaltendes Getriebe, *Motorwagen* 31, 1928, pp. 133–141.

Radzimovsky, E. I. A Simplified Approach for Determining Power Losses and Efficiency of Planetary Gear Drives, *Machine Design* 28, no. 3, 1956, pp. 101–110.

Radzimovsky, E. I. Planetary Gear Drives, *Machine Design* 31, June 11, 1959, pp. 144–153.

Radzimovsky, E. I. Planetary Gear Drives, *Machine Design* 32, September 15, 1960, pp. 190–197.

Rasmussen, A. C. Gear Calculation Based on Dynamic Loading and Wear, *Machinist (London)*, August, 1939, pp. 372–375.

Ravigneaux, P. 1re addition au brevet d'invention no. 823.757, no. 53.264, demandée le 15 octobre, 1943, Paris.

Reichenbaecher, K. Verzweigungsgetriebe und Leistungsregelung bei motorgetriebenen Fahrzeugen, *Z VDI* 87, 1943, pp. 700–714.

Reichenbaecher, H. *Gestaltung von Fahrzeug-Getrieben,* Berlin, Springer-Verlag, 1955.

Reimann, Ein Umlaufraederwerk als Antrieb fuer Hobelmaschinen, *Masch.-Bau* 6, 1927, pp. 357–360.

Reimann, Ein Umlaufraederwerk in Automaten, *Masch.-Bau* 6, 1927, p. 527.

Resetov, L. N. Design of Multiple Epicyclic Transmissions, *Russian Engng. J.* no. 10, 1964, pp. 9–13.

Reuleaux, F. *Lehrbuch der Kinematik,* Vol. 2, Braunschweig, 1900, pp. 1–140.

Richter, O. and Voss, R. von *Bauelemente der Feinmechanik,* 3rd printing. VDI-Verlag, Berlin, 1942.

Richter, W. Auslegung profilverschobener Aussenverzahnungen. *Konstruktion* 14, no. 5, 1962, pp. 189–196.

Richter, W. Auslegung von Innenverzahnungen und Planetengetrieben, *Konstruktion* 14, no. 12, 1962, pp. 489–497.

Rickli, H. Bestimmung des Wirkungsgrades von Zahnraedern, *Z VDI* 55, 1911, pp. 1435–1438.

Rideout, T. R. Why Gear Teeth Wear, *Machinist (London)*, July 15, 1939, pp. 425–427.

Roegnitz, H. *Die Getriebe der Werkzeugmaschinen,* Berlin, J. Springer, 1936.

Ross, W. A. Designing a Hydromechanical Transmission for Heavy Duty Trucks, *SEA-Paper* 720725.

Rossenbeck, *ATZ* 50, 1948, p. 76.

Ryba, A. and Moldin, O. Elektromagnetische Bandbremse und ihre Anwendung auf Wechselgetriebe, *Ma-Bau/Betrieb* 14, 1935, pp. 341–343.

Salamoun, C. and Valenta, J. Untersuchung des Wirkungsgrades von Planetengetrieben, Notizen, Puhu-Verl, (CSSR), September-October 1966.

Schachter, M. Hydroplanetary Transmission Units, *Automob. Eng.* 54, no. 7, 1964, pp. 278–283.

Schaefer, W. F. Einfaches Verfahren zur Ermittlung der Profilueberdeckung ε_α bei aussen- und innenverzahnten Getriebestufen, *Forschung im Ingenieurwesen* 36, VDI-Verlag, 1970, p. 5.

Schlesinger, G. Beitraege zur Berechnung von Durchbildung der Werkzeugmaschinen, Umlaufraederwerke, *Werkstattstechnik* 4, 1910, pp. 269–278.

Schlums, K. D. Hubwerksgetriebe fuer Eintrommelantriebe, *Foerdern und Heben* 16, no. 2, 1966, pp. 108–111.

Schmidt, K. A. Grundsaetzliche Fragen der Verstellschraubentechnik unter Beruecksichtigung der hauptsaechlichsten deutschen Bauarten, *Luftwissen* 6, 1939, pp. 267–273.

Schnetz, K. Optimierung zusammengesetzter Planetengetriebe. *Fortschrittsberichte* der VDI-Zeitschriften, Reihe 1, no. 30, VDI-Verlag, Duesseldorf, 1971.

Schoepke, H. Doppelt gebundene Norton-Getriebe, *Maschinenbau* 7, 1939, p. 513.

Schorsch, C. J. Formative Number of Teeth, *Design News,* Dec. 7, 1959, p. 74.

Schorsch, C. J. Nomogram for Choice of Single or Double Reduction, *Design News,* August 17, 1959, pp. 60–61.

Schreier, G. *Stirnrad-Verzahnungen,* Berlin, Verlag Technik, 1961.

Schulze-Allen, K. Schiffswendegetriebe, *Mitt. Forschungsanst.* Gutehoffnungshuette Konzern 5, 1937, p. 150.

Schulze-Allen, K. Schiffswendegetriebe, Konstruktiver Aufbau verschiedener Arten von Wendegetrieben, *MTZ* 4, 1942, pp. 323–325.

Schwerdhoefer, H. J. Planetengetriebe, Schaltsysteme und Freilaeufe in Fahrrad- und Motorradnaben, *Antriebstechnik* 4, no. 10, 1965, pp. 374–382.

Seeliger, K. Leistungsuebertragung mit Planetengetrieben, *Das Industrieblatt* 55, no. 12, 1955, pp. 591–595.

Seeliger, K. Planetenstirnradgetriebe fuer sehr hohe Uebersetzungen, *Industrie Anzeiger,* no. 10, 1955, pp. 15–16.

Seeliger, K. Planetengetriebe in Leistungsverzweigungen, *Werkstatt und Betrieb* 89, 1956, pp. 64–66.

Seeliger, K. Einsatzmoeglichkeiten fuer Planetengetriebe, *Maschine und Werkzeug* 59, no. 23, 1958, pp. 11–17.

Seeliger, K. Planetengetriebe in drehzahlveraenderlichen Antrieben, *VDI-Z* 101, no. 6, February 21, 1959, pp. 217–224.

Seeliger, K. Einsteg-Planetenschaltgetriebe, *Werkstatt und Betrieb* 93, no. 1, January 1960, pp. 12–14.

Seeliger, K. Betriebsverhalten der Planetengetriebe-Systeme, *die Maschine* 17, no. 9, 1963, pp. 53–55.

Seeliger, K. Das Mehrsteg-Getriebe, *Desch Antriebstechnik* 5, 1963, pp. 9–16, Herausgegeben von der H. Desch GmbH Neheim-Huesten.

Seeliger, K. Das einfache Planetengetriebe, *Antriebstechnik,* June 1964, pp. 216–221.

Seeliger, K. Das Leistungsverhalten von Getriebekombinationen, *VDI-Z* 106, no. 6, 1964, pp. 206–211.

Seeliger, K. Einsatzmoeglichkeiten fuer Planetengetriebe, *Maschine & Werkzeug* 65, no. 7, 1965, pp. 7–13.

Seeliger, K. Ueberlagerungsgetriebe fuer konstantes Abtriebsmoment, *Antriebstechnik* 4, no. 10, 1965, pp. 368–373.

Seeliger, K. Neue Planetengetriebe-Baureihe, *Die Maschine* 20, no. 4, Antriebs. & Getriebe-Techn., 1966, pp. 67–69.

Seeliger, K. Ueberlagerungsgetriebe, *Maschinenmarkt* 7, 1968, pp. 1567–1570.

Seifert, H. and Weber, C. Systematik der hydrodynamischen und der Riemen-Stellkoppelgetriebe, part 1 and 2, *Antriebstechnik* 22, no. 9, 1983, pp. 66–70, no. 10, pp. 47–52.

Shen, F. A. Planet-Pinion Spacing, *Machine Design,* June 12, 1958, pp. 147–150.

Shuster, J. Planetary Drives, *Machine Design,* July 7, 1960, pp. 147–149.

Simpson. Planetary Transmission, *SAE Quest. Trans.* 69, no. 3, 1949.

Smirra, J. Kolbentriebwerkseinheiten von 4000 PS, *Wissen,* 1942, pp. 210–215.

Snell, S. S. A bending stress criterion for idler gears, *Machinery, London*, 100, 1962, p. 2588.

Soderholm, L. G. Cycloidal Teeth Give Epicyclic Gear Reducer High Reduction Ratios, *Design News* 21, no. 9, 1966, pp. 24–27.

Sonderkonstruktion eines Gross-Planetengetriebes, *Antriebstechnik,* June 1964, pp. 235–237.

Spotts, M. F. Formula for Epicyclic Gear Trains of Large Reduction, *Product Design Handbook,* Mid-October, 1957.

Spotts, M. F. Adjustable-Speed Epicyclic-Gear Drives, *Machine Design,* February 1, 1962, pp. 119–122.

Stefanides, E. J. Stamped Planet-Carrier Cuts Cost of Multi-Stage Reducer, *Design News,* June 23, 1965, pp. 136–137.

Stiefermann, P. Planetengetriebe vereinigt mit Drehstrom-Kurzschlusslaeufer-Motor fuer drehzahl-veraenderliche Antriebe, *TZ fur praktische Metallbearbeitung* 59, no. 4, 1965, pp. 243–248.

Stoeckicht, W. G. DRP 737 886, DBP 814 981, DBP 858 185.

Stoeckicht, W. G. Some Advantages of Planetary Gears, *Journal of the American Society of Naval Engineers* 60, 1948, pp. 169–178.

Stoeckicht, W. G. Planetengetriebe, Kraftuebertragung bei grossen Leistungen, *VDI-Tagungsheft* 2, 1953, pp. 139–148.

Strauch, H. Das einfache Umlaufraedergetriebe und sein Wirkungsgrad, *Reuleaux-Mitteilungen* 2, 1935, pp. 105–106.

Strauch, H. Die Kratzenrauhmaschine und ihr getriebliches Abbild. *Monatsschrift fur Textilindustrie* 51, no. 1, 1936.

Strauch, H. Die Regulierung der Kratzenrauhmaschine, *Monatsschrift fuer Textilindustrie* 51, no. 2, 1936.

Strauch, H. Geomtrische Addition von Abtriebsbewegungen, *Masch.-Bau,* 1938, pp. 204–206.

Strauch, H. *Die Umlaufraedergetriebe,* Munich, Carl Hanser Verlag, 1950.

Strauch, H. Ueber die Selbsthemmung von Umlaufraedergetrieben, *Getriebetechnik I,* VDI-Verlag, 1953, pp. 77–79.

Strauch, H. Die Fundamentalgesetze der Planetengetriebe, *Antriebstechnik* 6, no. 8, 1967, pp. 265–270.

Strauch, H. Kreiskolbenmaschinen nach dem Orbit®-Prinzip, *Antriebstechnik* 6, no. 4, 1967, pp. 134–135.

Strauch, H. *Theorie und Praxis der Planetengetriebe,* Mainz, Krausskopf Verlag, 1970.

Stoelzle, K. Schnellaufende Planetengetriebe fuer schweren und stosshaften Betrieb, *VDI-Z* 106, no. 6, 1964, pp. 213–214.

Stroemblad, J. Beschleunigungsverlauf und Gleichgewichtsdrehzahlen einfacher Planetengetriebe nebst Selbsthemmungsversuche, *Transactions of Chalmers University of Technology* 226, Gothenburg, 1960, pp. 1–80.

Strohhaechker, Ausgleich-und Uebersetzungsgetriebe fuer mehrachsige Kraftfahrzeuge, *ATZ* 41, 1938, pp. 607– 09.

Strum, F. and Schirmer, H. Triebwerkanlage mit Vorlage im Luftschiff Graf Zeppelin, *Z. VDI* 75, 1931, p. 1189.

Stubbs, P. W. R. The Development of a Perbury Traction Transmission for Motor Car Applications, *Journal of Mechanical Design* 103, no. 1, 1981, pp. 29–40.

Stueper, J. *Automatische Automobilgetriebe,* Wien, Springer-Verlag, 1965.

Stueper, J. Selbsttaetige Automobilgetriebe in Europa, *Automobil Rev.* 61, no. 30, 1966, pp. 17, 19, 23.

Stuhr, H. W. Baueinheit aus hydrostatischen Getriebe und schaltbarem Vorgelege, *Antriebstechnik* 5, no. 3, 1966, pp. 107–110.

Sunaga, T. Innenverzahnungen mit geringer Zaehnezahldifferenz als Planeten-Reduktionsgetriebe. *Bulletin of Japan Society of Mechanical Engineers,* 1967.

Szenasy, St. v. Die amerikanischen Modelle 1940, *ATZ* 42, 1939, pp. 601–605.

Tank, G. Untersuchung von Beschleunigungs- und Verzoegerungsvorgaengen an Planetengetrieben, *VDI Z* 96, no. 10, April 1, 1954, p. 305.

Taylor, F. E. The Design of a Planetary Gear Box, *Engineering Materials and Design* 7, no. 10, 1964, pp. 674–679.

Ten Bosch, *Vorlesung ueber Maschinenelemente,* 2nd Edition, Berlin, 1940, pp. 383–391.

Terplan, Z. Der Wirkunsgrad eines Planetengetriebes mit Doppelantrieb, *Acta Technica Academiae Scientarum Hungaricae* 49, no. 1–2, 1964, pp. 207–217.

Terplan, Z. Verschiedene Methoden fuer die analytische Untersuchung der einfachsten Planetenradergetriebe, *Acta Technica Academiae Scientarum Hugaricae* 49, no. 3–4, 1964, pp. 437–452.

Terplan, Z. Einfache Umlaufgetriebe mit ausgeglichenem Gleiten, *Konstruktion* 18, no. 10, 1966, pp. 409–412.

Terplan, Z. *Dimensionierungsfragen der Zahnrad-Planetengetriebe*, Akademiaikiado, Budapest, 1974.

The De Normanvill Gear Box, *Automobile Engineer*, 1935, p. 245.

The Present Trend in Naval Propulsion, *The Marine Engineer and Naval Architect*, November 1961.

Thiemann, A. E. Schwungkraftstarter fuer Fahrzeugdieselmotoren, *ATZ* 44, 1941, p. 509.

Thiering, *Getriebe der Textiltechnik*, Berlin, 1926.

Thuengen, H. v. Grundlagen fuer die selbsttaetige Regelung von Kraftfahrzeuggetrieben, *Z VDI* 78, 1934, pp. 309–315.

Thuengen, H. v. Wesen der Kupplung und des Getriebes beim Kraftfahrzeug, *Z VDI* 81, 1937, pp. 645–651.

Thuengen, H. v. Leistungsverzweigung und Scheinleistung in Getrieben, *VDI Z* 83, no. 24, 1939, pp. 730–734.

Thuengen, H. v. Leistungsverzweigung in Getrieben, *ATZ* 54, no. 2, 1952.

Thuengen, H. v. Mechanische Leistungsverzweigung in Getrieben, *VDI-Berichte* 29, 1958, pp. 139–142.

Thuengen, H. v. Getriebe mit negativem Leistungsfluss, *ATZ* 67, no. 2, 1965, pp. 44–49.

Timperley, D. A. Epicyclic Gear Planet-Wheel Bearing Test-Rig, *Allen Engineering Rev.*, no. 53, 1964, pp. 22–24.

Titl, A. Getriebetechnik heute, *Technische Rundschau* 58, no. 39, 1966, pp. 1–5.

Toepler, H. Lastausgleich bei Zwillings- und Drillingsgetrieben, *Technische Mitteilungen* 54, no. 7, 1961, pp. 258–259.

Town, H. C. Analysis and Design of Planetary Gearing, *Engineering Materials and Design* 6, no. 1, 1963, pp. 12–18.

Trier, H. *Die Kraftuebertragung durch Zahnraeder*, Berlin, J. Springer, 1942, pp. 43–53, 58–64.

Trier, H. Berechnung von Umlaufgetrieben, Werkstattblatt 126/127/128, Carl Hanser Verlag, Munich.

Trylinski, W. Kraftmomentsschwankungen der Evolventenverzahnungen, *Feingeraetetechnik*, April 1962, pp. 146–150.

Tuplin, W. A. Designing Compound Epicyclic Gear Trans for Maximum Efficiency at High Velocity Ratio, *Machine Design*, April 4, 1957, pp. 100–104.

Tuplin, W. A. Stepless Speed Variation with Epicyclic Gearing, *Machinery (NY)* 66, no. 12, Aug., 1960, pp. 126–128.

Tuplin, W. A. Don't Overlook Basic Epicyclic Gear, *Machinery*, June 10, 1961, p. 122.

Tuplin, W. A. The Limitations of Epicyclic Gears, *Machine Design* 36, no. 3, 1964, pp. 124–126.

Tymann, H. Metallurgische Behandlung von Roheisenschmelzen durch gesteuerte vermischungsaktive Bewegungen, *DEMAG-Nachr* 180, 1966, pp. 22–26.

Upton, E. W. Application of Hydrodynamic Drive Units to Passenger Car Automatic Transmissions, 2nd Edition, *SAE*, New York, 1973, pp. 165–182.

Verhoeff, A. Epicyclic Gearing for Lower Powers, *Philips Tech. Rev.* 9, no. 9, 1947–1948, p. 285.

Voeroes, I. V-Null-Verzahnung fuer ausgeglichenes spezifisches Gleiten, *Maschinenbautechnik* 13, no. 2, 1964, pp. 102–106.

Voeroes, I. V-Verzahnung und Verzahnung mit Profilruecknahme fuer ausgeglichenes Gleiten, *Maschinenbautechnik* 13, no. 4, 1964, pp. 219–221.

Volmer, J. *Umlaufraedergetriebe,* Berlin, Verlag Technik, 1978.

Von Soden, E. Umlaufraederwerke, *Auto-Technik* 9, 1920, pp. 9–16.

Von Soden, E. Getriebe fuer Leichtmotoren, *Berichtsheft* 71. Hauptversammlung des VDI in Friedrichshafen, Berlin, VDI-Verlag, 1933, pp. 93–97.

Von Soden, E. Les boites de vitesses éléctromagnetiques, *Soc. Ingrs. Autom.* 13, 1940, pp. 21–26.

Waclawek, M. J. General Equations for Solving All Planetary Ratios, *Product Design Handbook,* 1956, p. E-12.

Wadman, B. Responder Automatic Transmission Ready for Market, *Diesel and Gas Turbine Progress,* June 1973.

Wahl, C. G. Getriebe fuer grosse Leistungen und hohe Drehzahlen mit variabler Sekundardrehzahl, *Antriebstechnik,* March 1964, pp. 98–100.

Walker, F. H. Multiturbine Torque Converters, Chapter 22, *Design Practices—Passenger Car Automatic Transmissions,* 2nd Edition, New York, 1973, pp. 227–240.

Walker, G. E. Graphical method for Epicyclic Gears, *Mach. World* 131, no. 3391, February 1952, p. 58.

Walter, H. Gear Tooth Deflection and Profile Modification, *The Engineer* 166, October 14 and 21, 1938, pp. 409–412, 434–436, respectively.

Walter, H. Gear Tooth Deflection and Profile Modification, *The Engineer* 170, August 16, 1940, pp. 102–103.

Wankel, F. *Einteilung der Rotationskolbenmaschinen,* Stuttgart, Deutsche Verlaganstalt, 1963.

Wear Life of Aluminum Gears, *Product Engineering,* September 1956, p. 160.Welbourn, D. B. Velocity Ratios of Epicyclic Gears, *Engineering* 176, no. 4568, August 14, 1953, pp. 198–199.

Welch and Voron Thermal Instability in High-Speed Gearing, *Transactions of the ASME, Series A, Journal of Engineering for Power* 83, 91961, no. 1.

Wellauer, E. J. Helical and Herringbone Gears, *Machine Design,* December 21, 1961, p. 125.

Werner, R. Flaschenzug mit neuerem Zahnradgetriebe, *VDI Z,* 1868, p. 27.

White, G. Compounded two-path variable ratio transmissions with coaxial gear trains, *Mechanisms and Machine Theory* 11, 227–240.

Whole Number Solution of Indeterminate Equations for Gearing, *Product Engineering 25,* May 1954, p. 198.

Wilkinson, W. H. Planetary Gear Trains, *Machine Design,* Jan. 7, 1960, pp. 155–159.

Willis, R. *Principles of Mechanisms,* London, Parker, 1841.

Willis, R., Jr. Stress Cycles in Planetary Gears, *Product Engineering,* January, 1957, p. 219.

Willis, R., Jr. Epicyclic Gear Train Calculations, *Product Engineering,* April 19, 1957, p. 186.

Willis, R., Jr. Formula for Epicyclic Gear Train of Large Reduction, *Product Design Handbook,* Mid-October 1957, p. E-12.

Willis, R., Jr. How Many Planets in an Epicyclic Gear?, *Product Engineering,* Nov. 24, 1958, pp. 57–59.

Willis, R., Jr. Volume Requirements of Epicyclic Gear Systems, *Product Engineering,* June 13, 1960, pp. 54–57.

Wilson, W. G. Epicyclic Gearing, *Proceedings of the Institution of Automobile Engineers* 26, 1931–32, pp. 216–257.

Wilson, W. G. Epicyclic Gearing, *Autom. Engr.* 22, 1932, pp. 139–146; Concerning the Manufacture of Wilson-Transmission, *Autom. Engr.* 23, 1933, pp. 375–382; *Vie Autom.* 29, 1933, pp. 172–176; *Autom.* 67, 1932, pp. 384–387; *Autom.* 36, 1933, p. 10.

Winter, F. Getriebe fuer verschiedene Zwecke, *Schweizerische Maschinenmarkt* 66, no. 6, 1966, pp. 57, 59, 61, 63, 65.

Wintermeyer, F. and Theil, W. *Untersuchungen ueber den Bau von Umlaufgetrieben,* Berlin, Verlag Technik, 1952.

Wissmann, K. Berechnung und Konstruktion von Zahnradern fuer Krane und aehnliche Maschinen, Dissertation, T. H. Berlin.

Wittekind, F. Verstellschrauben, *Motorschau* 3, 1939, pp. 805–809.

Wojnarowski, J. The Graph Method of Determining the Loads in Complex Gear Trains, *Mechanisms and Machine Theory* 11, pp. 103–121.

Wolf, A. Beitrag zur Berechnung von Planetengetrieben, *Z. der Osterr. Ing. u. Arch. Vereins* 77, 1925, pp. 224–226.

Wolf, A. Zur Weiterentwicklung der Umlaufraedergetriebe, BHS-Stoeckicht-Getriebe, *Masch.-Bau* 15, 1936, p. 338.

Wolf, A. Hemmwerk mit Umlaufraedergetriebe, *Masch.-Bau* 7, 1939, pp. 98–99.

Wolf, A. Die Umlaufgetriebe und ihre Berechnung, *Z. VDI* 91, no. 22, 1949, pp. 597–603; *Z. VDI* 92, 1950, p. 1032.

Wolf, A. *Die Grundgesetze der Umlaufgetriebe,* Dissertation, T. H. Stuttgart, 1953.

Wolf, A. *Die Grundgesetze der Umlaufgetriebe,* Braunschweig, F. Vieweg & Sohn, 1958.

Wolf, A. M. Automatic Transmissions, *ASE Journal* 41, 1937, pp. 311–313.

Wolf, O. Beitrag zur Berechnung von Planetenradgetrieben, *Z. oest. Ing. Arch. Verein* 77, 1925, pp. 224–226.

Wolf, O. Der Wirkungsgrad der Achsen und Wellen, *Werkzeugmaschine* 31, 1927, pp. 433–444.

Wolf, O. Konstruktive Entwicklung der Getriebetechnik unter besonderer Beruecksichtigung der Anwendung hochwertiger Werkstoffe, *Z. VDI* 80, 1936, pp. 1093–1098.

Wolfrom, U. Der Wirkungsgrad von Planetenraedergetrieben, *Werstattstechnik* 6, 1912, pp. 615–617.

Wolkenstein, R. Die Evolventen-Innenverzahnung, *Maschinenbau und Waermewirtschaft (Wien)* 12, no. 4, 1957, pp. 107–112; no. 5, pp. 135–140.

Wolkenstein, R. Zur Frage der konstruktiven Gestaltung von Planetengetrieben, *Z. VDI* 99, no. 25, 1957, pp. 1245–1249.

Wolkenstein, R. Ein neues, universell einsetzbares Industriegetriebe hoher Wirtschaftlichkeit, *Antriebstechnik* 8, 1969, special edition, Zurich, pp. 6–13.

Woerner, E. Neuzeitliche beschleunigungsfreie und ueberlastungssichere Antriebe fuer mehreckige Antriebsund Umlenksterne von Baggern, Absetzern und Becherwerken, *Foerdertechnik* 32, 1939, pp. 228–232, 241–247, 271–276.

Yu, D. The Optimizing Programming Design of the KHV Planetary Gear Driving, *Proceedings of the International Symposium on Gearing and Power Transmission,* Tokyo, 1981, pp. 295–299.

Yu, D. and Beachley, N. Alternative Split-Path Transmission Designs for Motor & Vehicle Applications, *Proceedings, First International Symposium on Design and Synthesis,* Tokyo, 1984, pp. 744–751.

Yu, D. and Beachley, N. On the Mechanical Efficiency of Differential Gearing, *Journal of Mechanisms, Transmissions and Automation in Design* 107, no. 1, 1985, pp. 61–67.

Zajonz, R. Die zeichnerische und rechnerische Untersuchung von Stirnrad-Umlaufgetrieben, Dissertation, Dresden, T. H., 1938.

Zajonz, R. Die zeichnerische und rechnerische Untersuchung von Stirnrad-Umlaufgetrieben, *ATZ*-Beiheft, Vol. 4, Stuttgart, 1939.

Zaporozec, O. L. Load Distribution Between Planet Gears in an Epicyclic Train, *Russian Engineering J.*, no. 10, 1965, pp. 30–31.

Zink, H. Lastdruckausgleich, Laufruhe und Konstruktion moderner Planetengetriebe, *Konstruktion* 16, 1964, part 1, pp. 41–47; part 2, pp. 81–8; part 3, pp. 188–191.

Index